Lecture Notes in Computer Science 13133

More information about this subseries at http://www.springer.com/series/7409

Hao-Ren Ke · Chei Sian Lee ·
Kazunari Sugiyama (Eds.)

Towards Open and Trustworthy Digital Societies

23rd International Conference on Asia-Pacific Digital Libraries, ICADL 2021
Virtual Event, December 1–3, 2021
Proceedings

Editors
Hao-Ren Ke
National Taiwan Normal University
Taipei, Taiwan

Chei Sian Lee
Nanyang Technological University
Singapore, Singapore

Kazunari Sugiyama
Kyoto University
Kyoto, Japan

ISSN 0302-9743 ISSN 1611-3349 (electronic)
Lecture Notes in Computer Science
ISBN 978-3-030-91668-8 ISBN 978-3-030-91669-5 (eBook)
https://doi.org/10.1007/978-3-030-91669-5

LNCS Sublibrary: SL3 – Information Systems and Applications, incl. Internet/Web, and HCI

This Springer imprint is published by the registered company Springer Nature Switzerland AG
The registered company address is: Gewerbestrasse 11, 6330 Cham, Switzerland

Preface

This volume contains the papers presented at the 23rd International Conference on Asia-Pacific Digital Libraries (https://icadl.net/icadl2021/). Due to the COVID-19 pandemic, the conference was organized as a virtual meeting which took place from December 1 to December 3, 2021. Since starting in Hong Kong in 1998, ICADL has become a premier international conference for digital library research. The conference series brings together researchers and practitioners from diverse disciplines within and outside the Asia-Pacific region to explore the present and future roles of digital libraries as we advance towards digital societies. The theme for ICADL 2021 was "Towards Open and Trustworthy Digital Societies."

ICADL 2021 was organized as a collaboration with the Asia-Pacific Chapter of iSchools (AP-iSchools). We believe that digital libraries and information schools play important roles in the development of an open and trustworthy digital society, and we hoped this conference would create a synergy for both groups to share and exchange common research interests and goals. Notably, organizing the conference virtually for a second time during the COVID-19 pandemic has important significances. The COVID-19 pandemic has not only disrupted our societies and caused upheavals in the economy, healthcare services, education, work, and employment but also underscored the importance of information-related research and practice. Specifically, the COVID-19 experience has reminded us of the need to stay resilient as societies look to rebound from the pandemic. New ideas, research paradigms, and information resources will be required as we prepare for the new normal in the post-pandemic era. Hence, ICADL 2021 was particularly timely and meaningful as it provided a forum to facilitate the sharing, exchange, and development of new ideas, research paradigms, and information resources to support ground-breaking innovations that will be needed in the digital information environment as we advance towards an open and trustworthy digital society in the post-pandemic era.

ICADL 2021 presented two keynotes – Wesley Teter from the United Nations Educational, Scientific and Cultural Organization (Bangkok) and Min-Yen Kan from National University of Singapore. Both academic research papers and practice papers were solicited for ICADL 2021. The integration of theory, research, and evidence-based policies and practice is critical to the discussion of research problems and future directions for digital library and digital society research.

Additionally, a doctoral consortium was also organized in collaboration with the Asia-Pacific Chapter of iSchools (AP-iSchools), the Asia-Pacific Library and Information Education and Practice (A-LIEP) Conference, and the Asia Library Information Research Group (ALIRG). Co-chaired by Atsuyuki Morishima and Di Wang, the consortium provided opportunities for doctoral students to connect and interact with other junior and senior research scholars from the Asia-Pacific region. The consortium also enabled senior research scholars in the region to participate in the mentoring and advising of doctoral students.

In response to the call for papers, 87 papers from authors in 28 countries were submitted to the conference and each paper was reviewed by at least three Program Committee (PC) members. Based on the reviews and recommendation from the PC, 17 full papers, 14 short papers, and 5 practice papers were selected and included in the proceedings. Collectively, these papers covered topics such as information technologies; data science; digital humanities; social informatics; digital heritage preservation; digital curation; models and guidelines; information retrieval, integration, extraction, and recommendation; education and digital literacy; open access and data; and information access design.

We would like to thank all those who contributed to ICADL 2021 – Emi Ishita and Sue Yeon Syn (Conference Chairs) as well as members of the Organizing Committee who lent their support: Atsuyuki Morishima and Di Wang (Doctoral Consortium Chairs), Ricardo Campos, Songphan Choemprayong, Akira Maeda, Maciej Ogrodniczuk, Natalie Pang, Rahmi, Nguyễn Hoàng Sơn, Suppawong Tuarob, Sohaimi Zakaria, and Lihong Zhou (Publicity Chairs), Shun-Hong Sie (Web Chair), Mitsuharu Nagamori (Registration Chair) and Hiroyoshi Ito (Session Management Chair). We are especially grateful to Shigeo Sugimoto for his relentless enthusiasm and effort in connecting the members of the Organizing Committee and to the ICADL Steering Committee Chair, Adam Jatowt, for his leadership and support. Finally, we would like to thank the researchers who contributed papers, the members of the Program Committee who generously offered their time and expertise to review the submissions, and the participants of ICADL 2021. Your contributions were vital in making ICADL 2021 a success.

December 2021

Hao-Ren Ke
Chei Sian Lee
Kazunari Sugiyama

Organization

Organizing Committee

Conference Co-chairs

Emi Ishita	Kyushu University, Japan
Sue Yeon Syn	Catholic University of America, USA

Program Committee Co-chairs

Hao-Ren Ke	National Taiwan Normal University, Taiwan
Chei Sian Lee	Nanyang Technological University, Singapore
Kazunari Sugiyama	Kyoto University, Japan

Doctoral Consortium Co-chairs

Atsuyuki Morishima	University of Tsukuba, Japan
Di Wang	Wuhan University, China

Publicity Co-chairs

Ricardo Campos	Polytechnic Institute of Tomar, Portugal
Songphan Choemprayong	Chulalongkorn University, Thailand
Akira Maeda	Ritsumeikan University, Japan
Maciej Ogrodniczuk	Institute of Computer Science, Polish Academy of Sciences, Poland
Natalie Pang	National University of Singapore, Singapore
Rahmi	Universitas Indonesia, Indonesia
Nguyễn Hoàng Sơn	Vietnam National University, Vietnam
Suppawong Tuarob	Mahidol University, Thailand
Sohaimi Zakaria	Universiti Teknologi MARA, Malaysia
Lihong Zhou	Wuhan University, China

Registration Chair

Mitsuharu Nagamori	University of Tsukuba, Japan

Web Chair

Shun-Hong Sie	National Taiwan Normal University, Taiwan

Session Management Chair

Hiroyoshi Ito	University of Tsukuba, Japan

Program Committee

Trond Aalberg	Norwegian University of Science and Technology, Norway
Biligsaikhan Batjargal	Ritsumeikan University, Japan
Chih-Ming Chen	National Chengchi University, Taiwan
Hung-Hsuan Chen	National Central University, Taiwan
Kun-Hung Cheng	National Chung Hsing University, Taiwan
Songphan Choemprayong	Chulalongkorn University, Thailand
Chiawei Chu	City University of Macau, Macau
Mickaël Coustaty	University of La Rochelle, France
Fabio Crestani	Università della Svizzera italiana, Switzerland
Yijun Duan	National Institute of Advanced Industrial Science and Technology, Japan
Edward Fox	Virginia Tech, USA
Liangcai Gao	Peking University, China
Dion Goh	Nanyang Technological University, Singapore
Simon Hengchen	University of Gothenburg, Sweden
Ji-Lung Hsieh	National Taiwan University, Taiwan
Jen Jou Hung	Dharma Drum Institute of Liberal Arts, Taiwan
Adam Jatowt	University of Innsbruck, Austria
Makoto P. Kato	University of Tsukuba, Japan
Marie Katsurai	Doshisha University, Japan
Yukiko Kawai	Kyoto Sangyo University, Japan
Christopher S. G. Khoo	Nanyang Technological University, Singapore
Yunhyong Kim	University of Glasgow, UK
Shaobo Liang	Wuhan University, China
Chern Li Liew	Victoria University of Wellington, New Zealand
Chung-Ming Lo	National Chengchi University, Taiwan
Akira Maeda	Ritsumeikan University, Japan
Muhammad Syafiq Mohd Pozi	Universiti Utara Malaysia, Malaysia
Jin-Cheon Na	Nanyang Technological University, Singapore
Preslav Nakov	Qatar Computing Research Institute, HBKU, Qatar
Thi Tuyet Hai Nguyen	University of La Rochelle, France
David Nichols	University of Waikato, New Zealand
Chifumi Nishioka	Kyoto University, Japan
Maciej Ogrodniczuk	Institute of Computer Science, Polish Academy of Sciences, Poland
Hiroaki Ohshima	University of Hyogo, Japan
Gillian Oliver	Monash University, Australia
Christos Papatheodorou	National and Kapodistrian University of Athens, Greece

Yohei Seki	University of Tsukuba, Japan
Catherine Smith	Kent State University, USA
Shigeo Sugimoto	University of Tsukuba, Japan
Yasunobu Sumikawa	Takushoku University, Japan
Masao Takaku	University of Tsukuba, Japan
Yuen-Hsien Tseng	National Taiwan Normal University, Taiwan
Suppawong Tuarob	Mahidol University, Thailand
Nicholas Vanderschantz	University of Waikato, New Zealand
Diane Velasquez	University of South Australia, Australia
Shoko Wakamiya	Nara Institute of Science and Technology, Japan
Di Wang	Wuhan University, China
Chiranthi Wijesundara	University of Colombo, Sri Lanka
Dan Wu	Wuhan University, China
Zhiwu Xie	Virginia Tech, USA
Sohaimi Zakaria	Universiti Teknologi MARA, Malaysia
Maja Žumer	University of Ljubljana, Slovenia

Additional Reviewers

Prashant Chandrasekar
Monica Landoni
Ivan Sekulic

Contents

Knowledge Discovery from Digital Collections

Diachronic Linguistic Periodization of Temporal Document Collections for Discovering Evolutionary Word Semantics

Yijun Duan[1(✉)], Adam Jatowt[2], Masatoshi Yoshikawa[3], Xin Liu[1], and Akiyoshi Matono[1]

[1] AIST, Tsukuba, Japan
{yijun.duan,xin.liu,a.matono}@aist.go.jp
[2] Department of Computer Science, University of Innsbruck, Innsbruck, Austria
jatowt@acm.org
[3] Graduate School of Informatics, Kyoto University, Kyoto, Japan
yoshikawa@i.kyoto-u.ac.jp

Abstract. Language is our main communication tool. Deep understanding of its *evolution* is imperative for many related research areas including history, humanities, social sciences, etc. To this end, we are interested in the task of segmenting long-term document archives into naturally coherent periods based on the evolving word semantics. There are many benefits of such segmentation such as better representation of content in long-term document collections, and support for modeling and understanding semantic drift. We propose a two-step framework for learning time-aware word semantics and periodizing document archive. Encouraging effectiveness of our model is demonstrated on the New York Times corpus spanning from 1990 to 2016.

Keywords: Dynamic word embedding · Temporal document segmentation · Knowledge discovery in digital history

1 Introduction

Language is an evolving and dynamic construct. The awareness of the necessity and possibilities of large scale analysis of the temporal dynamics on linguistic phenomena has increased considerably in the last decade [26,29,30]. Temporal dynamics play an important role in many time-aware information retrieval (IR) tasks. For example, when retrieving documents based on their embeddings, one needs accurate representations of content by temporal embedding vectors.

It is intuitive that, if an IR system is required to effectively return information from a target time period T_a in the past, it may fail to do so if it is unable to capture the change in context between T_a and the current time T_b. To which extent is the context of T_a different from that of T_b? Are there any turning points

H.-R. Ke et al. (Eds.): ICADL 2021, LNCS 13133, pp. 3–17, 2021.
https://doi.org/10.1007/978-3-030-91669-5_1

in the interval between T_a and T_b when a significant context change occurred, or rather do T_a and T_b belong to the same stage in the evolving process of language? Being capable of answering such questions is crucial for effective IR systems when coping with time-aware tasks. However, to the best of our knowledge, the research problem of *distinguishing key stages in the evolution's trajectory of language* still remains a challenge in the field of temporal IR and text mining.

Traditionally, a language's diachrony is segmented into pre-determined periods (e.g., the "Old", "Middle" and "Modern" eras for English) [24], which is problematic, since such an approach may yield results concealing the true trajectory of a phenomenon (e.g., false assumption on abrupt turning point about the data). Moreover, these traditional segments are very coarse as well as can be easily obscured and derived from arbitrary and non-linguistic features [7]. Thanks to accumulated large amounts of digitized documents from the past, it is possible now to employ large scale data-driven analyses for uncovering patterns of language change. In this study, we propose a data-driven approach for segmenting a temporal document collection (e.g., a long-term news article archive) into natural, linguistically coherent periods. Based on our method, we can both capture the features involved in diachronic linguistic change, as well as identify the time periods when the changes occurred. Our approach is generic and can be applied to any diachronic data set. The detected periods could be then applied in diverse time-aware downstream tasks, such as temporal analog retrieval, archival document recommendation, and summarization.

Our method is based on the computation of *dynamic word embeddings.* Semantic senses of words are subject to broadening, narrowing or other kinds of shifts throughout time. For instance, *Amazon* originally referred to mythical female warriors (in ancient Greek mythology), while it assumed a new sense of a rainforest in South Africa since around 16th century, and a large e-commerce company since middle 1990s. Additionally, different words may become conceptually equivalent or similar across time. For example, a music device *Walkman* played a similar role of mobile music playing device 30 years ago as *iPod* plays nowadays. Such phenomenon of evolving word semantics is however rarely considered in the existing corpus periodization schemes.

In this paper, we structure document collections by periodizing the evolving word semantics embodied in the corpus. Specifically, for a long-term document corpus, our goal is to split the entire time span into several consecutive periods, where we assume within the same period most words do not undergo significant fluctuations in term of their senses, while linguistic shifts are on the other hand relatively prevalent across different periods. In other words, *a word is represented by an identical vector in the same period, while it may have fairly different representations in different periods* (see Fig. 1).

The problem of document collection periodization based on evolving word semantics is however not trivial. In order to solve this problem, we address the following two research questions:

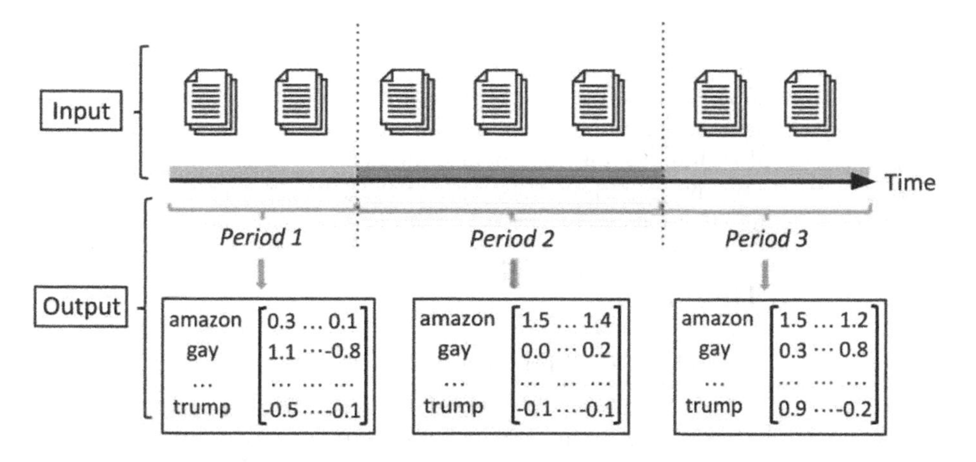

Fig. 1. Conceptual view of our task. Our goal is to identify *latent periods* in the input document collection, such that word semantics are relatively stable within the same period (i.e., a word is represented by the same embedding vector), and major linguistic shifts exist between different periods (i.e., a word may be represented by fairly different vectors in different periods).

a. How to compute temporal-aware word embeddings (**Task 1**)?
b. How to split the document collection based on learned word embeddings (**Task 2**)?

Our main technical contribution lies in a two-step framework for answering the above questions. First of all, we develop an *anchor-based joint matrix factorization* framework for computing time-aware word embeddings. More specifically, we concurrently factorize the time-stamped PPMI (positive pointwise mutual information) matrices, during which we utilize *shared frequent terms* (see Sect. 3) as anchors for aligning word embeddings of all time to the same latent space. Secondly, we formulate the periodization task as an optimization problem, where we aim to maximize the aggregation of differences between the word semantics of any two periods. To get the optimal solution, we employ three classes of algorithms which are based on *greedy splitting*, *dynamic programming* and *iterative refinement*, respectively.

In the experiments, we use the crawled and publicly released New York Times dataset [29], which contains a total of 99,872 articles published between January 1990 and July 2016. To evaluate the periodization effectiveness, we construct the test sets by utilizing New York Times article tags (see Sect. 5), and evaluate the analyzed methods based on two standard metrics: Pk [2] and WinDiff [22], which are commonly reported in text segmentation tasks.

In summary, our contributions are as follows:

- From a conceptual standpoint, we introduce a novel research problem of periodizing diachronic document collections for discovering the embodied evolutionary word semantics. The discovered latent periods and corresponding

temporal word embeddings can be utilized for many objectives, such as tracking and analyzing linguistic and topic shifts over time.
- From a methodological standpoint, we develop an anchor-based joint matrix factorization framework for computing time-aware word embeddings, and three classes of techniques for document collection periodization.
- We perform extensive experiments on the New York Times corpus, by which the encouraging effectiveness of our approach is demonstrated.

2 Problem Definition

We start by presenting the formal problem definition.

Input: The input are documents published across time. Formally, let $D = \{D_1, D_2, ..., D_N\}$ denote the entire article set where D_i represents the subset of documents belonging to the time unit t_i. The length of a time unit can be at different levels of granularity (months, years, etc.)

Task 1: Our first task is to embed each word in the corpus vocabulary $V = \{w_1, w_2, ..., w_{|V|}\}$ [1] into a d-dimensional vector, for each time unit $t_i (i = 1, ..., N)$, respectively. Thus, the expected output is a tensor of size $N \times |V| \times d$, which we denote by A. A_i the embedding matrix for t_i, thus A_i is of size $|V| \times d$.

Task 2: Based on Task 1, our second goal is to split the text corpus D into m latent periods $\Theta = (P_1, P_2, ..., P_m)$ and compute their corresponding word embedding matrix $E_i, i = 1, ..., m$. Each period $P_i = [\tau_b^i, \tau_e^i]$ is expressed by two time points representing its beginning date τ_b^i and the ending date τ_e^i. Let $L(\Theta) = (\tau_b^1, \tau_b^2, ..., \tau_b^m)$ denote the list of beginning dates of all periods, notice that searching for Θ is equivalent to searching for $L(\Theta)$.

3 Temporal Word Embeddings

In this section, we describe our approach for computing dynamic word embeddings (solving **Task 1** in Sect. 2), which captures word semantic evolution across time.

3.1 Learning Static Embeddings

The distributional hypothesis [10] states that semantically similar words usually appear in similar contexts. Let v_i denote the vector representing word w_i, then v_i can be expressed by the co-occurrence statistics of w_i. In this study, we compute the PPMI (positive pointwise mutual information) matrix for obtaining such inter-word co-occurrence information, following previous works [13,16,29]. Moreover, for word vectors v_i and v_j, we should have $\text{PPMI}[i][j] \approx v_i \cdot v_j$, thus static word vectors can be obtained through factorizing the PPMI matrix.

[1] The overall vocabulary V is the union of vocabularies of each time unit, and thus it is possible for some $w \in V$ to not appear at all in some time units. This includes emerging words and dying words that are typical in real-world news corpora.

3.2 Learning Dynamic Embeddings

We denote PPMI_i as the PPMI matrix for time unit t_i, then word embedding matrix A_i at t_i should satisfy $\text{PPMI}_i \approx A_i \cdot A_i^T$.

However, if A_i is computed separately for each time unit, due to the invariant-to-rotation nature of matrix factorization, these learned word embeddings A_i are non-unique (i.e., we have $\text{PPMI}_i \approx A_i \cdot A_i^T = (A_i W^T) \cdot (W A_i^T) = \tilde{A}_i \tilde{A}_i^T$ for any orthogonal transformation W which satisfies $W^T \cdot W = I$). As a byproduct, embeddings across time units may not be placed in the same latent space. Some previous works [13,15,30] solved this problem by imposing an alignment before any two adjacent matrices A_i and A_{i+1}, resulting in $A_i \approx A_{i+1}, i = 1, ..., N-1$.

Instead of solving a separate alignment problem for circumventing the non-unique characteristic of matrix factorization, we propose to learn the temporal embeddings across time *concurrently*. Intuitively, if word w did not change its meaning across time (or change its meaning to very small extent), we desire its vector to be close among all temporal embedding matrices. Such words are regarded as *anchors* for aligning various embedding matrices, in our joint factorization framework.

Essentially, we assume that very frequent terms (e.g., man, sky, one, water) did **not** experience significant semantic shifts in the long-term history, as their dominant meaning are commonly used in everyday life and used by so many people. This assumption is reasonable as it has been reported in many languages including English, Spanish, Russian and Greek [17,20]. We refer to these words as SFT, standing for *shared frequent terms*. Specifically, we denote by A_i^{SFT} the $|V| \times d$ embedding matrix whose i-th row corresponds to the vector of word w_i in A_i, if w_i is a shared frequent term, and corresponds to zero vector otherwise, for time unit t_i. Our joint matrix factorization framework for learning temporal word embeddings is then shown as follows (see Fig. 2 for an illustration):

$$\begin{aligned} A_1, ..., A_N = \arg\min \sum_{i=1}^{N} \left\| \text{PPMI}_i - A_i \cdot A_i^T \right\|_F^2 \\ + \alpha \cdot \sum_{i=1}^{N} \|A_i\|_F^2 + \beta \cdot \sum_{i=1}^{N-1} \sum_{j=i+1}^{N} \left\| A_i^{SFT} - A_j^{SFT} \right\|_F^2 \end{aligned} \tag{1}$$

Here $\|\cdot\|_F$ represents the Frobenius norm. $\left\| A_i^{SFT} - A_j^{SFT} \right\|_F^2$ is the key smoothing term aligning *shared frequent terms* in all time units, thus places word embeddings across time in the same latent space. The regularization term $\|A_i\|_F^2$ is adopted to guarantee the low-rank data fidelity for overcoming the problem of overfitting. α and β are used to control the weight of different terms to achieve the best factorization performance. Finally, we iteratively solve for A_i by fixing other embedding matrices as constants, and optimizing Eq. (1) using the block coordinate descent method [27].

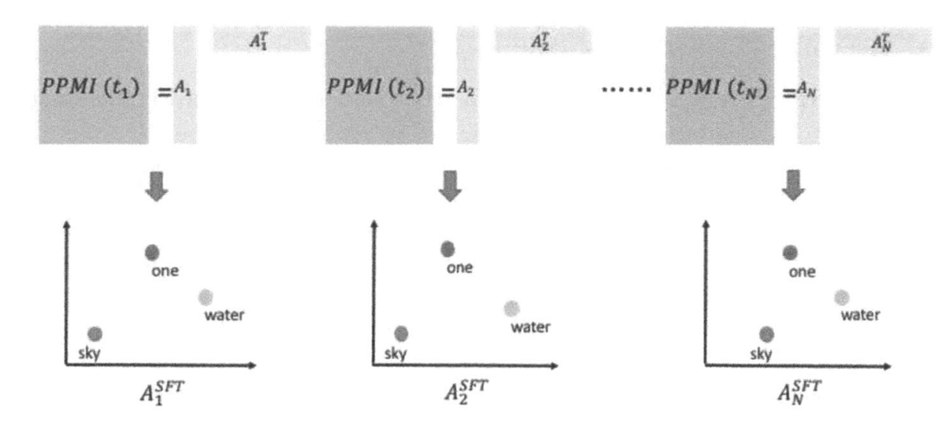

Fig. 2. Illustration of our joint matrix factorization model. *Shared frequent terms* (e.g., sky, one, water) in all time units are aligned to similar positions, which leads word embeddings across time in the same latent semantic space.

4 Document Collection Periodization

In this section, we explain how to split the document collection based on learned temporal word embeddings (solving **Task 2** in Sect. 2).

4.1 Scoring

In general, we prefer the embedding matrices of different periods to be characterized by *high inter-dissimilarity*. Thus, the objective $Obj(\Theta)$ for an overall segmentation is given by aggregating the dissimilarity between all pairs of period-specific embedding matrices, as follows:

$$Obj(\Theta) = Obj(L(\Theta)) = \sum_{i=1}^{m-1} \sum_{j=i+1}^{m} \|E_i - E_j\|_F^2 \tag{2}$$

Here E_i is measured as the average of embeddings in period P_i:

$$E_i = \frac{1}{\tau_e^i - \tau_b^i + 1} \sum_{t=\tau_b^i}^{\tau_e^i} A_t \tag{3}$$

The segmentation that achieves the highest score of Eq. (2) will be adopted.

4.2 Periodization

Greedy Algorithm Based Periodization. At each step, this algorithm inserts a new boundary (which is the beginning date of a new period) to the existing boundaries to locally maximize the objective function, until desired m periods are discovered. The process is formulated in Algorithm 1, where $L(\Theta)^i$ denotes the list of boundaries at the i-th step, and $L(\Theta)^0 = \{t_1\}$.

Algorithm 1: Greedy algorithm based periodization

```
   input : L(Θ)^0; m
   output: L(Θ)^{m-1}
1  for i ← 0 to m − 2 do
2  |   max_score ← 0;
3  |   next_boundary ← 0;
4  |   for t_p ← t_1 to t_N do
5  |   |   ▷Find the best local boundary;
6  |   |   if t_p ∈ L(Θ)^i then
7  |   |   |   continue
8  |   |   end
9  |   |   score ← Obj(L(Θ)^i ∪ {t_p});
10 |   |   if score > max_score then
11 |   |   |   max_score ← score;
12 |   |   |   next_boundary ← t_p;
13 |   |   end
14 |   end
15 |   L(Θ)^{i+1} ← L(Θ)^i ∪ {next_boundary};
16 end
```

Dynamic Programming Based Periodization. The core idea of this algorithm is to break the overall problem into a series of simpler smaller segmentation tasks, and then recursively find the solutions to the sub-problems. Let Θ_k^l denotes the segmentation of the first l time slices of the entire time span into k periods, the computational process of dynamic programming based periodization is expressed in Algorithm 2, where $\Theta_1^l = [t_1, t_l]$ and $L(\Theta_1^l) = \{t_1\}, l = 1, ..., N$.

Iterative Refinement Based Periodization. The iterative refinement framework starts with the greedy segmentation. At each step, after the best available boundary is found, a *relaxation* scheme which tries to adjust each segment boundary optimally while keeping the adjacent boundaries to either side of it fixed, is applied. This method can improve the performance of the greedy scheme, while at the same time partially retain its computational benefit. Let $L(\Theta)_G^i[j]$ denote the j-th element in $L(\Theta)^i$ after the i-th greedy search step, the refinement process for finding $L(\Theta)^i[j]$ is shown in Algorithm 3:

Algorithm 2: Dynamic programming based periodization

input : $L(\Theta_1^l)$, $l = 1, ..., N$; m
output: $L(\Theta_m^N)$

1 **for** $row \leftarrow 2$ **to** m **do**
2   **for** $col \leftarrow row$ **to** N **do**
3   `▷Recursively find the solutions to the sub-problems;`
4     $max\_score \leftarrow 0$;
5     $next\_boundary \leftarrow 0$;
6     $subtask \leftarrow 0$;
7   **for** $j \leftarrow row - 1$ **to** $col - 1$ **do**
8       $score \leftarrow Obj(L(\Theta_{row-1}^j) \cup \{t_{j+1}\})$;
9   **if** $score > max\_score$ **then**
10         $max\_score \leftarrow score$;
11         $next\_boundary \leftarrow t_{j+1}$;
12         $subtask \leftarrow j$;
13   **end**
14   **end**
15     $L(\Theta_{row}^{col}) \leftarrow L(\Theta_{row-1}^{subtask}) \cup \{next\_boundary\}$
16   **end**
17 **end**

Algorithm 3: Iterative refinement based periodization

input : $L(\Theta)^0$; m
output: $L(\Theta)^{m-1}$

1 **for** $i \leftarrow 0$ **to** $m - 2$ **do**
2   $next\_boundary, max\_score \leftarrow Greedy(L(\Theta)^i)$;
3   $L(\Theta)^{i+1} \leftarrow L(\Theta)^i \cup \{next\_boundary\}$;
4   **for** $j \leftarrow 1$ **to** i **do**
5   `▷Iteratively refine the previous boundaries;`
6     $new\_boundary \leftarrow L(\Theta)^{i+1}[j]$;
7     $t_{begin} \leftarrow L(\Theta)^{i+1}[j - 1]$;
8     $t_{end} \leftarrow L(\Theta)^{i+1}[j + 1]$;
9   **for** $t_p \leftarrow t_{begin}$ **to** t_{end} **do**
10       $score \leftarrow Obj(L(\Theta)^{i+1} - L(\Theta)^{i+1}[j] \cup \{t_p\})$;
11   **if** $score > max\_score$ **then**
12         $max\_score \leftarrow score$;
13         $next\_boundary \leftarrow t_p$;
14   **end**
15   **end**
16     $L(\Theta)^{i+1} \leftarrow (L(\Theta)^{i+1} - L(\Theta)^{i+1}[j]) \cup \{new\_boundary\}$;
17   **end**
18 **end**

4.3 Analysis of Time Complexity

For greedy periodization, it requires $m-1$ steps and the i-th step calls scoring function Eq. (2) $N-i$ times. In total, it is $O(Nm - N - m^2 + m/2)$. In the case of $N \gg m$, the greedy periodization algorithm takes $O(Nm)$. For dynamic programming based periodization, it requires $O(Nm)$ states and evaluating each state involves an $O(N)$ calling of Eq. (2). Then the overall algorithm would take $O(N^2m)$. Finally, for iterative refinement based periodization, an upper bound on its time complexity is $O(\sum_{i=1}^{m-1}(N-i)*i) = O(Nm^2)$.

5 Periodization Effectiveness

5.1 Datasets

News corpora, which maintain consistency in narrative style and grammar, are naturally advantageous to studying language evolution [29]. We thus perform the experiments on the New York Times Corpus, which has been frequently used to evaluate different researches on temporal information processing or extraction in document archives [4]. The dataset we use [29] is a collection of 99,872 articles published by the New York Times between January 1990 and July 2016. For the experiments, we first divide this corpus into 27 units, setting the length of time unit to be 1 year. Stopwords and rare words (which have less than 200 occurrences in entire corpus) were removed beforehand, following the previous work [29,30]. The basic statistics of our dataset are shown in Table 1.

Table 1. Summary of New York Times dataset.

#Articles	#Vocabulary	#Word Co-occurances	#Time units	Range
99,872	20,936	11,068,100	27	Jan. 1990 - Jul. 2016

5.2 Experimental Settings

For the construction of PPMI matrix, the length of sliding window and the value of embedding dimension is set to be 5 and 50, respectively, following [29]. During the training process of learning dynamic embeddings, the values of parameters α and β (see Eq. (1)) are set to be 20 and 100, respectively, as the result of a grid search. The selection of *shared frequent terms* used as anchors is set to be the top 5% most frequent words in the entire corpus, as suggested by [30].

5.3 Analyzed Methods

Baseline Methods. We test four baselines as listed below.

- **Random:** The segment boundaries are randomly inserted.
- **VNC** [12]**:** A bottom-up hierarchical clustering periodization approach.
- **KLD** [7]**:** An entropy-driven approach which calculates the Kullback-Leibler Divergence (KLD) between term frequency features to segment.

- **CPD** [15]: An approach which uses statistically sound change point detection algorithms to detect significant linguistic shifts.

Proposed Methods. We list three proposed methods below (see Sect. 4.2).

- **G-WSE:** Greedy periodization based on word semantic evolution.
- **DP-WSE:** Dynamic programming periodization based on word semantic evolution.
- **IR-WSE:** Iterative refinement based on word semantic evolution.

5.4 Test Sets

As far as we know, there is no standard testsets for New York Time Corpus, we then manually create test sets. The collected news articles dataset is associated with some metadata, including title, author, publish time, and topical section label (e.g., *Science*, *Sports*, *Technology*) which describes the general topic of news articles. Such section labels could be used to locate the boundaries.

Naturally, if a word w is strongly related to a particular section s in year t, we associate w, s and t together and construct a $<w, s, t>$ triplet. A boundary of w is registered if it is assigned to different sections in two adjacent years (i.e., both triplet $<w, s, t>$ and $<w, s', t+1>$ hold and $s \neq s'$). Some examples of words changing their associated section in adjacent years are shown in Table 2.

For each word w in the corpus vocabulary V, we compute its frequency in all sections for each year t, and w is assigned to the section in which w is most frequent. Note that this word frequency information is not used in our learning model. In this study we utilize the 11 most popular and discriminative sections [2] of the New York Times, following previous work [29].

Recall that parameter m denotes the number of predefined latent periods. For each different m, we first identify the set of words S_m characterized by the same number of periods. Then for each method and each value of m, we test the performance of such method by comparing the generated periods with the reference segments of each word in S_m, and then take the average. In this study, we experiment with the variation in the value of m, ranging from 2 to 10.

Table 2. Example words changing their associated section for evaluating *periodization effectiveness*.

Word	Year	Section	Year	Section
cd	1990	Arts	1991	Technology
seasoning	2002	Home and Garden	2003	Fashion and Style
zoom	2008	Fashion and Style	2009	Technology
roche	2009	Business	2010	Health
viruses	2009	Health	2010	Science
uninsured	2014	Health	2015	U.S.

[2] These sections are *Arts*, *Business*, *Fashion & Style*, *Health*, *Home & Garden*, *Real Estate*, *Science*, *Sports*, *Technology*, *U.S.*, *World*.

5.5 Evaluation Metrics

We evaluate the performance of analyzed methods with respect to two standard metrics commonly used in text segmentation tasks: Pk [2] and WinDiff [22]. Both metrics use a sliding window over the document and compare the machine-generated segments with the reference ones. Within each sliding window, if the machine-generated boundary positions are not the same as the reference, Pk will register an error. If the number of boundaries are different, WinDiff will register an error. Both Pk and WinDiff are scaled to the range [0, 1] and equal to 0 if an algorithm assigns all boundaries correctly. The **lower** the scores are, the better the algorithm performs.

5.6 Evaluation Results

Table 3 and Table 4 summarize the Pk and WinDiff scores for each method, respectively. Based on the experimental results we make the following analysis.

- The proposed methods exhibit the overall best performance regarding both Pk and WinDiff. More specifically, they outperform the baselines under 7 of 9 predefined numbers of periods in terms of Pk, and 6 of 9 in terms of WinDiff. Such encouraging observations demonstrate the effectiveness of our proposed periodization frameworks.
- Regarding baseline methods, Random achieves the worst performance. CPD and KLD show competitive performance under certain settings. CPD gets two wins in terms of Pk, and KLD obtains three wins in terms of WinDiff.
- DP-WSE is the best performer among all three proposed periodization algorithms. It contributes 6 best performance in terms of Pk, and 5 in terms of WinDiff. Moreover, when compared to G-WSE and IR-WSE, DP-WSE shows a 3.79% and 3.24% increase in terms of Pk, and a 7.77% and 6.46% increase in terms of WinDiff, respectively. This observation is in good agreement with the theoretical analysis, which states that dynamic programming based segmentation sacrifices certain computational efficiency for the globally optimal splitting.
- The operation of iterative refinement indeed improves the performance of greedy periodization in some cases, though many results generated by IR-WSE and by G-WSE are the same.

6 Related Work

6.1 Text Segmentation

The most related task to our research problem is text segmentation. Early text segmentation approaches include TextTiling [14] and C99 algorithm [5], which are based on some heuristics on text coherence using a bag of words representation. Furthermore, many attempts adopt topic models to tackle the segmentation task, including [9,23]. [1] is a segmentation algorithm based on time-agnostic

Table 3. Performance comparison by each method using Pk.

Acronym	Number of periods								
	2	3	4	5	6	7	8	9	10
Random	0.467	0.474	0.545	0.522	0.542	0.480	0.480	0.480	0.539
VNC	0.385	0.253	0.249	0.290	0.282	0.302	0.302	0.294	0.303
KLD	0.385	0.278	0.244	0.270	0.276	0.278	0.284	0.290	0.304
CPD	0.238	0.234	0.246	0.260	0.282	**0.263**	**0.249**	0.299	0.338
G-WSE	**0.115**	**0.201**	0.248	0.282	0.300	0.310	0.312	0.292	0.303
DP-WSE	**0.115**	0.230	**0.236**	**0.251**	**0.271**	0.290	0.291	**0.286**	**0.296**
IR-WSE	**0.115**	**0.201**	0.244	0.279	0.300	0.304	0.312	0.292	0.303

Table 4. Performance comparison by each method using WinDiff.

Acronym	Number of periods								
	2	3	4	5	6	7	8	9	10
Random	0.467	0.474	0.545	0.478	0.542	0.480	0.480	0.480	0.500
VNC	0.417	0.346	0.396	0.416	0.426	0.434	0.439	0.435	0.388
KLD	0.417	0.343	**0.383**	**0.384**	0.428	0.437	0.434	0.430	**0.384**
CPD	0.414	0.386	0.387	0.394	0.430	0.430	**0.430**	0.432	0.385
G-WSE	**0.383**	0.430	0.435	0.449	0.456	0.449	0.447	0.432	0.387
DP-WSE	**0.383**	**0.336**	0.387	0.403	**0.423**	**0.422**	**0.430**	0.431	0.388
IR-WSE	**0.383**	0.405	0.428	0.449	0.456	0.449	0.447	**0.421**	0.387

semantic word embeddings. Most text segmentation methods are unsupervised. However, neural approaches have recently been explored for domain-specific text segmentation tasks, such as [25]. Many text segmentation algorithms are greedy in nature, such as [5,6]. On the other hand, some works search for the optimal splitting for their own objective using dynamic programming [11,28].

6.2 Temporal Word Embeddings

The task of representing words with low-dimensional dense vectors has attracted consistent interest for several decades. Early methods are relying on statistical models [3,18], while in recent years neural models such as word2vec [19], GloVE [21] and BERT [8] have shown great success in many NLP applications. Moreover, it has been demonstrated that both word2vec and GloVE are equivalent to factorizing PMI matrix [16], which primarily motivates our approach.

The above methods assume word representation is static. Recently some works explored computing time-aware embeddings of words, for analyzing linguistic change and evolution [13,15,29,30]. In order to compare word vectors across time most works ensure the vectors are aligned to the same coordinate axes, by solving the least squares problem [15,30], imposing an orthogonal transformation [13] or jointly smoothing every pair of adjacent time slices [29]. Different from the existing methods, in this study we inject additional knowledge by using *shared frequent terms* as anchors to simultaneously learn the temporal word embeddings and circumvent the alignment problem.

7 Conclusion

This work approaches a novel and challenging research problem - *diachronic linguistic periodization of temporal document collections.* The special character of our task allows capturing evolutionary word semantics. The discovered latent periods can be an effective indicator of linguistics shifts and evolution embodied in diachronic textual corpora. To address the introduced problem we propose a two-step framework which consists of a joint matrix factorization model for learning dynamic word embeddings, and three effective embedding-based periodization algorithms. We perform extensive experiments on the commonly-used New York Times corpus, and show that our proposed methods exhibit superior results against diverse competitive baselines.

In future, we plan to detect correlated word semantic changes. We will also consider utilizing word sentiments in archive mining scenarios.

Acknowledgement. This paper is based on results obtained from a project, JPNP20006, commissioned by the New Energy and Industrial Technology Development Organization (NEDO).

References

1. Alemi, A.A., Ginsparg, P.: Text segmentation based on semantic word embeddings. arXiv preprint arXiv:1503.05543 (2015)
2. Beeferman, D., Berger, A., Lafferty, J.: Statistical models for text segmentation. Mach. Learn. **34**(1–3), 177–210 (1999)
3. Blei, D.M., Ng, A.Y., Jordan, M.I.: Latent Dirichlet allocation. J. Mach. Learn. Res. **3**(Jan), 993–1022 (2003)
4. Campos, R., Dias, G., Jorge, A.M., Jatowt, A.: Survey of temporal information retrieval and related applications. ACM Comput. Surv. (CSUR) **47**(2), 1–41 (2014)
5. Choi, F.Y.: Advances in domain independent linear text segmentation. arXiv preprint cs/0003083 (2000)
6. Choi, F.Y., Wiemer-Hastings, P., Moore, J.D.: Latent semantic analysis for text segmentation. In: Proceedings of the 2001 Conference on Empirical Methods in Natural Language Processing (2001)
7. Degaetano-Ortlieb, S., Teich, E.: Using relative entropy for detection and analysis of periods of diachronic linguistic change. In: Proceedings of the Second Joint SIGHUM Workshop, pp. 22–33 (2018)
8. Devlin, J., Chang, M.W., Lee, K., Toutanova, K.: BERT: pre-training of deep bidirectional transformers for language understanding. In: NAACL 2019, pp. 4171–4186 (2019)

9. Du, L., Buntine, W., Johnson, M.: Topic segmentation with a structured topic model. In: Proceedings of the 2013 Conference of the North American Chapter of the Association for Computational Linguistics: Human Language Technologies, pp. 190–200 (2013)
10. Firth, J.R.: Papers in Linguistics 1934–1951: Repr. Oxford University Press (1961)
11. Fragkou, P., Petridis, V., Kehagias, A.: A dynamic programming algorithm for linear text segmentation. J. Intell. Inf. Syst. **23**(2), 179–197 (2004)
12. Gries, S.T., Hilpert, M.: Variability-based neighbor clustering: a bottom-up approach to periodization in historical linguistics (2012)
13. Hamilton, W.L., Leskovec, J., Jurafsky, D.: Diachronic word embeddings reveal statistical laws of semantic change. In: Proceedings of the 54th Annual Meeting of the Association for Computational Linguistics (Volume 1: Long Papers), Berlin, pp. 1489–1501. Association for Computational Linguistics, August 2016. https://doi.org/10.18653/v1/P16-1141. https://www.aclweb.org/anthology/P16-1141
14. Hearst, M.A.: TextTiling: segmenting text into multi-paragraph subtopic passages. Comput. Linguist. **23**(1), 33–64 (1997)
15. Kulkarni, V., Al-Rfou, R., Perozzi, B., Skiena, S.: Statistically significant detection of linguistic change. In: Proceedings of the 24th International Conference on World Wide Web, pp. 625–635 (2015)
16. Levy, O., Goldberg, Y.: Neural word embedding as implicit matrix factorization. In: Advances in Neural Information Processing Systems, pp. 2177–2185 (2014)
17. Lieberman, E., Michel, J.B., Jackson, J., Tang, T., Nowak, M.A.: Quantifying the evolutionary dynamics of language. Nature **449**(7163), 713 (2007)
18. Lund, K., Burgess, C.: Producing high-dimensional semantic spaces from lexical co-occurrence. Behav. Res. Methods Instrum. Comput. **28**(2), 203–208 (1996)
19. Mikolov, T., Chen, K., Corrado, G., Dean, J.: Efficient estimation of word representations in vector space. arXiv:1301.3781 (2013)
20. Pagel, M., Atkinson, Q.D., Meade, A.: Frequency of word-use predicts rates of lexical evolution throughout Indo-European history. Nature **449**(7163), 717 (2007)
21. Pennington, J., Socher, R., Manning, C.D.: Glove: global vectors for word representation. In: Proceedings of the 2014 Conference on Empirical Methods in Natural Language Processing (EMNLP), pp. 1532–1543 (2014)
22. Pevzner, L., Hearst, M.A.: A critique and improvement of an evaluation metric for text segmentation. Comput. Linguist. 19–36 (2002)
23. Riedl, M., Biemann, C.: Text segmentation with topic models. J. Lang. Technol. Comput. Linguist. 47–69 (2012)
24. Schätzle, C., Booth, H.: DiaHClust: an iterative hierarchical clustering approach for identifying stages in language change. In: Proceedings of the 1st International Workshop on Computational Approaches to Historical Language Change, pp. 126–135 (2019)
25. Sehikh, I., Fohr, D., Illina, I.: Topic segmentation in ASR transcripts using bidirectional RNNs for change detection. In: 2017 IEEE Automatic Speech Recognition and Understanding Workshop (ASRU), pp. 512–518. IEEE (2017)
26. Tahmasebi, N., Borin, L., Jatowt, A.: Survey of computational approaches to diachronic conceptual change. arXiv preprint arXiv:1811.06278 (2018)
27. Tseng, P.: Convergence of a block coordinate descent method for nondifferentiable minimization. J. Optim. Theory Appl. **109**(3), 475–494 (2001)
28. Utiyama, M., Isahara, H.: A statistical model for domain-independent text segmentation. In: Proceedings of the 39th Annual Meeting of the Association for Computational Linguistics, pp. 499–506 (2001)

29. Yao, Z., Sun, Y., Ding, W., Rao, N., Xiong, H.: Dynamic word embeddings for evolving semantic discovery. In: Proceedings of the Eleventh ACM International Conference on Web Search and Data Mining, pp. 673–681 (2018)
30. Zhang, Y., Jatowt, A., Bhowmick, S., Tanaka, K.: Omnia Mutantur, Nihil Interit: connecting past with present by finding corresponding terms across time. In: Proceedings of the 53rd Annual Meeting of the Association for Computational Linguistics and the 7th International Joint Conference on Natural Language Processing (Volume 1: Long Papers), pp. 645–655 (2015)

Joint Model Using Character and Word Embeddings for Detecting Internet Slang Words

Yihong Liu[1] and Yohei Seki[2(✉)]

[1] Graduate School of Comprehensive Human Sciences, University of Tsukuba, Tsukuba, Japan
s2021716@s.tsukuba.ac.jp

[2] Faculty of Library, Information and Media Science, University of Tsukuba, Tsukuba, Japan
yohei@slis.tsukuba.ac.jp

Abstract. The language style on social media platforms is informal and many Internet slang words are used. The presence of such out-of-vocabulary words significantly degrades the performance of language models used for linguistic analysis. This paper presents a novel corpus of Japanese Internet slang words in context and partitions them into two major types and 10 subcategories according to their definitions. The existing word-level or character-level embedding models have shown remarkable improvement with a variety of natural-language processing tasks but often struggle with out-of-vocabulary words such as slang words. We therefore propose a joint model that combines word-level and character-level embeddings as token representations of the text. We have tested our model against other language models with respect to type/subcategory recognition. With fine-grained subcategories, it is possible to analyze the performance of each model in more detail according to the word formation of Internet slang categories. Our experimental results show that our joint model achieves state-of-the-art performance when dealing with Internet slang words, detecting semantic changes accurately while also locating another type of novel combinations of characters.

Keywords: Internet slang words · Joint embeddings · ELMo

1 Introduction

In natural-language processing, an online collection of digital objects is analyzed to identify metadata-related phrases. Examples include ontology mapping from clinical text fragments [5] and keyphrase extraction for a scholarly digital library [13].

To alleviate fixed vocabulary or terminology-resource constraints and expand the metadata to include informal content such as social tags [20], we propose a method for detecting unknown Internet slang words. In this study, we introduce a new resource containing Internet slang words specifiable at two levels: as a "major type" or as a "fine-grained subcategory" that subdivides words according to their semantics and construction. We aim to further analyze Internet slang words through fine-grained subcategories based on the internal construction and diversity of context which are learned with pre-trained embedding models.

H.-R. Ke et al. (Eds.): ICADL 2021, LNCS 13133, pp. 18–33, 2021.
https://doi.org/10.1007/978-3-030-91669-5_2

The richness of social media platforms, such as Twitter and Weibo, enables users from different countries, even from different races and using different languages, to communicate effectively. Such a large diversity has led to the semantic development of terminology on the Internet, including Internet slang words. Many Internet slang words have evolved from existing words (denoted "new semantic words") or are newly created (denoted "new blend words"). These are constantly being introduced by users via social media platforms and can quickly become popular. Some Internet slang words have become widely used and even become popular in our daily lives outside of the Internet. The semantic meanings of Internet slang words, however, have not been codified in a dictionary or a corpus, which is essential if human researchers or AI-based machines are to understand them. In addition, many of them share similar features, such as harmonic sounds, abbreviations, or derived meanings, and knowledge of such features may allow better analysis of such words. Although such new words (also known as "youth jargon" [8]) have been analyzed and investigated in detail in terms of their social and linguistic context, existing word-embedding methods have not focused on the differences between these features.

Our proposed method combines character and word embedding using embeddings from existing language models (ELMo) [3] and word2vec [12]. We then input the combination of the obtained token representations (or subword-based word embeddings), aiming to detect if the word in context is an Internet slang word.

Because no public Japanese slang-word datasets are available, we constructed a novel Internet slang word dataset that contains a total of 100 Internet slang words for each language, together with their meanings. We divided them manually in terms of their characteristics into two major types: "new semantic words" (SEM) and "new blend words" (BLN). We also identified 10 fine-grained subcategories based on word-formation features to help classify the details of lexical changes in the slang words. We then compared our model with existing models that use character embeddings only, word embeddings only, and subword embeddings. Experiments using the new dataset showed that our proposed joint embedding method performed better than the baseline methods with respect to the detection and classification of Internet slang words.

The contributions of this paper can be summarized as follows.

1. We have proposed an Internet slang word corpus with a novel approach that classifies Internet slang words into two main types, "new semantic words" and "new blend words," and 10 subcategories defined in terms of word creation and morphological features.
2. To obtain token representations for Internet slang words effectively, we propose a novel joint embedding method that combines character embedding and word embedding by utilizing ELMo and word2vec. To compare our method with a subword-based embedding method, we constructed a bidirectional encoder representations from transformers (BERT) model as one of the comparison models.
3. Our experimental results show that our proposed encoder can match the performance of the baseline approach, precisely discriminating the polysemy of SEM-type words but also distinguishing the BLN-type words based different ways of constructing combinations at the same time.

This paper is organized as follows. Section 2 describes related work. Section 3 details the Japanese Internet-slang-word dataset we constructed using tweets from Twitter[1]. In Sect. 4, we introduce our proposed model to address the identified problems. Section 5 gives the details of the comparison experiments. Finally, we draw conclusions and discuss future work in Sect. 6.

2 Related Works

2.1 Word and Character Embedding

Word-embedding methods, which learn embedding rules according to the external context of words, have been used in many natural-language processing (NLP) tasks. They contain some limited word-level contextual information [23], because of other word models' poor performance when handling out-of-vocabulary (OOV) words. This is caused by the sparseness of word distributions for such words. Many approaches segment the text into character units, with such character-based models consistently outperforming word-based models for deep learning of Chinese representations [10]. Lample et al. [9] demonstrated that character-based representations can be used to handle OOV words in a supervised tagging task. Pinter et al. [17] aggregated language-related characteristics from the perspective of individual hidden units within a character-level long short-term memory (LSTM). Chen et al. [2] proposed multiple-prototype character embeddings and an effective word selection method to address the issues of character ambiguity and noncompositional words, which are also problematic for Internet slang words. Using both character and word representation, language models should have a more powerful capability to encode internal contextual information.

2.2 ELMo

ELMo [15] creates contextualized representations of each token by concatenating the internal states of a two-layer biLSTM trained on a bidirectional language modeling task. In contrast to neural networks such as LSTM, which can only generate a fixed vector for each token, ELMo's main approach is to train a complete language model first and then use this language model to process the text to be trained to generate the corresponding word vectors. It also contains character-based information, which allows the model to form representations of OOV words [24]. With this type of character-based dynamic vector, ELMo considers both character information and the context between words, enabling it to recognize semantic differences most effectively.

2.3 BERT

BERT [3] is a bidirectional model that is based on a transformer architecture. It replaces the sequential nature of recurring neural networks with a much faster attention-based approach. As it is bidirectional, given the powerful capabilities of the encoder, BERT is quite effective in certain NLP cases [3]. The BERT model can interact fully with the characteristics of the deep neural network to improve the accuracy of the model.

[1] https://twitter.com/.

It employs a variety of subword tokenization methods, with byte-pair encoding [21] being the most popular approach to segmenting text into subword units. Although both character-based and subword-based BERTs can solve the OOV problem, the character-based model demonstrates stronger autocorrection once a word is judged to be a type error [11]. Therefore, we selected subwords as the token representation in our comparisons.

2.4 Detection of Internet Slang Words

Compared with English or other languages whose units are characters from the Latin alphabet and have space segments between words, Japanese languages not only have words built using characters or pseudonyms but also no obvious word delimiters. Although progress has been made in processing English Internet slang words [7], only a few works have focused on processing Japanese Internet slang.

The emergence of Internet slang words is the result of language evolution. Linguistic variation is not only a core concept in sociolinguistics [1], but a new type of language feature and a manifestation of the popular culture of social networks. In the information transmitted on social media, some Internet slang words have evolved meanings different from standard dictionaries and there are also words that are not listed at all in such dictionaries. To address this, we identify two types, "new semantic words" (different meanings from those recorded in dictionaries) and "new blend words" (not recorded in general dictionaries). In addition, with reference to the construction of NYTWIT [16], a dataset of novel English words in The New York Times, all of our 100 identified Internet slang words were partitioned into 10 subcategories, based on the nature of the word formation in existing Internet slang.

When dealing with new words in Japanese, considering the linguistic patterning of word formation enables better refinement of such words, in addition to helping to analyze how models learn various types of semantic variation from a linguistic perspective. Therefore, to analyze slang words in depth, it is necessary to both identify word components and the formation process involved in creating new words. Hida et al. [4] showed that words are made up of word bases and affixes. Yonekawa [25] concluded that the novel words either have no connection to existing words or are created by existing words. He also identified five types of word formation based on existing words: Borrowing, Synthesis, Transmutation, Analogy, and Abbreviation.

1. Borrowing: Borrowing from jargon, dialects, and foreign languages. It may be accompanied by the Japanization of pronunciation in foreign languages and a change of meaning.
2. Synthesis: Combining two or more words to create a new word.
3. Transmutation: Creating a new word by prefixing or suffixing with an existing word.
4. Analogy: A word made from an incorrect solution by analogy with an existing word form.
5. Abbreviation: A method of omitting parts of a word.

We investigated our collection of Internet slang words introduced in Sect. 3 from the viewpoint of word-formation types. We eliminated some non-Internet slang forms like words newly named in a professional field, and grouped words that did not belong to the types described above into novel categories. We describe the details in Sect. 3.3.

3 Internet Slang Corpus

In this section, we describe our proposed two types and 10 subcategories for labeling Internet slang words. We then explain the construction of our dataset.

3.1 New Semantic Words

"New semantic words" involve entries initially recorded in the dictionary and used daily. These words, however, now have new meanings that have become popular because of their similarities to other terms or popular iconic events. Although such vocabulary can be segmented directly, the context-based meaning is often very different from the original, and even the parts of speech can change. Examples are shown in Table 1.

Table 1. Examples of new semantic words.

New Semantic Words			
Word	**Common Usage**	**Internet Usage**	**Subcategory**
草	grass	interesting	Neologis
丸い	round	safe	Rhetoric
垢	dirt	account	Pun
		...	

3.2 New Blend Words

"New blend words" often borrow foreign words, dialects, numeric elements, and icons by focusing on sound similarity. They often combine definitions, homonyms, abbreviations, repetitions, and other word-formation methods. They can also involve unconventional grammars [7]. Internet language has achieved the effect of "novelty" through its unconventional nature and nonstandard usage. Examples are shown in Table 2.

Table 2. Examples of new blend words.

New Blend Words			
Word	**Common Usage**	**Internet Usage**	**Subcategory**
禿同	–	strongly agree	Pun
ふぁぼ	–	favourite	Japanese-English
過去1	–	Most or Best ... in the past	Pun
		...	

3.3 Word-Formation Subcategories

We identify subcategories by considering the formation of the Japanese Internet slang words. Examples for each subcategory are given in Table 3. Beside, a summary is given in Table 4.

1 *Gairaigo*[2]: the foreign words transliterated into Japanese and usually written in the "*katakana*" phonetic script.
2 Japanese–English: the Japanese words look or sound just like English, but have different meanings from their English origins.
3 Dialect Borrowing: the words are derived from nonstandard Japanese dialects but have different meanings and contexts from the original words.
4 Compound Word: words created by joining together two (or more) root words.
5 Derived Word: words created by attaching affixes (which cannot be used in isolation) to the root words.
6 Abbreviation: words that omit some characters of existing words to create a shorter word form.
7 Acronymic: words composed of acronyms from multiple words.
8 Pun: each word is replaced with another word that sounds very similar but has a different meaning, thereby making the expression humorous.
9 Rhetoric: words with new and more specific meanings based on figurative expressions.
10 Neologism: words whose compositional characteristics are difficult to recognize from existing root words and affixes according to word-formation subcategories in 1–9.

Table 3. Examples of internet slangs for each subcategory

Subcategory	Word	Etymology
Gairaigo	ミーム	From English Slang Word *meme*.
Japanese-English	ワンチャン	With the same pronunciation as *One Chance*.
Dialect Borrowing	してもろて	From the dialect in Kansai district of "してもらって" (let someone do).
Compound Word	秒で	Combination of the noun "秒" (second) and the auxiliary "で" (at).
Derived Word	わかりみ	A verb "わかる" (understand) with a noun suffix "み" (-ing) to nominalize the verb.
Abbreviation	そマ	An abbreviation of the sentence "それマジ" (Is that really?)
Acronymic	三密	*Three Cs* (crowded places, close contact settings, and closed spaces)
Pun	鯖	With the same pronunciation as *server*.
Rhetoric	世紀末	A decadent worldview extended from the end of the century.
Neologism	タヒる	Katakana characters combination "タヒ" is morphologically similar to "死" (death).

[2] "Foreign words" in Japanese.

Table 4. Subcategories of New Semantic Words (SEM) and New Blend Words (BLN).

Subcategory	Tag	SEM	BLN	Total
Gairaigo	g		3	3
Japanese-English	je		7	7
Dialect Borrowing	db	2		2
Compound Word	c		8	8
Derived Word	dw		5	5
Abbreviation	ab		19	19
Acronymic	ac	1	1	2
Pun	p	8	1	9
Rhetoric	r	38		38
Neologism	n	1	6	7

3.4 Corpus Construction

The dataset used in our experiments was constructed from Japanese-language tweets accessed via Twitter's application programmer's interface[3]. A preprocessing stage removed any emoji or *kaomoji* (emoticon) data in the tweets. We then selected 100 Japanese Internet slang words whose meanings in Internet usage were specified in an online Japanese slang words collection[4], with 50 of the words being identified as "new semantic words" and the remainder as "new blend words." Among these words, we eliminated some ineligible words, such as those that had been updated or added to the standard Japanese dictionary, those that did not originate on the Internet, and those that had extremely low usage. Then, using the word creation rules in Sect. 3.3, 10 subcategories were finally determined for this set of Japanese Internet slang words. Next, we collected 50 sentences containing Internet slang used as such. For comparison, we collected another 50 sentences containing the same words but used in a general sense. Please note that the texts in which contain the selected Internet slang words may also contain other Internet slangs, including those outside the 100 words we selected. We have also annotated those words, therefore the actual samples in our corpus contain more Internet slangs than we planned at first. In the annotation step, three native Japanese annotators identified Internet slang words in the sentences and labeled their types and subcategories. Most of the internet slang words can be clearly distinguished from the words with common usage, but in a few cases annotators were unable to agree the results, because of incorrect usage by the tweet users or insufficient information in the short tweets. Finally, we decided to exclude such ambiguous cases from the corpus.

The collected Japanese sentences were then segmented into character units, subword units (via the SentencePiece [6] algorithm), and word units (via MeCab[5]). Finally, we tagged the characters, using the *BIO (B-Begin I-Inside O-Others)* tagging style to represent positional information. Examples are given in Table 5.

[3] https://developer.twitter.com/en/docs.
[4] https://numan.tokyo/words/.
[5] https://www.mlab.im.dendai.ac.jp/~yamada/ir/MorphologicalAnalyzer/MeCab.html.

Table 5. Examples of annotations of Japanese internet slang words.

<table>
<tr><th>New Semantic Words</th></tr>
<tr><td>-Internet Usage
初/O 鯖/ B-sem | p の/O 初/O 心/O 者/O に/O 迷/O惑/O
か/Oけ/Oる/O な/O ！/O</td></tr>
<tr><td>-Common Usage
脂/O の/O 乗/O っ/O た/O 鯖/O の/O 塩/O 焼/O き/O と/O か/O
と/O 合/O わ/O せ/O た/O い/O</td></tr>
<tr><th>New Blend Words</th></tr>
<tr><td>-Internet Usage
そ/B-bln|ab マ/I-bln | ab ？/O 行/O け/O る/O 時/O 言/O っ/O て/O
バ/O イ/O ト/O 無/O け/O れ/O ば/O ワ/O イ/O も/O 行/O く/O わ/O</td></tr>
<tr><td>-Common Usage
今/O 日/O こ/O そ/O マ/O へ/O ラ/O し/O ま/O す/O ！/O</td></tr>
</table>

4 Method

We propose an encoder model that combines characters and word embeddings as the token representation for each token. We take them as the input to the two layers of a biLSTM network and perform a weighted sum of each biLSTM layer, following Peter et al. [14] with respect to weight values. The output embeddings are provided to downstream tasks as "features" of each token.

For the "new blend words" type, it is necessary to consider the connection between and the collocation of characters. We assumed that the performance in detecting "new blend words" would be improved by learning these characteristics. Considering the relationships between a word and its characters, the discriminative characters are crucial for distinguishing slight semantic differences [18]. In the joint model, word embedding can store contextual information among surrounding words, and character embeddings provide semantic information between characters in the word.

The word embedding obtained by word2vec is fixed, whereas ELMo [15] allows flexible changes according to the context. Therefore, using embeddings from word2vec that obtain static standard contextual information and accumulating them into the ELMo representation can enable semantic distinctions based on context from the dataset.

Following [19], we also use a "conditional random field" as the output layer to predict a character's tag. In this architecture, the last hidden layer is used to predict confidence scores for characters with each of the possible labels.

4.1 Joint Model Using Character and Word Embeddings

The general structure of our model is given in Fig. 1.

In Fig. 1, the parameter w_j is the word embedding of token j, N_j is the number of characters in this token, c_j^k is the embedding of the k-th character, and y_j^k is joint

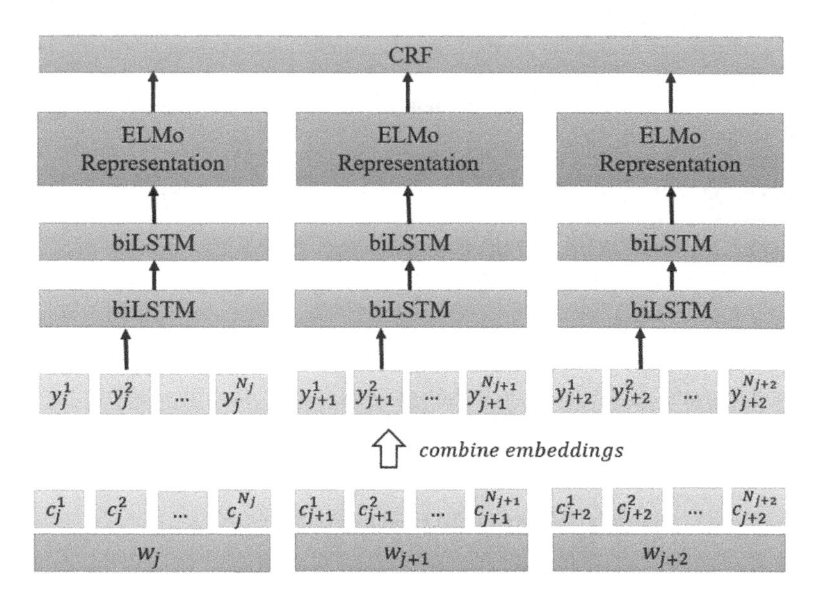

Fig. 1. Structure of our proposed model

embedding for the character-level unit annotation, for which the relationship between these parameters is given in Eq. (1).

$$y_j^k = c_j^k \oplus \frac{w_j}{N_j} \tag{1}$$

4.2 ELMo Representation

We also trained the ELMo model introduced by [15] using the Japanese Wikipedia Dataset[6], for which all of the layers were combined via a weighted-average pooling operation. The output of the ELMo model is given in Eq. (2).

$$\mathbf{ELMo}_k = \gamma \sum_{j=0}^{L} s_j \mathbf{h}_{k,j} \tag{2}$$

Here, k represents the position of each (character-level) token and j is the number of layers. $\mathbf{h}_{k,j}$ is the hidden-state output for each biLSTM layer. The parameter s represents the softmax-normalized weight, and the scalar parameter γ allows the task model to scale the entire ELMo vector. γ could enhance the optimization process.

[6] https://dumps.wikimedia.org/jawiki/latest/ [accessed on October 2020].

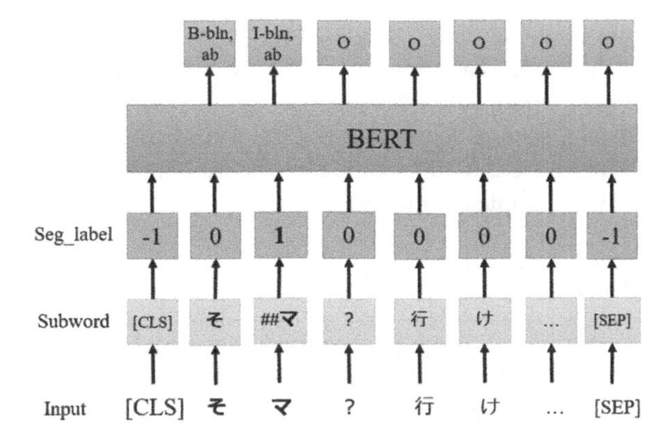

Fig. 2. Structure of the subword-based method on BERT.

4.3 Subword-Level Representation Based on BERT

According to the Japanese BERT model [22], we distinguish the first part of the non-head subwords after segmentation, then add "##" to the first character of all non-head subwords, and generate a seg_label. The subword-level BERT model is given in Fig. 2.

5 Experiments

We compared the proposed model with LSTM, pure ELMo from Tohoku NLP LAB[7], and BERT, with respect to the detection of our two types and 10 subcategories. The aim was to evaluate the impact of using a joint-embedding input instead of using common word embedding alone. In an attempt to dissociate this impact from any other effects that may be related to the training models used in our own specific settings, we trained our ELMo model using the Japanese Wikipedia Dataset with the same vector dimensionality as word2vec. We set the same parameters for the baseline models as used in our models (e.g., the number of epochs was set as 10 for all models; the batch size was set as 16 for ELMo and LSTM family models, and as 8 for BERT model by default values.)

5.1 Dataset and Settings

For learning the character and word-embedding of all Japanese terms, we chose the Japanese Wikipedia Dataset introduced in Sect. 4.2 as the pre-trained dataset, for which the number of words and characters are about 312,000 and 10,000, respectively.

We set the vector dimensionality as 200 and the context window size as 10 for character-based embedding and the vector dimensionality as 200 and window size as 5 for word-based embedding. We also adjusted the other settings to match those in the Japanese Wikipedia Entity Vector[8].

[7] https://github.com/cl-tohoku/elmo-japanese.

[8] http://www.cl.ecei.tohoku.ac.jp/~m-suzuki/jawiki_vector/.

We used five-fold cross-validation to split the Internet Slang Corpus into five small datasets for training the model; each dataset has 80% training data and 20% testing data. In addition, 20% of the training data was randomly selected as the validation set in order to adjust the parameters, such as learning rate (1e−2 for ELMo and LSTM; 2e−5 for BERT). As a result, there were 6,400 samples in the training set, 1,600 samples in the validation set, and 2,000 samples in the test set.

5.2 Baseline Methods

To investigate the performance of the joint embedding model of characters and words with respect to its ability to distinguish semantics with ELMo embedding, we compared it with three baseline methods based on LSTM, two ELMo methods, and one BERT method as follows[9].

1. Baseline1 (c): character-only embedding with LSTM network
2. Baseline2 (w): word-only embedding with LSTM network
3. Baseline3 (w+c): character and word embeddings with LSTM network
4. Pure-ELMo (c): character-only embedding with original ELMo method
5. BERT (subword): subword-based embedding with BERT
6. Our ELMo (c): character-only embedding with our proposed ELMo method.

We used the pure ELMo model from Tohoku NLP LAB but used only character-level embedding, for which the dimensionality was 512. We used the BERT model from BERT-base [27]. Note that both of them were trained via the Japanese Wikipedia Dataset.

We trained and tested the models with each of the five datasets and took the average value of the precision, recall and F1-score among the Internet slang words assigned to the type or subcategories. Finally, to conduct ablation study, we also prepared two sets of our models to compare: (1) excluding the second layer of ELMo output; and (2) excluding CRF layer. We also conducted two-tailed paired samples t-tests for the average value of the F1-score in terms of subcategories to investigate the statistical significance for our method against the other methods.

5.3 Experimental Results

From the results given in Table 6, all models were better at recognizing "new blend words" (BLN) than recognizing "new semantic words" (SEM). For the SEM recognition task, our model was better at recognizing Internet slang usage in different contexts. The BERT model was also effective especially for BLN. Among the LSTM family of models, using only a character-embedding input (Baseline1) is better than using only a word-embedding input (Baseline2) for SEM and BLN, and the joint-embedding models outperform these methods. This suggests that joint-embedding methods can extract Internet slang words more completely. In addition, character-vector-based methods contribute more than word-vector-based methods do by focusing on the connections between characters.

[9] "c" and "w" denote character and word embeddings, respectively.

Table 6. Results of detecting "New Semantic Words" and "New Blend Words".

Type	Measure	Baseline1	Baseline2	Baseline3	Pure-ELMo	BERT	Our ELMo (c)	Our model	-ELMo2nd	-CRF
SEM	Precision	0.381	0.357	0.371	0.446	0.376	0.412	**0.451**	0.326	0.337
	Recall	0.281	0.049	0.349	0.226	0.438	0.453	**0.470**	0.286	0325
	F1-score	0.303	0.086	0.359	0.297	0.405	0.429	**0.460**	0.303	0.330
BLN	Precision	0.482	0.465	0.485	0.589	0.576	0.579	**0.596**	0.452	0.452
	Recall	0.487	0.166	0.490	0.420	**0.612**	0.577	0.607	0.431	0.456
	F1-score	0.485	0.244	0.488	0.488	0.593	0.577	**0.602**	0.442	0.454

Table 7. F1-scores for detecting fine-grained subcategories. "**" and "*" denote cases where the difference in macro average between Our Model and other methods is statistically significant for $p < 0.01$ and $p < 0.05$ using two-tailed paired samples t-tests.

Subcategory	Baseline1	Baseline2	Baseline3	Pure-ELMo	BERT	Our ELMo (c)	Our model	-ELMo2nd	-CRF
Gairaigo	0.588	0.393	0.762	0.644	0.676	0.625	0.639	0.694	**0.797**
Japanese-English	0.492	0.150	0.596	0.570	0.627	0.545	0.600	0.594	**0.696**
Dialect Borrowing	0.682	0.896	**0.898**	0.604	0.593	0.616	0.596	0.693	0.795
Compound Word	0.593	0.363	0.596	0.556	**0.631**	**0.631**	0.627	0.490	0.495
Derived Word	0.483	0.334	0.486	0.512	**0.622**	0.572	0.598	0.470	0.588
Abbreviation	0.326	0.220	0.325	**0.399**	0.378	0.350	0.368	0.317	0.326
Acronymic	0.296	0.246	0.374	0.428	0.461	0.421	**0.486**	0.312	0.390
Pun	0.386	0.153	0.416	0.385	**0.560**	0.496	0.532	0.375	0.413
Rhetoric	0.312	0.062	0.361	0.360	0.414	0.458	**0.486**	0.293	0.355
Neologism	0.282	0.298	**0.371**	**0.371**	0.316	0.329	0.355	0.260	0.267
F1-score Avg.	0.444	0.312	0.519	0.483	0.528	0.504	**0.529**	0.450	0.512
Significance	**	*		*		*	–	*	

The experimental results with respect to "Fine-grained subcategories" are given in Table 7. Our model performs best overall in subcategory detection, and improved significantly against Baseline1, Baseline2, Pure-ELMo, Our ELMo (c), and ablation study excluding ELMo 2nd layer methods. Baseline3 performs better for some BLN words (Neologism subcategory) and also for SEM words (Dialect Borrowing subcategory). BERT also performs better for BLN words (Compound Word and Derived Word subcategories) and also for SEM words (Pun subcategory).

5.4 Discussion

Using a joint embedded model, both ELMo and LSTM (Our Model and Baseline3) are better able to recognize subcategories than models using character embeddings alone (Our ELMo (c) and Baseline1) or word embeddings alone (Baseline2) alone. This demonstrates that joint embeddings enable better learning of relationships between vocabulary and contexts and better identify feature correspondences between characters and the word that characters belong to.

The results of the subcategory experiment demonstrates that our proposed model is excellent at recognizing the subcategories associated SEM words. Moreover, although previous studies have shown that ELMo models can recognize semantic changes in words successfully, our joint embedded ELMo model outperforms the other two ELMo comparison methods (Pure-ELMo and Our ELMo (c)) with statistical significance. This suggests that our ELMo model can better capture subtle differences such as word constituents when learning the joint embedding of two granularities than can the original ELMo model.

Comparing to LSTM model (Baseline3), our model outperforms it for some subcategories (Abbreviation, Compound Word, Derived Word, etc.). The joint embedding ELMo model could generate the dynamic character and word representations. The former helps the model to learn the components of word formation and the latter aids the control of word boundaries. For the words Abbreviation, Compound Word, and Derived Word, whose constituents are relevant to the new meaning, our ELMo model is better than LSTM models thanks to the distinction of character-level and word-level semantics.

BERT can flexibly represent tokens corresponding to textual difference, and can therefore learn the polysemy through the "contextual pulse" of subwords. Subwords with less than word-level granularity, however, also inevitably misses word-to-word correspondence. So although it performs optimally on the subcategory recognition of BLN-type words, it does not perform as well when it encounters SEM-type words like Rhetoric, whose lexical changes do not involve the constituent elements.

Zhang et al. [26] pointed out that fine-tuning with BERT has instability problems when processing small datasets. The model will continue to oscillate at the beginning of training, which will reduce the efficiency of the entire training process, slow down the speed of convergence, and limit the model to a certain extent. In summary, it is significantly slower than the other models in terms of loss convergence during training and has the instability issues mentioned above.

Finally, for the ablation study, our model outperforms the excluding ELMo 2nd layer case significantly. Although the 2nd upper layer can be used to induce more contextual semantic content, the 1st lower layer focuses on local syntax. We suppose that this is the reason why our model with the 2nd layer outperforms for detecting Internet slang words.

Our model also outperforms the excluding CRF layer case, because the introduction of the CRF layer not only makes the algorithm complexity simple but also ensures the orderliness of the BIO tag. We also suppose that the orderliness is not essential for *Gairaigo* or Japanese-English subcategories.

Our results demonstrate that, for our proposed corpus for OOV-related slang word recognition and differentiation tasks, joint embedding in the ELMo model has achieved significant improvements.

6 Conclusion

We have introduced a new corpus of Japanese Internet slang words labeled with two major types and 10 subcategories based on semantic and morphological features. We have also proposed a novel encoder method for detecting Internet slang words, combining character and word embeddings through a deep bidirectional language model. The detection of major types demonstrates the superiority of our joint model in identifying Internet slang words. Complemented by fine-grained subcategories, we could analyze language models in detail from the viewpoints of word formation of Internet slang words. In addition, our experiments showed that the joint embedding of words and characters can enrich the information acquired during training because it can utilize both contextual semantic information and the diverse association between a word and its characters. In future work, considering the similarity between Chinese and Japanese Internet slang words, we will create a corpus of Chinese predictions for similar recognition tasks. We will also aim to disambiguate the word senses of Internet slang words using joint embedding models. Finally, we will apply our method to detect metadata phrases from informal social texts.

Acknowledgments. We are very grateful to Dr. Wakako Kashino at the *National Institute for Japanese Language and Linguistics* for her guidance and help in identifying and classifying Japanese Internet slang words. We are also grateful to the Japanese members of our research laboratory for their help in the annotation and checking of the dataset. This work was partially supported by a Japanese Society for the Promotion of Science Grant-in-Aid for Scientific Research (B) (#19H04420).

Ethics and Impact Statement. This research was conducted with the approval of the Ethics Review Committee of the Faculty of Library, Information and Media Science, the University of Tsukuba. The participants in the corpus creation experiment were asked to sign a consent form in advance and were allowed to quit the experiment at any time.

References

1. Chambers, J.K.: Sociolinguistic Theory, 3rd edn. Wiley-Blackwell (2008)
2. Chen, X., Xu, L., Liu, Z., Sun, M., Luan, H.: Joint learning of character and word embeddings. In: Proceedings of the 24th International Conference on Artificial Intelligence (IJCAI 2015), pp. 1236–1242, July 2015
3. Devlin, J., Chang, M.W., Lee, K., Toutanova, K.: BERT: pre-training of deep bidirectional transformers for language understanding. In: Proceedings of the 2019 Conference of the North American Chapter of the Association for Computational Linguistics: Human Language Technologies (Long and Short Papers), Minneapolis, Minnesota, vol. 1, pp. 4171–4186, June 2019
4. Hida, Y., Endo, Y., Kato, M., Sato, T., Hachiya, K., Maeda, T.: The research encyclopedia of Japanese linguistic. Jpn. Liguist. **3**(4), 125–126 (2007). (in Japanese)
5. Kersloot, M.G., van Putten, F.J.P., Abu-Hanna, A., Cornet, R., Arts, D.L.: Natural language processing algorithms for mapping clinical text fragments onto ontology concepts: a systematic review and recommendations for future studies. J. Biomed. Semant. **11** (2020)
6. Kudo, T., Richardson, J.: SentencePiece: a simple and language independent subword tokenizer and detokenizer for Neural Text Processing. In: Proceedings of the 2018 Conference on Empirical Methods in Natural Language Processing: System Demonstrations (EMNLP 2018), Brussels, Belgium, pp. 66–71. Association for Computational Linguistics, November 2018
7. Kundi, F.M., Ahmad, S., Khan, A., Asghar, M.Z.: Detection and scoring of internet slangs for sentiment analysis using SentiWordNet. Life Sci. J. **11**(9), 66–72 (2014)
8. Kuwamoto, Y.: A shift of morphological and semantic structures in ambiguous expression of Japanese Youth Jargons Wakamono-kotoba: approaching a diachronic study with a database of a TV drama. Natl. Inst. Technol. Akita Coll. **49**, 68–75 (2014). (in Japanese)
9. Lample, G., Ballesteros, M., Subramanian, S., Kawakami, K., Dyer, C.: Neural architectures for named entity recognition. In: Proceedings of the 2016 Conference of the North American Chapter of the Association for Computational Linguistics: Human Language Technologies, San Diego, California, pp. 260–270, June 2016. https://doi.org/10.18653/v1/N16-1030. https://www.aclweb.org/anthology/N16-1030

10. Li, X., Meng, Y., Sun, X., Han, Q., Yuan, A., Li, J.: Is word segmentation necessary for deep learning of Chinese representations? In: Proceedings of the 57th Annual Meeting of the Association for Computational Linguistics, Florence, Italy, pp. 3242–3452, July 2019
11. Ma, W., Cui, Y., Si, C., Liu, T., Wang, S., Hu, G.: CharBERT: character-aware pre-trained language model. In: Proceedings of the 28th International Conference on Computational Linguistics (COLING 2020), Barcelona, Spain, pp. 39–50, December 2020
12. Mikolov, T., Sutskever, I., Chen, K., Corrado, G.S., Dean, J.: Distributed representations of words and phrases and their compositionality. In: Burges, C., Bottou, L., Welling, M., Ghahramani, Z., Weinberger, K. (eds.) Advances in Neural Information Processing Systems 26, pp. 3111–3119. Curran Associates, Inc. (2013). http://papers.nips.cc/paper/5021-distributed-representations-of-words-and-phrases-and-their-compositionality.pdf
13. Patel, K., Caragea, C., Wu, J., Giles, C.L.: Keyphrase extraction in scholarly digital library search engines. In: IEEE International Conference on Web Services (ICWS 2020), pp. 179–196, October 2020
14. Peters, M., Ammar, W., Bhagavatula, C., Power, R.: Semi-supervised sequence tagging with bidirectional language models. In: Proceedings of the 55th Annual Meeting of the Association for Computational Linguistics (Volume 1: Long Papers), Vancouver, Canada, pp. 1756–1765, July 2017. https://doi.org/10.18653/v1/P17-1161. https://www.aclweb.org/anthology/P17-1161
15. Peters, M., et al.: Deep contextualized word representations. In: Proceedings of the 2018 Conference of the North American Chapter of the Association for Computational Linguistics: Human Language Technologies (Long Papers), New Orleans, Louisiana, vol. 1, pp. 2227–2237, June 2018. https://doi.org/10.18653/v1/N18-1202. https://www.aclweb.org/anthology/N18-1202
16. Pinter, Y., Jacobs, C.L., Bittker, M.: NYTWIT: a dataset of novel words in the New York times. In: Proceedings of the 28th International Conference on Computational Linguistics (COLING 2020), Barcelona, Spain, pp. 6509–6515. International Committee on Computational Linguistics, December 2020. https://www.aclweb.org/anthology/2020.coling-main.572
17. Pinter, Y., Marone, M., Eisenstein, J.: Character eyes: seeing language through character-level taggers. In: Proceedings of the 2019 ACL Workshop BlackboxNLP: Analyzing and Interpreting Neural Networks for NLP, Florence, Italy, pp. 95–102, August 2019. https://doi.org/10.18653/v1/W19-4811. https://www.aclweb.org/anthology/W19-4811
18. Qiao, X., Peng, C., Liu, Z., Hu, Y.: Word-character attention model for Chinese text classification. Int. J. Mach. Learn. Cybern. **10**(12), 3521–3537 (2019)
19. Rei, M., Crichton, G., Pyysalo, S.: Attending to characters in neural sequence labeling models. In: Proceedings of the 26th International Conference on Computational Linguistics: Technical Papers (COLING 2016), Osaka, Japan, pp. 309–318, December 2016. https://www.aclweb.org/anthology/C16-1030
20. Samanta, K.S., Rath, D.S.: Social tags versus LCSH descriptors: a comparative metadata analysis in the field of economics. J. Libr. Inf. Technol. **39**(4), 145–151 (2019)
21. Sennrich, R., Haddow, B., Birch, A.: Neural machine translation of rare words with subword units. In: Proceedings of the 54th Annual Meeting of the Association for Computational Linguistics (Volume 1: Long Papers), Berlin, Germany, pp. 1715–1725, August 2016
22. Shibata, T., Kawahara, D., Kurohashi, S.: Improved accuracy of Japanese parsing with BERT. In: Proceedings of 25th Annual Meeting of the Association for Natural Language Processing, pp. 205–208 (2019). (in Japanese)
23. Sun, Y., Lin, L., Yang, N., Ji, Z., Wang, X.: Radical-enhanced Chinese character embedding. In: Loo, C.K., Yap, K.S., Wong, K.W., Teoh, A., Huang, K. (eds.) ICONIP 2014. LNCS, vol. 8835, pp. 279–286. Springer, Cham (2014). https://doi.org/10.1007/978-3-319-12640-1_34

24. Ulčar, M., Robnik-Šikonja, M.: High quality ELMo embeddings for seven less-resourced languages. In: Proceedings of the 12th Language Resources and Evaluation Conference (LREC 2020), Marseille, France, pp. 4731–4738. European Language Resources Association, May 2020. https://aclanthology.org/2020.lrec-1.582
25. Yonekawa, A.: New Words and Slang Words. NAN'UN-DO Publishing (1989). (in Japanese)
26. Zhang, T., Wu, F., Katiyar, A., Weinberger, K.Q., Artzi, Y.: Revisiting few-sample BERT fine-tuning. In: Proceedings of International Conference on Learning Representations (ICLR 2021), May 2021. https://openreview.net/forum?id=cO1IH43yUF
27. Zhao, X., Hamamoto, M., Fujihara, H.: Laboro BERT Japanese: Japanese BERT Pre-Trained With Web-Corpus (2020). https://github.com/laboroai/Laboro-BERT-Japanese

Knowledge Discovery from the Digital Library's Contents: Bangladesh Perspective

Md. Habibur Rahman[1,2](✉), Sohaimi Zakaria[3], and Azree Ahmad[4]

[1] Chattogram Veterinary and Animal Sciences University, Chattogram, Bangladesh
librarian@cvasu.ac.bd
[2] Universiti Teknologi MARA (UiTM), Shah Alam, Malaysia
[3] Faculty of Information Management, Universiti Teknologi MARA (UiTM), Shah Alam, Malaysia
sohaimiz@uitm.edu.my
[4] Faculty of Information Management, Universiti Teknologi MARA (UiTM), Kedah Branch, Merbok, Malaysia
azree@uitm.edu.my

Abstract. The purpose of this study is to explore the present trends of knowledge discovery (KD) from digital library (DL) systems in Bangladesh. The main obstacles of KD from the contents of DL and ways to overcome the barriers are also described. This study uses both qualitative and quantitative approaches along with the review of related literature. The present scenario of the KD from the contents of DL in Bangladesh is presented by a survey with a structured questionnaire and reviewing related literature. This study identifies the challenges of KD from the contents of DL in Bangladesh, which are inaccurate bibliographic metadata, a lack of accurate holdings data, a lack of synchronized bibliographic metadata, etc. This study also suggests some suitable ways to overcome the existing challenges of KD from DL contents, such as providing high-quality data, preserving complete bibliographic metadata, preserving accurate holdings data, synchronizing bibliographic metadata and holdings data, and using consistent data formats.

Keyword: Knowledge discovery · Digital library · Contents of digital library · Information retrieval · Library of Bangladesh · Discovery tools · Metadata

1 Introduction

The contents of the digital library (DL) are critical for any twenty-first-century library user. Especially it is very important for students, faculty members, researchers, and scientists who want to get information in the quickest possible time. As a result of the advancement of ICT and the demand for current users, DL has been established all over the world to effectively and efficiently serve those communities. With the advent of modern science and technology, information management services that enable the creation of digital libraries have significantly improved [1]. As a consequence, the authorities of that DL would purchase a significant number of digital resources each year based on their ability to meet patrons' demands. However, due to a lack of good searching tools, many

H.-R. Ke et al. (Eds.): ICADL 2021, LNCS 13133, pp. 34–42, 2021.
https://doi.org/10.1007/978-3-030-91669-5_3

great library materials are overlooked. Libraries spend a significant amount of money each year growing their collections through various information resources or platforms, yet most of these resources are underutilized owing to a lack of a robust search engine with a single interface [2].

Knowledge discovery (KD) or retrieval of information from sources is very critical for any library user. KD tools are essential for any library to make its materials visible to users so that they can make appropriate use of them. KD tools use many approaches to search for and extract usable knowledge from data, databases, and documents [3]. However, KD tools can work effectively and efficiently if the library metadata and holdings data are properly managed. Discovery and access can be difficult or impossible without accurate bibliographic metadata and holdings data [4]. Nowadays, the most crucial responsibility of library professionals is to ensure KD from DL content is effectively and efficiently done in order to provide good library services to their users. As a consequence, they must rely on advanced technology, cutting-edge tools, and approaches that are critically needed in the DL to help make use of the contents of crucial information sources and locate the knowledge [5].

The purpose of this study is to explore the present trends of KD from DL systems in Bangladesh. It tried to identify the advantages and disadvantages, current barriers and challenges, and the tools and technologies required for KD from DL contents. It also attempted to offer effective strategies for overcoming the existing barriers and obstacles associated with KD. As a result, this research will be highly useful to all Bangladeshi and worldwide information scientists in understanding KD from the content of DL. It will serve as a guideline for library academicians, practitioners, and patrons.

The rest of the paper is structured as follows. The second section describes the conceptual map by reviewing related literature. The third section explains the aims and objectives of this study. The fourth section presents the research questions. The fifth section presents the research methodology. The sixth section describes the analysis of data, findings and discussion. The seventh section presents the technique of discovering knowledge as per the library professionals in Bangladesh. The eighth section mentions the limitations and direction for future research and the final and ninth section concludes the paper.

2 Literature Review

2.1 Concept of Knowledge Discovery

Knowledge discovery has been a hot topic not only in data mining and artificial intelligence but also in many other disciplines throughout the history of humanity in general and in particular over the past decades in the digital age [7]. Knowledge discovery is responsible for the quality of data entering discovery systems [8]. The high-level process of extracting effective, undiscovered, possibly valuable, and ultimately understood patterns from enormous amounts of data is referred to as knowledge discovery [9]. Web software that searches journal articles and library catalog metadata in a unified index and shows search results in a single interface are known as discovery tools [2]. The knowledge discovery process has some distinct characteristics, particularly when dealing with data of high velocity, variety, and volume [10]. Characteristics such as large data volume, knowledge discovery efficiency, the accuracy of discovered knowledge, and discovery automation are linked and must be addressed together for reliable knowledge discovery [10, 11]. Knowledge Discovery in databases is the process of extracting usable information from a huge data set using a data mining algorithm with specified metrics and thresholds and in more recent times, artificial intelligence approaches have been utilized to achieve this goal [11]. Because humans' intelligent actions include discovering knowledge, extracting knowledge, and mining knowledge, knowledge discovery is a type of intelligent behavior [7, 12].

2.2 Tools of Knowledge Discovery

A knowledge discovery tool is defined as a search engine that uses unified indexes of licensed scholarly information to search across multiple library databases provided by various vendors and can be customized for size, range, and comprehensiveness of data inclusion for targeted solutions [6]. Discovery tools that ingest metadata into a single index employ a single set of search algorithms to retrieve and rank results [2]. Although it is evident that a tool like Summon is required and value to library users, libraries must take care to position such devices in such a way that they can be a successful addition to the resource discovery dynamic [13]. Discovery tools, such as EBSCO Discovery Service, summon service (Serial Solution), Encore Discovery (Innovative interface), and Primo Central (Ex Libris Group), provides both opportunities and challenges for library instruction, depending on the academic discipline, users' knowledge, and information-seeking need [2]. A metasearch engine, sometimes known as a search aggregator, is an online information retrieval tool that generates its own results using the data of a web search engine. Metasearch engines take user input and query search engines for results right away. The users are supplied with enough data that has been acquired, ranked, and presented to them [14]. Federated search is a key component of an Information Portal, which serves as a gateway to a variety of information sources. When a user types a search query into the Information Portal's search box, the system uses federated search technology to send the search string to each resource included in the Portal [15].

2.3 The Current State of Digital Library Practice in Bangladesh

We have become comfortable with the DL, which is paperless, borderless, and always accessible from anywhere in the globe, due to the use of information and communication technology. Digital libraries and Institutional Repositories have grown increasingly popular as a means of gaining quick access to electronic information, however digital library activities in Bangladesh are still in their infancy [16]. In Bangladesh, the expansion of IRs is modest and it has a poor rate of IR development as compared to other Asian countries [17]. According to ROAR (Registry of Open Access Repositories), the number of repositories in Bangladesh is quite low, with only twelve [18]. They suggested that Bangladeshi librarians may require assistance in developing repositories in their own organizations. Islam & Naznin pointed out that the implementation of DL in Bangladesh is difficult, and specified library users should be knowledgeable of how to utilize it. It also necessitates an educated workforce, enhanced infrastructural facilities, and government attention [19]. Islam and Naznin; Rahman et al. noted significant progress in library digitization efforts in Bangladesh over the last decade, as well as issues like capacity building, shrinking funding, insufficient facilities, and traditional duties being supplanted by modern technologies. They also promoted joint digitalization efforts and provided helpful recommendations [19, 20].

3 Objectives of the Study

The main objective of this study is to explore the present trend of KD from digital library systems in Bangladesh. The more particular objectives are:

a) To determine the pros and cons of knowledge discovery from DL contents
b) To identify the existing barriers and challenges of DL
c) To identify the tools and technologies that are required for KD from DL contents
d) To offer effective ways to overcome the existing barriers and challenges of KD.

4 Research Questions (RQs)

Based on the above objectives, one major research question (MRQ) and three subsidiary research questions (SRQs) have been formulated that will guide the study.

MRQ1. What is the present trend of KD from the DL system in Bangladesh?
SRQ2. What difficulties and challenges are being faced in knowledge discovery from the DL content in Bangladesh?
SRQ3. What types of tools and technologies are required for knowledge discovery from DL contents?
SRQ4. How could the existing barriers and challenges of KD be overcome?

5 Research Methodology

Both quantitative and qualitative methods have been applied in this study. A structured questionnaire was designed and sent to the library professionals through emails for collecting data. The questionnaire was sent to 150 library professionals of different organizations, including universities, research institutes, etc. in Bangladesh. We received 85 filled questionnaires. Among 85 respondents 78 respondents, filled questionnaires appropriately and the remaining 7 respondents did not fill properly. Therefore, we used 78 responses in this study. Due to the current pandemic circumstances, many professionals are staying in remote areas where internet connectivity does not work properly. Hence, they could not participate in this study. We conducted telephone interviews with key professionals who are actively involved in directing DL. The quantitative data were analyzed using SPSS (version 20.0) software and qualitative data were analyzed thematically.

6 Findings and Discussion

6.1 Demographic Profile of the Respondents

This section focuses on the male and female ratio as well as the respondents' highest levels of education. Among the 78 respondents, 70.5% (55) respondents were male, and 29.5% (23) were female. The educational qualification of the respondents is seen that the highest 77% of respondents have a Master of Arts (M.A.) degree, while the lowest 4% have the Masters of Social Science (M.S.S.) degree. The second highest 13% has a master's of Philosophy (M.Phil) degree and 6% of respondents have a Doctor of Philosophy (Ph.D.) degree.

6.2 Designation of the Respondents

According to the survey, it is seen that the highest percentage of responses (36%) from the "Assistant librarian", followed by "Deputy Librarian" (32%), "Librarian" (17%), and "Library Officer" (4.5%) "Additional Librarian" (3%) Deputy Director (3%). We received only one response (1%) from "Chief librarian", "Principal Scientific Officer", "Assistant Library officer" and "Cataloguer" respectively.

6.3 Competencies of Respondents in Digital Library Management

This section summarizes the results of the survey on respondents' digital library management skills. The highest 44.9% of respondents have an intermediate level of competencies in the management of the digital library, on the other hand, 30.8% of respondents are beginners in the management of the digital library. 14.1% of respondents stated that they have excellent knowledge in the management of digital libraries, while 10.3% mentioned there have no setup of digital libraries.

6.4 The Tools and Technologies Needed to Discover Knowledge from the DL Contents

The respondents were asked to indicate their level of agreement with some specific tools and technologies which are required for Knowledge discovery from the content of digital libraries. The level of agreement of the respondent is analyzed on 1–5 Likert scales in the Sect. 6.4, 6.5 and 6.6. The mean and standard deviation of the responses were calculated according to the following scores: strongly disagree = 1.00, disagree = 2.00, neutral = 3.00, agree = 4.00, and strongly agree = 5.00 using the descriptive analysis techniques of SPSS.

The highest mean score was 4.49 for the statement of "OPAC," while the lowest mean score was 4.03 for the statement of "Apps." The second highest mean score was 4.42 for the statement of "Library Catalogue," followed by 4.35, 4.26, 4.21, 4.18, 4.15 and 4.14 for the statements of "Search engine," "Index," "Discovery software", "Document Object Identifier (DOI)", "Information of contents page", and "Metadata" respectively.

6.5 The Difficulties and Obstacles of Discovering Knowledge

The opinion of the respondents regarding the difficulties and obstacles of discovering knowledge from the content of a digital library is stated below.

Inaccurate Bibliographic Metadata
Without the proper bibliographic metadata, users won't be able to find the information they need [4]. The respondents agreed with the statement (with a mean score of 3.91) that content could not be traced due to inaccurate bibliographic metadata.

Lack of Accurate Holdings Data
The content from the digital library is unable to discover due to a lack of accurate holdings data. The respondents agreed (with a mean score of 3.83) that one of the most significant barriers to discovering knowledge from the digital library's contents is a lack of accurate holdings data.

Lack of Synchronized Bibliographic Metadata and Holdings Data
The respondents agreed with the statement (with a mean score of 4.24) that unsynchronized metadata and holdings data creating hindrance in discovering knowledge.

Libraries Receive Data in Multiple Formats
The respondents agreed with the statement (with a mean score of 4.21) that the libraries have to receive data in multiple formats, which causes difficulties discovering knowledge from the digital library's content.

6.6 Ways of Overcoming the Barriers and Challenges of KD from Contents of DL

The respondents were asked to specify how to overcome the barriers and challenges of KD from the digital library's content. The responses of the respondents are presented below.

The highest mean score was 4.60 for the statement of "Using Knowledge Bases And Related Tools (KBART) and MARC standards," while the lowest mean score was 4.12 for the statement of "Preserving complete bibliographic metadata." The second highest mean score was 4.37 for the statement of "Providing high-quality data," followed by the score of 4.32, 4.28, 4.24 for the following statements of "Using consistent data formats", "Preserving accurate holdings data" and "Synchronizing bibliographic metadata and holdings data" respectively.

6.7 The Most Commonly Used Discovery Tool/software in the Libraries of Bangladesh

The most commonly used discovery tool/software in Bangladeshi libraries. The highest 55 (66.7%) respondents agreed that "VuFind" is used to discover knowledge in Bangladeshi libraries, while the lowest 2 (2.6%) respondents believed that "Encore" and "Blacklight" are used for the same purpose. The second highest 37 (47.4%) respondents agreed that “EBSCO’s Discovery Service” are being used in libraries in Bangladesh followed by 16 (20.5%), 15 (19.2%), 9 (11.5%), 8 (10.3%), 6 (7.7%) for the tools/software “eXtensible Catalog”, “WorldCatLocal”, “BiblioCommons”, “ProQuest AquaBrowser”, “Ex Libris” respectively.

7 The Technique of Discovering Knowledge as Per the Library Professionals in Bangladesh

An open-ended question, “Would you please share your thoughts on how to effectively discover knowledge from the digital library’s content?” was asked through the questionnaire. The responses to this question were analyzed and categorized into themes as discussed below.

Creating systematic index and keywords: IP-1, IP-5, and IP-7 (Interview Participant) mentioned that the knowledge could be discovered from the content of the digital library by creating a systematic index of resources and searching knowledge applying appropriate keywords of desired information. Powerful search engine: IP-2 and IP-10 stated that a powerful search engine should be developed for discovering knowledge. A search engine can find information from any organized digital source. It is a popularly used tool all over the world for discovering knowledge. The resources should be collected from authentic sources: IP-50 and IP-72 emphasized the collection of resources from authentic, renowned, and standard organizations. They revealed that a world-famous organization generally maintains proper quality and standard system from creation to delivery of information. Hence the information provided by world-class organizations that information can be easily discovered using searching tools. Use of Artificial Intelligent and Big Data technology: IP-22, IP-66, and IP-71 stated that knowledge might be discovered effectively and efficiently from the contents of the digital libraries utilizing Artificial Intelligence (AI) and Big Data technologies. DL can leverage AI and Big Data technologies to discover knowledge for its databases, as they are used in almost every industry. Following standard metadata: IP-33, IP-77 and IP 80 stated that every document should be uploaded in the digital library with the following standard metadata and setting all the possible keywords and subjects should be added as much as possible. The database should be relational; the author, subject, place, publisher, supplier, etc. should be hyperlinked. Search can be filtered within the search result. The interface should be user-friendly.

As per the study respondents, KD from the contents of DL would be simple to discover using the approaches indicated.

8 Limitations and Direction for Future Research

In this pandemic circumstance, educational institutes and most research organizations in Bangladesh and the rest of the world are physically closed. The information and library professionals from those educational institutes and research organizations make up the research population for this study. We were unable to communicate with many professionals since they were not available at their workstations; as a result, we could not reflect their views in the article. Future research could involve a larger sample size and a look at the demographic information of the diverse responders. As a result of the limitations mentioned above, other scholars may be inspired to conduct further empirical research in this field.

9 Conclusion

The aim of the study was to explore the present trend of KD from digital library systems in Bangladesh. The findings showed that 70 respondents (out of 78) stated that they are successfully capable of running the DL and some of them are experts in DL management and only 8 respondents mentioned they have not existed the DL. Moreover, the study exposed that "VuFind" and "EBSCO's Discovery Service" software are being popularly used as KD tools in the libraries in Bangladesh. As a result, Bangladesh has a healthier overall scenario in terms of DL setup, skilled manpower, management, and KD from DL's content. The findings also revealed that the DL faces numerous challenges in discovering knowledge from its content, including inaccurate bibliographic metadata, a lack of accurate holdings data, a lack of synchronized bibliographic metadata and holdings data, and data in a variety of formats, among others. However, this study recommended that DL should maintain high-quality data, preserving complete bibliographic metadata and accurate holdings data, synchronizing bibliographic metadata and holdings data, using consistent data formats and using Knowledge Bases And Related Tools (KBART) and MARC standards. Every year, almost every library acquires valuable resources for library clients based on its financial capabilities. Nevertheless, library users may not be able to read all of these materials since adequate discovery techniques are not used to retrieve information from the sources. Therefore, discovery tools are playing a crucial role to solve that problem. We may conclude that KD from DL's content could be a success if data suppliers, service providers, and competent professionals work together to update and synchronize bibliographic and holdings data, as well as standardize data formats.

References

1. Pan, Z.: Optimization of information retrieval algorithm of digital library based on semantic search engine. In: 2020 International Conference on Computer Engineering and Application (ICCEA). IEEE (2020)
2. Karadia, A., Pati, S.: Discovery Tools and Services for Academic Libraries (2015)
3. Shi, H., He, W., Xu, G.: Workshop proposal on knowledge discovery from digital libraries. In: Proceedings of the 18th ACM/IEEE Joint Conference on Digital Libraries, JCDL 2018, Fort Worth, Texas USA, 3–7 June 2018, vol. 2 (2018)
4. Kemperman, et al.: Success Strategies for Electronic Content Discovery and Access: A Cross-Industry White Paper. OCLC, Dublin (2014). http://www.oclc.org/content/dam/oclc/reports/data-quality/215233-SuccessStrategies.pdf
5. Viet, N.T., Kravets, A.G.: Analyzing recent research trends of computer science from academic open-access digital library. In: 8th International Conference on System Modeling & Advancement in Research Trends, 22nd–23rd November 2019, Proceedings of the SMART–2019. IEEE (2019)

6. Shi, X., Levy, S.: An empirical review of library discovery tools. J. Serv. Sci. Manag. **8**, 716–725 (2015). https://doi.org/10.4236/jssm.2015.85073
7. Sun, Z., Stranieri, A.: A knowledge discovery in the digital age. PNG UoT BAIS **5**(1), 1–11 (2020)
8. Sharma, G., Tripathi, V.: Effective knowledge discovery using data mining algorithm. In: Fong, S., Dey, N., Joshi, A. (eds.) ICT Analysis and Applications. LNNS, vol. 154, pp. 145–153. Springer, Singapore (2021). https://doi.org/10.1007/978-981-15-8354-4_15
9. Zhu, H., Li, X.: Research on the application of knowledge discovery in digital library service. In: 7th International Conference on Social Network, Communication and Education (SNCE 2017), Advances in Computer Science Research, vol. 82 (2017)
10. Misra, S., Mukherjee, A., Roy, A.: Knowledge Discovery for Enabling Smart Internet of Things: A Survey. Wiley Periodicals, Inc. (2018)
11. Soundararajan, E., Joseph, J.V.M., Jayakumar, C., Somasekharan, M.: Knowledge discovery tools and techniques. In: Proceedings of the Conference on Recent Advances in Information Technology, Kalapakkam, India, vol. 141 (2005)
12. Russell, S., Norvig, P.: Artificial Intelligence: A Modern Approach, 3rd edn. Prentice Hall (2010)
13. Boyer, G.M., Besaw, M.: A study of librarians' perceptions and use of the summon discovery tool. J. Electron. Resour. Med. Libr. **9**(3), 173–183 (2012). https://doi.org/10.1080/15424065.2012.707056
14. Wikipedia: Metasearch Engine (2021). https://en.wikipedia.org/wiki/Metasearch_engine. Accessed 14 Sept 2021
15. Lingam, A.S.: Federated search and discovery solutions. IP Indian J. Libr. Sci. Inf. Technol. January-June **5**(1), 39–42 (2020)
16. Rahman, M.S.: Challenges and initiatives of digital library system and institutional repository: Bangladesh scenario. Eastern Libr. **25**(1), 1–23 (2020)
17. Elahi, M.H., Mezbah-ul-Islam, M.: Open access repositories of Bangladesh: an analysis of the present status. IFLA J. **44**(2), 132–142 (2018). https://doi.org/10.1177/0340035218763952
18. Chowdhury, M.H.H., Mannan, S.M.: Identifying the possible contents for university repositories of Bangladesh. Eastern Libr. **25**(2), 1–12 (2020)
19. Islam, M.S., Naznin, S.: Present status of digital library initiatives in Bangladesh. In: Proceedings of the 6th International Conference on Asia-Pacific Library and Information Education and Practice. Asia-Pacific LIS: Exploring Unity Amid Diversity, Philippine International Convention Center, Manila, Philippines, 28–30 October (2015)
20. Rahman, A.I.M.J., Rahman, M.M., Chowdhury, M.H.H.: Digital resources management in libraries: step towards digital Bangladesh. In: Proceedings of the National Seminar on Cross-Talk of Digital Resources Management: Step Towards Digital Bangladesh. Bangladesh Association of Librarians, Information Scientists, and Documentalists (BALID), Dhaka, pp. 1–24 (2015)

DataQuest: An Approach to Automatically Extract Dataset Mentions from Scientific Papers

Sandeep Kumar[1(✉)], Tirthankar Ghosal[2], and Asif Ekbal[1]

[1] Department of Computer Science and Engineering, Indian Institute of Technology Patna, Bihta, India
{19_11mc12,asif}@iitp.ac.in

[2] Institute of Formal and Applied Linguistics, Faculty of Mathematics and Physics, Charles University, Prague, Czech Republic
ghosal@ufal.mff.cuni.cz

Abstract. The rapid growth of scientific literature is presenting several challenges for the search and discovery of research artifacts. Datasets are the backbone of scientific experiments. It is crucial to locate the datasets used or generated by previous research as building suitable datasets is costly in terms of time, money, and human labor. Hence automated mechanisms to aid the search and discovery of datasets from scientific publications can aid reproducibility and reusability of these valuable scientific artifacts. Here in this work, utilizing the *next sentence prediction* capability of language models, we show that a BERT-based entity recognition model with POS aware embedding can be effectively used to address this problem. Our investigation shows that identifying sentences containing dataset mentions in the first place proves critical to the task. Our method outperforms earlier ones and achieves an F1 score of 56.2 in extracting dataset mentions from research papers on a popular corpus of social science publications. We make our codes available at https://github.com/sandeep82945/data_discovery.

Keywords: Dataset discovery · Dataset mention extraction · Publication mining · Deep learning

1 Introduction

Data is the new oil for as they say, and datasets are crucial for scientific research. There has been an enormous growth of data and rapid advancement in data science technologies a generation or two ago, which has opened considerable opportunities to conduct empirical research. Now the researchers can rapidly acquire and develop massive, rich datasets, routinely fit complex statistical models, and conduct their science in increasingly fine-grained ways. Finding a good dataset to support/carry out the investigation or creating a new one is crucial to research.

S. Kumar and T. Ghosal—Equal contribution.

H.-R. Ke et al. (Eds.): ICADL 2021, LNCS 13133, pp. 43–53, 2021.
https://doi.org/10.1007/978-3-030-91669-5_4

Faced with a never-ending stream of new findings and datasets generated using different code and analytical techniques, researchers cannot readily determine who has worked in an area before, what methods were used, what was produced, and where those products can be found. However, many datasets go unnoticed due to lack of proper dataset discovery tools, and hence many efforts are duplicated. A survey [16] even suggests that data users' and analysts' productivity grow less because more than a third of their time is spent finding out about data rather than in model development and production. The links from scientific publications to the underlying datasets and vice versa are helpful in many scenarios, including building a dataset recommendation system, determining the impact of a given dataset, or identifying the most used datasets in a given community, sharing available datasets through the research community.

Empirical researchers and analysts who want to use data for evidence and policy mostly face challenges in finding out who else worked with the data. Hence, good research is underused, great data go undiscovered and are undervalued, and time and resources are wasted redoing empirical work [1]. It will also help governments modernize their data management practices and building policies based on evidence and science [3]. Too often, scientific data and outputs cannot be easily discovered, even if publicly available, which leads to the reproducibility crisis of empirical science, thereby threatening its legitimacy and utility [12,22]. Automatically detecting dataset references is challenging even within one research community because of a wide variety of dataset citations and the variety of places in which datasets can be referenced in articles [14].

A significant effort towards this problem were made in the Rich Context Competition [4] (RCC). This paper improves the previously used state-of-the-art approaches for dataset extraction from scientific publications by proposing an end-to-end pipeline. Our approach consists of two stages: *(1) Dataset Sentence Classification, (2) Identification of Actual Dataset Mentions within that sentence.* To the best of our knowledge, our approach is novel in this domain.

2 Related Work

Researchers have long investigated extracting entities, artifacts from research paper full text to make knowledge computable [23,25,28]. However, here in this work, we concentrate on the investigations that specifically address dataset extraction and discovery. Recently Google released their Dataset Discovery engine [26] which relies on an open ecosystem, where dataset owners and providers publish semantically enhanced metadata on their sites. Singhal et al. [32] leverage on a user profile-based search and a keyword-based search from open-source web resources such as scholarly articles repositories and academic search engines to discover the datasets. Lu et al. [21] extracted dataset from publications using handcrafted features. Ghavimi et al. [15] proposed a semi-automatic three-step approach for finding explicit references to datasets in social sciences articles. To identify references to datasets in publications, Katarina Boland et al. [8] proposed a pattern induction approach to induce patterns iteratively using a bootstrapping strategy. The task of identifying biomedical dataset is addressed by [9] open source biomedical data discovery system called DataMed. Within the RCC challenge [2], the winner was the Semantic Scholar team from Allen AI [18]. They built a rule-based extraction system with Named Entity Recognition (NER) model using Bidirectional Long Short-Term Memory (Bi-LSTM) model with a conditional random field (CRF) decoding layer to predict dataset mentions. The honorable mention KAIST team [17] used a machine-learning-based question answering system for retrieving data sets by generating questions regarding datasets. Another finalist, team GESIS [27] also explored a named entity recognition (NER) approach using SPACY for full text. The DICE team [24] from Paderborn University trained an entity extraction model based on CRFs and combined it with a simple dataset mention search to detect datasets in an article. The team from Singapore Management University (SMU) [30] used SVM for dataset detection followed by rules to extract dataset names. The work reported in [29,33] by SU and NUS describes a method for extracting dataset-mentions using various BiLSTM variants with CRF attention models for the dataset extraction task.

The previous works have some limitations in generalizing unseen datasets, discriminating ambiguous names to datasets, and reducing noise. Our current work aims to tackle the limitation and improve the results by combining the transfer capabilities of Bi-Directional Encoder Representations from Transformers (BERT).

3 Methodology

RCC organizers provided a labeled corpus of 5000 publications with an additional development fold of 100 publications. Overall, there are around 8 lakhs and 32k sentences, not containing dataset mention and dataset mention, respectively. Each publication was labeled to indicate which of the datasets from the list were referenced within and what specific text was used to refer to each dataset. However, many of the listed datasets do not appear in the corpus. We consider only those publications that contain a mention of the dataset and filtered out the rest for training the dataset-mention extraction model.

We employ a pipeline of two tasks in sequence: Dataset Sentence Classification, followed by Dataset Mention Extraction, as shown in Fig. 1. The sentences that contain dataset mentions are considered further for the dataset mention extraction task. The first task helps us quickly filter out the sentences that do not refer to any dataset.

3.1 Dataset-Sentence Classification

We propose a SciBERT+MLP model (a sentence-level binary classifier), which encodes hidden semantics and long-distance dependency. In this module, the goal is to classify each sentence in a sequence of n sentences in a document to find out whether it contains a dataset reference or not. For this purpose, we

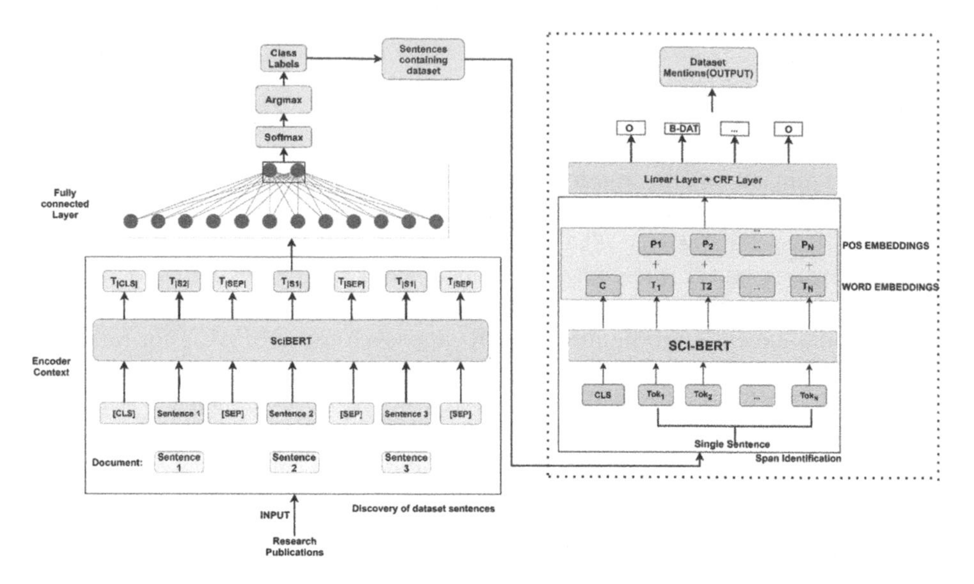

Fig. 1. Overall architecture diagram showing: (a) Dataset Sentence Classification (on the left), (b) Dataset Mention Extraction (on the right)

develop a technique based on the Sequential Sentence Classification [6] (SSC) model. The SSC model is based on SciBERT [7], a variant of BERT [10] pre-trained on a large multi-domain corpus of scientific publications. Figure 1(a) gives an overview of our dataset sentence identification module. Consider the training dataset as T = $D_1, D_2, .., D_i, .., D_Z$ comprising of Z documents. Each D_i can be represented as $D_i = s_{i1}, s_{i2}, .., s_{ij}, .., s_{iN}$ where N is the number of sentences in the document and s_{ij} is the jth sentence of document D_i. Each sentence is assigned a ground-truth label where label "1" represents a sentence containing dataset mention reference and label "0" a sentence doesn't contain dataset mention reference. The standard [CLS] is inserted as the first token of the sequence, and another delimiter token [SEP] is used for separating the segments. The initial input embedding (E_{Tok}) is calculated by summing up the token, sentence, and positional embedding. The transformer layers [11] allow the model to fine-tune the weights of these special tokens according to the task-specific training data (RCC corpus). We use a multi-layer feedforward network on top of each sentence's [SEP] representations to classify them to their corresponding categories (Has Dataset Mention or Not?). During fine-tuning, the model learns appropriate weights for the [SEP] token to capture contextual information and learns sentence structure and relations between continuous sentences (through the next sentence objective). Further, we use a softmax classifier on top of the MLP to predict the label's probability. The last linear layer consists of two units corresponding to label "0" and label "1". The final output label is the label whose corresponding unit has a higher score in the last linear layer. Our loss function is weighted binary cross entropy loss, whose weights are decided by the number of samples in each class. We use the AllenNLP [13] toolkit for the model implementation. As in prior work [10], for training we use dropout of 0.1, the Adam optimizer for 2–5 epochs, and learning rates of 5e−6, 1e−5, 2e−5, or 5e−5.

3.2 Dataset Mention Extraction

Dataset Mention Extraction is a binary sequence tagging task where we classified each token to indicate whether it is part of a dataset mention phrase fragment. Here, the goal is to extract the dataset mentions from the sentences which contain at least one mention of the dataset. To detect the boundary of a dataset mention, we use the BIO tagging scheme[1]. We finetune the pre-trained SciBERT model using the annotated corpus with the BIO-schema for dataset mention recognition. While BERT has its tokenization with Byte-Pair encoding and will assign tags to its extracted tokens, we should take care of it. BERT extracted tokens are always equal to or smaller than our main tokens because BERT takes tokens of our dataset one by one, as described by [31]. As a result, we will have intra-tokens that take X tag (meaning don't mention). We employ masking to ignore the padded elements in the sequences.

To add syntactic features to the BERT model, we create a syntax-infused vector for each word by adding a POS embedding vector of dimension d = D to the BERT embedding vector of the word. To determine the POS label of each word of a sentence, we use the pretrained spacy model [5]. We make a POS embedding vector from the BERT embedding of the POS label of the word. Here D is the input dimension of the encoder (D = 768). We add a token-level classifier on top of the BERT layer followed by a Linear-Chain CRF to classify the *dataset mention tokens*. For an input sequence of n tokens, BERT outputs an encoded token sequence with hidden dimension H. The classification model projects each token's encoded representation to the tag space, i.e. $\mathbb{R}^H$-> $\mathbb{R}^K$ where K is the number of tags and depends on the number of classes and the tagging scheme. The output scores $\mathbf{P} \in \mathbb{R}^{n \times K}$ of the classification model are then fed to the CRF layer. The matrix $\mathbf{A}$ is such that $A_{i,j}$ represents the score of transitioning from tag i to tag j including two more additional states representing start and end of sequence.

As described by [20] for an input sequence $\mathbf{X} = (\mathbf{x}_1, ..., \mathbf{x}_n)$ and a sequence of tag predictions $\mathbf{y} = (y_1, ..., y_n), y_i \in \{1, ..., K\}$ the score of the sequence is defined as:-

$$s(\mathbf{X}, \mathbf{y}) = \sum_{i=0}^{n} A_{y_i, y_{i+1}} + \sum_{i=1}^{n} P_{i, y_i} \tag{1}$$

where y_0 and y_{n+1} are the start and end tags. A softmax over all possible sequences yields the probability for sequence $\mathbf{y}$

$$p(\mathbf{y}|\mathbf{X}) = \frac{e^{s(\boldsymbol{X}, \boldsymbol{y})}}{\sum_{\tilde{\boldsymbol{y}} \in \boldsymbol{Y_X}} e^{s(\boldsymbol{X}, \tilde{\boldsymbol{y}})}} \tag{2}$$

[1] B, I, and O denote the beginning, intermediate, and outside of dataset mention.

The model is trained to maximize the log probability of the correct tag sequence:-

$$log(p(\mathbf{y}|\mathbf{X})) = s(\mathbf{X}, \mathbf{y}) - log(\sum_{\hat{y} \in Y_X} e^{s(\mathbf{X}, \hat{y})}) \tag{3}$$

where $\boldsymbol{Y_X}$ are all possible tag sequences. Equation 3 is computed using dynamic programming. During evaluation, the most likely sequence is obtained by Viterbi decoding. As per [10] we compute predictions and losses only for the first sub-token of each token. While we tried different batch sizes and learning rates for fine-tuning while we report the best. We use a learning rate of 5e-6 for the Adam optimizer [19], with a batch size of 16 for 10 epochs. Gradient clipping was used, with a max gradient of 1. This module's output will be the BIO-tagged sentence from where we can extract the B followed by I-tagged tokens signifying the dataset mention.

Table 1. Result of Dataset-Sentence Classification (P → Precision, R → Recall, F1 → F1-Score)

Result of Task-1	P	R	F1
Sentence not containing data	0.99	0.98	0.99
Sentence containing data	0.83	0.82	0.83
Macro average	0.92	0.91	0.91

Table 2. Result of Dataset Mention Extraction, Details of each of these comparison systems is described in Sect. 2

Model	Partial match			Exact match		
	P	R	F1	P	R	F1
SMU [30]	–	–	–	34.0	30.0	32.0
BiLSTM(NUS) [29]	71.4	64.4	67.7	31.3	34	32.6
SL-E-C(NUS) [29]	72.2	72.6	74.8	39.9	41.6	40.7
CNN-BiLSTM(NUS) [29]	77.5	75.5	76.5	41.4	44.6	43.0
CNN-BiLSTM-CRF(NUS) [29]	79.1	71.1	74.9	42.7	44.6	43.6
SU [33]	88.2	88.4	88.3	–	–	–
CNN-BiLSTM-Att-CRF(NUS) [29]	76.1	73.8	74.9	39.4	47.7	43.2
Allen AI [18]	–	–	–	52.4	50.3	51.8
Our model	**89.2**	88.1	**88.6**	**60.24**	**52.8**	**56.2**
GESIS [27]	93.0	95.0	93.8	80.0	81.0	80.4
Our model	**94.2**	**95.2**	**94.6**	**85.2**	**86.7**	**85.9**

4 Results and Analysis

Table 1 shows the result of Task-1 (Dataset Sentence Classification). Our model has reported a 0.91 macro average for Task 1. While Table 2 shows the result of

Task-2 (Dataset Mention Extraction) and the comparison with other baselines. We evaluate our model for strict and partial (relaxed) F1-score. While strict criterion contributes a true positive count if and only if the ground truth tokens are exactly predicted, whereas matched correctly predicted assigns the credit if and only if the exact boundaries are matched, for partial (or relaxed) criterion, a partial match to the ground truth is also treated as the true positive count.

As expected, our proposed model results are better for the partial match than the exact match, which means we can find the proper context with very high precision even if we could not match the full dataset mention in the text exactly. Results also show that our proposed system performs the best for both strict and relaxed evaluation metrics than the other existing methods. The closest system, AllenAI [18], reported having achieved the F1-scores of 51.8 for the strict. We observe a relative improvement of 6.4% F1-score compared to AllenAI wrt strict. The closest system, GESIS [27], reported having achieved the F1-scores of 80.4 for the strict and 93.8 for the relaxed criterion, respectively. We observe a relative improvement of 5.4% F1-score compared to GESIS wrt strict and almost equal F1-score to relaxed criterion (All results are on the development set while GESIS divided the training set into the split of 80:20, 80% for training and 20% for testing; we also report for the same set). The other participants, including AllenAI [18] and GESIS [27], have not tried transformer-based NER and have also performed NER on the paper's full context. In contrast, we filtered out the irrelevant sentences (not containing the dataset mention) and then used the relevant sentences for mention extraction. Also, the BERT-based NER understood the context better, resulting in better results. We also perform test-of-significance (T-test) and observe that the obtained results are statistically significant w.r.t. the state-of-the-art with p-values < 0.05.

Table 3. Ablation study

Model	Precision	Recall	F1
BERT+CRF (On full test set)	54.9	51.2	52.9
BERT+CRF (After dataset sentence classification)	58.2	51.3	54.4
BERT+CRF+POS (On full test set)	55.2	52.3	53.7
BERT+CRF+POS (After dataset sentence classification)	60.2	52.8	56.2

4.1 Analysis

Table 3 shows the ablation study examining our system's various components' importance. We observe dataset sentence classification before dataset mention extraction, and POS-aware BERT embedding for dataset mention extraction boosts the overall model's performance for this task.

Table 4. Examples of the dataset sentence identification task, where the red coloured text indicate sentences being filtered out whereas blue colored text indicate sentences passed for the next dataset mention extraction task.

Dataset Name: "SWAN' Title: "Study of Women's Health across the Nation"
1. Swans are bird of family Anatidae within the genus Cygnus (-) 2. Several enduring themes have emerged from SWAN that have associated certain patterns of hormones and symptoms with metabolic status. (+) 3. SWAN Energy LTd. is an emerging "green energy" company (-) 4. Weight gain has been observed to occur in conjunction with the menopausal transition, but studies prior to SWAN have concluded that weight gain is driven primarily by age (+)
Dataset Name: "SUPPORT' Title: "Study to understand Prognoses and Preferences for Outcomes and Risks of Treatments"
5. Peng's findings do not appear to support his conclusions (-) 6. Phase I of SUPPORT collected data from patients accessioned during 1989-1991 to characterize the care, treatment preferences, and pattern of decision-making among critically ill patients. (+)

Role of Sentence Identification. As the string may occur multiple times in the document, and all occurrences may or may not be correct dataset mentions; this is especially problematic when the string is a common word which may have multiple meanings in different contexts. As shown in Table 4, we provide some examples to show how the sentence identification task can overcome other participants' limitations, including that of GESIS. 'SWAN' is a dataset mention of a dataset with the title "Study of Women's Health Across Nation," which is also the name of a bird, company, etc. Similarly, 'SUPPORT' is a dataset mention of a dataset with the title "Study to Understand Prognoses and Preferences for Outcomes and Risks of Treatments." However, it is also a commonly used word in the English language with a different meaning. Using NER directly does not discriminate these confusion cases and mislabels all of them as dataset names. While the sentence identification task understands the context of the sentences and filters out these irrelevant sentences (red), and preserves only relevant sentences (blue) before feeding them to NER.

Table 5. Examples showing the use of adding POS embedding to word embedding (red: wrongly identified dataset mention, blue: correctly identified dataset mention)

Without POS embedding	**With POS embedding**
The data has been used from progress in international reading literacy study.... There is a high level of risk in the rise in the number of cases.	The data has been used from progress in international reading literacy study.... There is a high level of risk in the rise in the number of cases.
We combine two micro datasets provided by the deutsche bundesbank ,the its and the midi and complement them[..]	We combine two micro datasets provided by the deutsche bundesbank ,the its and the midi and complement them [..]

Role of POS Embedding. Dataset mentions are usually noun phrases, such as in Table 5 *"National health and educational survey"*, *"coastal erosion study"*, etc. The examples *"progress in"* and *"rise in"* are misclassified by the NER, as the dataset mentions. However, adding the POS embedding gives more weightage to the noun chunks. Hence, some misclassified verbs or other POS dataset phrases are reduced.

4.2 Error Analysis

- **Roman numbers:** Our model finds difficulty in determining full dataset names having roman names. For example *"[..]add health (waves i, ii, and iii) with obesity[..]"*, contains roman letters in the dataset name (*"add health and add health waves i ii and iii"*). However, the model predicts only *add health*, i.e., does not predict the full dataset name.
- **Too many numbers or punctuations:** Our model confuses when there are too many numbers or punctuations in the sentence. For example *"002 hospital beds per 100,000 population –0:002***[..] national profile of local health departments[..]"* shows the example having the dataset mentions *"national profile of local health departments,"* but the model fails to understand the context due to many punctuation or numbers, hence fails to predict the dataset name.

5 Conclusion and Future Work

In this work, we report a novel BERT-based model for extracting dataset mentions from scientific publications. Our model is simple and outperforms earlier approaches. Our overall goal is to understand the impact of any given dataset (*Data Impact Factor*) in the community. The critical observation we make here is that *identifying sentences containing the dataset-mentions are highly useful before proceeding with the task of dataset-mention extraction* and using BERT with POS embedding can enhance the task of dataset-mention extraction. In the future, we intend to explore extracting other helpful information (tasks, methods, metrics) from research publications to automate automated literature comparison.

Acknowledgement. Sandeep Kumar acknowledges the Prime Minister Research Fellowship (PMRF) program of the Government of India for its support. Asif Ekbal is a recipient of the Visvesvaraya Young Faculty Award and acknowledges Digital India Corporation, Ministry of Electronics and Information Technology, Government of India for supporting this research.

References

1. The coleridge initiative announces rich context competition—NYU cusp. https://cusp.nyu.edu/blog/the-coleridge-initiative-announces-rich-context-competition/. Accessed 14 July 2021
2. Github - rich-context-competition/rich-context-book-2019. https://github.com/rich-context-competition/rich-context-book-2019. Accessed 14 July 2021
3. Rich context project - coleridge initiative. https://coleridgeinitiative.org/rich-context-project/. Accessed 14 July 2021
4. Richcontextcompetition - coleridge initiative. https://coleridgeinitiative.org/richcontext/richcontextcompetition/. Accessed 14 July 2021
5. Spacy industrial-strength natural language processing in python. https://spacy.io/. Accessed 15 July 2021
6. Cohan, A., Beltagy, I., King, D., Dalvi, B., Weld, D.S.: Pretrained language models for sequential sentence classification. In: EMNLP (2019)

7. Beltagy, I., Lo, K., Cohan, A.: SciBERT: a pretrained language model for scientific text. In: Inui, K., Jiang, J., Ng, V., Wan, X. (eds.) Proceedings of the 2019 Conference on Empirical Methods in Natural Language Processing and the 9th International Joint Conference on Natural Language Processing, EMNLP-IJCNLP 2019, Hong Kong, China, 3–7 November 2019, pp. 3613–3618. Association for Computational Linguistics (2019). https://doi.org/10.18653/v1/D19-1371
8. Boland, K., Ritze, D., Eckert, K., Mathiak, B.: Identifying references to datasets in publications. In: Zaphiris, P., Buchanan, G., Rasmussen, E., Loizides, F. (eds.) TPDL 2012. LNCS, vol. 7489, pp. 150–161. Springer, Heidelberg (2012). https://doi.org/10.1007/978-3-642-33290-6_17
9. Chen, X., et al.: DataMed - an open source discovery index for finding biomedical datasets. J. Am. Medical Informatics Assoc. **25**(3), 300–308 (2018)
10. Devlin, J., Chang, M.W., Lee, K., Toutanova, K.: BERT: pre-training of deep bidirectional transformers for language understanding (2019)
11. Devlin, J., Chang, M., Lee, K., Toutanova, K.: BERT: pre-training of deep bidirectional transformers for language understanding. In: Burstein, J., Doran, C., Solorio, T. (eds.) Proceedings of the 2019 Conference of the North American Chapter of the Association for Computational Linguistics: Human Language Technologies, NAACL-HLT 2019, Minneapolis, MN, USA, 2–7 June 2019 (Long and Short Papers), vol. 1, pp. 4171–4186. Association for Computational Linguistics (2019). https://doi.org/10.18653/v1/n19-1423
12. Feger, S.S.: Interactive tools for reproducible science - understanding, supporting, and motivating reproducible science practices. CoRR abs/2012.02570 (2020). https://arxiv.org/abs/2012.02570
13. Gardner, M., et al.: AllenNLP: a deep semantic natural language processing platform (2018). http://arxiv.org/abs/1803.07640
14. Ghavimi, B., Mayr, P., Lange, C., Vahdati, S., Auer, S.: A semi-automatic approach for detecting dataset references in social science texts. Inf. Serv. Use **36**(3–4), 171–187 (2016)
15. Ghavimi, B., Mayr, P., Vahdati, S., Lange, C.: Identifying and improving dataset references in social sciences full texts. In: Loizides, F., Schmidt, B. (eds.) Positioning and Power in Academic Publishing: Players, Agents and Agendas, 20th International Conference on Electronic Publishing, Göttingen, Germany, 7–9 June 2016, pp. 105–114. IOS Press (2016). https://doi.org/10.3233/978-1-61499-649-1-105
16. Grover, M.: Amundsen - Lyft's data discovery & metadata engine—by mark grover—Lyft engineering, April 2019. https://eng.lyft.com/amundsen-lyfts-data-discovery-metadata-engine-62d27254fbb9. Accessed 31 Oct 2020
17. Hong, G., Cao, M.S., Puerto-San-Roman, H.: Rich text competition. In: Rich Search and Discovery for Research Datasets: Building the Next Generation of Scholarly Infrastructure. Sage, London (2020)
18. King, D., Ammar, W., Beltagy, I., Betts, C., Gururangan, S., van Zuylen, M.: The AI2 submission at the rich context competition. In: Rich Search and Discovery for Research Datasets: Building the Next Generation of Scholarly Infrastructure. Sage, London (2020)
19. Kingma, D.P., Ba, J.: Adam: a method for stochastic optimization (2015). http://arxiv.org/abs/1412.6980
20. Lample, G., Ballesteros, M., Subramanian, S., Kawakami, K., Dyer, C.: Neural architectures for named entity recognition. CoRR abs/1603.01360 (2016). http://arxiv.org/abs/1603.01360

21. Lu, M., Bangalore, S., Cormode, G., Hadjieleftheriou, M., Srivastava, D.: A dataset search engine for the research document corpus. In: 2012 IEEE 28th International Conference on Data Engineering, pp. 1237–1240. IEEE (2012)
22. Munafò, M., et al.: A manifesto for reproducible science. Nat. Hum. Behav. **1**, 0021 (2017). https://doi.org/10.1038/s41562-016-0021
23. Nasar, Z., Jaffry, S.W., Malik, M.K.: Information extraction from scientific articles: a survey. Scientometrics **117**(3), 1931–1990 (2018)
24. Ngonga, P.D.A., Srivastava, N., Jalota, R.: Dice @ rich context competition. In: Rich Search and Discovery for Research Datasets: Building the Next Generation of Scholarly Infrastructure. Sage, London (2020)
25. Nguyen, T.D., Kan, M.-Y.: Keyphrase extraction in scientific publications. In: Goh, D.H.-L., Cao, T.H., Sølvberg, I.T., Rasmussen, E. (eds.) ICADL 2007. LNCS, vol. 4822, pp. 317–326. Springer, Heidelberg (2007). https://doi.org/10.1007/978-3-540-77094-7_41
26. Noy, N., Burgess, M., Brickley, D.: Google dataset search: building a search engine for datasets in an open web ecosystem. In: 28th Web Conference (WebConf 2019) (2019)
27. Otto, W., Zielinski, A., Ghavimi, B., Dimitrov, D., Tavakolpoursaleh, N.: Rich context competition phase 2. In: Rich Search and Discovery for Research Datasets: Building the Next Generation of Scholarly Infrastructure. Sage, London (2020)
28. Peng, F., McCallum, A.: Information extraction from research papers using conditional random fields. Inf. Process. Manag. **42**(4), 963–979 (2006)
29. Prasad, A., Si, C., Kan, M.Y.: Dataset mention extraction and classification. In: Proceedings of the Workshop on Extracting Structured Knowledge from Scientific Publications, Minneapolis, Minnesota, pp. 31–36. Association for Computational Linguistics, June 2019. https://doi.org/10.18653/v1/W19-2604. https://www.aclweb.org/anthology/W19-2604
30. Prasetyo, P.K., Silva, A., Lim, E.P., Achananuparp, P.: Simple extraction for social science publications. In: Rich Search and Discovery for Research Datasets: Building the Next Generation of Scholarly Infrastructure. Sage, London (2020)
31. Shamsfard, M., Jafari, H.S., Ilbeygi, M.: Step-1: a set of fundamental tools for Persian text processing. In: Calzolari, N., et al. (eds.) Proceedings of the International Conference on Language Resources and Evaluation, LREC 2010, Valletta, Malta, 17–23 May 2010. European Language Resources Association (2010). http://www.lrec-conf.org/proceedings/lrec2010/summaries/809.html
32. Singhal, A., Srivastava, J.: Research dataset discovery from research publications using web context. In: Web Intelligence, vol. 15, pp. 81–99. IOS Press (2017)
33. Zeng, T., Acuna, D.: Dataset mention extraction in scientific articles using a BiLSTM-CRF model. In: Rich Search and Discovery for Research Datasets: Building the Next Generation of Scholarly Infrastructure. Sage, London (2020)

Embedding Transcription and Transliteration Layers in the *Digital Library of Polish and Poland-Related News Pamphlets*

Maciej Ogrodniczuk[1(✉)] and Włodzimierz Gruszczyński[2]

[1] Institute of Computer Science, Polish Academy of Sciences, Jana Kazimierza 5, 01-248 Warszawa, Poland
maciej.ogrodniczuk@ipipan.waw.pl

[2] Institute of Polish Language, Polish Academy of Sciences, al. Mickiewicza 31, 31-120 Kraków, Poland
wlodzimierz.gruszczynski@ijp.pan.pl

Abstract. The paper presents an experiment intended to overcome the problem of searching for different spelling variants in old Polish prints. In the case of *The Digital Library of Polish and Poland-related Ephemeral Prints from the 16th, 17th and 18th Centuries* two concurrent layers of text (transliteration and transcription) underlying selected digital library items are available in the related *Electronic Corpus of the 17th and 18th Century Polish Texts (until 1772)*. Both variants are retrieved and a double-hidden layer representation of a sample item is prepared and made available for textual searching in a PDF containing its scanned image. The experiment can be generalized to other libraries dealing with multiple concurrent textual interpretations of graphical items.

Keywords: Digital library · Transcription · Transliteration · Middle Polish

1 Introduction

The Digital Library of Polish and Poland-related Ephemeral Prints from the 16th, 17th and 18th Centuries (Pol. *Cyfrowa Biblioteka Druków Ulotnych polskich i Polski dotyczących z XVI, XVII i XVIII wieku*, hereafter abbreviated CBDU[1] [7–9] is a thematic digital library of approx. 2 000 Polish and Poland-related prepress documents (ephemeral prints—short, disposable and irregular informative publications) dated between 1501 and 1729 (all surviving documents of this kind described in scientific publications, particularly by Zawadzki [11]).

[1] See https://cbdu.ijp.pan.pl/.

The work was financed by a research grant from the Polish Ministry of Science and Higher Education under the National Programme for the Development of Humanities for the years 2019–2023 (grant 11H 18 0413 86, grant funds received: 1,797,741 PLN).

H.-R. Ke et al. (Eds.): ICADL 2021, LNCS 13133, pp. 54–60, 2021.
https://doi.org/10.1007/978-3-030-91669-5_5

The library is managed by EPrints [1], a database management system configured to using extended metadata: apart from the usual ones such as item title. author, publication date etc. CBDU defines extended metadata such as historical comments, glossaries of foreign interjections, explanations of lesser-known background details and relations between library objects (translations, adaptations, alterations of the base text, their alleged sources etc.) For 1404 prints (out of 2011 present in the digital library) scans of actual microfilmed documents are available, stored in multipage PDF files displayed on demand in a side pane.

The PDFs are currently graphical files only but textual transcriptions are planned to be soon added using the process described by [10]. The texts will be acquired from *The Electronic Corpus of the 17th and 18th Century Polish Texts (until 1772)* (Pol. *Elektroniczny Korpus Tekstów Polskich z XVII i XVIII w. (do roku 1772)*) [6], also referred to as the *Baroque Corpus* (Pol. *Korpus Barokowy* – hence its acronym KORBA[2] and merged into the PDFs as hidden text using the OCR tools.

The process is straightforward but not complete: the orthography of middle-Polish text may be much different from contemporary Polish, hindering usability of the searchable text even when it is properly embedded in the resulting PDF. In the paper we present the method of overcoming this inconvenience by offering double textual layer with spelling variants of the text.

2 Related Work and Background Information

Although the creators of CBDU intended to add two spelling versions of library items - transliterated and transcribed ones[3], the task proved impossible within the project timeframe and only one sample text was transcribed[4]. The implementation of KORBA corpus allowed for transcribing 40 more prints, amounting to over 200,000 tokens (words and punctuation).

Apart from the rich structural annotation available in the corpus (e.g. marked page numbers for text ranges or foreign language interjections), the texts have been encoded in both transcribed and transliterated forms, cf. *nazajutrz* (identical to a contemporary Polish word) and older *nazaiutrz*):

```
<fs type="morph">
  <f name="orth">
    <string>nazajutrz</string>
  </f>
  <f name="translit">
    <string>nazaiutrz</string>
  </f>
</fs>
```

[2] See also https://korba.edu.pl/overview?lang=en.

[3] The intention of transliteration is accurate representation of the graphemes of a text while transcription is concerned with representing its phonemes.

[4] See *Translation into Contemporary Polish* section of print 1264 at http://cbdu.ijp.pan.pl/12640/.

What must be noted, KORBA creators decided not to follow the extremely faithful transliteration featured e.g. in the IMPACT corpus [3][5] called *strict diplomatic* or *facsimile* transcription, preserving the distinctions irrelevant for many users. For example, minor differences in certain graphical or typographical features were neglected such as different variants of the letter ę, all Unicode-encoded as `LATIN SMALL LETTER E WITH OGONEK` (also used in comtemporary Polish) and not as other existing Unicode representations such as `LATIN SMALL LETTER E WITH STROKE`. Obviously, both transliterated and transcribed variants retain their capitalisation as per the original texts.

Both layers are available not only in the source corpus files but can also be queried in the corpus search engine[6], featuring e.g. an interface add-on for inputting letters which are absent from modern Polish orthography but which may appear in the transliteration layer, the switch between displayed layer mode (modernized or transliterated) or attributes for querying both layers (in our example, `[orth=nazajutrz"]` and `[translit=nazaiutrz"]` respectively) – see Fig. 1.

Still, the corpus search interface is separated from the digital library and offers access only to generated concordances of text, linked to the *Electronic Dictionary of the 17th–18th Century Polish* (Pol. *Elektroniczny słownik języka polskiego XVII i XVIII wieku*) [4,5] but not to actual scans. What could offer much more flexibility for researchers accessing both layers would be to enable search in them directly in the browser, most conveniently in the PDF file containing both the scanned version of an item and its transcribed/transliterated text.

Table 1 shows the importance of this issue: over 32% of words in the 40 prints from CBDU currently available in the corpus[7] are subject to variation between transcription and transliteration and over 20% of these differences are significant ones (going beyond accent variants corresponding to characters currently unused in Polish alphabet such as *a/á* and *e/é*, as e.g. in *potrzébná*).

Table 1. Counts of differences between transcription and transliteration layers

	Tokens	Percentage
No difference	135 101	67.11%
Punctuation	11 914	5.92%
Difference	66 199	32.89%
Significant difference	40 911	20.32%
All	201 300	100.00%

[5] *Improving Access to Texts* international project, see also http://www.impact-project.eu.

[6] See https://korba.edu.pl/.

[7] KORBA project is being continued until 2023 and several new texts from CBDU will be included in the corpus.

THE ELECTRONIC CORPUS OF 17TH- AND 18TH-CENTURY POLISH TEXTS (UP TO 1772)

Corpus
Automatically annotated corpus

Query
[translit="nazaiutrz"]

á é Á É

QUERY BUILDER DISCARD FOREIGN METADATA

Displayed layer
transliterated

Number of results per page
10

Search

104 results found.

Left context	Result	Right context	Text ID	Date
M. Pan Woiewodá Krákowski Hetman Polny cum eadem apparentia	nazaiutrz [nazajutrz:adv]	Die[...] ma praesentis wiachał. Woysko do Soboty przeszłey stało	AwLwow	1693

I. W. I. M. Pan Woiewodá Krákowski Hetman Polny cum eadem apparentia **nazaiutrz** Die[...] ma praesentis wiachał. Woysko do Soboty przeszłey stało pod Báriszem, w Niedźielę miało się daley ruszyć zá lázłowiec ku Wasiłowu, co ieśli się stało czekamy in momentá wiádomośći.

Text ID: AwLwow
Page: 1
Title: Awizy lwowskie z Krakowa
Author: Anonim
Place of publication: Kraków

Region: Lesser Poland
Rhymed/Non-rhymed: non-rhymed
Type of text: press releases and leaflets
Humorous: no
Release date: 1693

ELECTRONIC DICTIONARY OF THE 17TH- AND 18TH-CENTURY POLISH

NAZAJUTRZ

Part of speech: przysł.
Meanings:
1. »następnego dnia, na drugi dzień«

Reference to the dictionary

Fig. 1. Corpus search interface: querying transliteration layer and dictionary linking

3 Embedding Transcription and Transliteration Layers in a PDF File

Storing text in PDF along with the scanned version of an item is usually done using the invisible text rendering mode (either by drawing text in the background which is then covered by the scanned image or drawing the invisible text in the foreground of the scanned image). Embedding spelling variants is then possible by manipulating the hidden textual layer in various ways, e.g. by: adding word variants directly in the single hidden layer. This method proves inconvenient for the reader since hidden text is not aligned in the page view (see Fig. 2).

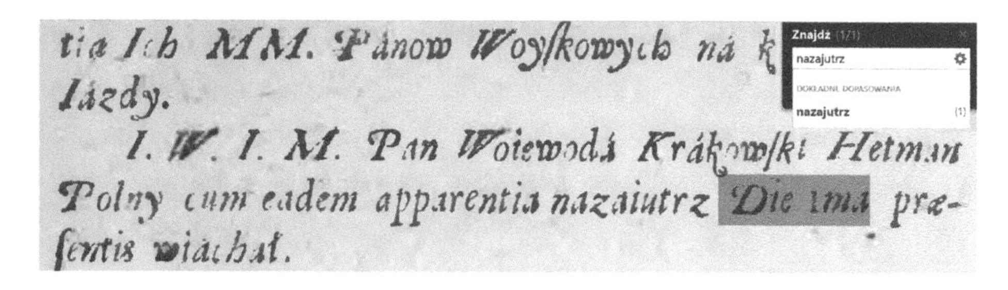

Fig. 2. Word variants added one by one in the hidden layer

A much better solution can be obtained by merging both hidden textual layers into a single file so that they can be searched concurrently. This function is easily obtained using many freely-available PDF manipulation libraries such as *PDFtk*, the PDF toolkit[8] or *CPDF*, Coherent PDF Command Line Toolkit[9] by merging two files each containing a separate hidden layer using *background* or *combine pages* options:

```
pdftk transcription.pdf multibackground transliteration.pdf
      output merged.pdf

cpdf  -combine-pages transcription.pdf transliteration.pdf
      -o merged.pdf
```

The result of this process can be found for print 1179 in CBDU[10] where both transcribed *nazajutrz* and transliterated *nazaiutrz* can be searched for, pointing at exactly the same scan segment as in Fig. 3.

The two tools tested in the process are just examples and several other PDF command-line utilities can be applied to merge multiple textual layers. However, it must be noted that the selected tool should be tested for compatibility with

[8] See https://www.pdflabs.com/tools/pdftk-the-pdf-toolkit/.
[9] See https://community.coherentpdf.com/.
[10] See https://cbdu.ijp.pan.pl/id/eprint/11790/.

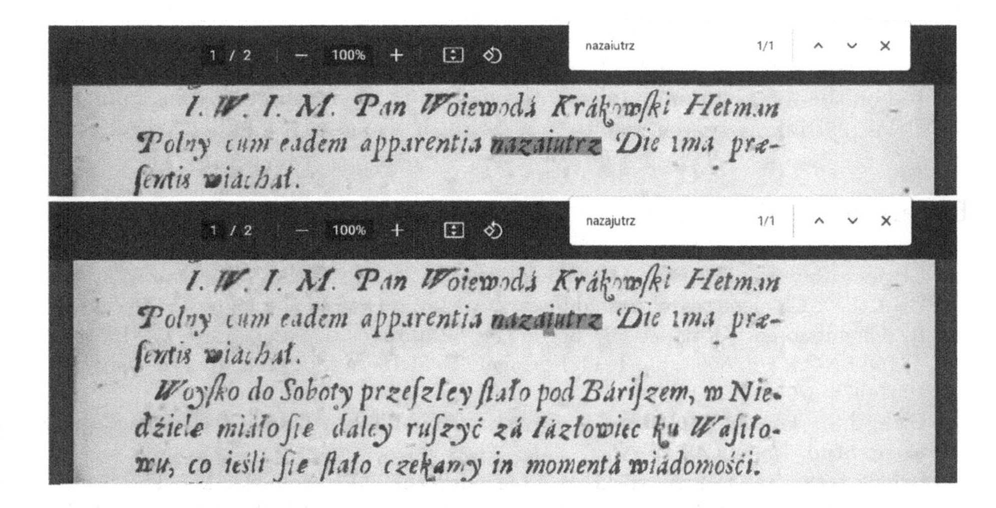

Fig. 3. Two spelling variants available for search in a single item

popular PDF viewers since e.g. files output by *PDFtk* can cause search problems when open offline in Acrobat Reader while they are perfectly processed with the default PDF viewer in the browser[11].

The hidden layers stored this way do not have any negative impact on copying of text from the PDF file (i.e. the text is not copied twice which happens with using the word variants method) and the layer available for copying can be selected by placing it in the foreground.

4 Conclusions and Future Plans

The presented experiment intended to show the method of supplementing CBDU but also similar libraries of old texts with several interpretation layers. Whenever suitable textual data is available, it can be encoded directly in the file to provide the best user experience and make the PDF file function independently on any digital library management system.

An additional experiment showed that embedding more than two layers in a single PDF file is also possible which gives many opportunities for encoding various interpretation over a single graphical layer of an item.

Another path of implementation could lead to integration of search in the digital library with both the corpus and the dictionary. From the point of view of the corpus search user it might be useful to view the retrieved concordance directly on the scan of the source document. Similarly, search in the dictionary could be illustrated with examples shown in actual context, on the respective page of the scanned item.

[11] Tested with Chrome 91.0.4472.124, Firefox 90.0 and Edge 91.0.864.67.

Last but not least, search in multiple PDF files in the form of so called graphical concordance (a Key Word in Context index with the scan snippets created on the fly) could be implemented following the method used by *Poliqarp for DjVu* [2], linking search results to a series of scans with highlighted hits.

References

1. EPrints Manual (2010). http://wiki.eprints.org/w/EPrints_Manual
2. Bień, J.S.: Efficient search in hidden text of large DjVu documents. In: Bernardi, R., Chambers, S., Gottfried, B., Segond, F., Zaihrayeu, I. (eds.) AT4DL/NLP4DL -2009. LNCS, vol. 6699, pp. 1–14. Springer, Heidelberg (2011). https://doi.org/10.1007/978-3-642-23160-5_1
3. Bień, J.S.: The IMPACT project Polish Ground-Truth texts as a Djvu corpus. Cogn. Stud. 75–84 (2014)
4. Bronikowska, R., Gruszczyński, W., Ogrodniczuk, M., Woliński, M.: The use of electronic historical dictionary data in corpus design. Stud. Pol. Linguist. **11**(2), 47–56 (2016). https://doi.org/10.4467/23005920SPL.16.003.4818
5. Gruszczyński, W. (ed.): Elektroniczny słownik języka polskiego XVII i XVIII w. (Electronic Dictionary of the 17th and the 18th century Polish, in Polish). Institute of Polish Language, Polish Academy of Sciences (2004). https://sxvii.pl/
6. Gruszczyński, W., Adamiec, D., Bronikowska, R., Wieczorek, A.: Elektroniczny Korpus Tekstów Polskich z XVII i XVIII w. - problemy teoretyczne i warsztatowe. Poradnik Językowy (8/2020 (777)), 32–51 (2020). https://doi.org/10.33896/porj.2020.8.3
7. Gruszczyński, W., Ogrodniczuk, M.: Cyfrowa Biblioteka Druków Ulotnych Polskich i Polski dotyczących z XVI, XVII i XVIII w. w nauce i dydaktyce (Digital Library of Poland-related Old Ephemeral Prints in research and teaching. In: Polish). In: Materiały konferencji Polskie Biblioteki Cyfrowe 2010 (Proceedings of the Polish Digital Libraries 2010 Conference), Poznań, Poland, pp. 23–27 (2010)
8. Ogrodniczuk, M., Gruszczyński, W.: Digital library of Poland-related old ephemeral prints: preserving multilingual cultural heritage. In: Proceedings of the Workshop on Language Technologies for Digital Humanities and Cultural Heritage, Hissar, Bulgaria, pp. 27–33 (2011). http://www.aclweb.org/anthology/W11-4105
9. Ogrodniczuk, M., Gruszczyński, W.: Digital library 2.0 – source of knowledge and research collaboration platform. In: Calzolari, N., et al. (eds.) Proceedings of the Ninth International Conference on Language Resources and Evaluation (LREC 2014), Reykjavík, Iceland, pp. 1649–1653. European Language Resources Association (2014). http://www.lrec-conf.org/proceedings/lrec2014/pdf/14_Paper.pdf
10. Ogrodniczuk, M., Gruszczyński, W.: Connecting data for digital libraries: the library, the dictionary and the corpus. In: Jatowt, A., Maeda, A., Syn, S.Y. (eds.) ICADL 2019. LNCS, vol. 11853, pp. 125–138. Springer, Cham (2019). https://doi.org/10.1007/978-3-030-34058-2_13
11. Zawadzki, K.: Gazety ulotne polskie i Polski dotyczące z XVI, XVII i XVIII wieku (Polish and Poland-related Ephemeral Prints from the 16th-18th Centuries, in Polish). National Ossoliński Institute, Polish Academy of Sciences, Wrocław (1990)

Search for Better User Experience

A Qualitative Evaluation of User Preference for Link-Based vs. Text-Based Recommendations of Wikipedia Articles

Malte Ostendorff[1(✉)], Corinna Breitinger[2], and Bela Gipp[2]

[1] Open Legal Data, Berlin, Germany
mo@openlegaldata.io
[2] University of Wuppertal, Wuppertal, Germany
{breitinger,gipp}@uni-wuppertal.de

Abstract. Literature recommendation systems (LRS) assist readers in the discovery of relevant content from the overwhelming amount of literature available. Despite the widespread adoption of LRS, there is a lack of research on the user-perceived recommendation characteristics for fundamentally different approaches to content-based literature recommendation. To complement existing quantitative studies on literature recommendation, we present qualitative study results that report on users' perceptions for two contrasting recommendation classes: (1) link-based recommendation represented by the Co-Citation Proximity (CPA) approach, and (2) text-based recommendation represented by Lucene's MoreLikeThis (MLT) algorithm. The empirical data analyzed in our study with twenty users and a diverse set of 40 Wikipedia articles indicate a noticeable difference between text- and link-based recommendation generation approaches along several key dimensions. The text-based MLT method receives higher satisfaction ratings in terms of user-perceived similarity of recommended articles. In contrast, the CPA approach receives higher satisfaction scores in terms of diversity and serendipity of recommendations. We conclude that users of literature recommendation systems can benefit most from hybrid approaches that combine both link- and text-based approaches, where the user's information needs and preferences should control the weighting for the approaches used. The optimal weighting of multiple approaches used in a hybrid recommendation system is highly dependent on a user's shifting needs.

Keywords: Information retrieval · Recommender systems · Human factors · Recommender evaluation · Wikipedia · Empirical studies

1 Introduction

The increasing volume of online literature has made recommendation systems an indispensable tool for readers. Over 50 million scientific publications are in circulation today [13], and approx. 3 million new publications are added annually

M. Ostendorff and C. Breitinger—Both authors contributed equally to this research.

H.-R. Ke et al. (Eds.): ICADL 2021, LNCS 13133, pp. 63–79, 2021.
https://doi.org/10.1007/978-3-030-91669-5_6

[15]. Encyclopedias such as Wikipedia are also subject to constant growth [39]. In the last decade, various recommendation approaches have been proposed for the literature recommendation use case. In a review of 185 publications, 96 different approaches for literature recommendation were identified. The majority of approaches (55%) continued to make use of content-based (CB) methods [4]. Only 18% of the surveyed recommendation approaches relied on collaborative filtering, and another 16% made use of graph-based recommendation methods, i.e. the analysis of citation networks, author, or venue networks [4]. A question that remains largely unexplored in today's literature is if these fundamentally different classes of recommendation algorithms are perceived differently by users. If a noticeable difference can be observed among users, across what dimensions do the end-users of such recommendation algorithms perceive that the approaches differ for a given recommendation use case? The majority of studies dedicated to evaluating LRS make use of offline evaluations using statistical accuracy metrics or error metrics without gathering any qualitative data from users in the wild [3]. More recently, additional metrics have been proposed to measure more dimensions of user-perceived quality for recommendations, e.g. novelty [11,29], diversity [22,41], serendipity [6,9,17,18], and overall satisfaction [14,27,43]. However, empirical user studies examining the perceived satisfaction with recommendations generated by different approaches remain rare. Given the emerging consensus on the importance of evaluating LRS from a user-centric perspective beyond accuracy alone [9], we identify a need for research to examine the user-perception of fundamentally different recommendation classes.

In this paper, we perform a qualitative study to examine user-perceived differences and thus highlight the benefits and drawbacks of two contrasting LRS applied to Wikipedia articles. We examine a text-based recommendation generation approach, represented by Lucene's MLT [1], and contrast this with a link-based approach, as implemented in the Citolytics recommendation engine [34], which uses co-citation proximity analysis (CPA) as a similarity measure [10]. Our study seeks to answer the following three research questions:

- RQ1: Is there a measurable difference in users' perception of the link-based approach compared to the text-based approach? If so, what difference do users perceive?
- RQ2: Do the approaches address different user information needs? If so, which user needs are best addressed by which approach?
- RQ3: Does one approach show better performance for certain topical categories or article characteristics?

Finally, we discuss how the evaluated recommendation approaches could be adapted in a hybrid system depending on the information needs of a user.

2 Background

2.1 Link-Based Similarity Measure

Co-citation proximity analysis (CPA; [10]) determines the similarity among articles by comparing the patterns of shared citations or links to other works within the full text of a document. The underlying concept of CPA originates from the Library Science field, where it takes inspiration from the co-citation (CoCit) measure introduced by Small [37]. Beyond co-citation, CPA additionally takes into account the positioning of links to determine the similarity of documents. When the links of co-linked articles appear in close proximity within the linking article, the co-linked articles are assumed to be more strongly related.

Schwarzer et al. [34] applied the concept of CPA to the outgoing links contained in Wikipedia articles. They introduced Citolytics as the first link-based recommendation system using the CPA measure and applied it to the Wikipedia article recommendation use case. To quantify the degree of relatedness of co-linked articles, CPA assigns a numeric value, the Co-Citation Proximity Index (CPI), to each pair of articles co-linked in one or more linking articles. Schwarzer et al. [35] derived a general-purpose CPI that is independent of the structural elements of academic papers, e.g., sections or journals, which were proposed in the original CPA concept. The general-purpose CPI is, therefore, more suitable for links found in the Wikipedia corpus. We define the co-link proximity $\delta_j(a,b)$ as the number of words between the links to article a and b in the article j. Equation 1 shows the CPI for article a being a recommendation for article b.

$$\mathrm{CPI}(a,b) = \sum_{j=1}^{|D|} \delta_j(a,b)^{-\alpha} * log(\frac{|D| - n_a + 0.5}{n_a + 0.5}) \quad (1)$$

The CPI consists of two components: First, the general co-link proximity of a and b which is the sum of all marker proximities $\delta_j(a,b)$ over all articles in the corpus D. The parameter α defines the non-linear weighting of the proximity δ. In general, co-links in close proximity should result in a higher CPI than co-links further apart. Thus, α must be greater or equal to zero. Moreover, the higher α, the closer the co-link proximity must be to influence the final CPI score. Prior to the user study, we conducted an offline evaluation similar to [35] and found that CPA achieves the best results with $\alpha = 0.9$.

The second component of CPI is a factor that defines the specificity of article a based on its in-links n_a. This factor is inspired by the Inverse Document Frequency of TF-IDF, whereby we adapted the weighting schema from Okapi BM25 [38]. Hence, we refer to the factor as Inverse Link Frequency (ILF). We introduced ILF to counteract the tendency of CPA to recommend more general Wikipedia articles, which we discovered in a manual analysis of CPA recommendations. ILF increases the recommendation specificity by penalizing articles with many in-links, which tend to cover broad topics.

2.2 Text-Based Similarity Measure

The text based *MoreLikeThis* (MLT) similarity measure from Elasticsearch [7] (based on Apache Lucence) differs fundamentally from the CPA approach. Instead of links, MLT relies entirely on the terms present in the article text to determine similarity. Using a Vector Space Model [33], MLT represents articles as sparse vectors in a space where each dimension corresponds to a separate index term. Term Frequency-Inverse Document Frequency (TF-IDF) proposed by [16] defines the weight of these index terms. Accordingly, MLT considers two articles similar the more terms they share and the more specific these terms are. Thus, MLT-based article recommendations are more likely to cover similar topics when the topic is defined by specific terms that do not occur in other topics. MLT's simplicity and ability to find similar articles has made it popular for websites. For instance, MLT is currently used by Wikipedia's MediaWiki software, as part of its *CirrusSearch* extension [28], to recommend articles to its users.

2.3 Related Work

Previously, Schwarzer et al. [35] performed an offline evaluation using the English Wikipedia corpus. They examined the offline performance of two link-based approaches, namely CPA and the more coarse CoCit measure, in addition to the text-based MLT measure. Schwarzer et al. made use of two quasi-gold standards afforded by Wikipedia. First, they considered the manually curated 'see also' links found at the end of Wikipedia articles as a quasi-gold-standard, which they used to evaluate 779,716 articles. Second, they used historical Wikipedia clickstream data in an evaluation of an additional 2.57 million articles. The results of this large-scale offline evaluation showed that the more fine-grained CPA measure consistently outperformed CoCit. This finding has also been validated by the research community in other recommendation scenarios [21,26].

Interestingly, this offline evaluation indicated that MLT performed better in identifying articles featuring a more narrow topical similarity with their source article. In contrast, CPA was better suited for recommending a broader spectrum of related articles [35]. However, prior evaluations by Schwarzer et al., using both 'See also' links (found at the bottom of Wikipedia articles) and clickstream data, were purely data-centric offline evaluations. This prohibits gaining in-depth and user-centric insights into the users' perceived usefulness of the recommendations shown. For example, click-through rates are a misleading metric for article relevance because users will click on articles with sensational or surprising titles before realizing that the content is not valuable to them. Accuracy and error metrics alone are not a reliable predictor of a user's perceived quality of recommendations [5]. Only user studies and online evaluations can reliably assess the effectiveness of real-world recommendation scenarios.

Knijnenburg et al. [20] proposed a user-centered framework that explains how objective system aspects influence subjective user behavior. Their framework is extensively evaluated with four trials and two controlled experiments and attempts to shed light on the interactions of personal and situational characteristics. Additionally, they take into account system aspects to explain the perceived user experience for movie recommendations.

Pu et al. [32] developed a user-centric evaluation framework termed ResQue (Recommender system's quality of user experience) consisting of 32 questions and 15 constructs to define the essential qualities of an effective and satisfying recommender system, including the recommendation qualities of accuracy, novelty, and diversity. Since their framework can be applied to article recommendations, we include several questions from the ResQue framework in our evaluation design (refer to Sect. 3.1 for details).

Despite the large size and popularity of Wikipedia, the potential of this corpus for evaluating LRS has thus far not been exploited by the research community. To the best of our knowledge, no prior work, aside from the initial offline study [35] and the work by [30], has made use of the Wikipedia corpus to evaluate the effectiveness of different recommendation approaches. The implications of studying the user-perceived recommendation effectiveness for Wikipedia articles may also be applicable to Wikimedia projects in a broader context, which tend to contain a high frequency of links.

3 Methodology

This section describes our study methodology and the criteria for selecting the Wikipedia articles used in our study. The Wikipedia encyclopedia is one of the most prominent open-access corpora among online reference literature. As of June 2021, the English Wikipedia contains approximately 6.3 million articles [39]. Its widespread use and accessibility motivated us to use the English Wikipedia to source the articles for our live user study. We consulted the same English Wikipedia corpus as in [35] with the pre-processing as in [34].

3.1 Study Design

Prior to our study, we created a sample of 40 seed articles covering a diverse spectrum of article types in Wikipedia. When selecting these seed articles, our aim was to achieve a diversity of topics, which nonetheless remained comprehensible to a general audience. To ensure comprehensibility, we excluded topics that would require expert knowledge to judge the relevance of recommendations, e.g., articles on mathematical theorems. Moreover, the seed articles featured diverse article characteristics, such as article length and article quality[1].

We distinguished seed articles into four categories. First, according to their *popularity* (measured by page views) into either niche or popular articles, and second, according to the content of the article into either *generic*, i.e., reference articles typical of encyclopedias, or *named entities*, i.e., politicians, celebrities, or locations. We choose popularity as a criterion because, on average, popular articles receive more in-links from other articles. Schwarzer et al. [35] found that the number of in-links affected the performance of the link-based CPA approach. Moreover, we expect study participants to be more familiar with popular topics compared to niche articles. Therefore users will be better able to verbalize their spontaneous information needs when examining a topic. The 'article type' categories were chosen to study the effect that articles about named entities may have on MLT. Names of entities tend to be more unique than terms in articles

[1] We judge the article quality using Wikipedia's vital article policy [40].

on generic topics. Therefore, we expect that specific names may affect MLT's performance. Likewise, due to the nature of Wikipedia articles linking to generic topics, they may appear in a broader context than links to named entities. Thus, CPA's performance may also be affected. These considerations resulted in four article categories: (A) niche generic articles, (B) popular generic articles, (C) niche named entities, and (D) popular named entities. Table 1 shows these four categories and the 40 seed articles selected for recommendation generation.

To perform our qualitative evaluation of user-perceived recommendation effectiveness, we recruited 20 participants. Participants were students and doctoral researchers from several universities in Berlin and the University of Konstanz. The average age of participants was 29 years. 65% of our participants said they spend more than an hour per month on Wikipedia, with the average being 4.6 h spent on Wikipedia.

Our study contained both qualitative and quantitative data collection components. The quantitative component was in the form of a written questionnaire. This questionnaire asked participants about each recommendation set separately and elicited responses on a 5-point Likert scale. Some questions were tailored to gain insights on the research questions we defined for our study. The remainder of the questions adhered to the ResQue framework for user-centric evaluation [32]. The qualitative data component was designed as a semi-structured interview. The interview contained open-ended questions that encouraged participants to verbally compare and contrast the two recommendation sets. The participants were also asked to describe their perceived satisfaction. Resulting from this mixed methods study design, we could use the findings from the qualitative interviews to interpret and validate the results from the quantitative questionnaires. All interviews were audio-recorded with the permission of our participants.

In the study, each participant was shown four Wikipedia articles, one at a time. For each article, two recommendation sets, each containing five recommended articles, were displayed. One set was generated using CPA, i.e., the link-based Citolytics implementation [34], while the other was generated using the MLT algorithm. Each set of four Wikipedia articles was shown to a total of two participants to enable checking for the presence of inter-rater agreement. Participants were aware that recommendation sets had been generated using different approaches, but they did not know the names of the approaches or the method behind the recommendations. We alternated the placement of the recommendation sets to avoid the recognition of one approach over the other and forming a potential bias based on placement. The seed Wikipedia articles were shown to participants via a tablet or a laptop. The participants were asked to read and scroll through the full article so that the exploration of the article's content was a natural as possible. We have made the complete questionnaire and the collected data publicly available on GitHub[2].

[2] https://github.com/malteos/wikipedia-article-recommendations.

Table 1. Overview of seed articles selected for the study.

#	Article (Quality (See footnote 1))	Words	#	Article (Quality (See footnote 1))	Words
A	*Niche generic topics*		*C*	*Niche named entities*	
1	Babylonian mathematics (B)	3,825	21	Mainau (S)	567
2	Water pollution in India (S)	1,697	22	Lake Constance (C)	7,079
3	Transport in Greater Tokyo (C)	3,046	23	Spandau (C)	599
4	History of United States cricket (S)	3,610	24	Appenzell (C)	2,667
5	Firefox for Android (C)	4,821	25	Michael Müller (politician) (Stub)	602
6	Chocolate syrup (Stub)	391	26	Olympiastadion (Berlin) (C)	3,360
7	Freshwater snail (C)	1,757	27	Theo Albrecht (S)	929
8	Touring car racing (S)	2,550	28	ARD (broadcaster) (S)	2,397
9	Mudflat (C)	787	29	Kaufland (Stub)	680
10	Philosophy of healthcare (B)	3,804	30	Sylt Air (Stub)	110
B	*Popular generic topics*		*D*	*Popular named entities*	
11	Fire (C)	4,297	31	Albert Einstein (GA)	15,071
12	Basketball (C)	11,172	32	Hillary Clinton (FA)	28,645
13	Mandarin Chinese (C)	698	33	Brad Pitt (FA)	9,955
14	Cancer (B)	16,300	34	New York City (B)	30,167
15	Vietnam War (C)	32,847	35	India (FA)	16,861
16	Cat (GA)	17,009	36	Elon Musk (C)	11,529
17	Earthquake (C)	7,541	37	Google (C)	16,216
18	Submarine (C)	11,968	38	Star Wars (B)	16,046
19	Rock music (C)	19,833	39	AC/DC (FA)	10,442
20	Wind power (GA)	15,761	40	FIFA World Cup (FA)	7,699

4 Results

In this section, we summarize and discuss the empirical data collected. First, we present the primary findings, in which we provide answers to the three research questions specified in the Introduction, and illustrate them with participants' quotes. Second, we discuss secondary findings that arose from coding the participants' responses, which go beyond the research questions we set out to answer.

4.1 Primary Findings

Our study found several differences in reader's perception of the link-based approach compared to the text-based approach. A notable difference could be identified especially in the perceived degree of 'similarity' of the recommendations. Participants were significantly more likely to agree with the statement 'the recommendations are more similar to each other' (see 1.6 in Fig. 1) for the MLT approach. 73% of responses 'agreed' or 'strongly agreed' (58 out of 80 responses) with this statement, compared to only 36% of the responses for the CPA approach (29 out of 80). Keep in mind each of the 40 seed articles was examined by two participants resulting in 80 responses in total. A question about whether the articles being recommended 'matched with the content' of the source article (see 1.1) was answered with a similar preference, with a significantly higher portion of

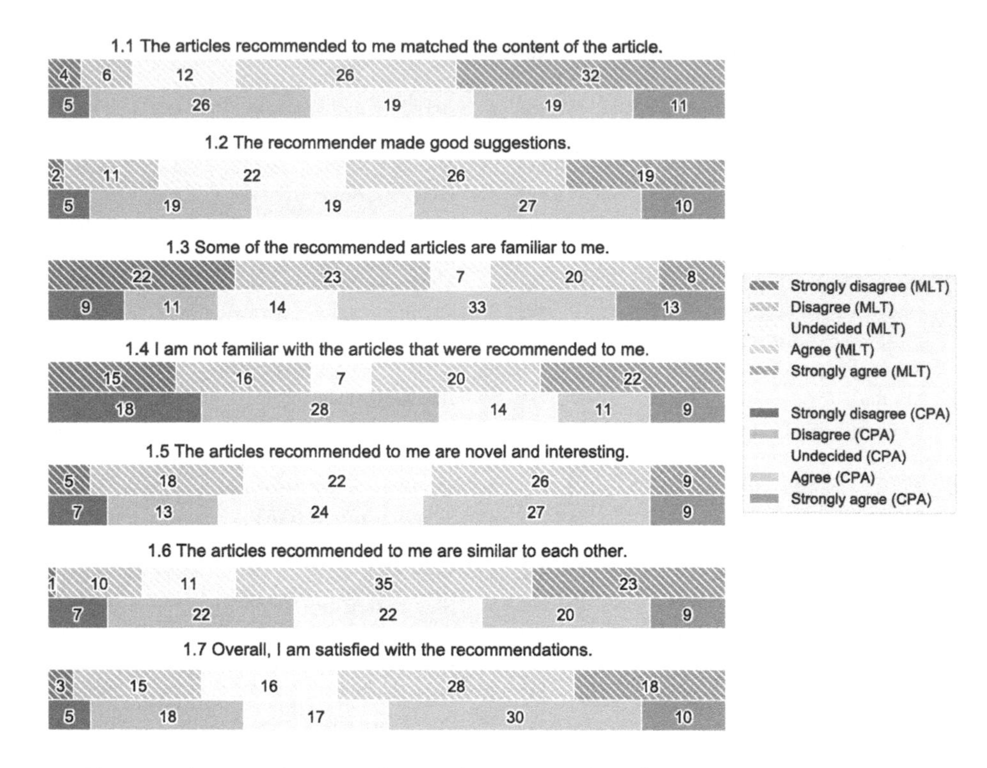

Fig. 1. Responses for MLT (dashed) and CPA (solid) on a 5-point Likert scale.

the responses indicating 'strongly agree' or 'agree' for the MLT approach (73%) and only 38% of responses choosing the same response for the CPA approach.

Overall, users perceived recommendations of CPA as more familiar (see 1.3). They felt less familiar (1.4) with the recommendations made by MLT. We found that this difference was observed by nearly all participants and can be attributed to how MLT considers textual similarity. In general, MLT focuses on overlapping terms, while CPA utilizes the co-occurrence of links. The quantitative results in [35] already suggested that this leads to diverging recommendations.

Perceived Difference between CPA and MLT. The participants observed that the methodological difference between the approaches affected their recommendations. In the questionnaire, participants expressed 48 times that the articles recommended by CPA are more diverse, i.e., less similar, compared to the seed article (Fig. 2a). MLT's recommendations were found to be diverse only 13 times. Regarding the similarity of recommendations, the outcome is the opposite.

The participants' answers also indicate the difference between MLT's and CPA's recommendations. Participant P20 explained that *"approach A [CPA] is more an overview of things and approach B [MLT] is focusing on concrete data or issues and regional areas"*. CPA providing an "overview of things" is not favorable for all participants as they describe different information needs.

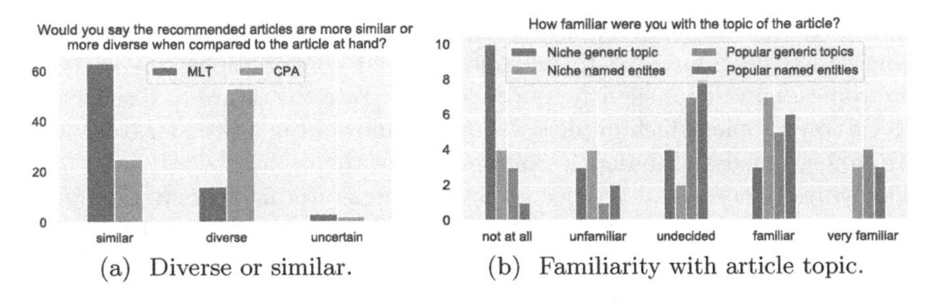

(a) Diverse or similar. (b) Familiarity with article topic.

Fig. 2. Quantitative answers

For example, participant P20 prefers MLT's recommendation since *"it is better to focus on the details"*. Some participants attributed the recommendations' similarity (or diversity) to terms co-occurring in the title of the seed and recommended articles. For instance, participant P15 found MLT's recommendations for *Star Wars* to be more similar because *"Star Wars is always in the title [of MLT's recommendations]"*. Participant P17 also assumes a direct connection between the seed and MLT's recommendations *"I'd guess recommendations of A [MLT] are already contained as link in the source article"*.

Nonetheless, participants struggled to put the observed difference between MLT and CPA in words, although they noticed categorical differences in the recommendation sets. Participant P19 said *"I can see a difference but I don't know what the difference is"*. Similarly, participant P20 found that *"they [MLT and CPA] are both diverse to the same extent but within a different scope"*.

Concerning the overall relevancy of the recommendations, MLT outperformed CPA. In total, the participants agreed or strongly agreed 45 times that MLT made 'good suggestions' (see 1.2), whereas only 37 times the same was stated for CPA. Similarly, the overall satisfaction was slightly higher for MLT (46 times agree or strongly agree) compared to CPA (40 times; see 1.7).

Information Need. The participants are also aware of relevancy depending on their individual information need. When asked about the 'most relevant recommendation' the participants' answers contained the words 'depends' or 'depending' ten-times. Participant P15 states that *"if I want a broader research I'd take B [CPA] but if I'd decide for more punctual research I would take A [MLT] because it is more likely to be around submarine and because in B [CPA] I also get background information"*. Similarly, participant P13 would click on a recommendation as follows: *"if you're looking for a specific class/type of snails then this [MLT] could be one, but if you're just looking to get an overview of aquatic animals, then probably you would click on the other approach [CPA]"*. In summary, the participants agreed on CPA providing 'background information' that is useful to 'get an overview of a topic', while MLT's recommendations were perceived as 'more specific' and having a 'direct connection' to the seed article.

For articles on science and technology, the most commonly expressed information needs were understanding how a technology works or looking up a definition. For articles about individuals, participants expressed the need to find dates relating to an individual and to understand their contributions to society. For 'niche' topics, users were slightly more likely to state the desire to discover subcategories on a topic, which implies wishing to move from a broader overview to a more fine-grained and in-depth examination of the topic.

The subjectiveness is also reflected by the inter-rater agreement. The participants who reviewed the same articles had a Cohen's kappa of $\kappa = 0.14$ on average, which corresponds to slight agreement. The inter-rater agreement increases to a "fair agreement" (see [24]) when we move from a 5-point to a 3-point Likert scale, i.e., possible answers are 'agree', 'undecided', or 'disagree'. Low agreement indicates that the perception of recommendation highly depends on the individual's prior knowledge and information needs.

Article Characteristics. The article 'types', which we defined as described in the methodology section according to article popularity, length and breadth into the four categories 'popular generic', 'niche generic', 'popular named entities', and 'niche named entities' had no observable impact on user's preference for one recommendation approach over the other.

However, we found that the user-expressed information need, for example, the desire to identify related articles that were either more broadly related or were more specialized, did have a measurable impact on the user's preference for the recommendation method. For *popular generic articles* on science and technology, e.g., the article on wind power, the most frequently expressed information needs were understanding how a technology works or looking up definitions.

For articles in the categories 'popular generic' and 'niche generic', we could observe that the information needs expressed by our readers were more *broad.* For example, they wanted to find definitions for the topic at hand, more general information to understand a topic in its wider context, or examples of subcategories on a topic. There was no observable difference between the specified categories of information need for 'popular' vs. 'niche' generic articles.

Resulting from our initial classification of the 40 Wikipedia articles selected, the empirical questionnaire data showed that 'niche' entities were on average more familiar to the participants than we initially expected (Fig. 2b). This was especially the case for niche named entities, many of which were rated as being familiar to the participants. The reason for this may be that our participants were from Germany and were thus familiar with many of these articles, despite the articles reporting on regional German topics, e.g., Spandau, Mainau. On the other hand, users rated niche generic topics as being less familiar, which was in line with what we expected.

Furthermore, both popular generic topics and popular named entities were less often classified as 'unfamiliar' by the participants than they were classified as 'neutral or familiar'. Lastly, one notable finding is that Wikipedia listings, e.g., *List of rock genres* or *List of supermarket chains in Germany*, were found to be the most relevant recommendations in some cases. The CPA implementation [34] intentionally excludes Wikipedia listings from its recommendation sets. Thus, the implementation needs to be revised accordingly.

4.2 Secondary Findings

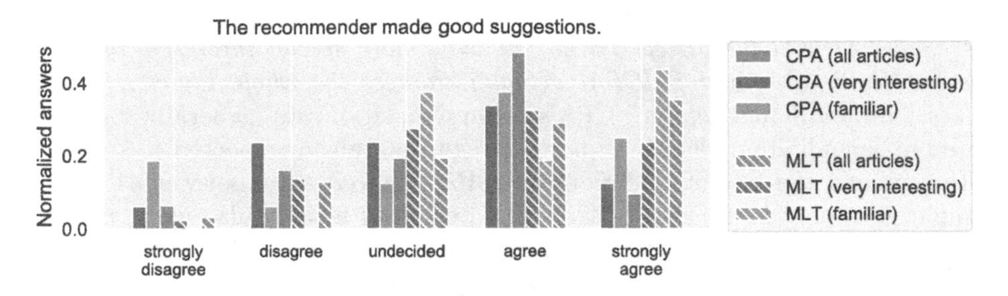

Fig. 3. User's satisfaction depending on interest and familiarity, i.e., for all articles or only the articles which are very interesting or familiar to the user.

Effect of User's Interest on Recommendations. Figure 3 shows that MLT outperforms CPA in recommendation satisfaction if participants 'strongly agreed' with the article topic being (a) interesting or (b) being at least 'familiar' to them. This is likely the case because users are more versed in judging the relevance of the text-based recommendations of the MLT approach if they already have more in-depth knowledge of a topic. For example, one participant observed regarding CPA's recommendations of *Renewable energy* for *Wind power* as the seed article: *"Renewable energy is least relevant because everybody knows something about it [Renewable energy]"*. Sinha and Swearingen [36] have already shown previous familiarity with an item as a confounding factor on a user 'liking' a recommendation. Interestingly, this trend was no longer observable in cases when participants only 'agreed' but without strong conviction that the articles were interesting or familiar to them. In these cases, the MLT and CPA approaches were seen as more equal, with the CPA approach taking a slight lead.

User-Based Preferences. Our findings confirm the subjectiveness of recommendation performance, since we observe user-based preferences. For instance, participant P2 only agreed or strongly agreed for CPA on MC 1.2 'The recommender made good suggestions' and 1.7 'Overall, I am satisfied with the recommendations' but never gave the same answers for MLT. However, participant P3 showed the opposite preference, i.e., only MLT made good suggestions according to P3. The remaining participants had more balanced preferences. In terms of MC 1.2 and 1.7, nine participants had a tendency to prefer MLT, while six participants preferred CPA, and five participants did not show any particular preference for one of the two recommendation approaches.

Perception of Novelty and Serendipity of Recommendations. To gain insights on the user-perceived novelty, we asked participants to rate the following statements: 'I am not familiar with the articles that were recommended to me' and

'The articles recommended to me are novel and interesting'. While novelty determines how unknown recommended items are to a user, serendipity is a measure of the extent to which recommendations positively surprise their users [9,18]. For instance, participant P19 answered *"approach A [MLT] shows me some topics connected with the article I read but with more special interest - they are about "Healthcare", and B [CPA] actually changes the whole topic. B [CPA] offers totally different topics."* CPA's recommendations are generally found to be more serendipitous. For the question regarding an 'unexpected recommendation among the recommendation set' CPA received a yes answer 41 times, compared to only 23 times for MLT. The perceived novelty also made participants click on recommendations. Among others, participant P1 explained that *"there are more [MLT] articles that I would personally click on, because they are new to me."* Similarly, participant P16 stated that they would *"click first on Star Wars canon, because I don't know what it is"*.

Trust and (Missing) Explanations. Although the questionnaire is not designed to investigate the participants' trust in the recommendations, many answers addressed this topic. When users were asked for the relevancy of recommendations, some participants expressed there "must be a connection" between the article at hand and a recommendation and that they just "do not know what is has to do with it". Others were even interested in topically irrelevant recommendations. For example, they expressed *"it interests me why this is important to the article I am reading"*. Similarly, a participant said they might click on a recommendation *"because I do not know what it has to do with [the seed article]"*. Such answers were more often found for CPA recommendations since they tend to be more broadly related than MLT's more narrow topical similarity. In some cases, there is no semantic relatedness. Yet, even then, participants often do not recognize a recommendation as irrelevant. Instead, they say it is their fault for not knowing how the recommendation is relevant to the seed. This behavior indicates a high level of trust from the participants placed in the recommender system.

5 Discussion

The experimental results demonstrate that MLT and CPA differ in their ability to satisfy specific user information needs. Furthermore, our study participants were capable of perceiving a systematic difference between the two approaches.

CPA was found to provide an 'overview of things' with recommendations more likely to be unfamiliar to the participants and less likely to match with the content of the seed article. In contrast, MLT was found to 'focus on the details'. Participants also felt that MLT's recommendations matched the content of the seed article more often. At the same time, participants perceive CPA's recommendations as more diverse, while MLT's recommendations are more similar to each other. So CPA and MLT, being conceptually different approaches and relying on different data sources, lead to unique differences in how the recommendations were perceived. In terms of the overall satisfaction with recommendations, most participants expressed a preference for MLT over CPA. MLT is based on TF-IDF and, therefore, its recommendations are centered around specific terms (e.g., P15: *"Star Wars is always in the title"*). In contrast, CPA relies on the co-occurrence of links. According to CPA, two articles are considered related when they are mentioned in the same context. Our results show that this leads to more distantly related recommendations, which do not necessarily share the same terminology. Given that the participants experience the two recommendation approaches differently, a hybrid text- and link combination, depending on the context, is preferable (as demonstrated in [8,19,31]).

Moreover, the differently perceived recommendations show the shortcoming of the notion of similarity. Both approaches, CPA and MLT, were developed to retrieve semantically similar documents, which they indeed do [10,16]. However, their recommendations are 'similar' within different scopes. A recommended article that provides an 'overview' can be considered similar to the seed article. Equivalently, a 'detailed' recommendation can also be similar to the seed but in a different context. Our qualitative interview data could show how users perceive these two similarity measures differently. These findings are aligned with [2], which found that text similarity inherits different dimensions.

We also found that either CPA's or MLT's recommendations are liked or disliked depending on the individual participant preferences. Some participants even expressed a consistent preference for one method over the other. However, a strict preference was the exception. We could also not identify any direct relation between the user or article characteristics and the preference for one method. At this point, more user data as in a user-based recommender system would be needed to tailor the recommendations to the user's profile. Purely content-based approaches such as CPA and MLT lack this ability [4,12,25]. The only option would be to allow users to select their preferred recommendation approach through the user interface depending on their information need.

The participants' answers also revealed a trust in the quality of the recommendations that was not always justified. Participants would assume a connection between the seed article and the recommended article just because it was recommended by the system. Instead of holding the recommender system accountable for non-relevant recommendations, participants found themselves responsible for not understanding a recommendations relevance. To not disappoint this trust, recommender systems should provide explanations that help users understand why a particular item is recommended. Also, explanations would help users to understand connections between seed and recommendations. Explainable recommendations are a subject of active research [23,42]. However, most research focuses on user-based approaches, while content-based approaches like CPA or MLT could also benefit from explanations.

Despite the insights of our qualitative study to elicit user's perceived differences in recommendation approach performance, the nature of our evaluation has several shortcomings. With 20 participants, the study is limited in size. Consequently, our quantitative data points suggest a difference that is not statistically significant. Large-scale offline evaluations (e.g., [35]) are more likely to produce statistically significant results. For this reason, and for not requiring participants, such offline evaluations are more commonly used in recommender system research [4]. But offline evaluations only provide insights in terms of performance measures. Our study shows that this can be an issue. When consulting only our quantitative data, one could assume that MLT and CPA are comparable in some aspects since their average scores are similarly high. The discrepancies between CPA and MLT only become evident when analyzing the written and oral explanations of live users. This highlights that recommender system research should not purely rely on offline evaluations [3].

Moreover, we acknowledge that recommendations of Wikipedia articles differ from recommendations of other literature types. It is thus uncertain whether our findings relating to encyclopedic recommendations in Wikipedia can be directly transferred to other domains. Recent advancements in recommender system research are focused on neural-based approaches, which may lead to the belief that the examined methods, MLT and CPA, are dated. This, however, is not the case since they are still used in practice, as Wikipedia's MLT deployment shows (Sect. 2.2), and the intuition of CPA is the basis for neural approaches like in Virtual Citation Proximity [30].

6 Conclusion

We elicit the user-perceived differences in performance for two well-known recommendation approaches. With the text-based MLT and the link-based CPA, we evaluate complementary content-based recommender system implementations. In a study with 20 participants, we collect qualitative and quantitative feedback on recommendations for 40 diverse Wikipedia articles.

Our results show that users are generally more satisfied with the recommendations generated by text-based MLT, whereas CPA's recommendation are perceived as more novel and diverse. The methodological difference of CPA and MLT, i.e., being based on either text or links, is reflected in their recommendations and noticed by the participants. Depending on information needs or user-based preferences, this leads to one recommendation approach being preferred over the other. Thus, we suggest combining both approaches in a hybrid system, since they both address different information needs.

As a result of the insights gained from our study, we plan to continue research on a hybrid approach tailored to the recommendation of literature, which accounts for diverse information needs. Moreover, we will investigate how content-based features can be utilized to provide explanation such that users can understand why a certain item is recommend to them. Lastly, we make our questionnaires, participants' answers, and code publicly available (See footnote 2).

Acknowledgements. We thank the anonymous reviewers, all participants, and especially Olha Yarikova for her assistance in the completion of the interviews.

References

1. Apache Lucence - More Like This. https://lucene.apache.org/core/7_2_0/queries/org/apache/lucene/queries/mlt/MoreLikeThis.html. Accessed 16 June 2021
2. Bär, D., Zesch, T., Gurevych, I.: A reflective view on text similarity. In: Proceedings of the International Conference Recent Advances in Natural Language Processing 2011, pp. 515–520. Association for Computational Linguistics, Hissar (2011). https://www.aclweb.org/anthology/R11-1071
3. Beel, J., Breitinger, C., Langer, S., Lommatzsch, A., Gipp, B.: Towards reproducibility in recommender-systems research. User Model. User-Adap. Inter. **26**(1), 69–101 (2016). https://doi.org/10.1007/s11257-016-9174-x
4. Beel, J., Gipp, B., Langer, S., Breitinger, C.: Research-paper recommender systems: a literature survey. Int. J. Digit. Libr. **17**(4), 305–338 (2015). https://doi.org/10.1007/s00799-015-0156-0
5. Cremonesi, P., Garzotto, F., Negro, S., Papadopoulos, A.V., Turrin, R.: Looking for "Good" recommendations: a comparative evaluation of recommender systems. In: Campos, P., Graham, N., Jorge, J., Nunes, N., Palanque, P., Winckler, M. (eds.) INTERACT 2011. LNCS, vol. 6948, pp. 152–168. Springer, Heidelberg (2011). https://doi.org/10.1007/978-3-642-23765-2_11
6. De Gemmis, M., Lops, P., Semeraro, G., Musto, C.: An investigation on the serendipity problem in recommender systems. Inf. Process. Manag. **51**(5), 695–717 (2015). https://doi.org/10.1016/j.ipm.2015.06.008
7. ElasticSearch - More like this query. https://www.elastic.co/guide/en/elasticsearch/reference/current/query-dsl-mlt-query.html. Accessed 16 June 2021
8. Färber, M., Sampath, A.: HybridCite: a hybrid model for context-aware citation recommendation. arXiv (2020). http://arxiv.org/abs/2002.06406
9. Ge, M., Delgado-Battenfeld, C., Jannach, D.: Beyond accuracy: evaluating recommender systems by coverage and serendipity. In: Proceedings of the Fourth ACM Conference on Recommender Systems, RecSys '10, pp. 257–260. Association for Computing Machinery, New York (2010). https://doi.org/10.1145/1864708.1864761
10. Gipp, B., Beel, J.: Citation Proximity Analysis (CPA) – a new approach for identifying related work based on co-citation analysis. In: Proceedings of the 12th International Conference on Scientometrics and Informetrics (ISSI '09), 2 July, pp. 571–575 (2009). https://ag-gipp.github.io/bib/preprints/gipp09a.pdf
11. Gravino, P., Monechi, B., Loreto, V.: Towards novelty-driven recommender systems. C R Phys. **20**(4), 371–379 (2019). https://doi.org/10.1016/j.crhy.2019.05.014
12. Jannach, D., Zanker, M., Felfernig, A., Friedrich, G.: Recommender Systems - An Introduction, 1st edn. Cambridge University Press, London (2010)
13. Jinha, A.E.: Article 50 million: an estimate of the number of scholarly articles in existence. Learn. Publish. **23**(3), 258–263 (2010). https://doi.org/10.1087/20100308
14. Joachims, T., Granka, L., Pan, B.: Accurately interpreting clickthrough data as implicit feedback. In: Proceedings of the 28th Annual International ACM SIGIR Conference on Research and Development in Information Retrieval (2005). https://doi.org/10.1145/1076034.1076063
15. Johnson, R., Watkinson, A., Mabe, M.: The STM report: an overview of scientific and scholarly publishing. Int. Assoc. Sci. Tech. Med. Publish. **5**(October), 212 (2018)

16. Jones, K.S.: Index term weighting. Inf. Storage Retrieval **9**(11), 619–633 (1973). https://doi.org/10.1016/0020-0271(73)90043-0
17. Kaminskas, M., Bridge, D.: Measuring surprise in recommender systems. In: RecSys REDD 2014: International Workshop on Recommender Systems Evaluation: Dimensions and Design, vol. 69, pp. 2–7 (2014)
18. Kaminskas, M., Bridge, D.: Diversity, serendipity, novelty, and coverage: a survey and empirical analysis of beyond-accuracy objectives in recommender systems. ACM Trans. Interact. Intell. Syst. **7**(1), 1–42 (2017). https://doi.org/10.1145/2926720
19. Kanakia, A., Shen, Z., Eide, D., Wang, K.: A scalable hybrid research paper recommender system for microsoft academic. In: The World Wide Web Conference (WWW 2019), pp. 2893–2899. ACM Press, New York (2019). https://doi.org/10.1145/3308558.3313700
20. Knijnenburg, B.P., Willemsen, M.C., Gantner, Z., Soncu, H., Newell, C.: Explaining the user experience of recommender systems. User Model. User-Adap. Inter. **22**(4–5), 441–504 (2012). https://doi.org/10.1007/s11257-011-9118-4
21. Knoth, P., Khadka, A.: Can we do better than co-citations? - Bringing citation proximity analysis from idea to practice in research article recommendation. In: 2nd Joint Workshop on Bibliometric-enhanced Information Retrieval and Natural Language Processing for Digital Libraries, Tokyo, Japan (2017)
22. Kunaver, M., Požrl, T.: Diversity in recommender systems - a survey. Knowl.-Based Syst. **123**, 154–162 (2017). https://doi.org/10.1016/j.knosys.2017.02.009
23. Kunkel, J., Donkers, T., Michael, L., Barbu, C.M., Ziegler, J.: Let me explain: impact of personal and impersonal explanations on trust in recommender systems. In: Conference on Human Factors in Computing Systems - Proceedings (2019). https://doi.org/10.1145/3290605.3300717
24. Landis, J.R., Koch, G.G.: The measurement of observer agreement for categorical data. Biometrics **33**(1), 159–174 (1977). http://www.jstor.org/stable/2529310
25. Lenhart, P., Herzog, D.: Combining content-based and collaborative filtering for personalized sports news recommendations. In: Proceedings of the 3rd Workshop on New Trends in Content-Based RecommenderSystems (CBRecSys '16) at RecSys'16, vol. 1673, pp. 3–10 (2016)
26. Liu, S., Chen, C.: The effects of co-citation proximity on co-citation analysis. In: Proceedings of the 13th Conference of the International Society for Scientometrics and Informetrics (ISSI 2011) (2011)
27. Maksai, A., Garcin, F., Faltings, B.: Predicting online performance of news recommender systems through richer evaluation metrics. In: Proceedings of the 9th ACM Conference on Recommender Systems, pp. 179–186. ACM, New York (2015). https://doi.org/10.1145/2792838.2800184
28. Media Wiki Extension - CirrusSearch. https://www.mediawiki.org/wiki/Extension:CirrusSearch. Accessed 16 June 2021
29. Mendoza, M., Torres, N.: Evaluating content novelty in recommender systems. J. Intell. Inf. Syst. **54**(2), 297–316 (2019). https://doi.org/10.1007/s10844-019-00548-x
30. Molloy, P., Beel, J., Aizawa, A.: Virtual Citation Proximity (VCP): learning a hypothetical in-text citation-proximity metric for uncited documents. In: Proceedings of the 8th International Workshop on Mining Scientific Publications, pp. 1–8 (2020). https://doi.org/10.31219/osf.io/t5aqf

31. Ostendorff, M., Ash, E., Ruas, T., Gipp, B., Moreno-Schneider, J., Rehm, G.: Evaluating document representations for content-based legal literature recommendations. In: Proceedings of the Eighteenth International Conference on Artificial Intelligence and Law, ICAIL '21, pp. 109–118. Association for Computing Machinery, New York (2021). https://doi.org/10.1145/3462757.3466073
32. Pu, P., Chen, L., Hu, R.: A user-centric evaluation framework for recommender systems. In: Proceedings of the Fifth ACM Conference on Recommender Systems - RecSys '11, p. 157. ACM Press, New York (2011). https://doi.org/10.1145/2043932.2043962
33. Salton, G., Wong, A., Yang, C.S.: Vector space model for automatic indexing. Information retrieval and language processing. Commun. ACM **18**(11), 613–620 (1975)
34. Schwarzer, M., Breitinger, C., Schubotz, M., Meuschke, N., Gipp, B.: Citolytics: a link-based recommender system for wikipedia. In: Proceedings of the Eleventh ACM Conference on Recommender Systems - RecSys '17, pp. 360–361. ACM, New York (2017). https://doi.org/10.1145/3109859.3109981
35. Schwarzer, M., Schubotz, M., Meuschke, N., Breitinger, C., Markl, V., Gipp, B.: Evaluating Link-based recommendations for wikipedia. In: Proceedings of the 16th ACM/IEEE-CS on Joint Conference on Digital Libraries - JCDL '16, pp. 191–200. ACM Press, New York (2016). https://doi.org/10.1145/2910896.2910908
36. Sinha, R., Swearingen, K.: The role of transparency in recommender systems. In: CHI '02 Extended Abstracts on Human Factors in Computing Systems - CHI '02, p. 830. ACM Press, New York (2002)
37. Small, H.: Co-citation in the scientific literature: a new measure of the relationship between two documents. J. Am. Soc. Inf. Sci. **24**(4), 265–269 (1973). https://doi.org/10.1002/asi.4630240406
38. Sparck Jones, K., Walker, S., Robertson, S.E.: A probabilistic model of information retrieval: development and comparative experiments, Part 2. Inf. Process. Manag. **36**(6), 809–840 (2000). https://doi.org/10.1016/S0306-4573(00)00015-7
39. Wikipedia - Size of Wikipedia. https://en.wikipedia.org/wiki/Wikipedia:Size_of_Wikipedia. Accessed 16 June 2021
40. Wikipedia - Vital articles. https://en.wikipedia.org/wiki/Wikipedia:Vital_articles. Accessed 16 June 2021
41. Yu, C., Lakshmanan, L., Amer-Yahia, S.: It takes variety to make a world: diversification in recommender systems. In: Edbt, pp. 368–378 (2009). https://doi.org/10.1145/1516360.1516404
42. Zhang, Y., Chen, X.: Explainable recommendation: a survey and new perspectives. Now Publish. (2020). https://doi.org/10.1561/9781680836592
43. Zhao, Q., Harper, F.M., Adomavicius, G., Konstan, J.A.: Explicit or implicit feedback? Engagement or satisfaction? a field experiment on machine-learning-based recommender systems. In: Proceedings of the 33rd Annual ACM Symposium on Applied Computing, SAC '18, pp. 1331–1340. Association for Computing Machinery, New York (2018). https://doi.org/10.1145/3167132.3167275

Narrative Query Graphs for Entity-Interaction-Aware Document Retrieval

Hermann Kroll(✉), Jan Pirklbauer, Jan-Christoph Kalo, Morris Kunz, Johannes Ruthmann, and Wolf-Tilo Balke

Institute for Information Systems, TU Braunschweig, Braunschweig, Germany
{kroll,kalo,balke}@ifis.cs.tu-bs.de,
{j.pirklbauer,morris.kunz,j.ruthmann}@tu-bs.de

Abstract. Finding relevant publications in the scientific domain can be quite tedious: Accessing large-scale document collections often means to formulate an initial keyword-based query followed by many refinements to retrieve a *sufficiently complete, yet manageable* set of documents to satisfy one's information need. Since keyword-based search limits researchers to formulating their information needs as a set of unconnected keywords, retrieval systems try to guess each user's intent. In contrast, distilling short narratives of the searchers' information needs into simple, yet precise entity-interaction graph patterns provides all information needed for a precise search. As an additional benefit, such graph patterns may also feature variable nodes to flexibly allow for different substitutions of entities taking a specified role. An evaluation over the PubMed document collection quantifies the gains in precision for our novel entity-interaction-aware search. Moreover, we perform expert interviews and a questionnaire to verify the usefulness of our system in practice.

Keywords: Narrative queries · Graph-based retrieval · Digital libraries

1 Introduction

PubMed, the world's most extensive digital library for biomedical research, consists of about 32 million publications and is currently growing by more than one million publications each year. Accessing such an extensive collection by simple means such as keyword-based retrieval over publication texts is a challenge for researchers, since they simply cannot read through hundreds of possibly relevant documents, yet cannot afford to miss relevant information in retrieval tasks. Indeed, there is a dire need for retrieval tools tailored to specific information needs in order to solve the above conflict. For such tools, deeper knowledge about the particular task at hand and the specific semantics involved is essential. Taking a closer look at the nature of scientific information search, interactions between entities can be seen to represent a short narrative [8], a short story of interest: how or why entities interact, in what sequence or roles they occur, and what the result or purpose of their interaction is [3,8].

H.-R. Ke et al. (Eds.): ICADL 2021, LNCS 13133, pp. 80–95, 2021.
https://doi.org/10.1007/978-3-030-91669-5_7

Indeed, an extensive query log analysis on PubMed in [4] clearly shows that researchers in the biomedical domain are often interested in interactions between entities such as drugs, genes, and diseases. Among other results, the authors report that a) on average significantly more keywords are used in PubMed queries than in typical Web searches, b) result set sizes reach an average of (rather unmanageable) 14,050 documents, and c) keyword queries are on average 4.3 times refined and often include more specific information about the keywords' intended semantic relationships, e.g., *myocardial infarction AND aspirin* may be refined to *myocardial infarction prevention AND aspirin*. Given all these observations, native support for entity-interaction-aware retrieval tasks can be expected to be extremely useful for PubMed information searches and is quite promising to generalize to other kinds of scientific domains, too. However, searching scientific document collections curated by digital libraries for such narratives is tedious when being restricted to keyword-based search, since the same narrative can be paraphrased in countless ways [1,4].

Therefore, we introduce the novel concept of *narrative query graphs for scientific document retrieval* enabling users to formulate their information need as entity-interaction queries explicitly. Complex interactions between entities can be precisely specified: Simple interactions between two entities are expressed by a basic query graph consisting of two nodes and a labeled edge between them. Of course, by adding more edges and entity nodes, these basic graph patterns can be combined to form arbitrarily complex graph patterns to address highly specialized information needs. Moreover, narrative query graphs support *variable nodes* supporting a far broader expressiveness than keyword-based queries. As an example, a researcher might search for treatments of *some disease* using *simvastatin*. While keyword-based searches would broaden the scope of the query far in excess of the user intent by just omitting any specific disease's name, narrative query graphs can focus the search by using a variable node to find documents that describe treatments of *simvastatin* facilitated by an entity of the type *disease*. The obtained result lists can then be clustered by possible node substitutions to get an entity-centric literature overview. Besides, we provide provenance information to explain why a document matches the query.

In summary, our contributions are:

1. We propose narrative query graphs for scientific document retrieval enabling fine-grained modeling of users' information needs. Moreover, we boost query expressiveness by introducing variable nodes for document retrieval.
2. We developed a prototype that processes arbitrary narrative query graphs over large document collections. As a showcase, the prototype performs searches on six million PubMed titles and abstracts in real-time.
3. We evaluated our system in two ways: On the one hand, we demonstrated our retrieval system's usefulness and superiority over keyword-based search on the PubMed digital library in a manual evaluation including practitioners from the pharmaceutical domain. On the other hand, we performed interviews and a questionnaire with eight biomedical experts who face the search for literature on a daily basis.

2 Related Work

Narrative query graphs are designed to offer complex querying capabilities over scientific document collections aiming at high precision results. Focusing on retrieving entity interactions, they are a subset of our conceptual overlay model for representing narrative information [8]. We discussed the first ideas to bind narratives against document collections in [9]. This paper describes the complete retrieval method and evaluation of narrative query graphs for document retrieval. In the last decade three major research areas were proposed to improve text-based information retrieval.

Machine Learning for Information Retrieval. Modern personalized systems try to guess each user's intent and automatically provide more relevant results by query expansion, see [1] for a good overview. Mohan et al. focus on information retrieval of biomedical texts in PubMed [13]. The authors derive a training and test set by analyzing PubMed query logs and train a deep neural network to improve literature search. Entity-based language models are used to distinguish between a term-based and entity-based search to increase the retrieval quality [16]. Yet, while a variety of approaches to improve result rankings by learning how a query is related to some document [13,19,20], have been proposed, gathering enough training data to effectively train a system for all different kinds of scientific domains seems impossible. Specialized information needs, which are not searched often, are hardly covered in such models.

Graph-Based Information Retrieval. Using graph-based methods for textual information retrieval gained in popularity recently [3,17,18,20]. For instance, Dietz et al. discuss the opportunities of entity linking and relation extraction to enhance query processing for keyword-based systems [3] and Zhao et al. demonstrate the usefulness of graph-based document representations for precise biomedical literature retrieval [20]. Kadry et al. also include entity and relationship information from the text as a learning-to-rank task to improve support passage retrieval [5]. Besides, Spitz et al. build a graph representation for Wikipedia to answer queries about events and entities more precisely [17]. But in contrast to our work, the above approaches focus on unlabeled graphs or include relationships only partially.

Knowledge Bases for Literature Search. GrapAl, for example, a graph database of academic literature, is designed to assist academic literature search by supporting a structured querying language, namely Cypher [2]. GrapAl mainly consists of traditional metadata like authors, citations, and publication information but also includes entities and relationship mentions. However, complex entity interactions are not supported, as only a few basic relationships per paper are annotated. As a more practical system that extracts facts from text to support question answering, QKBfly has been presented [14]. It constructs a knowledge base for ad-hoc question answering during query time that provides journalists with the latest information about emergent topics. However, they focus on retrieving relevant facts concerning a single entity. In contrast, our focus is on document retrieval for complex entity interactions.

3 Narrative Query Graphs

Entities represent things of interest in a specific domain: Drugs and diseases are prime examples in the biomedical domain. An entity $e = (id, type)$, where id is a unique identifier and $type$ the entity type. To give an example, we may represent the drug *simvastatin* by its identifier and entity type as follows: $e_{simvastatin} = (D019821, Drug)$. Typically, entities are defined by predefined ontologies, taxonomies, or controlled vocabularies, such as NLM's MeSH or EMBL's ChEBI. We denote the set of known entities as $\mathcal{E}$. Entities might also be classes as well, e.g., the entity *diabetes mellitus* (Disease) refers to a class of specialized diabetes diseases such as DM type 1 and DM type 2. Thus, these classes can be arranged in subclass relations, i.e., DM type 1 is the subclass of general diabetes mellitus. Since we aim to find entity interactions in texts, we need to know where entities are mentioned. In typical natural language processing, each sentence is represented as a sequence of tokens, i.e., single words. Therefore, an **entity alignment** maps a token or a sequence of tokens to an entity from $\mathcal{E}$ if the tokens refer to it.

We call an interaction between two entities a **statement** following the idea of knowledge representation in the Resource Description Framework (RDF) [12]. Hence, a **statement** is a triple (s, p, o) where $s, o \in \mathcal{E}$ and $p \in \Sigma$. Σ represents the set of all interactions we are interested in. We focus only on interactions between entities, unlike RDF, where objects might be literals too. For example, a *treatment* interaction between *simvastatin* and *hypercholesterolemia* is encoded as $(e_{simvastatin}$, *treats*, $e_{hypercholesterolemia})$. We call a set of extractions from a single document a so-called **document graph**.

Document graphs support narrative querying, i.e., the query is answered by matching the query against the document's graph. Suppose a user formulates a query like $(e_{simvastatin}$, *treats*, $e_{hypercholesterolemia})$. In that case, our system retrieves a set of documents containing the searched statement. Narrative query graphs may include typed variable nodes as well. A user might query $(e_{simvastatin}$, *treats*, *?X(Disease)*), asking for documents containing *some* disease treatment with *simvastatin*. Hence, all documents that include *simvastatin* treatments for diseases are proper matches. Formally, we denote the set of all variable nodes as $\mathcal{V}$. Variable nodes consist of a name and an entity type to support querying for entity types. We also support the entity type *All* to query for arbitrary entities. We write variable nodes by a leading question mark. Hence, a narrative query graph might include entities stemming from $\mathcal{E}$ and variable nodes from $\mathcal{V}$. Formally, a **fact pattern** is a triple $fp = (s, p, o)$ where $s, o \in (\mathcal{E} \cup \mathcal{V})$ and $p \in \Sigma$. A **narrative query graph** q is a set of fact patterns similar to SPARQL's basic graph patterns [15]. When executed, the query produces one or more matches μ by binding the variable symbols to actual entities, i.e., $\mu : \mathcal{V} \rightarrow \mathcal{E}$ is a partial function. If several fact patterns are queried, all patterns must be contained within a document forming a proper query answer. If queries include entities that are classes and have subclasses, then the query will be expanded to also query for these subclasses, i.e., direct and transitive subclasses.

4 Narrative Document Retrieval

In the following section we describe our system for narrative query graph processing. First, we perform a pre-processing that involves entity linking, information extraction, cleaning, and loading. It extracts document graphs from text and stores them in a structured repository. Then, a query processing that matches a user's query against the document graphs takes place. In this way, we can return a structured visualization of matching documents. An overview of the whole system is depicted in Fig. 1.

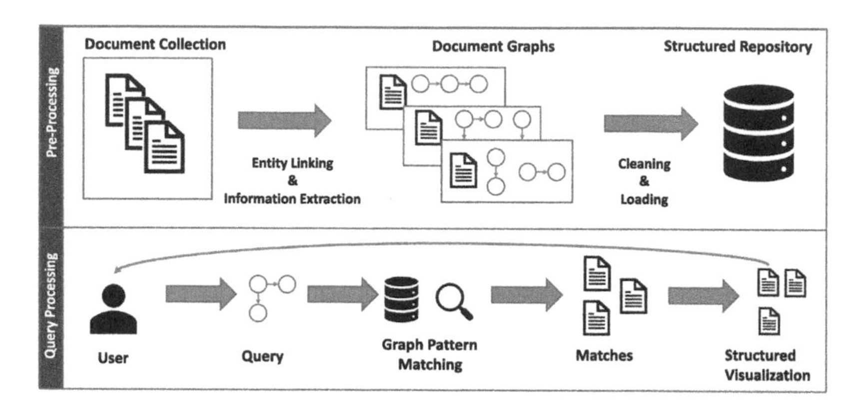

Fig. 1. System Overview: Document graphs are extracted from texts, cleaned, indexed, and loaded into a structured repository. Then, narrative query graphs can be matched against the repository to retrieve the respective documents.

4.1 Document Graph Extraction

The pre-processing step, including entity linking and information extraction, utilizes our toolbox for the nearly-unsupervised construction of knowledge graphs [10]. The toolbox requires the design of two different vocabularies: 1. An entity vocabulary that contains all entities of interest. An entry consists of a unique entity id, an entity name, and a list of synonyms. 2. A relation vocabulary that contains all relations of interest. An entry consists of a relation and a set of synonyms.

For this paper, we built an entity vocabulary that comprises *drugs, diseases, dosage forms, excipients, genes, plant families, and species*. Next, we wanted to extract interactions between these entities from texts since interactions between entities are essential to support retrieval with narrative query graphs. Although the quality of existing open information extraction like OpenIE 6 sounded promising [6], we found that open information extraction methods highly lack recall when processing biomedical texts, see the evaluation in [10]. That is why we developed a recall-oriented extraction technique **PathIE** in [10] that flexibly extracts interactions between entities via a path-based method. This method was evaluated and shared in our toolbox as well.

PathIE yields many synonymous predicates (treats, aids, prevents, etc.) that represent the relation *treats*. The relation vocabulary must have clear semantics and was built with the help of two domain experts. We designed a relation vocabulary comprising 60 entries (10 relations plus 50 synonyms) for the cleaning step. This vocabulary enables the user to formulate her query based on a well-curated vocabulary of entity interactions in the domain of interest. We applied our semi-supervised predicate unification algorithm to clean the extractions. To increase the quality of extractions, we introduced type constraints by providing fixed domain and range types for each interaction. Extracted interactions that did not meet the interaction's type constraints were removed. For example, the interaction *treats* is typed, i.e., the subject must be a drug, and the object must be a disease or species. Some interactions in our vocabulary like *induces* or *associated* are more general and thus were not annotated with type constraints.

4.2 Document Retrieval

Finally, the extracted document graphs had to be stored in a structured repository for querying purposes. For this paper, we built upon a relational database, namely PostgresV10. Relational databases support efficient querying and allowed us to provide additional provenance information and metadata for our purposes. For example, our prototype returned document titles, sentences, entity annotations, and extraction information to explain matches to the user. Due to our focus on pharmaceutical and medical users, we selected a PubMed subset that includes drug and excipient annotations. Therefore, we annotated the whole PubMed collection with our entity linking component, yielding 302 million annotations. Around six million documents included a drug or excipient annotation. Performing extraction and cleaning on around six million documents yielded nearly 270 million different extractions. Hence, the current prototype's version comprises about six million documents. We incrementally have increased the available data, but we entirely covered the relevant pharmaceutical part (drug and excipient).

As a reminder, a narrative query graph consists of fact patterns following simple RDF-style basic graph patterns. Our system automatically translates these narrative query graphs into a structured query language: They are translated into SQL statements for querying the underlying relational database. A single fact pattern requires a selection of the extraction table with suitable conditions to check the entities and the interaction. Multiple fact patterns require self-joining of the extraction table, and adding document conditions in the where clause, i.e., the facts matched against the query must be extracted from the same document. We developed an in-memory and hash-based matching algorithm that quickly combines the results. Another point to think about were ontological subclass relations between entities. For example, querying for treatments of *Diabetes Mellitus* would require to also search for the subclasses *Diabetes Mellitus Type 1* and *Diabetes Mellitus Type 2*. Query rewriting is necessary to compute complete results for queries that involve entities with subclasses [11]. We rewrite queries that include entities with subclasses to also query for these subclasses. Due to the long-standing development of databases, such a query processing can be performed very quickly when using suitable indexes. We computed an inverted index, i.e., each extraction triple was mapped to a set of document ids. Besides, we implemented some optimization strategies to accelerate the query processing, e.g., match fact patterns with concrete entities first and fact patterns with variable nodes afterward. We remark on our system's query performance in our evaluation.

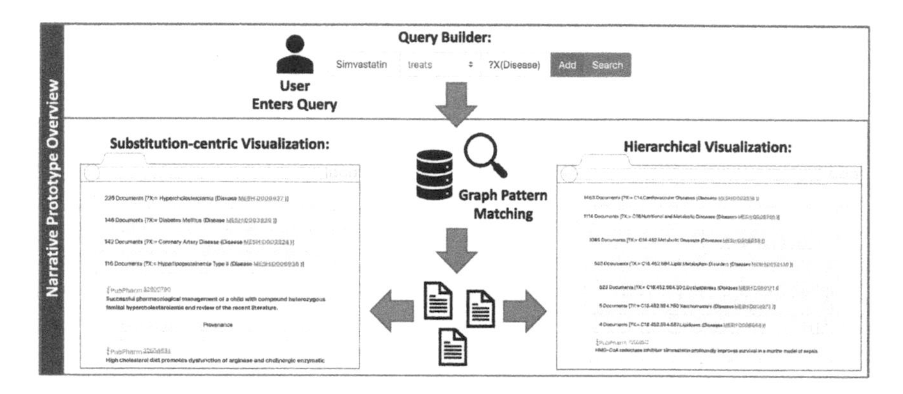

Fig. 2. A schematic overview of our prototype implementation. A query builder helps the users to formulate their information need. If the narrative query involves variable nodes, the results can be visualized in a substitution-centric visualization (left side) or in a hierarchical visualization (right side).

4.3 Prototype Design

We present a prototype resulting from joint efforts by the university library, the institute for information systems, and two pharmaceutical domain experts who gave us helpful feedback and recommendations. The prototype[1] offers precise biomedical document retrieval with narrative query graphs. A general overview of our prototype is shown in Fig. 2. We implemented a REST service handling queries and performing the query processing on the backend side. Furthermore, we supported the user with a query builder and suitable result visualization on the frontend side. In an early prototype phase, we tested different user interfaces to formulate narrative query graphs, namely, 1. a simple text field, 2. a structured query builder, and 3. a graph designer tool. We found that our users preferred the structured query builder which allows them to formulate a query by building a list of fact patterns. For each fact pattern, the users must enter the query's subject and object. Then, they can select an interaction between both in a predefined selection. The prototype assists the user by suggesting around three million terms (entity names plus synonyms). Variable nodes can be formulated, e.g., by writing *?X(Drug)* or just entering the entity type like *Drug* in the subject or object field. When users start their search, the prototype sends the

[1] http://www.pubpharm.de/services/prototypes/narratives/.

query to the backend and visualizes the returned results. The returned results are sorted by their corresponding publication date in descending order. The prototype represents documents by a document id (PubMedID), a title, a link to the digital library entry (PubMed), and provenance information. Provenance includes the sentence in which the matching fact was extracted. We highlight the linked entities (subject and object) and their interaction (text term plus mapping to the interaction vocabulary). Provenance may be helpful for users to understand why a document is a match. If a query contains multiple fact patterns, we attach a list of matched sentences in the visualization. Visualizing document lists is comparable to traditional search engines, but handling queries with variable nodes requires novel interfaces. We will discuss such visualizations for queries, including variable nodes, subsequently.

4.4 Retrieval with Variable Nodes

Variable nodes in a narrative query graph may be restricted to specific entity types like *Disease*. We also allow a general type *All* to support querying for arbitrary entities. For example, a user might formulate the query (*Simvastatin*, *treats*, *?X(Disease)*). Several document graphs might match the query with different variable substitutions for $?X$. A document d_1 with the substitution $\mu_1(?X) =$ *hypercholestorelemia* as well as a document d_2 with $\mu_2(?X) =$ *hyperlipidemia* might be proper matches to the query. How should we handle and present these substitutions to the users? Discussions with domain experts led to the conclusion that aggregating documents by their substitution seems most promising. Further, we present two strategies to visualize these document result groups in an user interface: *substitution-centric* and *hierarchical visualization*.

Substitution-Centric Visualization. Given a query with a variable node, the first strategy is to aggregate by similar variable substitutions. We retrieve a list of documents with corresponding variable substitutions from the respective document graph. Different substitutions represent different groups of documents, e.g., one group of documents might talk about the treatment of *hypercholestorelemia* while the other group might talk about *hypertriglyceridemia*. These groups are sorted in descending order by the number of documents in each group. Hence, variable substitutions shared by many documents appear at the top of the list. Our query prototype visualizes a document group as a collapsible list item. A user's click can uncollapse the list item to show all contained documents. Provenance information is used to explain why a document matches her query, i.e., the prototype displays the sentences in which a query's pattern was matched. Provenance may be especially helpful when working with variable nodes.

Hierarchical Visualization. Entities are arranged in taxonomies in many domains. Here, diseases are linked to MeSH (Medical Subject Heading) descriptors arranged in the MeSH taxonomy. The hierarchical visualization aims at showing document results in a hierarchical structure. For example, *hypercholestorelemia* and *hypertriglyceridemia* share the same superclass in MeSH, namely *hyperlipidemias.* All documents describing a treatment of *hypercholestorelemia* as well as *hypertriglyceridemia* are also matches to *hyperlipidemias.* Our prototype visualizes this hierarchical structure by several nested collapsible lists, e.g., *hyperlipidemias* forms a collapsible list. If a user's click uncollapses this list, then the subclasses of *hyperlipidemias* are shown as collapsible lists as well. We remove all nodes that do not have any documents attached in their node or all successor nodes to bypass the need to show the whole MeSH taxonomy.

5 System Evaluation and User Study

Subsequently, we analyze our retrieval prototype concerning two research questions: *Do narrative query graphs offer a precise search for literature? And, do variable nodes provide useful entity-centric overviews of literature?* We performed three evaluations to answer the previous questions:

1. Two pharmaceutical experts created test sets to quantify the retrieval quality (100 abstracts and 50 full-text papers). Both experts are highly experienced in pharmaceutical literature search.
2. We performed interviews with eight pharmaceutical experts who search for literature in their daily research. Each expert was interviewed twice: Before testing our prototype to understand their information need and introducing our prototype. After testing our prototype, to collect feedback on a qualitative level, i.e., how they estimate our prototype's usefulness.
3. Finally, all eight experts were asked to fill out a questionnaire. The central findings are reported in this paper.

5.1 Retrieval Evaluation

After having consulted the pharmaceutical experts, we decided to focus on the following typical information needs in the biomedical domain: I1: Drug-Disease treatments (*treats*) play a central role in the mediation of diseases. I2: Drugs might decrease the effect of other drugs and diseases (*decrease*). I3: Drug treatments might increase the expression of some substance or disease (*induces*). I4: Drug-Gene inhibitions (*inhibits*), i.e., drugs disturb the proper enzyme production of a gene.I5: Gene-Drug metabolisms (*metabolizes*), i.e., gene-produced enzymes metabolize the drug's level by decreasing the drug's concentration in an organism. Narrative query graphs specify the exact interactions a user is looking for. For each information need (I1-5), we built narrative query graphs with well-known entities from the pharmaceutical domain: Q1: *Metformin treats Diabetes Mellitus (I1)*, Q2: *Simvastatin decreases Cholesterol (I2)*, Q3: *Simvastatin induces Rhabdomyolysis (I3)*, Q4: *Metformin inhibits mtor (I4)*, Q5: *CYP3A4 metabolizes Simvastatin AND Erythromycin inhibits CYP3A4 (I4/5)*, and Q6: *CYP3A4 metabolizes Simvastatin AND Amiodarone inhibits CYP3A4 (I4/5).*

Further, we used the entities for each query to search for document candidates on PubMed, e.g., for Q1 we used *metformin diabetes mellitus* as the PubMed

Table 1. Expert evaluation of retrieval quality for narrative query graphs in comparison to PubMed and a MeSH-based search on PubMed. Two experts have annotated PubMed samples to estimate whether the information need was answered. Then, precision, recall and F1-measure are computed for all systems.

Query	#Hits	#Sample	#TP	PubMed	MeSH Search			Narrative QG		
				Prec.	Prec.	Rec.	F1	Prec.	Rec.	F1
Q1	12.7K	25	19	0.76	0.82	0.47	0.60	**1.00**	0.42	0.59
Q2	5K	25	16	0.64	**0.73**	0.50	0.59	0.66	0.25	0.36
Q3	427	25	17	0.68	0.77	0.59	0.67	**1.00**	0.35	0.52
Q4	726	25	16	0.64	**0.78**	0.44	0.56	0.71	0.31	0.43
Q5	397	25	6	0.24	–	–	–	**1.0**	0.17	0.25
Q6	372	25	5	0.20	–	–	–	**1.0**	0.20	0.33

query. We kept only documents that were processed in our pipeline. Then, we took a random sample of 25 documents for each query. The experts manually read and annotated these sample documents' abstracts concerning their information need (true hits / false hits). Besides, we retrieved 50 full texts documents of PubMed Central (PMC) for a combined and very specialized information need (Q5 and Q6). The experts made their decision for PubMed documents by considering titles and abstracts, and for PMC documents, the full texts. Subsequently, we considered these documents as ground truth to estimate the retrieval quality. We compared our retrieval to two baselines, 1) queries on PubMed and 2) queries on PubMed with suitable MeSH headings and subheadings.

PubMed MeSH Baseline. PubMed provides so-called MeSH terms for documents to assists users in their search process. MeSH (Medical Subject Headings) is an expert-designed vocabulary comprising various biomedical concepts (around 26K different headings). These MeSH terms are assigned to PubMed documents by human annotators who carefully read a document and select suitable headings. Prime examples for these headings are annotated entities such as drugs, diseases, etc., and concepts such as study types, therapy types, and many more. In addition to headings, MeSH supports about 76 subheadings to precisely annotate how a MeSH descriptor is used within the document's context. An example document might contain the subheading *drug therapy* attached to *simvastatin*. Hence, a human annotator decided that *simvastatin* is used in drug therapy within the document's context. The National Library of Medicine (NLM) recommends subheadings for entity interactions such as treatments and adverse effects. In cooperation with our experts who read the NLM recommendations, we selected suitable headings and subheadings to precisely query PubMed concerning the respective entity interaction for our queries.

Results. The corresponding interaction and the retrieval quality (precision, recall, and F1-score) for each query are depicted in Table 1. The sample size and the number of positive hits in the sample (TP) are reported for each query. The PubMed search contains only the entities as a simple baseline, and hence, achieved a recall of 1.0 in all cases. PubMed search yielded a precision of around 0.64 up to 0.76 for abstracts and 0.2 up to 0.24 for full texts. The PubMed MeSH search achieved a moderate precision of about 0.73 to 0.82 and recall of about 0.5 for PubMed titles and abstracts (Q1-Q4). Unfortunately, the important MeSH annotations were missing for all true positive hits for Q5 and Q6 in PMC full texts. Hence, the PubMed MeSH search did not find any hits in PMC for Q5 and Q6. Narrative query graphs (Narrative QG) answered the information need with good precision: Q1 (*treats*) and Q3 (*induces*) were answered with a precision of 1.0 and a corresponding recall of 0.42 (Q1) and 0.47 (Q3). The minimum achieved precision was 0.66, and the recall differed between 0.17 and 0.42. Our prototype could answer Q5 and Q6 on PMC full texts: One correct match was returned for Q5 as well as for Q6, leading to a precision of 1.0.

5.2 User Interviews

The previous evaluation demonstrated that our system could achieve good precision when searching for specialized information needs. However, the next questions are: How does our prototype work for daily use cases? And, what are the prototype's benefits and limitations in practice? Therefore, we performed two interviews with each of the eight pharmaceutical experts who search for literature in their daily work. All experts had a research background and worked either at a university or university hospital.

First Interview. In the first interview, we asked the participants to describe their literature search. They shared two different scientific workflows that we have analyzed further: 1. searching for literature in a familiar research area, and 2. searching for a new hypothesis which they might have heard in a talk or read in some paper. We performed think-aloud experiments to understand both scenarios. They shared their screen, showed us at least two different literature searches, and how they found relevant documents answering their information need. For scenario 1), most of them knew suitable keywords, works or journals already. Hence, they quickly found relevant hits using precise keywords and sorting the results by their publication date. They already had a good overview of the literature and could hence answer their information need quickly. For scenario 2), they guessed keywords for the given hypothesis. They had to refine their search several times by varying keywords, adding more, or removing keywords. Then, they scanned titles and abstracts of documents looking for the given hypothesis. We believe that scenario 1) was recall-oriented: They did not want to miss important works. Scenario 2) seemed to be precision-oriented, i.e., they quickly wanted to check whether the hypothesis may be supported by literature. Subsequently, we gave them a short introduction to our prototype. We highlighted two features: The precision-oriented search and the usage of variable nodes to get entity-centric literature overviews. We closed the first interview and gave them three weeks to use the prototype for their literature searches.

Second Interview. We asked them to share their thoughts about the prototype: What works well? What does not work well? What could be improved? First, they considered querying with narrative query graphs, especially with variable nodes, different and more complicated than keyword-based searches. Querying with variable nodes by writing *?X(Drug)* as a subject or an object was deemed too cryptic. They suggested that using *Drug*, *Disease*, etc. would be easier. Another point was that they were restricted to a fixed set of subjects and objects (all known entities in our prototype). For example, querying with pharmaceutical methods like *photomicrography* was not supported. Next, the interaction vocabulary was not intuitive for them. Sometimes they did not know which interaction would answer their information need. One expert suggested to introduce a hierarchical structure for the interactions, i.e., some general interactions like *interacts* that can be specified into *metabolizes* and *inhibits* if required. On the other side, they appreciated the prototype's precise search capability. They all agreed that they could find precise results more quickly using our prototype than other search engines. Besides, they appreciated the provenance information to estimate if a document match answers their information need. They agreed that variable nodes in narrative query graphs offered completely new search capabilities, e.g., *In which dosage forms was Metformin used when treating diabetes?* Such a query could be translated into two fact patterns: (*Metformin, administered, ?X(DosageForm)* and (*Metformin, treats, Diabetes Mellitus*). The most common administrations are done *orally* or via an *injection*. They agreed that such information might not be available in a specialized database like DrugBank. DrugBank covers different dosage forms for Metformin but not in combination with diabetes treatments. As queries get more complicated and detailed, such information can hardly be gathered in a single database. They argued that the *substitution-centric visualization* helps them to estimate which substitutions are relevant based on the number of supporting documents. Besides, they found the *hierarchical visualisation* helpful when querying for diseases, e.g., searching for (*Metformin, treats, ?X(Disease)*). Here, substitutions are shown in an hierarchical representation, e.g., *Metabolism Disorders, Glucose Disorders, Diabetes Mellitus, Diabetes Mellitus Type 1*, etc. They liked this visualization to get a drug's overview of treated disease classes. All of them agreed that searches with variable nodes were helpful to get an entity-structured overview of the literature. Four experts stated that such an overview could help new researchers get better literature overviews in their fields.

5.3 Questionnaire

We asked each domain expert to answer a questionnaire after completing the second interview. The essential findings and results are reported subsequently. First, we asked to choose between precision and recall when searching for literature. Q1: *To which statement would you rather agree when you search for related work?* The answer options were (rephrased): A1a: *I would rather prefer a complete result list (recall). I do not want to miss anything.* A2a: *I would rather prefer precise results (precision) and accept missing documents.* Six of

Table 2. Questionnaire Results: Eight participants were asked to rate the following statements about our prototype on a Likert scale ranging from 1 (disagreement) to 5 (agreement). The mean ratings are reported.

Statement about the Prototype	Mean
The prototype allows me to formulate precise questions by specifically expressing the interactions between search terms	4.0
The formulation of questions in the prototype is understandable for me	4.0
The displayed text passage from the document (Provenance) is helpful for me to understand why a document matches my search query	5.0
The prototype provides precise results for my questions (I quickly find a relevant match)	3.5
Basically, grouping results is helpful for me when searching for variable nodes	4.5
When searching for related work, I would prefer the prototype to a search using classic search tools (cf. PubPharm, PubMed, etc.)	2.8
When searching for or verifying a hypothesis, I would prefer the prototype to a search using classic search tools (cf. PubPharm, PubMed, etc.)	3.4
I could imagine using the prototype in my literature research	3.9

eight experts preferred recall, and the remaining two preferred precision. We asked a similar question for the second scenario (hypothesis). Again, we had let them select between precision and recall (A1a and A1b). Seven of eight preferred precision, and one preferred recall when searching for a hypothesis. Then, we asked Q3: *To which statement would you rather agree for the vast majority of your searches?* Again, seven of eight domain experts preferred precise hits over complete result lists. The remaining one preferred recall. The next block of questions was about individual searching experiences with our prototype: different statements were rated on a Likert scale ranging from 1 (disagreement) to 5 (full agreement). The results are reported in Table 2. They agreed that the prototype allows to formulate precise questions (4.0 mean rating), and the formulation of questions was understandable (4.0). Besides, provenance information was beneficial for our users (5.0). They could well imagine using our prototype in their literature research (3.9) and searching for a hypothesis (3.4). Still, users were reluctant to actually switch to our prototype for related work searches (2.8). Finally, the result visualization of narrative query graphs with variables was considered helpful (4.5).

5.4 Performance Analysis

The query system and the database ran on a server, having two Intel Xeon E5-2687W (3,1 GHz, eight cores, 16 threads), 377 GB of DDR3 main memory, and SDDs as primary storage. The preprocessing took around one week for our six million documents (titles and abstracts). We randomly generated 10k queries asking for one, two, and three interactions. We measured the time of query execution on a single thread. Queries that are not expanded via an ontology took in average 21.9 ms (1-fact)/52 ms (2-facts)/51.7 ms (3-facts). Queries that are expanded via an ontology took in average 54.9 ms (1-fact)/158.9 ms (2-facts)/158.2 ms (3-facts). However, the query time heavily depends on the interaction (selectivity) and how many subclasses are involved. In sum, our system can retrieve documents with a quick response time for the vast majority of searches.

6 Discussion and Conclusion

In close cooperation with domain experts using the PubMed corpus, our evaluation shows that overall document retrieval can indeed decisively profit from graph-based querying. The expert evaluation demonstrates that our system achieves a moderate up to good precision for highly specialized information needs in the pharmaceutical domain. Although the precision is high, our system has only a moderate recall. Moreover, we compared our system to manually curated annotations (MeSH and MeSH subheadings), which are a unique feature of PubMed. Most digital libraries may support keywords and tags for documents but rarely support how these keywords, and primarily, how entities are used within the document's context. Therefore, we developed a document retrieval system with a precision comparable to manual metadata curation but without the need for manual curation of documents.

The user study and questionnaire reveal a strong agreement for our prototype's usefulness in practice. In summary, the user interface must be intuitive to support querying with narrative query graphs. Further enhancements are necessary to explain the interaction vocabulary to the user. We appreciate the idea of hierarchical interactions, i.e., showing a few basic interactions that can be specified for more specialized needs. Especially the search with variable nodes in detailed narrative query graphs offers a new access path to the literature. The questionnaire reveals that seven of eight experts agreed that the vast majority of their searches are precision-oriented. Next, they agreed that they prefer our prototype over established search engines for precision-oriented searches. The verification of hypotheses seems to be a possible application because precise hits are preferred here. We believe that our prototype should not replace classical search engines because there are many recall-oriented tasks like related work searches. The recall will always be a problem by design when building upon error-prone natural language processing techniques and restricting extractions to sentence levels. Although the results seem promising, there are still problems to be solved in the future, e.g., improve the extraction and the user interface.

Conclusion. Entity-based information access catering even for complex information needs is a central necessity in today's scientific knowledge discovery. But while structured information sources such as knowledge graphs offer *high query expressiveness* by graph-based query languages, scientific document retrieval is severely lagging behind. The reason is that graph-based query languages allow to describe the desired characteristics of and interactions between entities in sufficient detail. In contrast, document retrieval is usually limited to simple keyword queries. Yet unlike knowledge graphs, scientific document collections offer *contextualized knowledge*, where entities, their specific characteristics, and their interactions are connected as part of a coherent argumentation and thus offer a clear advantage [7,8]. The research in this paper offers a novel workflow to bridge the worlds of structured and unstructured scientific information by performing graph-based querying against scientific document collections. But as our current workflow is clearly precision-oriented, we plan to improve the recall without having to broaden the scope of queries in future work.

References

1. Azad, H.K., Deepak, A.: Query expansion techniques for information retrieval: a survey. Inf. Process. Manag. **56**(5), 1698–1735 (2019)
2. Betts, C., Power, J., Ammar, W.: GrapAL: connecting the dots in scientific literature. In: Proceedings of the 57th Annual Meeting of the Association for Computational Linguistics: System Demonstrations, pp. 147–152. Association for Computational Linguistics, Florence (2019). https://doi.org/10.18653/v1/P19-3025
3. Dietz, L., Kotov, A., Meij, E.: Utilizing knowledge graphs for text-centric information retrieval. In: The 41st International ACM SIGIR Conference on Research & Development in Information Retrieval, SIGIR '18, pp. 1387–1390. Association for Computing Machinery, New York (2018). https://doi.org/10.1145/3209978.3210187
4. Herskovic, J.R., Tanaka, L.Y., Hersh, W., Bernstam, E.V.: A day in the life of PubMed: analysis of a typical day's query log. J. Am. Med. Inform. Assoc. **14**(2), 212–220 (2007)
5. Kadry, A., Dietz, L.: Open relation extraction for support passage retrieval: merit and open issues. In: Proceedings of the 40th International ACM SIGIR Conference on Research and Development in Information Retrieval, SIGIR '17, pp. 1149–1152. Association for Computing Machinery, New York(2017). https://doi.org/10.1145/3077136.3080744
6. Kolluru, K., Adlakha, V., Aggarwal, S., Mausam, Chakrabarti, S.: OpenIE6: iterative grid labeling and coordination analysis for open information extraction. In: Proceedings of the 2020 Conference on Empirical Methods in Natural Language Processing (EMNLP), pp. 3748–3761. ACL (2020). https://doi.org/10.18653/v1/2020.emnlp-main.306
7. Kroll, H., Kalo, J.-C., Nagel, D., Mennicke, S., Balke, W.-T.: Context-compatible information fusion for scientific knowledge graphs. In: Hall, M., Merčun, T., Risse, T., Duchateau, F. (eds.) TPDL 2020. LNCS, vol. 12246, pp. 33–47. Springer, Cham (2020). https://doi.org/10.1007/978-3-030-54956-5_3
8. Kroll, H., Nagel, D., Balke, W.-T.: Modeling narrative structures in logical overlays on top of knowledge repositories. In: Dobbie, G., Frank, U., Kappel, G., Liddle, S.W., Mayr, H.C. (eds.) ER 2020. LNCS, vol. 12400, pp. 250–260. Springer, Cham (2020). https://doi.org/10.1007/978-3-030-62522-1_18

9. Kroll, H., Nagel, D., Kunz, M., Balke, W.T.: Demonstrating narrative bindings: linking discourses to knowledge repositories. In: Fourth Workshop on Narrative Extraction From Texts, Text2Story@ECIR2021. CEUR Workshop Proceedings, vol. 2860, pp. 57–63. CEUR-WS.org (2021). http://ceur-ws.org/Vol-2860/paper7.pdf
10. Kroll, H., Pirklbauer, J., Balke, W.T.: A toolbox for the nearly-unsupervised construction of digital library knowledge graphs. In: Proceedings of the ACM/IEEE Joint Conference on Digital Libraries in 2021. JCDL '21, Association for Computing Machinery, New York (2021)
11. Krötzsch, M., Rudolph, S.: Is your database system a semantic web reasoner? KI - Künstliche Intelligenz **30**(2), 169–176 (2015). https://doi.org/10.1007/s13218-015-0412-x
12. Manola, F., Miller, E., McBride, B., et al.: RDF primer. W3C Recommend. **10**(1–107), 6 (2004)
13. Mohan, S., Fiorini, N., Kim, S., Lu, Z.: A fast deep learning model for textual relevance in biomedical information retrieval. In: Proceedings of the 2018 World Wide Web Conference, WWW '18, pp. 77–86. International World Wide Web Conferences Steering Committee, Republic and Canton of Geneva (2018). https://doi.org/10.1145/3178876.3186049
14. Nguyen, D.B., Abujabal, A., Tran, N.K., Theobald, M., Weikum, G.: Query-driven on-the-fly knowledge base construction. Proc. VLDB Endow. **11**(1), 66–79 (2017). https://doi.org/10.14778/3151113.3151119
15. Pérez, J., Arenas, M., Gutierrez, C.: Semantics and complexity of SPARQL. ACM Trans. Database Syst. **34**(3), 1–45 (2009). https://doi.org/10.1145/1567274.1567278
16. Raviv, H., Kurland, O., Carmel, D.: Document retrieval using entity-based language models. In: Proceedings of the 39th International ACM SIGIR Conference on Research and Development in Information Retrieval, SIGIR '16, pp. 65–74. Association for Computing Machinery, New York (2016). https://doi.org/10.1145/2911451.2911508
17. Spitz, A., Gertz, M.: Terms over load: leveraging named entities for cross-document extraction and summarization of events. In: Proceedings of the 39th International ACM SIGIR Conference on Research and Development in Information Retrieval, SIGIR '16, pp. 503–512. Association for Computing Machinery, New York(2016). https://doi.org/10.1145/2911451.2911529
18. Vazirgiannis, M., Malliaros, F.D., Nikolentzos, G.: GraphRep: boosting text mining, nlp and information retrieval with graphs. In: Proceedings of the 27th ACM International Conference on Information and Knowledge Management, CIKM '18, pp. 2295–2296. Association for Computing Machinery, New York (2018). https://doi.org/10.1145/3269206.3274273
19. Xiong, C., Power, R., Callan, J.: Explicit semantic ranking for academic search via knowledge graph embedding. In: Proceedings of the 26th International Conference on World Wide Web, WWW '17, pp. 1271–1279. International World Wide Web Conferences Steering Committee, Republic and Canton of Geneva (2017). https://doi.org/10.1145/3038912.3052558
20. Zhao, S., Su, C., Sboner, A., Wang, F.: Graphene: a precise biomedical literature retrieval engine with graph augmented deep learning and external knowledge empowerment. In: Proceedings of the 28th ACM International Conference on Information and Knowledge Management, CIKM '19, pp. 149–158. Association for Computing Machinery, New York (2019). https://doi.org/10.1145/3357384.3358038

BookReach-UI: A Book-Curation Interface for School Librarians to Support Inquiry Learning

Shuntaro Yada[1(✉)], Takuma Asaishi[2], and Rei Miyata[3]

[1] Nara Institute of Science and Technology, 8916-5, Takayama-cho, Ikoma, Nara 630-0192, Japan
s-yada@is.naist.jp

[2] Nanzan University, 18 Yamazato-cho, Showa-ku, Nagoya 466-8673, Japan
tasaishi@nanzan-u.ac.jp

[3] Nagoya University, Furo-cho, Chikusa-ku, Nagoya 464-8601, Japan
miyata@nuee.nagoya-u.ac.jp

Abstract. As a way of collaborating with teachers, school librarians select books that are useful for inquiry-based or exploratory learning classes. Although this method of teaching-material curation has gained importance as teachers aim to adopt more inquiry learning, it has not been well supported by existing tools, resources, and systems designed for librarians. In Asia, especially, the development of school librarianship is not well institutionalised either, and significant practical experience is required to select the books useful for given inquiry classes. To enable even less experienced school librarians to easily curate appropriate books, we developed a graphical user interface that directly shows the candidate books that are topically relevant to the inquiry-based class' subject, by making use of decimal classification classes assigned to books. This interface, BookReach-UI, naturally follows the standard workflow of teaching-material curation described in prior studies. We evaluated its usability by asking school librarians to curate books for mock inquiry classes via BookReach-UI. The results of this preliminary experiment showed substantially high levels of satisfaction and adoption.

Keywords: School library · Exploratory learning · Units of teaching · Book curation · Decimal classification system

1 Introduction

Because one of the major missions of school libraries is to collaborate with teachers [5], it is expected that school librarians can select or 'curate'[1] a set of books that would be useful for classes, especially those classes involving student-driven inquiries. Teachers worldwide are paying increasing attention to such

[1] This paper simply refers to the term 'curation' as a creative activity of reorganising items for a certain use-case, generalised from the classical definition in museology.

H.-R. Ke et al. (Eds.): ICADL 2021, LNCS 13133, pp. 96–104, 2021.
https://doi.org/10.1007/978-3-030-91669-5_8

inquiry-based learning [7,8,16], wherein "learners are motivated to find out about something that interests them by asking authentic questions, investigating multiple and diverse sources to find answers, making sense of the information to construct new understandings, drawing conclusions and forming opinions based on the evidence, and sharing their new understandings with others" [14, p. 51]. In response to this situation, the American Association of School Librarians declared that qualified school librarians should foster inquiry-based learning [1]. Curating books as teaching materials for inquiry-based learning, or *teaching-material curation*, has recently garnered greater importance in school librarianship.

Nonetheless, selecting appropriate books for inquiry-based learning classes can be difficult, especially if the school librarian is unfamiliar with the class subject in question. In Asian countries and regions, such as Singapore, Hong Kong, and Japan, the available human resources of professional school librarians are not sufficient given the larger number of students and professional training and development for school librarianship is less institutionalised by governments [9]. Because teaching-material curation requires expert knowledge and skills with practical experience [12], external support is required, especially for less-experienced school librarians.

School librarians often use Open Public Access Catalog (OPAC) systems for their work, ranging from managing the collections of their school libraries to teaching-material curation. OPAC systems in general, however, are designed for fine-grained search based on bibliographic fields to locate books on library shelves. With respect to teaching-material curation, school librarians need to translate the units of teaching (or curriculum units)[2] targeted in the given inquiry-based class into bibliographic field-based search queries, wherein the 'units' basically correspond to the goals or themes of the textbook chapters (e.g. "investigate the characteristics of your local towns and people" from geography and "living things and environments" from biology). Thus, for instance, a book entitled *Professor, a Chipmunk is Biting the Head of a Snake!* may not be retrieved via OPAC's book title search using obvious queries for the unit of 'animal ecology'.

We propose **BookReach-UI**, a graphical user interface (UI) that helps school librarians with their teaching-material curation for inquiry-based learning classes. Based on the findings of related work, we implemented an optimal workflow to curate books into this UI, wherein selecting textbook chapters directly shows topically relevant books. We evaluated the usability by asking school librarians to curate books for a mock class with BookReach-UI, the results of which showed reasonably high satisfaction with the system.

2 Related Work

2.1 Teaching-Material Curation by School Librarians

Although teaching-material curation is a fundamental mission of school librarians, few studies have examined the actual practices of teaching-material

[2] In this paper, the term 'unit' always mean a unit (or module) of teaching defined in the school curricula or textbooks.

curation. Some prior work revealed the detailed characteristics of teaching-material curation in Japanese school libraries. Miyata et al. [12] investigated teaching-material curation in Japan, which have been voluntarily stored in an online database named "The practice-case database of school libraries for teachers".[3] One of the notable findings of this study is that Nippon Decimal Classification (NDC)[4] classes assigned to the curated books generally corresponded to class subjects (e.g. science, mathematics, and national language).

To examine the book selection criteria in teaching-material curation directly, Asaishi et al. [2] conducted semi-structured interviews with two school librarians regarding a teaching-material curation task for a mock science class, and they were asked to do the curation test in advance of the interviews. That study reported the following procedure, which can be regarded as the standard workflow of teaching-material curation: (1) search for candidate books that contain some keywords of the curriculum unit (e.g. the term 'science' and some scientist names) in their titles or summaries, wherein they have generated these keywords by themselves; and (2) manually filter out books that are outdated or are substantially too difficult for the target students by referring to book covers.

2.2 Support Tools for School Librarians

Only a few tools dedicated to school librarians are currently available. In fact, searching system-oriented journals in the library and information science domain (e.g. *Library Hi Tech* and *Code4Lib Journal*) for keywords such as 'school library' and 'teaching support' does not return relevant research, at the time of this writing.

LibGuides [4] is a commercial content-management software dedicated for creating 'pathfinders', in which librarians can curate website links and library collections useful for students investigating a certain topic. It is popular among academic (university) libraries in the US and is versatile enough to publish curated book lists for arbitrary purposes. However, no automation feature for teaching-material curation was implemented in this software.

In Japan, a book database specialised for school libraries named 'TOOLi-S' is provided by a bookseller for libraries, TRC Library Service, Inc.,[5] which enables school librarians to purchase books relevant to major curriculum units in elementary school textbooks. Another related database is "the practice-case database of school libraries for teachers", as mentioned earlier, managed by Tokyo Gakugei University. It stores practice cases of teaching-material curation by Japanese school librarians, aiming to encourage teachers to make use of school libraries for teaching classes, rather than just for book loans for students. As of July 2021, approximately 400 cases had been voluntarily submitted by school librarians and organised manually by the university. However, these curation cases do not exhaustively cover all the typical curriculum units adopted in Japanese schools. Although the database enables the filtering of cases by school type and subject, curated books are not searchable across cases.

[3] http://www.u-gakugei.ac.jp/~schoolib/.
[4] NDC is a Japanese adaptation of Dewey Decimal Classification (DDC).
[5] https://www.trc.co.jp/school/tooli_s.html.

3 Proposed User Interface: BookReach-UI

To support school librarians in their teaching-material curation, we propose BookReach-UI, an alternative graphical UI that compensates for what the current general-purpose OPAC systems do not provide well. This UI is primarily intended to assist a school librarian in selecting relevant books available in the school library and inter-library loan (ILL)-partnered libraries in response to a request from a teacher who plans an inquiry-based learning class. We first summarise the UI requirements and then elaborate on its implemented features.

3.1 Overall Design

To design the book-curation UI, we examined the standard workflow of teaching-material curation as reported by Asaishi et al. [2] (see Sect. 2.2), which allowed us to derive the following two notable points:

(i) The judgement on which book is relevant to the class relies on how much school librarians understand the class content
(ii) Book covers help determining the relevance of the book to the class

Point (i) can be derived because school librarians may find it difficult to select class-relevant books if they are not familiar with the class subject. For example, a school librarian who know less about physics cannot construct specific search queries such as theories in physics and physicists. This can be compensated for by successful collaboration between teachers and school librarians, although not many schools have yet established this collaboration [10,13,15].

Point (ii) means that book covers include relevant content information pertaining to its targeted readers (e.g. children's books are likely to have more graphical covers) and the type of book (e.g. certain book series have iconic cover designs). This has been traditionally achieved by browsing physical shelves, while standard OPAC systems may require several more operations to show the book covers of a search result.

Considering these points, we designed the following features as essential:

(a) to support finding relevant books to the class subject even if the school-librarian user does not know the subject well
(b) to display book covers of a book collection so that users can perceive books as if they were browsing shelves

Whereas achieving Feature (b) is trivial given a database that provides book covers, there are several approaches to realise Feature (a). As a simple but effective solution based on Miyata et al. [12] (see Sect. 2.1), we utilised decimal classification systems (such as DDC and NDC), which label books with decimal-numbered classes of subjects, to bridge the gap between textbook chapters (or units of teaching) and books. That is, by using a pre-defined map between textbook chapters and their topically corresponding decimal classes (i.e. a *unit-to-decimal map*), we can provide relevant candidate books by selecting textbook chapters targeted in the inquiry-based class. This map enables the following workflow, which fits the typical teaching-material curation well:

1. Select a textbook used in the inquiry-based class
2. Select the chapters (curriculum units) targeted in the inquiry-based class
3. (The candidates of books relevant to the class subject, retrieved via the unit-to-decimal map, are displayed along with their book covers)
4. Select as many useful books as the user chooses
5. (Selected books are automatically organised into a shareable list)

We implemented this workflow in our UI, as shown and annotated in Fig. 1.

3.2 Detailed Features

Following the use-case workflow above, we elaborate on the detailed features of the BookReach-UI. After the user selected a unit of teaching from the *textbook/chapter selectors*, our implementation retrieves the candidate books for curation that are relevant to the curriculum unit using a unit-to-decimal map that we manually created by assigning the classes of NDC to Japanese textbook chapters, because we began by targeting Japanese school libraries first. When the retrieve button is clicked, the UI accesses to a database of books containing bibliographic information with cover images and NDC classes.[6] Suppose, hereafter, that all books displayed in the UI belong to the collections of the user's school library or ILL-partnered libraries.

The *book browser* (middle of Fig. 1) displays the cover images of the candidate books retrieved from the database, replicating the physical browsing of bookshelves. These books are grouped by tabs of the NDC classes relevant to the curriculum unit, achieving coarse-grained filtering of the book contents. Several buttons are provided for filtering which book to display by targeted school grade,[7] past curation use,[8] library location (in the own library or ILL libraries), and full-text search for the textual contents

When a book-cover image is clicked, a modal window appears over the book browser, as shown in Fig. 2. This view provides the detailed bibliographic information of the book, such as the authors, publishers, and total number of pages. If available, it also shows the book-content summary, table of contents, and its history of past curation use. The user can select this book for curation or close the modal window by clicking the buttons at the bottom.

Selected books are automatically organised into a *curation list* below the book browser. Users can access the major bibliographic information using the cover images in this list. The systems allows users to export the listed books as a teaching-material curation list. The available export formats are printing and tab-separated value (TSV) text; the former allows school librarians to share curated books immediately with teachers, whereas the latter is reusable in office spreadsheet applications to incorporate the list into another document.

---

[6] We adopted machine-readable cataloguing (MARC) records provided by TRC Library Service Inc. (Tokyo, Japan) for the current implementation.

[7] The adopted MARC records include targeted school grades as a bibliographic field.

[8] Past curation use for each book is retrieved from "the practice-case database of school libraries for teachers" if the book is listed in any cases [12] (see Sect. 2.2).

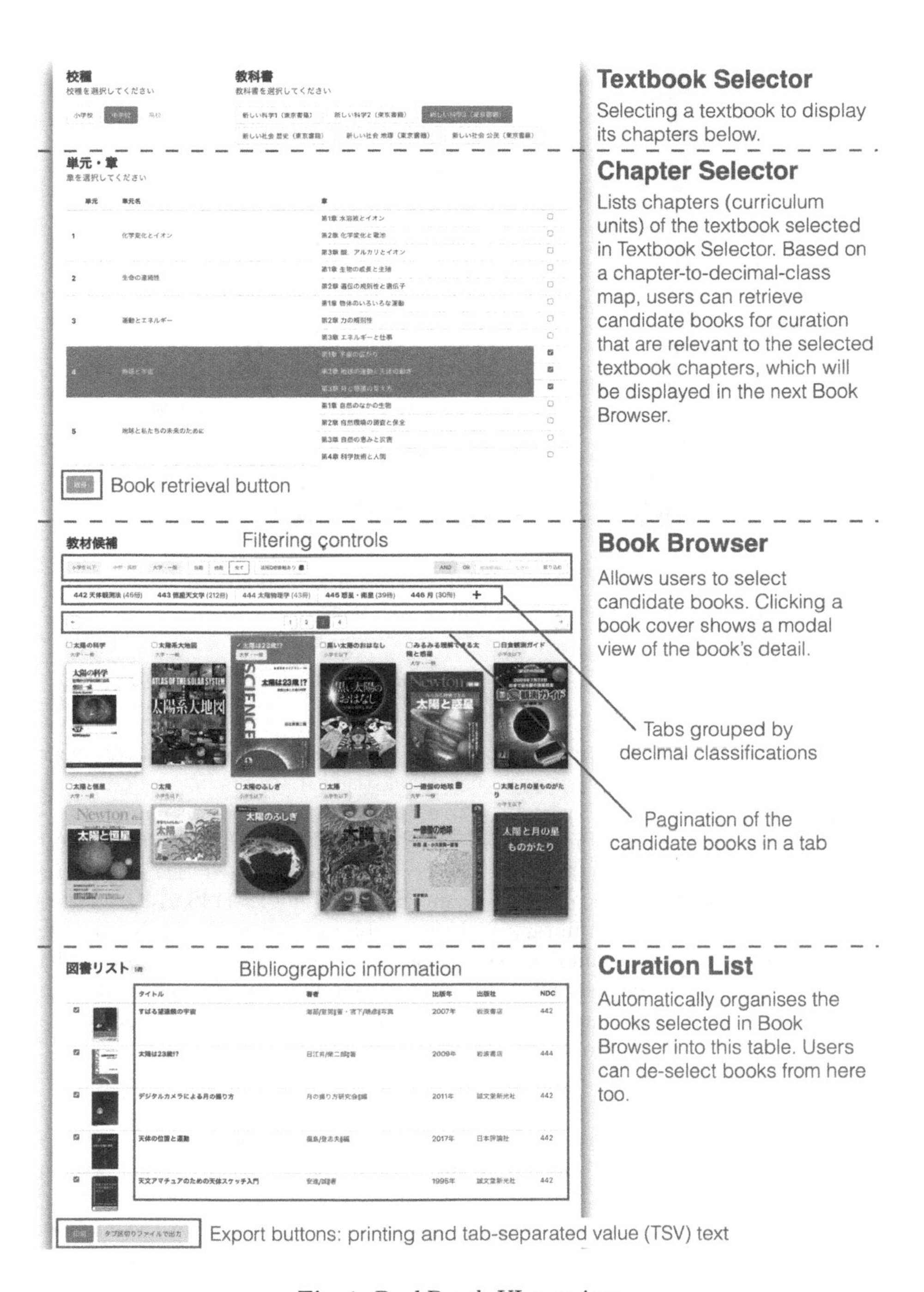

Fig. 1. BookReach-UI overview.

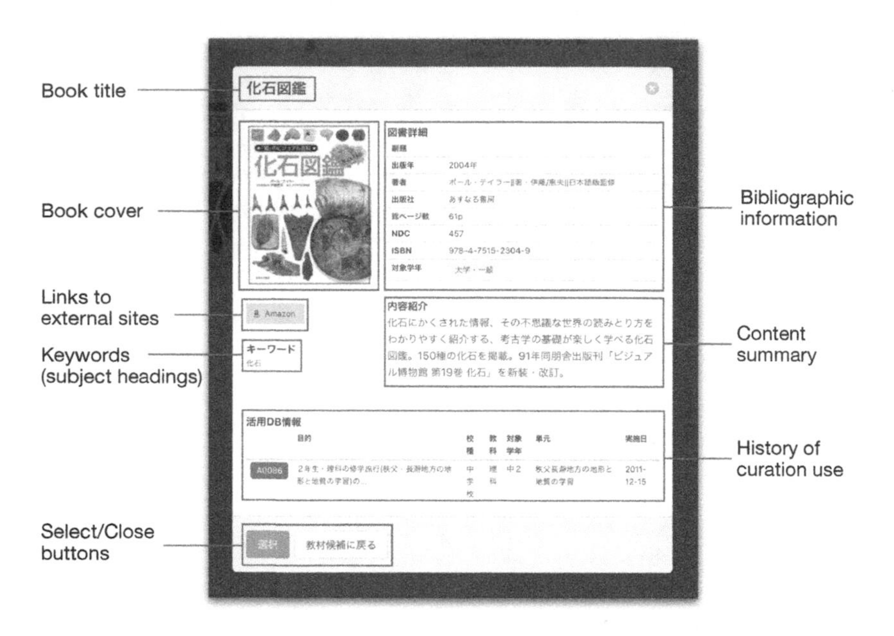

Fig. 2. Modal window view of a book within the browsing-and-filtering module, providing bibliographic information and content descriptions.

4 Experiment

To evaluate the usability of the proposed UI, we carried out a preliminary user study that asked school librarians to use BookReach-UI for mock requests for teaching-material curation. Each of the five Japanese school librarians (elementary: 3; junior high: 2), who used OPAC systems for daily library work, curated approximately 30 books for a single subject that we assigned (geography or civics), by browsing a book database containing all books (146,914 in total) for each participant's school library and two public libraries. For each subject, we set similar curriculum units across school types (elementary or junior high) and assigned eight to ten relevant NDC classes manually in advance. After the task, we asked the librarians to score the usability of the UI with the Japanese translation [11] of the System Usability Scale (SUS) [3].

The SUS results for each librarian are shown in Table 1. The overall SUS score ranging from 0 to 100 in 2.5-point increments was calculated as follows [3]: subtracted 1 from the scores of the odd-numbered questions, subtracted the scores of the even-numbered questions from 5, summed these modified scores, and multiplied the sum by 2.5 (see the bottom row). The mean score was 69, which was reasonably high. Specifically, Questions 3 and 10 gave the highest mean scores per question, indicating that the UI was intuitive enough for the first use. We observed 126 operations (i.e. UI element clicks) per session as a school-wise median value, which may include casual trials of our UI.

Table 1. System Usability Scale (SUS) results of five school librarians (A–E) using BookReach-UI for the book curation task (A and D: civics; others: geography).

SUS Question		Elementary			Junior high		Mean
		A	B	C	D	E	
1	I think that I would like to use this system frequently	2	5	3	4	4	3.6
2	I found the system unnecessarily complex	2	2	2	3	1	2.0
3	I thought the system was easy to use	4	4	4	3	5	4.0
4	I think that I would need the support of a technical person to be able to use this system	1	3	2	5	1	2.4
5	I found the various functions in this system were well integrated	2	2	4	3	4	3.0
6	I thought there was too much inconsistency in this system	2	2	1	1	2	1.6
7	I would imagine that most people would learn to use this system very quickly	4	4	4	3	4	3.8
8	I found the system very cumbersome to use	1	2	3	3	2	2.2
9	I felt very confident using the system	3	4	3	3	3	3.2
10	I needed to learn a lot of things before I could get going with this system	1	2	3	1	2	1.8
SUS Score		**70**	**70**	**67.5**	**57.5**	**80**	**69**

5 Conclusion

To enhance librarian-teacher collaboration, we developed the BookReach-UI, which supports school librarians in their selection of library books for inquiry-based learning classes (teaching-material curation). Based on prior work analysing this activity, we defined two key features: (1) to show candidate books for curation that are relevant to class subjects, and (2) to display book covers as the main source to judge useful books for the class. The implemented UI allows users to browse relevant books by selecting the targeted textbook chapter (curriculum unit) via pre-defined topically corresponding decimal-classification classes. Our preliminary usability experiment showed reasonably high satisfaction with the standard measurement (SUS).

We plan to expand this experiment to a broader range of school librarians to validate and improve the UI. One limitation of the approach is the necessity to create the unit-to-decimal map manually, which can be automated in the future by textual similarity [6] between chapter names and heading terms. Our implementation may contribute to related software and services for school librarians such as curation-case databases and book-recommendation systems.

Acknowledgements. This research was funded by Nanzan University Pache Research Subsidy I-A-2 for the 2019 academic year. We would like to express the deepest appreciation to TRC Library Service Inc., for offering the MARC records of a public library.

References

1. American Association of School Librarians: National school library standards for learners, school librarians, and school libraries. ALA Editions, Chicago (2018)
2. Asaishi, T., Miyata, R., Yada, S.: How librarians select learning resources: a preliminary analysis for developing the learning resource search system. Acad. Human. Nat. Sci. **20**, 99–112 (2020). (in Japanese)
3. Brooke, J.: SUS - a quick and dirty usability scale. Usability Eval. Ind. **189**(194), 4–7 (1996)
4. Bushhousen, E.: LibGuides. J. Med. Libr. Assoc.: JMLA **97**(1), 68 (2009)
5. IFLA and UNESCO: IFLA/UNESCO School Library Manifesto. Technical report, The International Federation of Library Associations and Institutions (1999). https://www.ifla.org/publications/iflaunesco-school-library-manifesto-1999
6. Jiang, H., He, P., Chen, W., Liu, X., Gao, J., Zhao, T.: SMART: robust and efficient fine-tuning for pre-trained natural language models through principled regularized optimization. In: Proceedings of the 58th Annual Meeting of the Association for Computational Linguistics, pp. 2177–2190. Association for Computational Linguistics, Online (2020)
7. Johnston, M.P.: Preparing teacher librarians to support STEM education. In: IASL Annual Conference Proceedings, pp. 1–11 (2017). https://doi.org/10.29173/iasl7145
8. Kim, S.Y.: School libraries as old but new supports for education in Japan: a review of Japan's national curriculum for elementary schools. Libri **61**(2), 143–153 (2011). https://doi.org/10.1515/libr.2011.012
9. Loh, C.E., Tam, A., Okada, D.: School library perspectives from Asia: trends, innovations and challenges in Singapore, Hong Kong and Japan. In: IASL Annual Conference Proceedings, pp. 1–9 (2019). https://doi.org/10.29173/iasl7258
10. Mandrell, J.C.: The effect of professional development on teachers' perceptions of the role of school librarians: an action research study. In: ProQuest LLC. ProQuest LLC (2018)
11. Miyata, R., Hartley, A., Kageura, K., Paris, C.: Evaluating the usability of a controlled language authoring assistant. Prague Bull. Math. Linguist. **108**(1), 147–158 (2017)
12. Miyata, R., Yada, S., Asaishi, T.: Analysing school librarians' practices in providing useful materials for classes to support teachers. J. Japan Soc. Libr. Inf. Sci. **64**(3), 115–131 (2018). https://doi.org/10.20651/jslis.64.3_115. (in Japanese)
13. Stafford, T., Stemple, J.: Expanding our understanding of the inquiry process. Libr. Media Connect. **30**(1), 34–37 (2011)
14. Stripling, B.K.: Empowering students to inquire in a digital environment. In: Alman, S.W. (ed.) School Librarianship: Past, Present, and Future, Beta Phi Mu Scholars Series, pp. 51–63. Rowman & Littlefield (2017)
15. Sturge, J.: Assessing readiness for school library collaboration. Knowl. Quest **47**(3), 24–31 (2019)
16. Young, T.E., Jr.: 24/7 STEMulation: reinventing discovery. Libr. Media Connect. **31**(6), 20–22 (2013)

SmartReviews: Towards Human- and Machine-Actionable Representation of Review Articles

Allard Oelen[1,2](✉), Markus Stocker[2], and Sören Auer[1,2]

[1] L3S Research Center, Leibniz University of Hannover, Hannover, Germany
oelen@l3s.de
[2] TIB Leibniz Information Centre for Science and Technology, Hannover, Germany
{markus.stocker,soeren.auer}@tib.eu

Abstract. Review articles are a means to structure state-of-the-art literature and to organize the growing number of scholarly publications. However, review articles are suffering from numerous limitations, weakening the impact the articles could potentially have. A key limitation is the inability of machines to access and process knowledge presented within review articles. In this work, we present SmartReviews, a review authoring and publishing tool, specifically addressing the limitations of review articles. The tool enables community-based authoring of living articles, leveraging a scholarly knowledge graph to provide machine-actionable knowledge. We evaluate the approach and tool by means of a SmartReview use case. The results indicate that the evaluated article is successfully addressing the weaknesses of the current review practices.

Keywords: Article authoring · Digital libraries · Living review documents · Semantic publishing

1 Introduction

As more scholarly articles are published every year [9], methods and tools to organize published articles are becoming increasingly important [12]. Traditionally, review (or survey) articles are used to organize information for a particular research domain [28]. Research articles, also referred to as primary sources, present original research contributions. Review articles, or secondary sources, organize the research presented in the primary sources [23]. The importance of review articles becomes apparent in the fact that these articles are often highly cited [31] which indicates that they are valuable for the community. Although reviews are important, they suffer from several major weaknesses, which affect the potential impact review articles can have. For example, once review articles are published, they are generally not updated when new research articles become available. This results in reviews that are outdated soon after publication. Furthermore, scholarly articles are not machine-actionable, which prevents machines from processing the contents.

H.-R. Ke et al. (Eds.): ICADL 2021, LNCS 13133, pp. 105–114, 2021.
https://doi.org/10.1007/978-3-030-91669-5_9

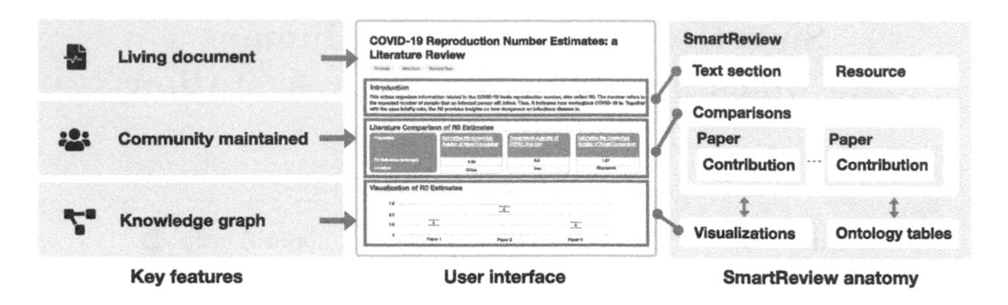

Fig. 1. Illustration of key features and anatomy of SmartReviews. They are composed of several building blocks, including natural text, comparisons and visualizations.

In this work, we present SmartReviews, a novel tool to author and publish review articles. The tool implements the requirements from the equally named SmartReview approach [21] which addresses the weaknesses from which current review articles are suffering. Reviews are authored in a community-based manner and are represented as living documents, meaning that they can be updated whenever deemed necessary by the community. SmartReviews are implemented within an existing scholarly knowledge graph called Open Research Knowledge Graph (ORKG) [8]. The key features and anatomy of SmartReviews are depicted in Fig. 1. In summary, this article provides the following research contributions: (i) Detailed description of authoring and publishing semantic review articles using knowledge graphs. (ii) Implementation of SmartReview authoring tool. (iii) Presentation and evaluation of an original SmartReview article.

2 Related Work

The current review authoring and publishing method faces numerous limitations and weaknesses [19]. In recent work [21], we identified these limitations and described them in detail. Table 1 summarizes them and includes an extended list of supporting related work. Based on those weaknesses, we devised an approach to address them. The two most pressing weaknesses relate to the inability to update articles once published and to the machine-inactionability of the presented knowledge. Both of these topics are extensively discussed in the literature.

Shanahan advocates for "living documents" and to move away from the traditional and obsolete print model in which articles are sealed after publishing [25]. The living documents concept also provides opportunities for article retractions and corrections [2]. This gives the possibility to embrace the features the modern web has to offer, including semantic web technologies [26]. Berners-Lee et al. used to term Linked Data to describe the interlinking of resources (i.e., data) by means of global identifiers, which constitutes the semantic web [3]. The Resource Description Framework (RDF) is the language used to represent the resources and provides an actionable format for machines [14]. RDF can be queried using the SPARQL query language [22]. The use of these technologies improves the

Table 1. Summarized weaknesses of the current review and their respective related work. A detailed list of the weaknesses is presented in previous work [21].

Weakness	Definition	Related work
Lacking updates	Published articles are generally not updated due to technical limitations or lacking author incentives	[15,16,19]
Lacking collaboration	Only the viewpoint from the review authors is reflected and not from the community as a whole	[19,24]
Limited coverage	Reviews are only conducted for popular fields and are lacking for less popular ones	[19,27,28]
Lacking machine-actionability	The most frequently used publishing format is PDF, which hinders machine-actionability	[4,7,10,12,13, 17,19]
Limited accessibility	The articles in PDF format are often inaccessible for readers with disabilities	[1,5,18]
Lacking overarching systematic representation	Web technologies are not used to their full potential because systematic representations are often lacking	[19,26]

machine-actionability of data and provides a means to make data FAIR (Findable, Accessible, Interoperable, Reusable) [29]. Semantic web technologies also play a key role in the living documents concept presented by Garcia-Castro et al. [6]. This type of document supports tagging and interlinking of individual article components and embeds ontologies in the core of their approach.

3 Approach

Our approach addresses the previously listed weaknesses. Accordingly, we introduce dimensions to address each weakness individually. The dimensions comprise: (i) Article updates (ii) Collaboration (iii) Coverage (iv) Machine-actionability (v) Accessibility (vi) Systematic representation. The approach leverages the SmartReview requirements as presented in [21].

The ORKG is used at the core of our approach. The use of knowledge graphs enables the reuse of existing ontologies, thus improving the machine-actionability of the data. To this end, the article has to be represented in a structured and semantic manner. Research articles are generally composed of multiple (non-structured) artifacts, among others this includes natural text sections, figures, tables, and equations. Review articles, in particular, do often include an additional artifact in the form of comparison tables. These tables present the reviewed work in a structured manner and compare the work based on a set of predefined properties. A previous study indicated that approximately one out of five review articles contains such tables [20]. Due to the structured nature of comparison tables, they can be processed more easily by machines. Complemented with semantic descriptions of the data, the comparisons can become FAIR data [19]. Therefore, we use comparison tables as the basis of our SmartReview approach. We leverage the comparisons tables within the ORKG which are specifically designed to be machine-actionable.

4 Implementation

The interface is implemented in JavaScript using the React framework, the source code is available online[1]. Additionally, a feature demonstration video is available.[2] The knowledge graph is built on top of a Neo4j property graph and SPARQL support is provided via a Virtuoso endpoint.

4.1 Section Types

The main building blocks of SmartReviews are sections. Each section has a section type that describes the section's content and its relation to the knowledge graph. The article writer has been implemented on top of the ORKG which allows for reusing artifacts already present in the graph. When adding a section, the type can be selected (Fig. 2, node 6). The section types comprise:

- **Natural language text** sections support markup text via a Markdown editor. References are supported via a custom syntax using the same notation as R Markdown [32].
- **Comparison** sections consist of tabular overviews of scientific contributions from papers that are being compared based on a selected set of properties. Comparison sections form the core of each review article.
- **Visualization** sections provide visual views of comparison data.
- **Ontology table** sections list descriptions of the properties and resources used within comparisons.
- **Resource and property** sections show a tabular representation of used resources and properties and their definitions from the knowledge graph.

4.2 Implementation per SmartReview Dimension

The implementation consists of various components to provide a comprehensive authoring interface. Among other things, this includes support for in-text citations, an interactive article outline, and reading time estimation. These features are ordinary functionalities for authoring tools and are therefore not discussed in detail. In the remainder of this section, we specifically focus on the dimensions of the SmartReview approach since they form the basis of the implementation.

[1] https://gitlab.com/TIBHannover/orkg/orkg-frontend/-/tree/master/src/components/SmartReview.

[2] https://doi.org/10.5446/53601.

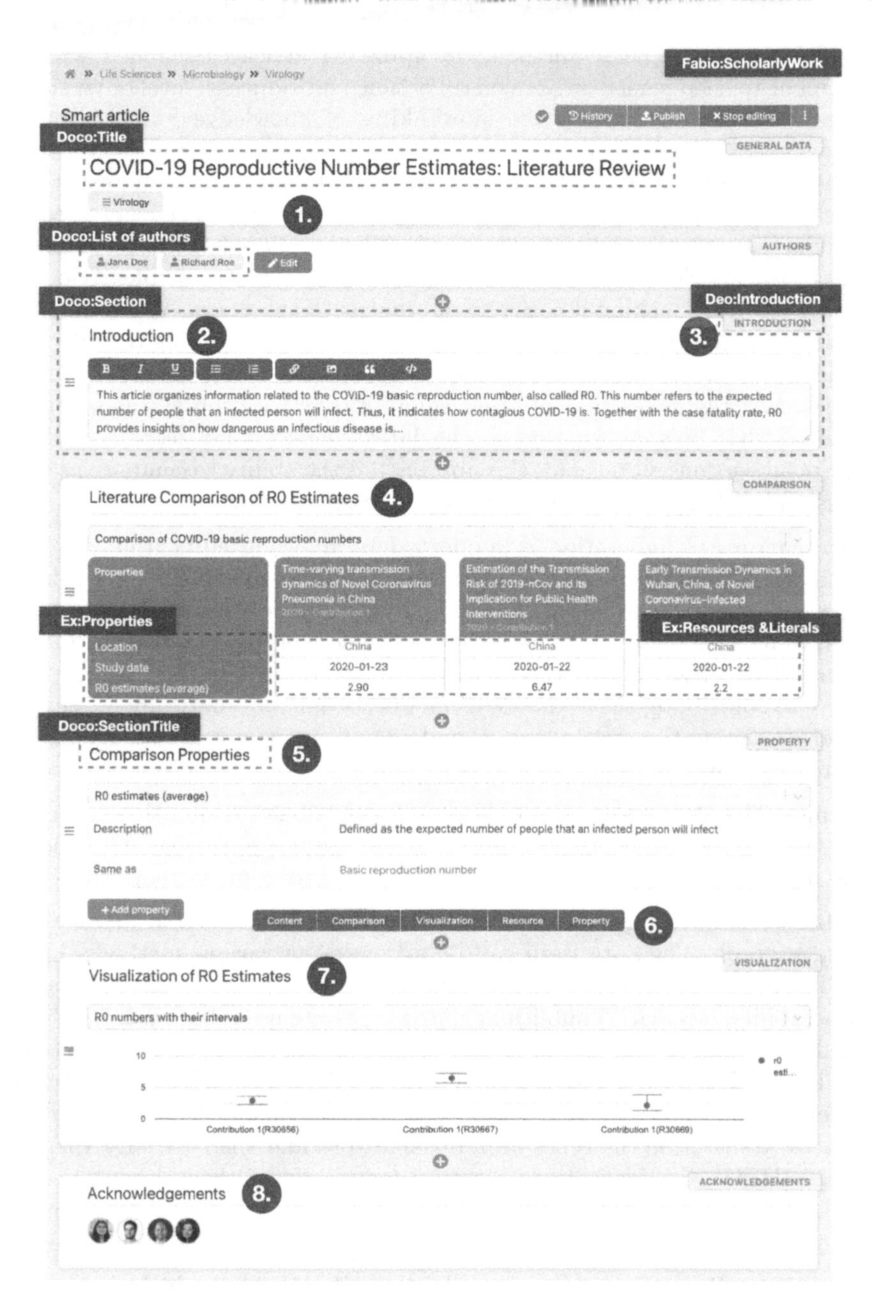

Fig. 2. Screenshot of the implemented interface. Black labels represent the RDF types. Types prefixed with "Ex" are from the scholarly graph used for the implementation. Node 1 relates to the metadata of the article. Node 2 is the natural text content section and its Markdown editor. Node 3 shows the DEO type, which can be selected by the users when clicking on the label. Node 4 is a comparison section and node 5 is a property section. Node 6 shows the type selection for a new section. Node 7 is the visualization of the comparison shows in node 4. Finally, node 8 lists the contributors of the article.

Article Updates. The requirement of updating articles combined with the requirement to keep persistent versions of articles introduces a level of versioning complexity. Especially due to the interlinking of knowledge graph resources, persistency is a complex endeavor that requires versioning at the resource level. To reduce the complexity, we added the constraint that only the latest version of an article can be updated, which we call the *head version*. The head version is the only version that is stored in the graph. This implies that always the latest version of the article is present in the graph, leaving version complexity outside the graph and thus making it easier to interact with the graph. When an article is published within the system (not to be confused with publishing the article via a publisher), a snapshot is created of the subgraph used to generate the article. This approach resembles that of other collaborative curation systems (such as Wikipedia) that only allow edits of the latest version and keep a persistent history of all versions. Crucial for this approach is the ability to compare previous versions and to track individual changes (i.e., the diff view).

Collaboration. Collaboration is supported by allowing edits from any user. As with the article updates, this resembles the approach Wikipedia takes to support collaborative authoring. In Wikipedia, this has resulted in high-quality articles, which is popularly explained by the "wisdom of the crowd" principle [11]. To acknowledge researchers who contributed to the article, and to create an incentive to contribute, the acknowledgements section automatically lists anyone involved in writing the article (Fig. 2, node 8). The list of acknowledgements is generated by traversing the article subgraph.

Coverage. The only prerequisite to be able to contribute to an article is the need for a user account. Authentication serves for tracking provenance data (needed for the acknowledgements) and as a basic abuse prevention system.

Machine-Actionability. As described, the article content is available in the knowledge graph. The data itself can be accessed via various methods, including a SPARQL endpoint, RDF dump, and REST interface. To enhance machine interoperability, (scholarly) publishing ontologies were used. In Fig. 2, RDF types prefixed with their respective ontologies are displayed next to system components. This includes the Document Components Ontology (DOCO)[3] to describe documents components. The FRBR-aligned Bibliographic Ontology (Fabio) (See footnote 3) to describe the types of published work and the Discourse Elements Ontology (DEO) (See footnote 3) ontology for specifying the section types. For the latter ontology, the article authors are responsible to select the appropriate type from a list of all DEO types for natural text sections (Fig. 2, node 3).

Accessibility. Review articles are available as HTML files, which makes them by design more accessible than their PDF counterpart. Furthermore, WCAG guidelines are followed to enhance accessibility. In particular, semantic HTML tags are used as well as hierarchical headings. Finally, articles are responsive (i.e., support different screen sizes) making them suitable for high browser zoom levels and large font settings.

[3] http://purl.org/spar/doco,fabio,deo.

Systematic Representation. Comparison tables form the main component to support systematic representations (Fig. 2, node 4). The tables are created in a spreadsheet-like editor. The papers used within the comparison are represented as structured data in the graph, including the metadata such as title, authors, and publication date. Furthermore, the properties and their corresponding values are stored in the graph. When creating the comparison table, users are recommended to use existing properties and resources to further enhance interlinking.

5 Evaluation

To evaluate our approach, we now present a use case with an original SmartReview article to demonstrate how SmartReviews look like and how they differ from regular reviews. Afterwards, we demonstrate how data presented within the article can be accessed in a machine-actionable manner.

The SmartReview presents a selective literature review, titled "Scholarly Knowledge Graphs", and it published online[4]. It consists of three comparisons and reviews in total 14 articles related to various types of scholarly knowledge graphs (i.e., identifier, bibliographic, domain-specific systems). This use case highlights the differences with regular static review articles. While regular review articles generally review the literature in comprehensive (and possibly lengthy) text sections, the SmartReview example shows how, instead, comparison tables are used to compare literature. Due to the interactive nature of the tables, they can contain more information than tables presented in static PDF files. Another notable difference is the presence of ontology tables within the article. The benefit of such tables is twofold: They improve machine-readability by linking the used properties to existing ontologies and improve human comprehension by textually describing the meaning of the property.

To demonstrate the machine-actionability of SmartReviews, we now present four SPARQL queries that are used to query the underlying data (cf. Query 1.1, 1.2, 1.3, and 1.4). The first query is for metadata, whereas the other queries are for the actual knowledge presented in the respective articles. The prefixes *orkgc*, *orkgp* and *orkgr* represent the class, predicate and resource URIs respectively.

```
SELECT DISTINCT ?smartReview
WHERE {
  ?smartReview a orkgc:SmartReview;
        orkgp:P30 orkgr:R278.
}
```

Query 1.1. Return all SmartReviews with research field (P30) information science (R278).

```
SELECT DISTINCT ?paper
WHERE {
  ?contrib a orkgc:Contribution;
        orkgp:P32 orkgr:R49584.
  ?paper orkgp:P31 ?contrib.
}
```

Query 1.2. Return paper contributions (P31) addressing Scholarly Communication (R49584) as research problem (P32).

[4] https://www.orkg.org/orkg/smart-review/R135360.

```
SELECT DISTINCT ?section
WHERE {
  ?review a orkgc:SmartReview;
        orkgp:P27 orkgr:R8193;
        orkgp:P31 ?contrib.
  ?contrib orkgp:HasSection ?section.
  ?section a orkgc:Introduction.
}
```

Query 1.3. Return all introduction sections from SmartReviews related to information science (R278).

```
SELECT DISTINCT ?paper
WHERE {
  ?contrib a orkgc:Contribution;
        orkgp:P32 orkgr:R49584;
        orkgp:P7009 "T"^^xsd:string.
  ?paper orkgp:P31 ?contrib.
}
```

Query 1.4. Return all scholarly communication systems (R49584) with RDF support (P7009).

6 Discussion

We acknowledge that our proposed approach is radical and will unlikely be immediately adopted in every aspect by the research community. While some of the weaknesses originate from technology limitations, the main challenge is not technological in nature. Rather it is rooted in researchers' habits and mindsets and being comfortable with familiar methods. This relates to the open access movement [30] which does not face a technical challenge but complex change that involves many aspects of traditional publishing.

Our proposed approach does not solely address review authoring but also impacts the publication and dissemination process. Articles can be published and accessed via the platform's user interface or directly via the graph. Therefore, the platform serves as a digital library for review articles. As discussed, any user can author new articles and contribute to existing articles. This means that articles are not peer-reviewed in the traditional sense, rather a community-based continuous review method is performed. However, traditional peer-review is still possible. For example, as soon as an article is mature enough (which is decided by the authors), it can be published with traditional publishing means. However, we want to stress that a traditional publishing body is optional and is therefore not part of our approach.

An extensive user evaluation is required to access the interactions and actual use of the system. Additionally, this user evaluation can focus on the usability aspects of the system. For future work, we have planned an evaluation with domain experts who will be asked to create a SmartReview for their field of expertise. This includes the creation of relevant comparisons and visualizations.

Our approach can be generalized to research articles. Concretely it means that the article writer can be used to author any type of scholarly article. We focused on review articles because several of the weaknesses are most apparent for this type of article. Furthermore, we deem the limitation of static non-updated articles as a key limitation for reviews.

7 Conclusions

We presented the SmartReview tool, an application to author and publish scholarly review articles in a semantic and community-maintained manner. With the implementation, we address the current weaknesses of review article authoring and demonstrate a possible future of publishing review articles. A scholarly knowledge graph is used at the core of our approach, which increases the machine-actionability of the presented knowledge. The presented use case demonstrates how SmartReviews look like and it shows that the contents within articles is published in a machine-actionable manner.

Acknowledgements. This work was co-funded by the European Research Council for the project ScienceGRAPH (Grant agreement ID: 819536) and the TIB Leibniz Information Centre for Science and Technology.

References

1. Ahmetovic, D., et al.: Axessibility: a LaTeX package for mathematical formulae accessibility in PDF documents. In: ASSETS 2018 - Proceedings of the 20th International ACM SIGACCESS Conference on Computers and Accessibility, pp. 352–354 (2018). https://doi.org/10.1145/3234695.3241029
2. Barbour, V., Bloom, T., Lin, J., Moylan, E.: Amending published articles: time to rethink retractions and corrections? F1000Research **6**, 1960 (2017). https://doi.org/10.12688/f1000research.13060.1
3. Berners-Lee, T., Hendler, J., Lassila, O.: The semantic web. Sci. Am. **284**(5), 34–43 (2001)
4. Corrêa, A.S., Zander, P.O.: Unleashing tabular content to open data, pp. 54–63 (2017). https://doi.org/10.1145/3085228.3085278
5. Darvishy, A.: PDF accessibility: tools and challenges. In: Miesenberger, K., Kouroupetroglou, G. (eds.) ICCHP 2018. LNCS, vol. 10896, pp. 113–116. Springer, Cham (2018). https://doi.org/10.1007/978-3-319-94277-3_20
6. Garcia-Castro, A., Labarga, A., Garcia, L., Giraldo, O., Montaña, C., Bateman, J.A.: Semantic web and social web heading towards living documents in the life sciences. J. Web Semant. **8**(2–3), 155–162 (2010). https://doi.org/10.1016/j.websem.2010.03.006
7. Heidorn, P.B.: Shedding light on the dark data in the long tail of science. Libr. Trends **57**(2), 280–299 (2008)
8. Jaradeh, M.Y., et al.: Open research knowledge graph: next generation infrastructure for semantic scholarly knowledge. In: Proceedings of the 10th International Conference on Knowledge Capture, pp. 243–246 (2019). https://doi.org/10.1145/3360901.3364435
9. Jinha, A.: Article 50 million: an estimate of the number of scholarly articles in existence. Learn. Publish. **23**(3), 258–263 (2010). https://doi.org/10.1087/20100308
10. Jung, D., et al.: ChartSense: interactive data extraction from chart images. In: Conference on Human Factors in Computing Systems - Proceedings 2017-May, pp. 6706–6717 (2017). https://doi.org/10.1145/3025453.3025957
11. Kittur, A., Kraut, R.E.: Harnessing the wisdom of crowds in wikipedia: quality through coordination (2008)
12. Klampfl, S., Granitzer, M., Jack, K., Kern, R.: Unsupervised document structure analysis of digital scientific articles. Int. J. Digit. Libr. **14**(3–4), 83–99 (2014). https://doi.org/10.1007/s00799-014-0115-1
13. Lipinski, M., Yao, K., Breitinger, C., Beel, J., Gipp, B.: Evaluation of header metadata extraction approaches and tools for scientific PDF documents. In: Proceedings of the ACM/IEEE Joint Conference on Digital Libraries, pp. 385–386 (2013). https://doi.org/10.1145/2467696.2467753

14. Manola, F., Miller, E., McBride, B., et al.: RDF primer. W3C Recommend. **10**(1–107), 6 (2004)
15. Mendes, E., Wohlin, C., Felizardo, K., Kalinowski, M.: When to update systematic literature reviews in software engineering. J. Syst. Softw. **167**, 110607 (2020). https://doi.org/10.1016/j.jss.2020.110607
16. Meyer, C.A.: Distinguishing published scholarly content with CrossMark. Learn. Publish. **24**(2), 87–93 (2011). https://doi.org/10.1087/20110202
17. Nasar, Z., Jaffry, S.W., Malik, M.K.: Information extraction from scientific articles: a survey. Scientometrics **117**(3), 1931–1990 (2018). https://doi.org/10.1007/s11192-018-2921-5
18. Nganji, J.T.: The Portable Document Format (PDF) accessibility practice of four journal publishers. Libr. Inf. Sci. Res. **37**(3), 254–262 (2015). https://doi.org/10.1016/j.lisr.2015.02.002
19. Oelen, A., Jaradeh, M.Y., Stocker, M., Auer, S.: Generate FAIR literature surveys with scholarly knowledge graphs. In: JCDL '20: Proceedings of the ACM/IEEE Joint Conference on Digital Libraries in 2020, pp. 97–106 (2020). https://doi.org/10.1145/3383583.3398520
20. Oelen, A., Stocker, M., Auer, S.: Creating a scholarly knowledge graph from survey article tables. In: Ishita, E., Pang, N.L.S., Zhou, L. (eds.) ICADL 2020. LNCS, vol. 12504, pp. 373–389. Springer, Cham (2020). https://doi.org/10.1007/978-3-030-64452-9_35
21. Oelen, A., Stocker, M., Auer, S.: SmartReviews: towards human- and machine-actionable reviews. In: Berget, G., Hall, M.M., Brenn, D., Kumpulainen, S. (eds.) TPDL 2021. LNCS, vol. 12866, pp. 181–186. Springer, Cham (2021). https://doi.org/10.1007/978-3-030-86324-1_22
22. Prudhommeaux, E., Seaborne, A.: SPARQL query language for RDF (2008). http://www.w3.org/TR/rdf-sparql-query/
23. Randolph, J.J.: A guide to writing the dissertation literature review. Pract. Assess. Res. Eval. **14**(1), 13 (2009)
24. Schmidt, L.M., Gotzsche, P.C.: Of mites and men: reference bias in narrative review articles; a systematic review. J. Fam. Pract. **54**(4), 334–339 (2005)
25. Shanahan, D.R.: A living document: reincarnating the research article. Trials **16**(1), 151 (2015). https://doi.org/10.1186/s13063-015-0666-5
26. Shotton, D.: Semantic publishing: the coming revolution in scientific journal publishing. Learn. Publish. **22**(2), 85–94 (2009). https://doi.org/10.1087/2009202
27. Webster, J., Watson, R.T.: Analyzing the past to prepare for the future: writing a literature review. MIS Q. **26**(2), xiii–xxiii (2002). https://doi.org/10.1.1.104.6570
28. Wee, B.V., Banister, D.: How to write a literature review paper? Transp. Rev. **36**(2), 278–288 (2016). https://doi.org/10.1080/01441647.2015.1065456
29. Wilkinson, M.D., et al.: Comment: the FAIR guiding principles for scientific data management and stewardship. Sci. Data **3**, 1–9 (2016). https://doi.org/10.1038/sdata.2016.18
30. Willinsky, J.: The access principle: the case for open access to research and scholarship. Lhs **2**(1), 165–168 (2006)
31. Wolmark, Y.: Quality assessment. Gerontol. Soc. **99**(4), 131–146 (2001). https://doi.org/10.3917/gs.099.0131
32. Xie, Y., Allaire, J.J., Grolemund, G.: R Markdown: The Definitive Guide. CRC Press, Boco Raton (2018)

A Large-Scale Collection Reselection Method Oriented by Library Space Rebuilding Based on the Statistics of Zero-Borrowing Books

Lei Huang(✉)

Library, Shanghai University of Electric Power, Shanghai 200090, China
huanglei@shiep.edu.cn

Abstract. Taking the relocation of Pudong branch Library of Shanghai University of Electric Power and comparing the difference between the goals of collection reselection oriented by space rebuilding and traditional collection weeding, this paper explores a large-scale collection reselection method based on the statistics of zero-borrowing books in recent *X* years, which takes "volume" as the unit, pointing out that the *X* value depends on the Library's reserved space and disciplines nature first, and discussing the using precautions of this method. The practice shows that the core collection selected by this method can meet the readers borrowing needs, ensure the reasonable discipline collection structure and the professional literature charactered by energy and electricity, which has some value for university libraries' large-scale collection adjustment oriented by space rebuilding in the digital transformation of libraries.

Keywords: Space rebuilding · Collection reselection · Zero-borrowing · Circulation statistics

1 Introduction

With the rapid development of the Internet technologies, significant changes have occurred in readers' reading behavior and learning habits: from paper reading to electronic reading, from individual learning to collaborative learning [1]. In the face of this change, university libraries in China are actively carrying out space rebuilding to meet the changing needs of readers [2, 3], such as Tongji University's continuous space rebuilding of its Sino-German Library [4], science and technology reading room [5], Hubei branch library [6]; and phased transformation of Library of Shanghai University of Finance and Economics [7]. The library of Shanghai Electric Power University has also actively responded to this trend. In the construction of its new library in Pudong Campus (after this referred to as the "new library"), it has fully considered the readers' demand for space and made a large-scale adjustment to the collection distribution: taking its Pudong branch as an example, it has a collection of more than 300000 volumes Chinese books (October 2017 data). We planned to put 150000 volumes Chinese books into the compact stack room on the first floor of the Pudong branch to release more space

H.-R. Ke et al. (Eds.): ICADL 2021, LNCS 13133, pp. 115–123, 2021.
https://doi.org/10.1007/978-3-030-91669-5_10

for collaborative learning and group discussion. Therefore, the Pudong branch library should select 150 thousand volumes as core collections from more than 30000 volumes Chinese books and allocate them to the "core stack of Pudong" when moving to the new library. The remaining 150 thousand Chinese books need to be drawn off and assigned to the "compact stack of Pudong".

For the collection construction of Pudong New Library of Shanghai Electric Power University, the selected collection is expected to achieve the following goals: 1) Meeting the readers borrowing needs basically; 2) Ensuring the reasonable distribution of disciplines in the collection structure of the new library; 3) Ensuring energy and power literatures as much as possible which are the key professional characters of the university; 4) The reselection method should be feasible and conducive to large-scale operations.

In this paper, we propose a large-scale collection reselection method and reach the above goals at the same time.

The rest of this paper is organized as follows: Sect. 2 surveys related work on collection reselection methods. Then, we introduce our method including the establishment of the conditions for selecting books and book extraction data generation in Sect. 3. In Sect. 4, we analyze whether the results can meet the expected goals followed by discussion in Sect. 5. A conclusion is drawn in Sect. 6.

2 Related Work

Through literature researching, it is found that most studies on the reselection of library collections are aimed at book weeding and collection optimizing [8], but not on the selection methods aiming at the rebuilding of digital library space. The emphases of the two are different because of their different goals. Book weeding is a process of weeding from the whole collection, and more focus on security [9]; The space rebuilding is the re-layout of the collection, emphasizing efficiency. The reselection method to optimize the collection by weeding has high costs of time and labor, and the processes are complex [10], which can not meet the requirements of large-scale selection guided by space rebuilding in library digitalization.

Based on the practice of moving the Pudong branch of Shanghai Electric Power University Library to the new library in 2018, this paper attempts to achieve a method suitable for large-scale operation through the statistics of zero-borrowing books(volume) in recent X years and can meet the above objectives at the same time.

3 Condition Establishment and Data Generation

3.1 The Establishment of the Conditions for Selecting Books

Considering the predicted learning space and the future collection growth rate, the new library should have 150000 books estimatedly, not more to ensure students learning space and not less to meet students reading needs, and these 150000 books also should meet the four selected goals in the introduction at the same time. In practice, the library of Shanghai University of Electric Power has adopted the "reverse selection" operation to extend the opening hours of the library as long as possible, that is, the selected core collections will continue to be on the shelves for readers to borrow, while the non-core collections will be removed from the shelves. Therefore, the extraction data generated in this paper is the "non-core collection" data, and the corresponding operation objects based on the extraction data are also the "non-core collection" books. Each library can flexibly select the operation objects according to the specific situation of the library.

Through a variety of experiments, it is found that the data based on the statistical method of zero-borrowing books (volume) in recent *X* years can better meet the four selection goals. In contrast, other methods, such as trying to take the total number of borrowing times or borrowing rate of a kind of book lower than a "certain value" as the condition of book extraction, have the following disadvantages: it doesn't consider the time factor - for example, a certain book used to be hot, but now no one is interested in it—although this kind of book borrowing times are high, it should be removed from the shelves; second, the "certain value" is difficult to be defined objectively. However, other more complex methods of book extraction by subject division are not feasible and difficult to large-scale operate. More importantly, after many experiments, the number of books extracted by other methods is far less than the target value of 150000.

To sum up, on the basis of comparing various methods, the conditions for book selection of the "non-core collection" are finally determined as follows: "zero-borrowing books(volume) in recent *X* years, excluding the books collected after the year of *Y*", and the schematic diagram is shown in Fig. 1.

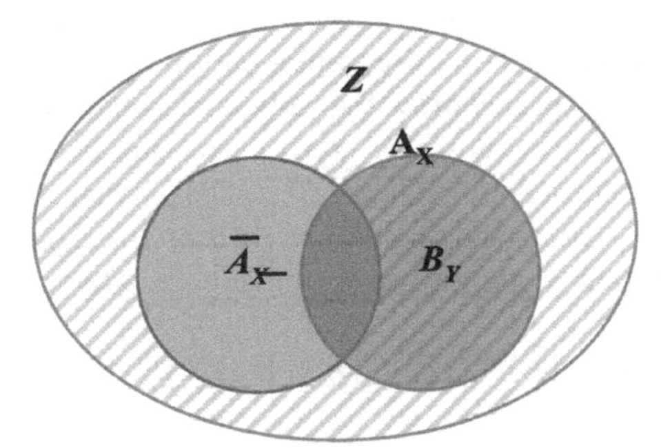

Fig. 1. Non-core collections (books to be drawn) diagram.

In Fig. 1, *Z* is the "non-core collection", that is, books to be drawn out, A_X is "zero-borrowing books in recent*X* years", and B_Y is "books collected after the year of *Y*". As can be seen from Fig. 1, the formula of book selection can be written as follows:

$$Z = A_X - B_Y \tag{1}$$

The Formula1 shows that the books to be withdrawn should consider the condition of zero borrowing quantity as well as consider the factor of collection time. The "books collected after the year of *Y*" should be eliminated so as to ensure that the new books collected after the year of *Y* will not be withdrawn from the shelves, and the values of *X* and *Y* mainly depend on the predicted collection space of stack rooms and the distribution of the disciplines.

3.2 Book Extraction Data Generation

According to Formula1, the extraction data is generated as follows:

1) First, setting the book extraction conditions for the overall collections as "zero-borrowing books in recent 5 years, excluding the books collected after January 1st, 2013", namely, $Z_O = A_5 - B_{2013}$. At this time, *X* is 5, *Y* is 2013, and 150000 books drawn list Z_O are generated;
2) Second, optimizing some Classes with large quantity by CLC (Chinese Library Classification, CLC) according to the different disciplines. For example, in this university characterized by energy and power, the books of Class T are industrial and technical books accounting for a largest collections, its conditions(the values of *X*, *Y*) for drawing book should be different, set as "zero-borrowing books in recent 5 years, excluding the books collected after January 1st, 2016", that is, $Z_T = A_5 - B_{2016}$. At this time, *X* is 5, *Y* is 2016, and the book extraction list Z_T is generated. Besides, Class TP3 computer books in Class T are further optimized according to their subject half-life [11, 12] and the expected growth of this kind of book collection. The Class TP3 book extraction condition is set as "zero-borrowing books in recent 3 years, excluding the books collected after January 1st, 2016", that is, $Z_{TP3} = A_3 - B_{2016}$. At this time, *X* is 3, *Y* is 2016, and the book extraction list Z_{TP3} is generated. After Class T, Class O and Class H are further optimized too, which is not repeated here.

After the above operation, the book extraction list of 150000 "non-core collection" books is generated.

4 Results and Analysis

4.1 Analysis of Readers Borrowing Needs Guarantee

As mentioned above, the book extraction lists generated can meet the space requirements of the new library firstly, but whether it can meet the 4 selection goals in the introduction still needs further verification. Taking the book list Z_O as an example, the condition of book selection is "zero-borrowing books in recent 5 years, excluding the books collected after January 1st, 2013". In other words, as long as the books borrowed within the past five years will be retained, and the books collected after January 1st, 2013 (inclusive) will also be retained on the shelves, that is, the book list itself is generated from the perspective of meeting the readers borrowing needs. From the standpoint of probability, books borrowed within recent 5 years are more likely to be borrowed again due to the similar professional distribution and similar needs of students in the same school. In addition, all the newer books are reserved. Therefore, the selection goal of "Meeting the borrowing needs of readers basically" can be achieved by the book extraction lists.

4.2 Analysis of Disciplines Distribution Reasonability

Before verifying whether the book list can "guarantee the reasonable distribution of disciplines in the collection structure of the new library", many staff members expressed concern: according to experience, the most books borrowed by students are art novels, while the number of professional books seems to be small. If we use the book list as the condition of the new book collection, there will be a large number of professional books be taken off the shelves. However, a large number of literary novels are kept on shelves, which affects the disciplines distribution of the new library and leads to the unreasonable collection structure for our school characterized by energy and power. Therefore, before the implementation of book selection, we comparatively analyzed the distribution of different disciplines before and after book selection. The results are shown in Fig. 2.

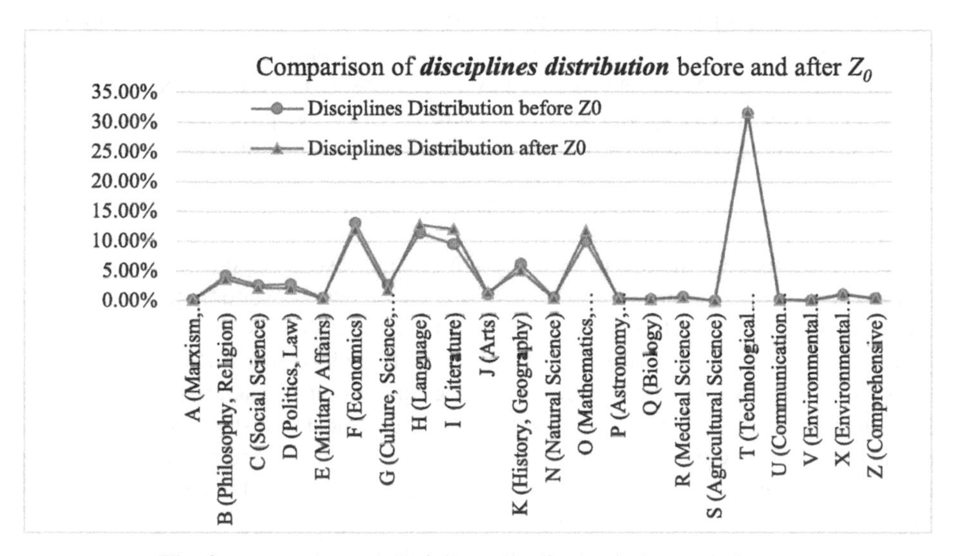

Fig. 2. Comparison of disciplines distribution before and after Z_0

Figure 2 shows that the distribution of all disciplines in the collection after the book extraction is basically the same as before. The major categories of books with a significant increase after the book extraction are Class H Language, I Literature, O Mathematical and Physical Sciences and Chemistry. This shows that the students have a high borrowing amount in these major categories of books because according to the book extraction condition Z_0, all the books borrowed in the past five years will be kept on the shelves. Therefore, which kind of books account for a high proportion after drawing books indicates that students have a high borrowing amount and a great demand for such books. However, the proportion of Class I literature and art novels increased the most, which is in line with people's experience expectations. Still, the ratio of Class T industrial technology with school characteristics has not been weakened at all, which is quite different from the previous staff's worry or feeling based on experience alone, which also shows that the school students' demand for professional books is beyond expectations.

The distribution of disciplines in the new library is almost consistent with that in the original library: the difference between the distribution of disciplines in the new library and the original library is in 22 categories, 14 categories fluctuate between 0–0.5%, only 3 categories fluctuate between 0.5–1%, and 5 categories fluctuate between 1–2.5%. Assuming that the distribution of disciplines in the original collection is reasonable, it can show that the book extraction list Z_0 meets the target of "Ensuring the reasonable distribution of disciplines in the collection structure of the new library".

4.3 Analysis of Key Disciplines Guarantee

Shanghai University of Electric Power is a University with the characteristics of energy and electric power. This is not only reflected in the collection structure, that is, the absolute advantage of the collection proportion of the Class T industrial technology books, but also reflected in the distribution of the key disciplines that can best represent the characteristics of energy and electric power in Class T. To verify whether the book list can guarantee the energy and power characteristic key disciplines of the school, we take the original collection as the reference and introduce the retention rate η,

$$\eta = q_i(\text{volumes on shelves})/Q_i(\text{total volumes}) \tag{2}$$

Further subdividing the book extraction list Z_T, the results are shown in Table 1.

Table 1 presents that the retention rate of TK energy and power engineering and TL atomic energy technology, the two most representative disciplines with energy and power characteristics, is in the front row, and TK is in the first place. The retention rate of the books that are highly related to the school's energy and power characteristics, such as the mechanical industry, TM electrical technology, TQ chemical industry, and other books, is more than 50%. In comparison, the retention rate of TK energy and power engineering books is 70%, followed by TL atomic energy technology books, which is also the micro reflection of the school's increasing research on nuclear power engineering in terms of borrowing quantity. To sum up, the book list can meet the selection goal of "Ensuring the key characteristic professional literature of energy and power".

5 Discussion

5.1 The Book Extraction Data by the Unit of "volume" Can Automatically Generate the Number of Copies Reserved for Each Kind of Book

Considering the given selection goals, this paper has tried many other methods and finds that, besides the convenience of data processing, this method processes data by the unite of "volume" instead of "kind" has an important but not easily perceived advantage: the

Table 1. Detailed data in retention rate of Z_T.

Subclasses in class T	Total number of volumes	Volumes to be drawn	Volumes retained on shelves	Retention rate η
T	28	17	11	**39.29%**
TB	4556	2181	2375	**52.13%**
TD	70	32	38	**54.29%**
TE	155	82	73	**47.10%**
TF	118	70	48	**40.68%**
TG	3025	1521	1504	**49.72%**
TH	5312	2323	2989	**56.27%**
TJ	63	24	39	**61.90%**
TK	1459	439	1020	**69.91%**
TL	211	66	145	**68.72%**
TM	12826	5646	7180	**55.98%**
TN	19677	10514	9163	**46.57%**
TP	39715	19477	20238	**50.96%**
TQ	1649	666	983	**59.61%**
TS	1520	781	739	**48.62%**
TU	5380	2918	2462	**45.76%**
TV	215	152	63	**29.30%**
SUM	95979	46909	49070	**51.13%**

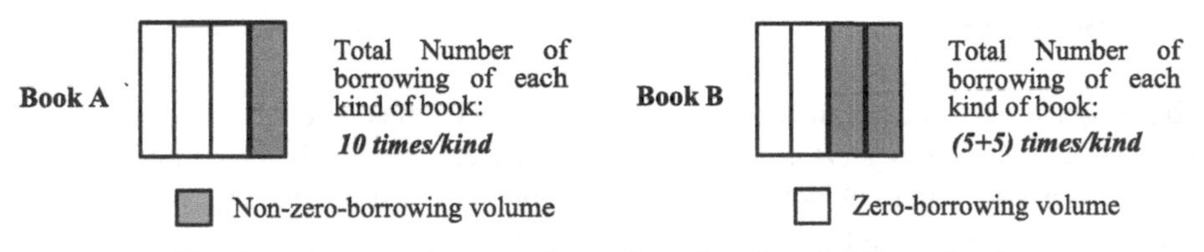

Fig. 3. Schematic diagram of retention of copies of different books.

number of copies of each kind of book will be automatically generated. As shown in Fig. 3.

As shown in Fig. 3, assuming that there are four copies of book A, should one or two be retained? If only one volume has been borrowed 10 times in the past five years, and the other three volumes have not been borrowed, then the three volumes will be withdrawn, and the only one volume will be retained. Another case: there are four copies of book B, two of which have been borrowed five times, while the other two volumes have not been borrowed, two volumes will be retained. Therefore, although the total borrowing times of the two kinds of books are the same, the probability of book A being borrowed only by one reader is higher, while the probability of book B being borrowed by different readers is higher, for which it is speculated that the reason may be related to whether the different kinds of books are of professional concentration or universal dispersion by their reader groups. Therefore, even if the total borrowing times of the two kinds of books are equal, the number of copies retained could be different, which skillfully avoids the unreasonable "one size fits all" decision in the number of copies of different kinds of books.

5.2 The Book Extraction Data Should Be Verified by the Rationality of Collection Disciplines Structure

The actual verification of the book extraction data generated by this method shows that, except for Class A books of Marx, Lenin, Mao Zedong, and Deng Xiaoping monographs, we decided to keep almost all of them because of their essential political core status and made necessary manual intervention. For other categories, the core collection selected by this method can meet readers borrowing needs and ensure the reasonable structure of the subject collection and the characteristics of the school energy and power professional literature. Therefore, it can be inferred that the seemingly simple condition of zero-borrowing book (in volume) actually contains multiple laws, which reveals the objective relationship between the borrowing rate and the disciplines distribution and the guarantee of key disciplines from a simple perspective. However, with the rapid development of the Internet, the impact of electronic literature, the migration of students' reading habits, whether the borrowing rate can truly reflect the students' reading needs should be further paid close attention to. In other words, the collection data generated by the statistical method of zero-borrowing books (in volume) can only be put into practice after the rationality verification of the collection disciplines structure and manual spot check.

6 Conclusion

Library space rebuilding is an active measure taken by China university libraries to meet the changing needs of readers and the changes of readers' reading behavior and learning habits in the Internet era, which makes it inevitable to adjust the existing collection on a large scale in the digital transformation of libraries.

Taking the relocation of the Pudong branch of Shanghai Electric Power University Library, this paper proposed a large-scale collection selection method based on the statistics of zero-borrowing books (in volume) in recent$X$ years. It points out that the X value firstly depends on the predicted collection space of stack rooms, then the nature of the disciplines such as half-life period etc., and also discusses what should be paid attention to in using this method. The practice shows that the core collection selected by this method can meet the readers borrowing needs and ensure the reasonable disciplines distribution and the characteristic professional literature of energy and power.

It should be pointed out that, before the large-scale implementation of this method, it is necessary to verify the rationality of disciplines distribution of the core collection and manual sampling check. On the long run, multiple methods should be taken to avoid the potential risk of unbalancing core collection, such as searching books on APP, e-display of non-core books information, assigning any returned books to the core collection including non-core books borrowed before, etc.

References

1. Su, L., Fang, F.: Hot spot and trend analysis of collaborative learning in China: a study of scientific knowledge map based on core journals. Adult Educ. **12**, 7–12 (2019)
2. Zhengqiang, L., Qiao, L.: A review of the research on library space reconstruction. Library (10), 55–59 (2018)
3. Yongjie, Z.: Review of the research on the space reconstruction and innovation of domestic library. J. Univ. Libr. Inf. **37**(3), 114–120 (2019)
4. Case introduction of German Library of Tongji University Library Homepage. https://www.lib.tongji.edu.cn/index.php?classid=11977&newsid=19664&t=show. Accessed 15 Jan 2021
5. Space transformation of science and technology reading room of Tongji University Library Homepage. https://www.lib.tongji.edu.cn/index.php?classid=11977&newsid=19668&t=show. Accessed 15 Jan 2021
6. Library space transformation (under construction) of Shanghai North Campus of Tongji University Library Homepage. https://www.lib.tongji.edu.cn/index.php?classid=11977&newsid=19669&t=show. Accessed 15 Jan 2021
7. Linlin, S., Li, Z.: Opening ceremony of academic database, academic exhibition, sharing space opening ceremony and Millennium Village Survey exhibition of Shanghai University of Finance and economics, and world top 500 enterprise exhibition held in the library. Res. Libr. Inf. Work Shanghai Univ. **29**(4), 100–101 (2019)
8. Na, X., Hong, Y., Limei, X.: Introduction and characteristics analysis of the old document removal project of Wesley University Library in USA. Libr. Inf. Work **59**(8), 38–41 (2015)
9. Yingchun, C.: The construction and empirical study of the collection reselection system. Libr. Constr. (3), 36–40 (2010)
10. Hailing, Y.: A new approach to the elimination of old books in library collection. Libr. Constr. (1), 66–72 (2015)
11. Liya, W.: Half life analysis of computer science based on CNKI. Libr. Inf. (1), 100–105 (2015)
12. Lijun, Z.: Statistics and analysis of zero lending books -- Taking the Chinese books in the branch library of South China Agricultural University as an example. Libr. Circles (3), 66–70 (2014)

Information Extraction

Automatic Cause-Effect Relation Extraction from Dental Textbooks Using BERT

Terapat Chansai[1], Ruksit Rojpaisarnkit[1], Teerakarn Boriboonsub[1], Suppawong Tuarob[1](✉), Myat Su Yin[1], Peter Haddawy[1,4], Saeed-Ul Hassan[2], and Mihai Pomarlan[3]

[1] Faculty of Information and Communication Technology, Mahidol University, Nakhon Pathom, Thailand
{terapat.chn,ruksit.roj,teerakarn.bor}@student.mahidol.edu, {suppawong.tua,myatsu.yin,peter.had}@mahidol.ac.th

[2] Department of Computing and Mathematics, Manchester Metropolitan University, Manchester, UK
S.Ul-Hassan@mmu.ac.uk

[3] Faculty of Linguistics, University of Bremen, Bremen, Germany
mihai.pomarlan@uni-bremen.de

[4] Bremen Spatial Cognition Center, University of Bremen, Bremen, Germany

Abstract. The ability to automatically identify causal relations from surgical textbooks could prove helpful in the automatic construction of ontologies for dentistry and building learning-assistant tools for dental students where questions about essential concepts can be auto-generated from the extracted ontologies. In this paper, we propose a neural network architecture to extract cause-effect relations from dental surgery textbooks. The architecture uses a transformer to capture complex causal sentences, specific semantics, and large-scale ontologies and solve sequence-to-sequence tasks while preserving long-range dependencies. Furthermore, we have also used BERT to learn word contextual relations. During pre-training, BERT is trained on enormous corpora of unannotated text on the web. These pre-trained models can be fine-tuned on custom tasks with specific datasets. We first detect sentences that contain cause-effect relations. Then, cause and effect clauses from each cause-effect sentence are identified and extracted. Both automatic and expert-rated evaluations are used to validate the efficacy of our proposed models. Finally, we discuss a prototype system that helps dental students learn important concepts from dental surgery textbooks, along with our future research directions.

Keywords: Relation extraction · Textbook mining · Deep learning

1 Introduction

Intelligent tutoring systems have been applied to a wide variety of domains, including clinical medicine. A particular challenge in medical applications is acquiring

H.-R. Ke et al. (Eds.): ICADL 2021, LNCS 13133, pp. 127–138, 2021.
https://doi.org/10.1007/978-3-030-91669-5_11

the structured knowledge needed to teach the required concepts. The teaching of clinical and surgical procedures requires providing students with a thorough understanding of causal relations between actions and their possible effects in the context of various states of the patient so that students can generalize beyond the particular scenario being presented. Such knowledge is available in fairly structured form in medical textbooks. If it could be automatically extracted and suitably structured, this would address a major bottleneck in building intelligent tutoring systems for clinical medicine and surgery. Yet, identifying causal relations in textbooks is challenging since causality can be linguistically expressed in many, sometimes ambiguous and wordy, ways. Furthermore, building a precise model to automatically extract causal relations requires good quality datasets and experts to analyze results concerning a large variety of sentences. For example, in the sentence "if undetected, transportation may lead to ledging, zipping, gouging, or even perforation of the root canal wall", the event "transportation" causes the events "ledging, zipping, gouging, or even perforation of the root canal wall". In a more complicated scenario, "gouging the root canal wall" can also be the effect of "forcing rotary instruments such as Gates-Glidden burs beyond the resistance level", therefore there are several potential causes for an effect, as well as more complex chains of causation, which might lead to the entity linking problem which will be discussed in a later section. The sentences above are from one endodontics textbook commonly used for teaching dental students.

Many approaches have been employed to extract machine-readable information from natural language text. Some of these used rule-based extraction models, while some used traditional machine learning algorithms. Regardless, most of the existing methods expect the participants to be denoted by one name each in one sentence expressing the relationship, limiting their applicability to natural text. Novel techniques like deep learning have only begun to be applied to identify and extract causal relations.

In this work, we address the problem of extracting causal relations from textbooks on dental endodontic procedures. We choose endodontics because it is one of the more challenging areas of dentistry and thus more difficult to learn. Our approach combines recent strategies to deal with NLP problems. Specifically, we employ BERT [4] and Transformers [18] to address the causal relation extraction problem. We propose a neural network architecture to capture such complex sentences in dental surgery textbooks and use BERT to learn contextual causal relations by training it with a rich dataset. We identify two subproblems in this task, cause-effect sentence classification (recognize when a sentence expresses a causal relation) and cause-effect relation extraction (once the relation is identified, identify its participants). Both of these are discussed in Sect. 3. The source code is made available for research purposes at: https://gitlab.com/dentoi/casual-extraction. Although our application domain in this paper is endodontics, the presented approach is quite general and applicable to other clinical and surgical domains.

2 Background and Related Work

While machine learning techniques have been extensively used in text mining to extract meaningful information from large-scale corpora [14,15], we briefly describe concepts and techniques from previous research that we use in this work. An *ontology* represents a formalization of a conceptualization, i.e., a description in some formal language of entities and relations between them. It is used for knowledge representation for discrete intelligent reasoning and to exchange such knowledge across systems that may use it. *Information extraction* is a task for extracting knowledge from unstructured text data, then converting it into a structured, machine-readable representation such as a semantic graph. The process consists of several steps such as Named Entity Recognition, Named Entity Linking, and Relation Extraction. *BERT (Bidirectional Encoder Representation from Transformer)* is a recent but prevalent technique in NLP to implement a language model, i.e., a model capable of assessing how likely a particular text is. Furthermore, BERT has proven to be applicable for a whole range of NLP tasks, including question answering and relation extraction. BERT is based on the Transformer model architecture, which uses attention mechanisms that learn contextual relationships between each token in a text. A transformer consists of an encoder to read the text input and a decoder to produce a prediction for the task.

Previous work has used several approaches to extract causal relations from the text in a variety of domains. Dasgupta et al. [3] used a recursive neural network architecture (Bi-directional LSTM) with word-level embeddings to detect causal relations in a sentence. A conditional random field (CRF) model was used to evaluate performance. They used a diverse trained dataset, including drug effect, BBC News Article, and SemEval.

Su Yin et al. [13] explored the possibility of automatically extracting causal relations from textbooks on dental procedures. They used pattern-matching over dependency parse trees produced by the spaCy NLP tool to identify causal relation assertions in a collection of fifteen textbooks on endodontic root canal treatment. Since their primary purpose was to extract knowledge for teaching, they focused on surgical mishaps and their possible causes. They achieved a precision of 95.7% but a recall of only 41.6%.

Zhao et al. [20] sought to extract causal transitive relations in medical text data, where one effect is a cause in another relation. First, they examined how such chains of causal relations are expressed over several sentences. Their approach used causal triads, which are defined as associations of co-occurring medical entities that are likely to be causality connected. Then, the authors tried to discover implicit causal relations. The result showed that the causal triad model is an appropriate means to extract textual medical causal relations.

Hybrid techniques have also been used to automatically extract causal relations from text [12]. The hybrid system combines both rule-based and machine learning methodologies; it extracts cause-effect pairs utilizing a set of rules for Simple Causative Verbs, rules for Phrasal Verbs/Noun, etc. Using the dataset SemEval-2010 (Task 8), they obtained 87% precision and 87% recall. Therefore, the rule-based model can also be adapted for other types of information, such as the dental textbooks that we use in this work.

3 Methodology

Our proposed methodology is composed of three modules: a) data preprocessing, b) cause-effect sentence classification, and c) cause-effect relation extraction. Each of the modules is described in the following sub-sections.

3.1 Dataset and Data Preprocessing

We used 16 dental surgery textbooks as our dataset to train the model. We first converted the PDF file to a text file and cleaned non-ASCII characters out of the text file for our data preparation processes. We then split each text file into several chapters using *chapterize*[1]. As a result, each chapter's text file was produced, consisting of several paragraphs and sentences, ready to use for model training. The dataset contains 5,642 annotated sentences, of which 2,032 sentences express a cause-effect relation. The sentences describe endodontic procedures and are labeled using the pattern-based method proposed by Su Yin, et al. [13], shown to yield high precision of 95.7%, but low recall of 41.57%.

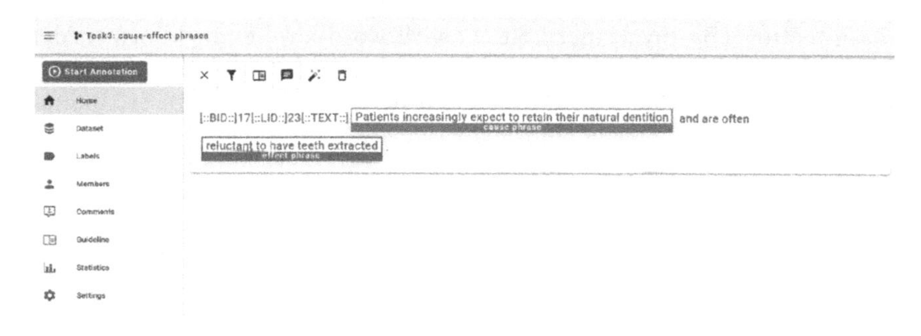

Fig. 1. Annotation tool for the cause-effect relation extraction task.

For each sentence that expresses a cause-effect relation, human annotators were asked to label parts of the sentence (word sequences) which represent cause and effect clauses. Doccano[2], an open source document annotation framework, was used to facilitate the annotation, as presented in Fig. 1.

3.2 Cause-Effect Sentence Classification

Cause-effect sentence classification is the method to identify whether or not a sentence expresses a cause-effect relation. Each sentence in the textbook is encoded before being used to train the BERT model. For the implementation of the cause-effect sentence classification model, there are four architectures. First, the input encoder consists of input_ids, input_mask, and segment_ids. Second, the BERT model consists of 24 layers and 1,024 hidden units in each layer.

[1] https://github.com/JonathanReeve/chapterize.

[2] https://github.com/doccano.

Third, a linear layer for the sigmoid activation function is used for formatting BERT's output then passed to the classifier, the last architecture in this model. The classifier uses the sigmoid activation function for binary classification. The 5,642 sentences used to develop this model are separated into 70% for training, 10% for validation, and 20% for testing.

Figure 2 illustrates the pipeline for model training and classification process, along with the model's optimal hyperparameters. First the annotated training data is used to train a BERT model, which is then fine-tuned with the validation dataset. The trained classifier takes a sentence as the input and outputs the probability of being a causal sentence.

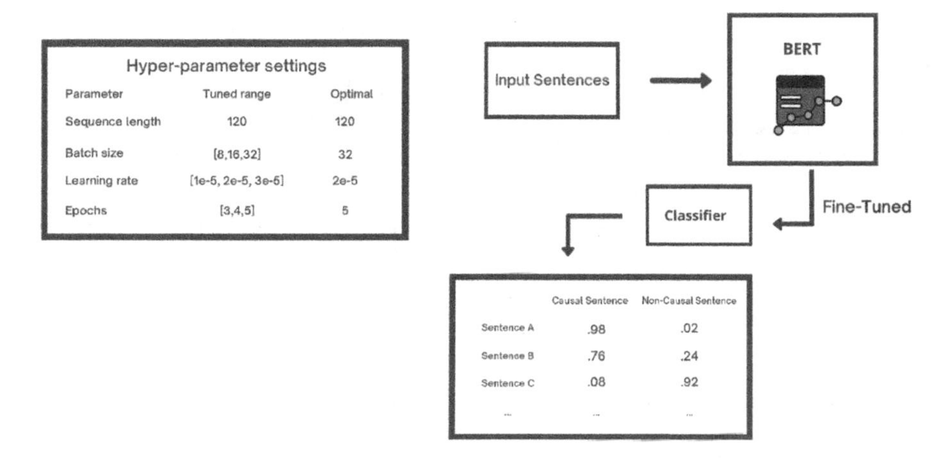

Fig. 2. The processing pipeline for the sentence classification task.

As baselines to BERT, we have also tried other deep learning based (i.e., CNN for sentence classification [7]) and traditional machine learning classifiers such as Naive Bayes [19], Support Vector Machine (SVM) [8], and Random Forest [1] with default hyperparameter settings. For traditional machine learning methods, each sentence is represented with a bag-of-word vector using TF-IDF weights.

3.3 Cause-Effect Relation Extraction

Cause-Effect relation extraction is the method to extract the cause and the effect from sentences expressing causal relations in textbooks. The causal sentence is tokenized into three types; C stands for Cause, E stands for Effect, and O stands for Other. Each token is labeled and encoded before training in the BERT model. For the implementation of the Cause-Effect Relation Extraction model, there are four architectures. The first and second architectures are the same as in the Cause-Effect Sentence Classification model. The third is the dropout layer, which differs from the Cause-Effect Sentence Classification model because the binary classification model is valid only for the first index of BERT's output. The last architecture is the classifier, which consists of three classes: cause, effect, and other, as mentioned earlier. The dataset for this model is the same that for the

Cause-Effect Sentence Classification model. As a baseline, we also compare the performance of the BERT model with a Bi-LSTM CRF model proposed in [10], due to its reported effectiveness in sequence tagging tasks such as named entity recognition [6].

Figure 3 illustrates the pipeline for model training and extraction process along with the model's optimal hyperparameters. We framed the cause-effect relation extraction task as a token classification problem, where each token is classified into Cause, Effect, or Other. Therefore, the BERT model has been configured for the multiclass classification task.

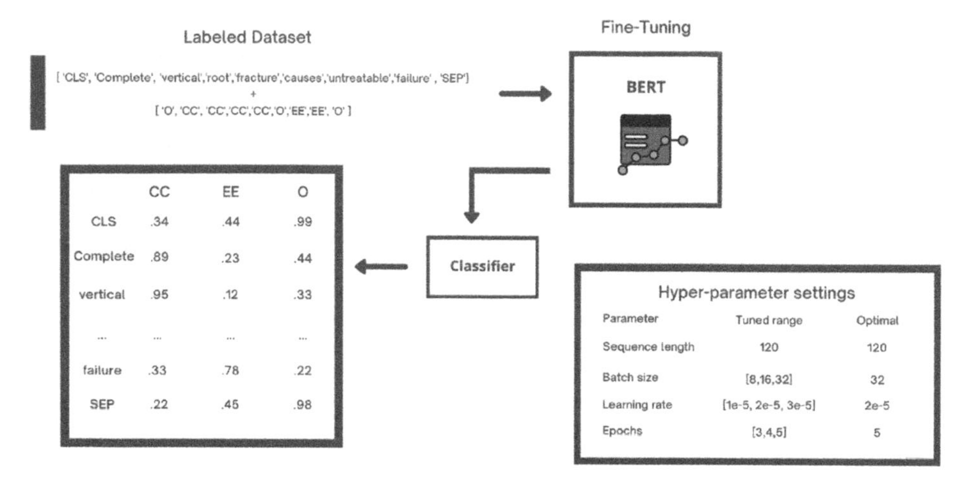

Fig. 3. The processing pipeline for the cause-effect extraction task.

4 Experiments, Results, and Discussion

All the experiments were conducted on a Linux machine with 16 CPU cores (32 threads), two RTX 2080 Super GPUs, and 128 GB of RAM. This section discusses the results.

4.1 Cause-Effect Sentence Classification

In cause-effect classification, each sentence is labeled as 0 or 1 for Non-causal or Causal sentence, respectively. Table 1 shows examples of causal and non-causal sentences taken from an endodontics textbook.

Table 1. Examples of causal and non-causal sentences, taken from [11].

Sentence	Label
Repeated recapitulation or remaining stationary in the canal with a non-landed rotary instrument can lead to apical transportation	1
A deep palate allows much greater vertical access when using a palatal approach	0
Furthermore, this pain can be further exacerbated by incorrect or unnecessary treatments, often resulting in the establishment of chronic pain pathways	1
This can be started 24 h prior to surgery but can also be swilled for 1 min prior to placement of anaesthetic	0

The precision, recall, and F1 scores for each class (Causal, Non-Causal, Weighted Average) predicted by BERT and other classifiers are reported in Table 2. It is apparent that deep learning based methods such as CNN and BERT outperform the traditional machine learning methods (i.e., Naïve Bayes, SVM, and Random Forest) in all aspects. While the CNN has a slightly better recall for the Causal Sentence class compared to BERT, BERT yields the highest F1 in both the Non-Causal and Causal Sentence classes. This may be due to BERT's ability to encode proximity semantics in a sentence into the embedding, making it suitable for tasks that require an understanding of word sequences and language syntax.

Table 2. Performance of the causal sentence classification task.

Model	Non causal sentence			Causal sentence			Weighted average		
	Precision	Recall	F1	Precision	Recall	F1	Precision	Recall	F1
Naive Bayes	0.81	0.98	0.89	0.86	0.31	0.45	0.82	0.78	0.82
SVM	0.95	0.94	0.95	0.84	0.86	0.85	0.93	0.93	0.93
Random Forest	0.95	0.94	0.95	0.84	0.86	0.85	0.93	0.92	0.92
CNN	**0.98**	0.95	0.96	0.83	**0.94**	0.88	0.95	0.95	0.95
BERT	0.97	**0.99**	**0.98**	**0.98**	0.90	**0.93**	**0.97**	**0.97**	**0.97**

4.2 Cause-Effect Relation Extraction

In the cause-effect relation extraction process, sentences are tokenized, and special tokens ("CLS" and "SEP") are added to the sentences. The "CLS" token annotates the sentences' starting point. The "SEP" token represents the ending of the sentences. The model's output is the vector representation of each sentence, which is later converted to each token label. The labels include 'C', 'E', and 'O', which represent Cause token, Effect token, and Other token, as illustrated in Table 3.

Table 3. Example labeling of cause and effect tokens.

Sentence	['CLS', 'Failure', 'to', 'achieve', 'patency', 'during', 'preparation', 'can', 'result', 'in', 'inadequate', 'penetration', 'of', 'irrigants', 'SEP']
Label	['O', 'C', 'C', 'C', 'C', 'C', 'C', 'O', 'O', 'O', 'E', 'E', 'E', 'E', 'O']

The precision, recall, and F1 of the cause-effect extraction task using Bi-LSTM CRF and BERT are reported in Table 4. It is evident that BERT outperforms Bi-LSTM CRF (baseline) in all metrics, both in the Cause Token and Effect Token classification tasks. Bi-LSTM CRF performs relatively well on extracting Cause tokens, but its performance drops when extracting the Effect tokens. Furthermore, Bi-LSTM CRF only yields F1 of 0.49 on identifying the Effect tokens. In contrast, BERT performs well on identifying both the Cause and Effect tokens, yielding an F1 of 0.89 on average.

Table 4. Performance of the cause-effect extraction task.

Model	Cause token			Effect token			Weighted average		
	Precision	Recall	F1	Precision	Recall	F1	Precision	Recall	F1
Bi-LSTM CRF	0.77	0.80	0.79	0.61	0.37	0.49	0.71	0.64	0.66
BERT	**0.89**	**0.91**	**0.90**	**0.89**	**0.90**	**0.89**	**0.89**	**0.89**	**0.89**

5 Future Directions

This paper has reported a primary investigation on extracting an ontology of causal relations from endodontics textbooks. Our overarching goal is to develop a toolbox capable of automatically extracting and structuring essential causal information from textbooks on clinical and surgical procedures. As future work, we plan to evaluate the result of the proposed approach by asking experts to assess our model's outcome regarding whether the extracted sentences contain a causal event or not. Subsequently, machine learning models could be further enhanced for textbooks in other languages using language-agnostic BERT-based models [5]. Besides improving the models, the subsections discuss our path forward in terms of implementation.

5.1 Relation Linking

Illustrated in Fig. 4 is a partial snapshot of extracted relations from individual causal sentences, visualized by *react-force-graph*[3]. Notable extracted relations include `post preparation` that leads to `root structures`, that further leads to `pulp therapy`. On the other branch, `access cavity` also leads to `pulp therapy` and `line access factors`. Linking these cause-effect relations together would give rise to many useful applications. Relation linking is a method to link each relevant entity to construct a knowledge graph, where several knowledge graph

[3] https://github.com/vasturiano/react-force-graph.

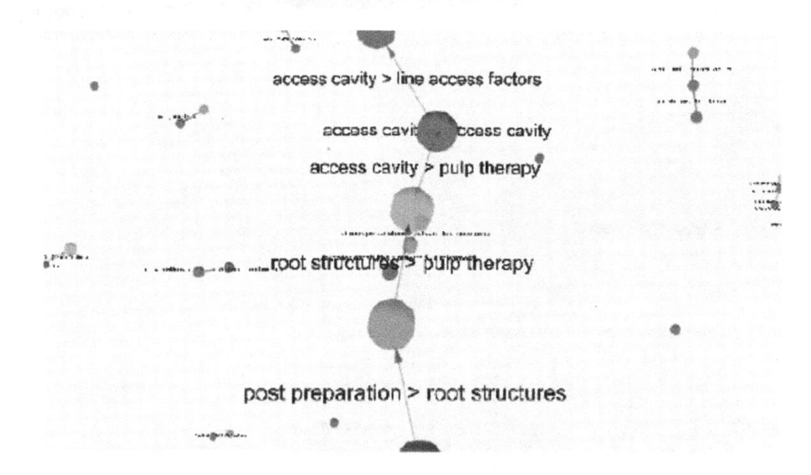

Fig. 4. Examples of extracted relations.

mining techniques can be applied to extract meaningful patterns. Our model might identify several cause-effect pairs that are similar or occur in a chain of causation. For example, if two cause-effect pairs have a similar cause, the effects of that two causes might be related. If we could link each relevant entity or cause together, it would be an efficient structure that effectively illustrates the knowledge and information. For example, linking causal relations and the conditions under which they hold with actions could enable reasoning about possible consequences of actions. Furthermore, it could be used for knowledge visualization or ontology visualization in a subsequent application.

5.2 Application in Dental Quiz Generation

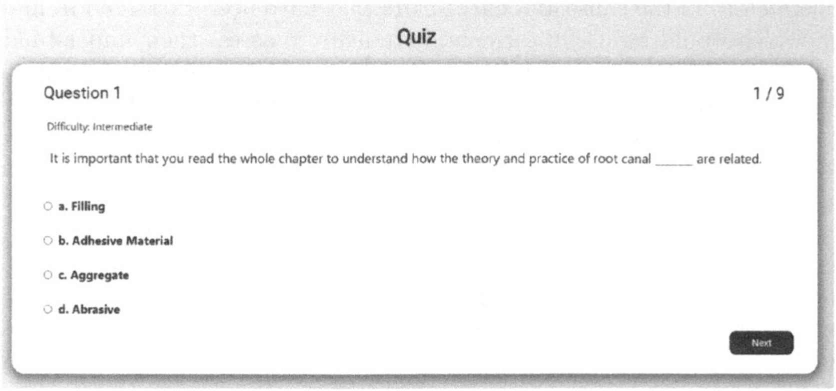

Fig. 5. Example of a dental quiz question.

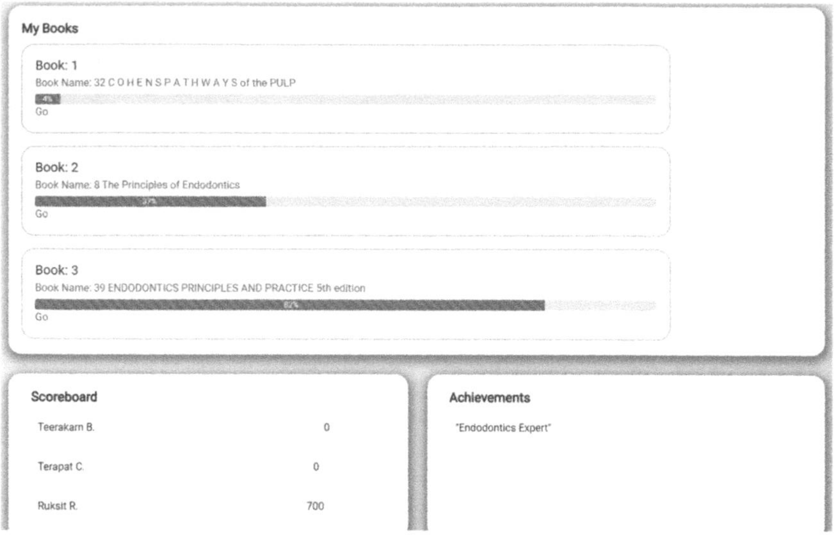

Fig. 6. Example dashboard showing a student' progress on completing quizzes from each textbook.

As dental students must learn from many textbooks, there are some difficulties distinguishing events and understanding information. Further, the method to assess the understanding of each topic is somewhat technical and requires an expert to construct the test. From the results of our proposed model, cause and effect pairs can be an efficient knowledge base that we could apply to generate a question and answering system, which could serve as a personal learning assistant to increase students' understanding and enhance their performance. Implementing a web-based application (similar to the work of Budovec et al. [2] in the field of radiology) could be an effective way for accessibility and quality. Quizzes can be generated from the cause and effect pairs that have been extracted from textbooks, which would be useful for dental students to assess their understanding. Automatic question-answer pair generation [9] could be adopted for this task. Figure 5 illustrates an example quiz question that could be helpful to students to assess their understanding of the textbook's material.

Dental instructors could also benefit from a system that allows them to automatically analyze each textbook and generate quizzes for each chapter. These quizzes can be assigned to students where the system can keep track of each student's progress on their self-study on each textbook, as illustrated in Fig. 6.

Recently Vannaprathip et al. [16,17] have presented an intelligent tutoring system for teaching surgical decision making, with application in the domain of endodontics. The intelligent tutor is integrated with a VR dental simulator. The tutor intervenes during the surgical procedure by pointing our errors, providing positive feedback, and asking a variety of types of questions. To do this, it makes use of causal knowledge represented in terms of conditional effects of actions using an adaptation of the PDDL AI planning language. The coding of this domain knowledge was a major bottleneck in building the automated tutoring system. The techniques presented in this paper represent a step toward automating that knowledge representation process.

6 Conclusions

This paper has proposed a neural network architecture to extract cause-effect relations from endodontics textbooks and built a precise learning-assistant tool for dental students. The ultimate goal is to teach students to make decisions in novel situations by providing them with a complete understanding of causal relations occurring in dental procedures, relations which are described in textbooks. This understanding can be assessed through quizzes that are automatically generated from the ontologies extracted by our approach.

Acknowledgments. This work was partially supported by a grant from the Mahidol University Office of International Relations to P. Haddawy in support of the MIRU joint unit. S. Tuarob is supported by Mahidol University (Grant No. MU-MiniRC02/2564). We also appreciate the computing resources from Grant No. RSA6280105, funded by Thailand Science Research and Innovation (TSRI) and the National Research Council of Thailand (NRCT).

References

1. Breiman, L.: Random forests. Mach. Learn. **45**(1), 5–32 (2001)
2. Budovec, J.J., Lam, C.A., Kahn, C.E., Jr.: Informatics in radiology: radiology gamuts ontology: differential diagnosis for the semantic web. Radiographics **34**(1), 254–264 (2014)
3. Dasgupta, T., Saha, R., Dey, L., Naskar, A.: Automatic extraction of causal relations from text using linguistically informed deep neural networks. In: Proceedings of the 19th Annual SIGdial Meeting on Discourse and Dialogue, pp. 306–316 (2018)
4. Devlin, J., Chang, M.W., Lee, K., Toutanova, K.: BERT: pre-training of deep bidirectional transformers for language understanding. arXiv preprint arXiv:1810.04805 (2018)
5. Feng, F., Yang, Y., Cer, D., Arivazhagan, N., Wang, W.: Language-agnostic BERT sentence embedding. arXiv preprint arXiv:2007.01852 (2020)
6. Huang, Z., Xu, W., Yu, K.: Bidirectional LSTM-CRF models for sequence tagging. arXiv preprint arXiv:1508.01991 (2015)
7. Kim, Y.: Convolutional neural networks for sentence classification. In: Proceedings of the 2014 Conference on Empirical Methods in Natural Language Processing (EMNLP), pp. 1746–1751. Association for Computational Linguistics, Doha, October 2014. https://doi.org/10.3115/v1/D14-1181. https://www.aclweb.org/anthology/D14-1181
8. Noble, W.S.: What is a support vector machine? Nat. Biotechnol. **24**(12), 1565–1567 (2006)
9. Noraset, T., Lowphansirikul, L., Tuarob, S.: WabiQA: a Wikipedia-based Thai question-answering system. Inf. Process. Manag. **58**(1), 102431 (2021)

10. Panchendrarajan, R., Amaresan, A.: Bidirectional lstm-crf for named entity recognition. In: PACLIC (2018)
11. Rhodes, J.S.: Advanced Endodontics: Clinical Retreatment and Surgery. CRC Press (2005)
12. Sorgente, A., Vettigli, G., Mele, F.: A hybrid approach for the automatic extraction of causal relations from text. In: Lai, C., Giuliani, A., Semeraro, G. (eds.) Emerging Ideas on Information Filtering and Retrieval. SCI, vol. 746, pp. 15–29. Springer, Cham (2018). https://doi.org/10.1007/978-3-319-68392-8_2
13. Su Yin, M., et al.: Automated extraction of causal relations from text for teaching surgical concepts. In: Proceedings of 8th IEEE International Conference on Healthcare Informatics, November 2020
14. Tuarob, S., Mitrpanont, J.L.: Automatic discovery of abusive Thai language usages in social networks. In: Choemprayong, S., Crestani, F., Cunningham, S.J. (eds.) ICADL 2017. LNCS, vol. 10647, pp. 267–278. Springer, Cham (2017). https://doi.org/10.1007/978-3-319-70232-2_23
15. Tuarob, S., et al.: DAViS: a unified solution for data collection, analyzation, and visualization in real-time stock market prediction. Financ. Innov. **7**(1), 1–32 (2021)
16. Vannaprathip, N., Haddawy, P., Schultheis, H., Suebnukarn, S.: Intelligent tutoring for surgical decision making: a planning-based approach. Int. J. Artif. Intell. Educ. (2021, to appear)
17. Vannaprathip, N., et al.: A planning-based approach to generating tutorial dialog for teaching surgical decision making. In: Nkambou, R., Azevedo, R., Vassileva, J. (eds.) ITS 2018. LNCS, vol. 10858, pp. 386–391. Springer, Cham (2018). https://doi.org/10.1007/978-3-319-91464-0_44
18. Vaswani, A., et al.: Attention is all you need. arXiv preprint arXiv:1706.03762 (2017)
19. Zhang, H.: Exploring conditions for the optimality of naive bayes. Int. J. Pattern Recognit. Artif. Intell. **19**(02), 183–198 (2005)
20. Zhao, S., Jiang, M., Liu, M., Qin, B., Liu, T.: CausalTriad: toward pseudo causal relation discovery and hypotheses generation from medical text data. In: Proceedings of the 2018 ACM International Conference on Bioinformatics, Computational Biology, and Health Informatics, pp. 184–193 (2018)

Multilingual Epidemic Event Extraction

Stephen Mutuvi[1,2](✉), Emanuela Boros[1], Antoine Doucet[1], Gaël Lejeune[3], Adam Jatowt[4], and Moses Odeo[2]

[1] L3i, University of La Rochelle, 17000 La Rochelle, France
{steve.mutuvi,emanuela.boros,antoine.doucet}@univ-lr.fr
[2] Multimedia University of Kenya, Nairobi, Kenya
{smutuvi,modeo}@mmu.ac.ke
[3] STIH Lab, Sorbonne University, 75006 Paris, France
gael.lejeune@paris-sorbonne.fr
[4] University of Innsbruck, 6020 Innsbruck, Austria
adam.jatowt@uibk.ac.at

Abstract. In this paper, we focus on epidemic event extraction in multilingual and low-resource settings. The task of extracting epidemic events is defined as the detection of disease names and locations in a document. We experiment with a multilingual dataset comprising news articles from the medical domain with diverse morphological structures (Chinese, English, French, Greek, Polish, and Russian). We investigate various Transformer-based models, also adopting a two-stage strategy, first finding the documents that contain events and then performing event extraction. Our results show that error propagation to the downstream task was higher than expected. We also perform an in-depth analysis of the results, concluding that different entity characteristics can influence the performance. Moreover, we perform several preliminary experiments for the low-resourced languages present in the dataset using the mean teacher semi-supervised technique. Our findings show the potential of pre-trained language models benefiting from the incorporation of unannotated data in the training process.

Keywords: Epidemiological surveillance · Multilingualism · Semi-supervised learning

1 Introduction

The ability to detect disease outbreaks early enough is critical in the deployment of measures to limit their spread and it directly impacts the work of health authorities and epidemiologists throughout the world. While disease surveillance has in the past been a critical component in epidemiology, conventional surveillance methods are limited in terms of both promptness and coverage, while at

This work has been supported by the European Union's Horizon 2020 research and innovation program under grants 770299 (NewsEye) and 825153 (Embeddia). It has also been supported by the French Embassy in Kenya and the French Foreign Ministry.

H.-R. Ke et al. (Eds.): ICADL 2021, LNCS 13133, pp. 139–156, 2021.
https://doi.org/10.1007/978-3-030-91669-5_12

the same time requiring labor-intensive human input. Often, they rely on information and data from past disease outbreaks, which more often than not, is insufficient to train robust models for extraction of epidemic events.

Epidemic event extraction from archival texts, such as digitized news reports, has also been applied for constructing datasets and libraries dedicated to tracking and understanding epidemic spreads in the past. Such libraries leverage the technology advantage of digital libraries by storing, processing, and disseminating data about infectious disease outbreaks. The work presented by Casey et al. [5] is an example of such an initiative aiming at analyzing outbreak records of the third plague pandemic in the period 1894 to 1952 in order to digitally map epidemiological concepts and themes related to the pandemic. Although the authors used semi-automatic approaches in their work, the discovery of documents related to epidemic outbreaks was done manually and the entity extraction was largely performed through manual annotation or the use of gazetteers, which have their own limitations. Other works devoted to the studies of past epidemics (e.g., the analysis of bubonic plague outbreak in Glasgow (1900) [10]) fully rely on manual efforts for data collection and preprocessing. We believe that automatic approaches to epidemic information extraction could also enhance this kind of scientific study.

The field of research focusing on data-driven disease surveillance, which has been shown to complement traditional surveillance methods, remains active [1,7]. This is majorly motivated by the increase in the number of online data sources such as online news text [14]. Online news data contains critical information about emerging health threats such as what happened, where and when it happened, and to whom it happened [35]. When processed into a structured and more meaningful form, the information can foster early detection of disease outbreaks, a critical aspect of epidemic surveillance. News reports on epidemics often originate from different parts of the world and events are likely to be reported in other languages than English. Hence, efficient multilingual approaches are necessary for effective epidemic surveillance [4,27].

Moreover, the large amounts of continuously generated unstructured data, for instance, in the ongoing COVID-19 epidemic, are often challenging and difficult to process by humans without leveraging computational techniques. With the advancements in natural language processing (NLP) techniques, processing such data and applying data-driven methods for epidemic surveillance has become feasible [2,30,40]. Although promising, the scarcity of available annotated corpora for data-driven epidemic surveillance is a major hindrance. Obtaining large-scale human annotations is a time-consuming and labor-intensive task. The challenge is more pronounced when dealing with neural network-based methods [18], where massive amounts of labeled data play a critical role in reducing generalization error.

Another specific challenge for the extraction of epidemic events from news text is class imbalance [19]. The imbalance exists between the disease and location entities, which when paired characterize an epidemic event. The large difference in the number of instances from different classes can negatively impact the performance of the extraction models. Another challenge relates to data sparsity where some languages in the multilingual setup have few annotated data [15], barely sufficient to train models that achieve satisfactory performance.

In this study, we use a multilingual dataset comprising news articles from the medical domain with diverse morphological structures (Chinese, English, French, Greek, Polish, and Russian). In this dataset, an epidemic event is characterized by the references to a disease name and the reported locations that are relevant to the disease outbreak. We evaluate a specialized baseline system and experiment with the most recent Transformer-based sequence labeling architectures. Additionally, error propagation from the classification task that affects the event extraction task is also evaluated since the event extraction task is a multi-step task, comprising various sub-tasks [13,22,30]. The classification task filters the epidemic-related documents from the large collection of online news articles, prior to the event extraction phase. We also perform a detailed analysis of various attributes (sentence length, token frequency, entity consistency among others) of the data and their impact on the performance of the systems.

Thus, considering the aforementioned challenges, our contributions are the following:

- We establish new performance scores after the evaluation of several pre-trained and fine-tuned Transformer-based models on the multilingual data and by comparing with a specialized multilingual news surveillance system;
- We perform a generalized, fine-grained analysis of our models with regards to the results on the multilingual epidemic dataset. This enables us to more comprehensively assess the proposed models, highlighting the strengths and weaknesses of each model;
- We show that semi-supervised learning is beneficial to the task of epidemic event extraction in low-resource settings by simulating different few-shot learning scenarios and applying self-training.

The remainder of this paper is organized as follows. Section 2 describes the related work. Section 3 presents the multilingual dataset utilized in our study. In Sect. 4, we discuss our experimental methodology and empirical results. Finally, Sect. 5 concludes this paper and provides suggestions for future research.

2 Related Work

Several works tackled the detection of events related to epidemic diseases. Some approaches include external resources and features at a sub-word representation level. For example, the Data Analysis for Information Extraction in any Language (DAnIEL) system was proposed as a multilingual news surveillance system that leverages repetition and saliency (salient zones in the structure of a news article), properties that are common in news writing [26]. By avoiding the usage of language-specific NLP toolkits (e.g., part-of-speech taggers, dependency parsers) and by focusing on the general structure of the journalistic writing genre [21], the system is able to detect key event information from news articles in multilingual corpora. We consider it as a baseline multilingual model.

Models based on neural network architectures and which take advantage of the word embeddings representations have been used in monitoring social media content for health events [25]. Word embeddings capture semantic properties of words, and thus the authors use them to compute the distances between relevant concepts for completing the task of flu event detection from text. Another type of approach is based on long short-term memory (LSTM) [47] models that approach the epidemic detection task from the perspective of classification of tweets to extract influenza-related information.

However, these approaches, especially the recent deep learning methods such as the Transformer-based model [44], remain largely unexplored in the context of epidemic surveillance using multilingual online news text. Transformer language models can learn powerful textual representations, thus they have been effective across a wide variety of NLP downstream tasks [3,9,23,46].

Despite the models requiring a large amount of data to train, annotated resources are generally scarce, especially in digital humanities [34,39]. Having sufficient data is essential for the performance of event extraction models since it can help reduce overfitting and improve model robustness [12]. To address the challenges associated with scarcity of large-scale labeled data, various methods have been proposed [6,12,15].

Among them is semi-supervised learning, where data is only partially labeled. The semi-supervised approaches permit harnessing unlabeled data by incorporating the data into the training process [43,50]. One type of semi-supervised learning method is self-training, which has been successfully applied to text classification [42], part-of-speech (POS) tagging [48] and named entity recognition (NER) [24,37]. Semi-supervised learning methods can utilize a teacher-student method where the teacher is trained on labeled data that generates pseudo-labels for the unlabeled data, and the pseudo-labeled examples are iteratively combined with the clean labels by the student [49]. These previous attempts in addressing the problem of limited labeled data have focused on resource-rich languages such as English [6,12,15]. In this study, we increase coverage to other languages, and most importantly languages with limited available training data.

3 Dataset

Due to the lack of dedicated datasets for epidemic event extraction from multilingual news articles, we adapt a freely available epidemiological dataset[1], referred to as DAnIEL [26]. The corpus was built specifically for the DAnIEL system [26,28], containing articles in six different languages: English, French, Greek, Russian, Chinese, and Polish. However, the dataset is originally annotated at the document level. We annotate the dataset to token-level annotations [31], a

[1] The DAnIEL dataset is available at https://daniel.greyc.fr/public/index.php?a=corpus.

Table 1. Statistical description of the DAnIEL partitions. DIS and LOC stand for the number of disease and location mentions, respectively.

	Partition	Documents	Sentences	Tokens	Entities	DIS	LOC
French	Train	2,185	62,748	1,786,077	2,677	1,438	1,239
	Dev	273	7,625	231,165	337	206	131
	Test	273	7,408	214,418	300	177	123
	Total	2,731	77,781	2,231,660	3,314	1,821	1,493
English	Train	379	7,312	204,919	524	319	205
	Dev	48	857	24,990	5	3	2
	Test	47	921	25,290	34	27	7
	Total	474	9,090	255,199	563	349	214
Greek	Train	312	4,947	151,959	259	144	115
	Dev	39	924	23,980	15	10	5
	Test	39	531	15,951	26	12	14
	Total	390	6,402	191,890	300	166	134
Chinese	Train	354	6,309	193,453	67	57	10
	Dev	44	838	26,720	16	14	2
	Test	44	624	19,767	7	5	2
	Total	442	7,771	239,940	90	76	14
Russian	Train	341	5,250	112,714	258	170	88
	Dev	43	618	14,168	30	27	3
	Test	42	547	11,514	39	27	12
	Total	426	6,415	138,396	327	224	103
Polish	Train	281	7,288	126,696	498	352	146
	Dev	35	954	17,165	73	40	33
	Test	36	998	17,026	67	52	15
	Total	352	9,240	160,887	638	444	194

common format utilized in research for the event extraction task. The token-level dataset is made freely and publicly available[2].

Typically in event extraction, this dataset is characterized by class imbalance. Only around 10% of the documents are relevant to epidemic events, which is very sparse. The number of documents in each language is rather balanced, except for French, having about five times more documents compared to the rest of the languages. More statistics on the corpus can be found in Table 1.

In this dataset, a document generally talks about an epidemiological event and the task of extracting the event comprises the detection of all the occurrences of a disease name and the locations of the reported event, as shown in Fig. 1. The document talks about the ending of a *polio* outbreak in *India*, more exactly

[2] The token-level annotated dataset is available at https://bit.ly/3kUQcXD.

Today marks one year since the last case of polio was recorded in India when the virus paralysed an 18-month-old girl in Howrah, near Kolkata. If pending test results return absent of the virus in coming weeks, India will be removed from the list of endemic polio countries. But India still remains at serious risk of fresh outbreaks if the virus is brought back into the country from overseas, and polio experts say the country's massive immunisation regimen must be maintained.

Fig. 1. Excerpt from an English article in the DAnIEL dataset that was published on January 13th, 2012 at http://www.smh.com.au/national/health/polio-is-one-nation-closer-to-being-wiped-out-20120112-1pxho.html.

in *Howrah* and *Kolkata*. An event extraction system should detect all the *polio* event mentions, along with the aforementioned locations.

4 Experiments

Our experiments are performed in two setups:

1. Supervised learning experiments:
   - Our first experiments focus on the *epidemic event extraction* utilizing the entire dataset.
   - Next, like most approaches for text-based disease surveillance [22], we follow a two-step process by first applying *document classification* into either relevant (documents that contain event mentions) or irrelevant (documents without event mentions) and then performing the *epidemic event extraction* task through the detection and extraction of the disease names and locations from these documents.
2. Semi-supervised learning experiments:
   - For these experiments, we simulate several few-shot scenarios for the low-resourced languages in our dataset, and we apply semi-supervised training with the mean teacher method in order to assess the ability of the models to alleviate the challenge posed by the lack of annotated data.

Models. We evaluate the pre-trained model BERT (Bidirectional Encoder Representations from Transformers) proposed by [11] for token sequential classification[3]. We decided to use BERT not only because it is easy to fine-tune, but it has also proved to be one of the most performing technologies in multiple NLP tasks [9,11,38]. Due to the multilingual characteristic of the dataset, we use the multilingual BERT pre-trained language models and fine-tune them on our epidemic-specific labeled data. We will refer to these models as BERT-multilingual-cased[4]

[3] For this model, we used the parameters recommended in [11].

[4] https://huggingface.co/bert-base-multilingual-cased. This model was pre-trained on the top 104 languages having the largest Wikipedia edition using a masked language modeling (MLM) objective.

and BERT-multilingual-uncased[5]. We also experiment with the XLM-RoBERTa-base model [8] that has shown significant performance gains for a wide range of cross-lingual transfer tasks. We consider this model appropriate for our task due to the multilingual nature of our dataset[6].

Evaluation. The epidemic event extraction evaluation is performed in a coarse-grained manner, with the entity as the reference unit [29]. We compute precision (P), recall (R), and F1-measure (F1) at the micro-level (error types are considered over all documents).

4.1 Supervised Learning Experiments

We chose DAnIEL [26] as a baseline model for epidemic event extraction. This is an unsupervised method that consists of a complete pipeline that first detects the relevant documents and then extracts the event triggers. The system considers text as a sequence of strings and does not depend on language-specific grammar analysis, hence can easily be adapted to a variety of languages. This is an important attribute of epidemic extraction systems for online news text, as the text is often heterogeneous in nature. Figure 2 presents the full procedure for the supervised learning experiments.

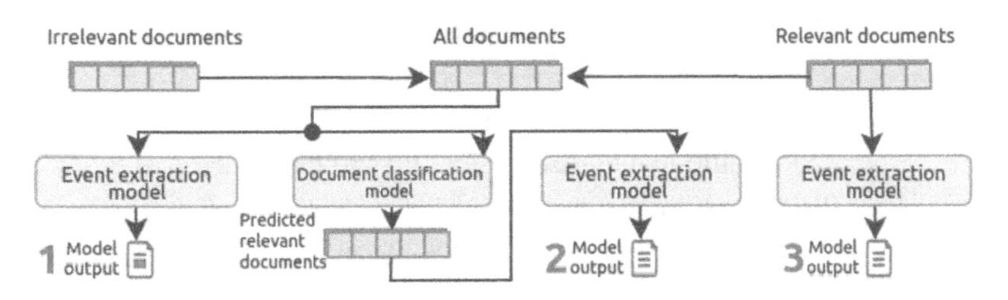

Fig. 2. Illustration of the types of experiments carried out: (1) using all data instances (*relevant and irrelevant documents*), (2) testing on the *predicted relevant documents* provided by the document classification step, (3) using only the *ground-truth relevant documents*.

For document classification, we chose the fine-tuned BERT-multilingual-uncased [11,30] whose performance on text classification is a F1 of 86.25%. The performance in F1 with regards to the relevant documents per language is 28.57% (Russian), 87.10% (French), 50% (English), 100% (Polish), and 50% (Greek). One drawback of this method is the fact the none of the Chinese relevant documents was found by the classification model, and thus, none of the events will be further detected.

[5] https://huggingface.co/bert-base-multilingual-uncased. This model was pre-trained on the top 102 languages having the largest Wikipedia editions using a masked language modeling (MLM) objective.

[6] XLM-RoBERTa-base was trained on 2.5 TB of newly created clean CommonCrawl data in 100 languages.

Holistic Analysis. We now present the results of the evaluated models, namely the DAnIEL system and the Transformer-based models. We first observe in Table 2 that all the models significantly outperform our baseline, DAnIEL. As it can be seen in Table 2, under *relevant and irrelevant documents* (1), when the models are trained on the entire dataset (i.e., the relevant and irrelevant documents), the BERT-multilingual-uncased model recorded the highest scores, with a very small margin when compared to the other two fine-tuned models, the cased BERT and XLM-RoBERTa-base.

Table 2. Evaluation results for the detection of disease names and locations on all languages and all data instances (relevant and irrelevant documents).

Models	P	R	F1
DAnIEL Baseline	38.97	47.32	42.74
Relevant and irrelevant documents (1)			
BERT-multilingual-cased	80.66	79.72	80.19
BERT-multilingual-uncased	82.25	**79.77**	**80.99**
XLM-RoBERTa-base	**82.41**	76.81	79.52
Predicted relevant documents (2)			
BERT-multilingual-cased	52.13	89.43	65.87
BERT-multilingual-uncased	**53.66**	**92.28**	**67.86**
XLM-RoBERTa-base	53.10	90.65	66.97
Ground-truth relevant documents (3)			
BERT-multilingual-cased	85.40	**90.95**	88.08
BERT-multilingual-uncased	87.16	89.79	88.46
XLM-RoBERTa-base	**88.53**	89.56	**89.04**

In Table 2, under *ground-truth relevant documents* (3), when evaluating the ground-truth relevant examples only, the task is obviously easier, particularly in terms of precision, while, when we test on the predicted relevant documents in Table 2, under *predicted relevant documents* (2), the amount of errors that are being propagated to the event extraction step is extremely high, reducing all the F1 scores by over 20% points for all models. Since there is a considerable reduction in the number of relevant instances after the classification step, this step alters the ratio between the relevant instances and the retrieved instances. Thus, not only in F1 but also a significant drop in precision is observed across all the models, when compared with the ground-truth results. The drop in precision is due to a number of relevant documents being discarded by the classifier.

Since our best results were not obtained after applying document classification for the relevant article detection, we consider that our best models are those applied on the initial dataset comprised of relevant and irrelevant documents. Thus, we continue by presenting the performance of these models for each language in the dataset.

Table 3. Evaluation scores (F1%) of the analyzed models for the predicted relevant documents per language, found by the classification model. The Chinese language was not included in the table because the classification model did not detect any relevant Chinese document.

Model	French	English	Greek	Chinese	Russian	Polish
BERT-multilingual-cased	83.60	65.52	**75.00**	**80.00**	**63.64**	82.35
BERT-multilingual-uncased	84.17	**80.70**	73.47	50.00	60.27	**84.62**
XLM-RoBERTa-base	**84.67**	52.00	72.73	66.67	61.11	81.90

As shown in Table 3, BERT-multilingual-uncased obtained the highest scores for three out of the four low-resource languages, while BERT-multilingual-cased was more fitted for Polish. The reason for the higher results in the case of the low-resourced Greek, Chinese, and Russian languages could be motivated by considering the experiments performed in the paper that describes the XLM-RoBERTa model [8]. The authors concluded that, initially, when training on a relatively small amount of languages (between 7 and 10), XLM-RoBERTa is able to take advantage of positive transfer which improves performance, especially on low resource languages. On a larger number of languages, the *curse of multilinguality* [8] degrades the performance across all languages due to a trade-off between high-resource and low-resource languages. As pointed out by Conneau et al. [8], adding more capacity to the model can alleviate this *curse of multilinguality*, and thus the results for low-resource languages could be improved when trained together.

Model-Wise Analysis. As demonstrated in the results, different models perform differently on different datasets. Thus, we move beyond the holistic score assessment (entity F1-score) and compare the strengths and weaknesses of the models at a fine-grained level.

We analyzed the individual performance of the models and the intersections of their predicted outputs by visualizing them in several UpSet plots[7]. As seen in Fig. 3(a), there are approximately 70 positive instances that none of the systems was able to find. The highest intersection, approximately 340 instances, represents the true positives found by the three systems. BERT-multilingual-cased was able to find a higher number of unique true positive instances, instances not detected by the other models.

BERT-multilingual-uncased had the highest number of true positive instances cumulatively, the second-highest number of unique true positives, and the lowest number of false positive instances. This reveals the ability of the BERT-multilingual-uncased model to find the relevant instances in the dataset and to correctly predict a large proportion of the relevant data points, thus the high recall and precision, and overall F1 performance.

[7] https://jku-vds-lab.at/tools/upset/.

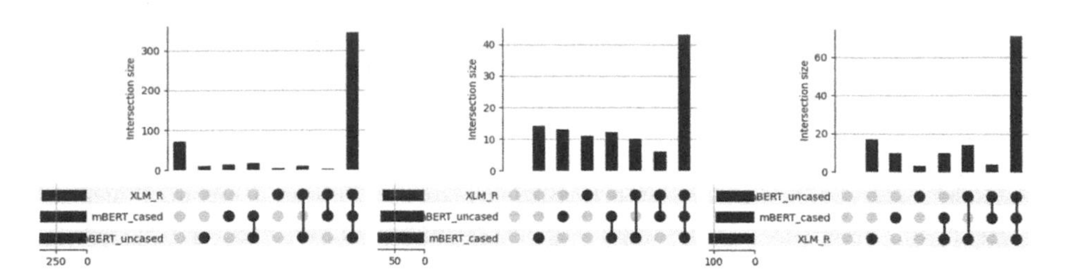

Fig. 3. Intersection of models predictions. The figures represent (from left) the true positive, false positive and false negative intersection sizes. The x-axis is interpreted as follows; from left to right, the first bar represents the number of instances that no system was able to find, the next three bars show the instances found by the respective individual models, the next three denote instances found by a pair of systems, while the last bar (the highest intersection) represents instances jointly found by all systems.

The overall performance is, generally, affected by the equally higher number of false positive and false negative results, as presented in Figs. 3(b, c). XLM-RoBERTa-base recorded the highest false negative rate and the lowest number of true positive instances, which explains the low recall and F1 scores.

Attribute-Wise Analysis. We chose to utilize an evaluation framework for interpretable evaluation for the named entity recognition (NER) task [16] that proposes a fine-grained analysis of entity attributes and their impact on the overall performance of the information extraction systems[8].

We conduct an attribute-wise analysis that compares how different attributes affect performance on the DAnIEL dataset, (e.g., how entity or sentence length correlates with performance). The entity attributes considered are entity length (eLen), sentence length (sLen), entity frequency (eFreq), token frequency (tFreq), out-of-vocabulary density (oDen), entity density (eDen) and label consistency. The label consistency describes the degree of label agreement of an entity on the training set. We consider both entity and token label consistencies, denoted as eCon and tCon. eCon represents the number of entities in a sentence. To perform the attribute-wise analysis, bucketing is applied, a process that breaks down the performance into different categories [16,17,33].

The process involves partitioning the attribute values into $m = 4$ discrete parts, whose intervals were obtained by dividing the test entities equally, with in some cases the interval method being customized depending on the individual characteristics of each attribute [16]. For example, in the entity length (eLen), entities in the test set with lengths of $\{1, 2, 3\}$ and >4 are partitioned into four buckets corresponding to the lengths. Once the buckets are generated, we calculate the F1 score with respect to the entities of each bucket.

The results in Table 4 illustrate that for our dataset the performance of all models varies considerably and it is highly correlated with oDen, eCon, tCon,

[8] The code [16] is available here: https://github.com/neulab/InterpretEval..

Table 4. Attribute-wise F1 scores (%) per bucket for the following entity attributes: entity length (eLen), sentence length (sLen), entity frequency (eFreq), token frequency (tFreq), out of vocabulary density (oDen), entity density (eDen), entity consistency (eCon) and token consistency (tCon).

Model	F1	Bucket	F1							
			eDen	oDen	eCon	tCon	tFreq	sLen	eFreq	eLen
BERT-multilingual-cased	80.19	1	84.15	86.00	59.11	18.18	74.76	74.16	76.78	81.12
		2	84.15	83.54	85.62	84.07	86.59	77.35	90.47	79.24
		3	88.03	70.32	100	87.94	83.52	85.58	85.50	92.30
		4	89.88	53.33	100	96.15	84.26	88.23	81.96	0
Standard deviation			**2.48**	**12.97**	16.69	**31.14**	4.48	5.76	**4.99**	**36.80**
BERT-multilingual-uncased	80.99	1	86.13	84.09	59.75	31.25	77.72	72.50	77.51	80.61
		2	87.12	84.61	84.60	81.22	85.17	78.67	86.40	76.36
		3	88.67	68.88	100	87.35	83.01	81.19	82.60	83.33
		4	82.75	55.88	100	94.33	83.00	90.36	81.30	0
Standard deviation			2.17	11.90	16.45	24.85	2.74	6.42	3.17	**34.77**
XLM-RoBERTa-base	79.52	1	81.24	84.28	53.06	100	72.27	72.80	76.40	79.73
		2	84.57	80.00	85.15	81.05	84.04	74.99	86.88	76.00
		3	85.43	63.52	87.50	87.60	84.04	81.73	81.20	92.30
		4	87.35	57.57	100	87.50	81.25	89.77	81.60	0
Standard deviation			2.20	11.10	**17.32**	6.86	**4.83**	**6.61**	3.70	**36.30**

and eLen. This proves that the prediction difficulty of en event mention is influenced by label consistency, entity length, out-of-vocabulary density, and sentence length. Regarding the entity length, the third bucket had fewer entities among the first three buckets and the highest F1 score among the four buckets, an indication that a majority of entities were correctly predicted. A very small number of entities had a length of size 4 or more, and at the same time, those entities were poorly predicted by the evaluated models (F1 of zero).

Moreover, the standard deviation values observed for BERT-multilingual-uncased are the lowest when compared with the other two models across the majority of the attributes (except for tCon, oDen, and sLen), which can be an indication that this model is not only the best performing, but it is also the most stable, thus being particularly robust.

4.2 Semi-supervised Learning Experiments

Due to the limited availability of annotated datasets in epidemic event extraction, we employ the self-training semi-supervised learning technique in order to analyze whether our dataset and models could benefit from having relevant unannotated documents. We then experiment with the mean teacher (MT) training method, a semi-supervised learning method where a target-generating teacher model is updated using the exponential moving average weights of the student model [41]. As such, the approach can handle a wide range of noisy input such as digitized documents, which often are susceptible to optical character recognition (OCR) errors [32,36,45].

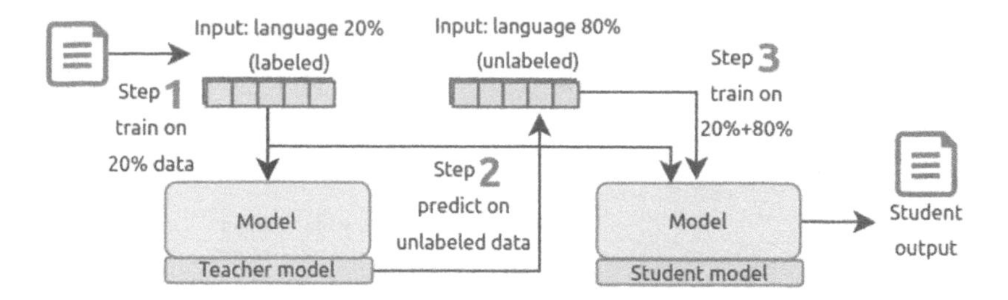

Fig. 4. The self-training process in the *20% for training and 80% unannotated data* few-shot setting.

Table 5. The four few-shot scenarios with a comparison between their increasing number of training sentences and the amounts of DIS and LOC per scenario and per language.

Model	Greek			Chinese			Russian			Polish		
	Sent	DIS	LOC	Sent	DIS	LOC	Sent	DIS	LOC	Sent	DIS	LOC
20%	854	35	19	1,280	–	–	1,115	44	22	1,486	56	28
30%	1,315	53	32	1,904	17	0	1,617	62	35	2,501	64	29
40%	1,801	69	46	2,427	22	1	2,031	72	40	3,078	83	47
50%	2,228	87	58	3,230	22	1	2,488	95	50	3,697	112	56
100%	4,947	144	115	6,309	57	10	5,250	170	88	7,288	352	146

First, we consider the documents in four low-resource languages from the DAnIEL dataset: Greek, Chinese, Russian, and Polish. These languages are around 80% less represented in this dataset when compared to French. For these experiments, we simulate several few-shot scenarios by implementing a strategy in which we split our training data into annotated and unannotated sets, starting from 20%, and increasing iteratively by 10% points until 50%. Thus, we obtain four few-shot learning scenarios as detailed in Table 5. For example, in the 20% scenario, 20% of the data is considered annotated and 80% unannotated. The process of self-training, as presented in Fig. 4, has the following steps:

- Step 1: Each of our models will be, at first, the teacher, trained and fine-tuned on the event extraction task using a cross-entropy loss and a small percentage of the DAnIEL dataset (i.e., we keep 20% for training). The rest of the data is considered unlabeled (i.e., the rest of 80%).
- Step 2: This data (80%) is annotated using the teacher model generating in this manner the pseudo labels which are added to the annotated percentage of data (20%) to form the final dataset.
- Step 3: Next, each of the models will be the student, trained and fine-tuned on this dataset using KL-divergence consistency cost function.

Table 6. The results for the low-resourced languages from DAnIEL when all data for all languages is trained together, and when the languages are trained separately.

Model	Greek	Chinese	Russian	Polish
Data instances trained on all languages				
BERT-multilingual-cased	73.47	50.00	60.27	84.62
BERT-multilingual-uncased	75.00	80.00	63.64	82.35
XLM-RoBERTa-base	72.73	66.67	61.11	81.90
Data instances trained per language				
BERT-multilingual-cased	80.77	**85.71**	60.00	86.67
BERT-multilingual-uncased	**81.56**	80.00	**64.00**	**86.79**
XLM-RoBERTa-base	80.77	80.00	63.16	86.54

Table 7. The results for the low-resourced languages in DAnIEL in the four few-shot scenarios (F1%).

Language	100%	Baseline				Self-training (MT)			
		20%	30%	40%	50%	20%	30%	40%	50%
BERT-multilingual-cased									
Greek	80.77	58.54	72.73	77.55	83.33	55.56	**77.19**	60.47	77.27
Chinese	80.00	0.0	80.00	80.00	66.67	0	80.00	72.73	**80.00**
Russian	63.16	41.79	50.67	54.84	53.73	34.15	47.06	53.85	40.82
Polish	86.54	74.51	74.23	78.10	78.50	**75.25**	73.47	**80.39**	76.47
BERT-multilingual-uncased									
Greek	81.56	51.43	72.73	72.73	80.77	45.71	72.73	70.83	78.26
Chinese	80.00	0	80.00	80.00	80.00	0	80.00	80.00	80.00
Russian	64.00	36.11	48.00	52.78	52.17	31.37	**55.38**	**53.52**	50.00
Polish	86.79	72.55	73.47	77.36	78.90	**75.25**	69.31	75.23	76.19
XLM-RoBERTa-base									
Greek	76.36	50.00	74.51	72.00	75.47	**53.33**	**75.00**	**72.34**	**77.55**
Chinese	85.71	0.0	66.67	72.73	66.67	0	66.67	72.73	66.67
Russian	60.00	32.50	47.62	53.66	53.16	**34.38**	**49.28**	**55.74**	**53.52**
Polish	86.67	65.55	70.59	75.00	75.23	**67.83**	**73.27**	**77.67**	**79.61**

Holistic Analysis. In Table 6, we compare the results obtained when the languages were all trained and tested together and when the languages were trained separately. One can notice that higher scores were obtained in the second case, showing the positive impact of fine-tuning one model per language. This could also be explained by the *curse of multilinguality* [8] that degrades the performance across all languages due to a trade-off between high-resource and low-resource languages when the languages are trained together. Meanwhile, the advantages of training them separately considerably increase the performance for each of the languages.

Table 7 presents the four few-shot scenarios, the F1 score when the models are trained on the entire language data, the F1 scores for the baselines (the models trained in a supervised manner on the few samples), and with the self-training using the mean teacher method. For the latter, we fine-tune all our models on between 800–3000 sentences of training data for each language (as shown in Table 5) and use it as a teacher model. Larger improvements in performance were noticed in the case of the XLM-roBERTa-base model, where self-training leads to 2.29% average gains on Greek (from 67.99% to 69.55%), 4.19% on Polish (from 71.59% to 74.59%), and on Russian, 3.21% (from 46.73% to 48.23% while remaining unchanged for Chinese.

In the majority of the cases and for all the models, the performance improvements can also be due to the fact that, because of the few-shot scenarios that are created from our initial dataset, the simulated unannotated data remains in-domain with the labeled data. It was proven that using biomedical papers for a downstream named entity recognition (NER) biomedical task considerably improves the performance of NER compared to using unannotated news articles [20]. Meanwhile, for the cases where we observed a decrease in the performance after self-training, it would mean that the teacher model was not that strong, leading to noisier annotations compared to the full or baseline dataset setup.

5 Conclusions

In this study, we evaluated supervised and semi-supervised learning methods for multilingual epidemic event extraction. First, with supervised learning, we observe low precision values when training and testing on all data instances and predict relevant documents. This is not surprising since the number of negative examples, with potential false positives, rises up to around 90%.

While the task of document classification, prior to event extraction, was expected to result in performance gains, our results reveal a significant drop in performance. This can be attributed to error propagation to the downstream task. Further, the fine-grained error analysis provides a comprehensive assessment and better understanding of the models. This facilitates the identification of the strengths of a model and aspects that can be enhanced to improve the performance of the model.

Regarding the semi-supervised experiments, we show that the mean teacher self-training technique can potentially improve the model results, by utilizing the fairly readily available unannotated data. As such, the self-training method can be beneficial to low-resource languages by alleviating the problems associated with the scarcity of labeled data.

In future work, we propose to focus on the integration of real unannotated data to improve our overall performance scores on the low-resourced languages. Also, since directly applying self-training on pseudo labels results in gradual drifts due to label noises, we propose to study in future work a judgment model to help select sentences with high-quality pseudo labels that the model predicted with high confidence. Further, we intend to explore the semi-supervised method under different noise levels and types to determine the robustness of our models to noise.

References

1. Aiello, A.E., Renson, A., Zivich, P.N.: Social media-and internet-based disease surveillance for public health. Ann. Rev. Public Health **41**, 101–118 (2020)
2. Bernardo, T.M., Rajic, A., Young, I., Robiadek, K., Pham, M.T., Funk, J.A.: Scoping review on search queries and social media for disease surveillance: a chronology of innovation. J. Med. Internet Res. **15**(7), e147 (2013)
3. Bosselut, A., Rashkin, H., Sap, M., Malaviya, C., Celikyilmaz, A., Choi, Y.: COMET: commonsense transformers for automatic knowledge graph construction. arXiv preprint arXiv:1906.05317 (2019)
4. Brixtel, R., Lejeune, G., Doucet, A., Lucas, N.: Any language early detection of epidemic diseases from web news streams. In: 2013 IEEE International Conference on Healthcare Informatics, pp. 159–168. IEEE (2013)
5. Casey, A., et al.: Plague dot text: text mining and annotation of outbreak reports of the Third Plague Pandemic (1894–1952). J. Data Min. Digit. Humanit. HistoInf. (2021). https://jdmdh.episciences.org/7105
6. Chen, S., Pei, Y., Ke, Z., Silamu, W.: Low-resource named entity recognition via the pre-training model. Symmetry **13**(5), 786 (2021)
7. Choi, J., Cho, Y., Shim, E., Woo, H.: Web-based infectious disease surveillance systems and public health perspectives: a systematic review. BMC Public Health **16**(1), 1–10 (2016)
8. Conneau, A., et al.: Unsupervised cross-lingual representation learning at scale. In: Proceedings of the 58th Annual Meeting of the Association for Computational Linguistics, ACL 2020, 5–10 July 2020, pp. 8440–8451. Association for Computational Linguistics (2020). https://www.aclweb.org/anthology/2020.acl-main.747/
9. Conneau, A., Lample, G.: Cross-lingual language model pretraining. In: Wallach, H., Larochelle, H., Beygelzimer, A., d' Alché-Buc, F., Fox, E., Garnett, R. (eds.) Advances in Neural Information Processing Systems, vol. 32, pp. 7059–7069. Curran Associates, Inc. (2019). http://papers.nips.cc/paper/8928-cross-lingual-language-model-pretraining.pdf
10. Dean, K., Krauer, F., Schmid, B.: Epidemiology of a bubonic plague outbreak in Glasgow, Scotland in 1900. R. Soc. Open Sci. **6**, 181695 (2019). https://doi.org/10.1098/rsos.181695
11. Devlin, J., Chang, M.W., Lee, K., Toutanova, K.: BERT: pre-training of deep bidirectional transformers for language understanding. In: Proceedings of the 2019 Conference of the North American Chapter of the Association for Computational Linguistics: Human Language Technologies, Volume 1 (Long and Short Papers), pp. 4171–4186. Association for Computational Linguistics, Minneapolis, June 2019. https://doi.org/10.18653/v1/N19-1423
12. Ding, B., et al.: DAGA: data augmentation with a generation approach for low-resource tagging tasks. arXiv preprint arXiv:2011.01549 (2020)
13. Doan, S., Ngo, Q.H., Kawazoe, A., Collier, N.: Global health monitor-a web-based system for detecting and mapping infectious diseases. In: Proceedings of the Third International Joint Conference on Natural Language Processing: Volume-II (2008)

14. Dórea, F.C., Revie, C.W.: Data-driven surveillance: effective collection, integration and interpretation of data to support decision-making. Front. Vet. Sci. **8**, 225 (2021)
15. Feng, X., Feng, X., Qin, B., Feng, Z., Liu, T.: Improving low resource named entity recognition using cross-lingual knowledge transfer. In: IJCAI, pp. 4071–4077 (2018)
16. Fu, J., Liu, P., Neubig, G.: Interpretable multi-dataset evaluation for named entity recognition. In: Proceedings of the 2020 Conference on Empirical Methods in Natural Language Processing (EMNLP), pp. 6058–6069 (2020)
17. Fu, J., Liu, P., Zhang, Q.: Rethinking generalization of neural models: a named entity recognition case study. In: Proceedings of the AAAI Conference on Artificial Intelligence, vol. 34, pp. 7732–7739 (2020)
18. Glaser, I., Sadegharmaki, S., Komboz, B., Matthes, F.: Data scarcity: Methods to improve the quality of text classification. In: ICPRAM, pp. 556–564 (2021)
19. Grancharova, M., Berg, H., Dalianis, H.: Improving named entity recognition and classification in class imbalanced Swedish electronic patient records through resampling. In: Eighth Swedish Language Technology Conference (SLTC). Förlag Göteborgs Universitet (2020)
20. Gururangan, S., et al.: Don't stop pretraining: adapt language models to domains and tasks. arXiv preprint arXiv:2004.10964 (2020)
21. Hamborg, F., Lachnit, S., Schubotz, M., Hepp, T., Gipp, B.: Giveme5W: main event retrieval from news articles by extraction of the five journalistic W questions. In: Chowdhury, G., McLeod, J., Gillet, V., Willett, P. (eds.) iConference 2018. LNCS, vol. 10766, pp. 356–366. Springer, Cham (2018). https://doi.org/10.1007/978-3-319-78105-1_39
22. Joshi, A., Karimi, S., Sparks, R., Paris, C., Macintyre, C.R.: Survey of text-based epidemic intelligence: a computational linguistics perspective. ACM Comput. Surv. (CSUR) **52**(6), 1–19 (2019)
23. Joshi, M., Chen, D., Liu, Y., Weld, D.S., Zettlemoyer, L., Levy, O.: SpanBERT: improving pre-training by representing and predicting spans. Trans. Assoc. Comput. Linguist. **8**, 64–77 (2020)
24. Kozareva, Z., Bonev, B., Montoyo, A.: Self-training and co-training applied to Spanish named entity recognition. In: Gelbukh, A., de Albornoz, Á., Terashima-Marín, H. (eds.) MICAI 2005. LNCS (LNAI), vol. 3789, pp. 770–779. Springer, Heidelberg (2005). https://doi.org/10.1007/11579427_78
25. Lampos, V., Zou, B., Cox, I.J.: Enhancing feature selection using word embeddings: the case of flu surveillance. In: Proceedings of the 26th International Conference on World Wide Web, pp. 695–704 (2017)
26. Lejeune, G., Brixtel, R., Doucet, A., Lucas, N.: Multilingual event extraction for epidemic detection. Artif. Intell. Med. **65** (2015). https://doi.org/10.1016/j.artmed.2015.06.005
27. Lejeune, G., Brixtel, R., Lecluze, C., Doucet, A., Lucas, N.: Added-value of automatic multilingual text analysis for epidemic surveillance. In: Peek, N., Marín Morales, R., Peleg, M. (eds.) AIME 2013. LNCS (LNAI), vol. 7885, pp. 284–294. Springer, Heidelberg (2013). https://doi.org/10.1007/978-3-642-38326-7_40
28. Lejeune, G., Doucet, A., Yangarber, R., Lucas, N.: Filtering news for epidemic surveillance: towards processing more languages with fewer resources. In: Proceedings of the 4th Workshop on Cross Lingual Information Access, pp. 3–10 (2010)
29. Makhoul, J., Kubala, F., Schwartz, R., Weischedel, R., et al.: Performance measures for information extraction. In: Proceedings of DARPA Broadcast News Workshop, Herndon, VA, pp. 249–252 (1999)

30. Mutuvi, S., Boros, E., Doucet, A., Lejeune, G., Jatowt, A., Odeo, M.: Multilingual epidemiological text classification: a comparative study. In: COLING, International Conference on Computational Linguistics (2020)
31. Mutuvi, S., Boros, E., Doucet, A., Lejeune, G., Jatowt, A., Odeo, M.: Token-level multilingual epidemic dataset for event extraction. In: Berget, G., Hall, M.M., Brenn, D., Kumpulainen, S. (eds.) TPDL 2021. LNCS, vol. 12866, pp. 55–59. Springer, Cham (2021). https://doi.org/10.1007/978-3-030-86324-1_6
32. Mutuvi, S., Doucet, A., Odeo, M., Jatowt, A.: Evaluating the impact of OCR errors on topic modeling. In: Dobreva, M., Hinze, A., Žumer, M. (eds.) ICADL 2018. LNCS, vol. 11279, pp. 3–14. Springer, Cham (2018). https://doi.org/10.1007/978-3-030-04257-8_1
33. Neubig, G., et al.: compare-MT: a tool for holistic comparison of language generation systems. arXiv preprint arXiv:1903.07926 (2019)
34. Neudecker, C., Antonacopoulos, A.: Making Europe's historical newspapers searchable. In: 2016 12th IAPR Workshop on Document Analysis Systems (DAS), pp. 405–410. IEEE (2016)
35. Ng, V., Rees, E.E., Niu, J., Zaghool, A., Ghiasbeglou, H., Verster, A.: Application of natural language processing algorithms for extracting information from news articles in event-based surveillance. Can. Commun. Dis. Rep. **46**(6), 186–191 (2020)
36. Nguyen, N.K., Boros, E., Lejeune, G., Doucet, A.: Impact analysis of document digitization on event extraction. In: 4th Workshop on Natural Language for Artificial Intelligence (NL4AI 2020) co-located with the 19th International Conference of the Italian Association for Artificial Intelligence (AI* IA 2020), vol. 2735, pp. 17–28 (2020)
37. Pan, X., Zhang, B., May, J., Nothman, J., Knight, K., Ji, H.: Cross-lingual name tagging and linking for 282 languages. In: Proceedings of the 55th Annual Meeting of the Association for Computational Linguistics (Volume 1: Long Papers), pp. 1946–1958 (2017)
38. Radford, A., Narasimhan, K., Salimans, T., Sutskever, I.: Improving language understanding by generative pre-training. Arxiv (2018)
39. Riedl, M., Padó, S.: A named entity recognition shootout for German. In: Proceedings of the 56th Annual Meeting of the Association for Computational Linguistics (Volume 2: Short Papers), pp. 120–125 (2018)
40. Salathé, M., Freifeld, C.C., Mekaru, S.R., Tomasulo, A.F., Brownstein, J.S.: Influenza a (H7N9) and the importance of digital epidemiology. N. Engl. J. Med. **369**(5), 401 (2013)
41. Tarvainen, A., Valpola, H.: Mean teachers are better role models: weight-averaged consistency targets improve semi-supervised deep learning results. arXiv preprint arXiv:1703.01780 (2017)
42. Van Asch, V., Daelemans, W.: Predicting the effectiveness of self-training: application to sentiment classification. arXiv preprint arXiv:1601.03288 (2016)
43. van Engelen, J.E., Hoos, H.H.: A survey on semi-supervised learning. Mach. Learn. **109**(2), 373–440 (2019). https://doi.org/10.1007/s10994-019-05855-6
44. Vaswani, A., et al.: Attention is all you need. In: Advances in Neural Information Processing Systems, pp. 5998–6008 (2017)
45. Walker, D., Lund, W.B., Ringger, E.: Evaluating models of latent document semantics in the presence of OCR errors. In: Proceedings of the 2010 Conference on Empirical Methods in Natural Language Processing, pp. 240–250 (2010)
46. Wang, A., Singh, A., Michael, J., Hill, F., Levy, O., Bowman, S.R.: GLUE: a multitask benchmark and analysis platform for natural language understanding. arXiv preprint arXiv:1804.07461 (2018)

47. Wang, C.K., Singh, O., Tang, Z.L., Dai, H.J.: Using a recurrent neural network model for classification of tweets conveyed influenza-related information. In: Proceedings of the International Workshop on Digital Disease Detection Using Social Media 2017 (DDDSM-2017), pp. 33–38 (2017)
48. Wang, W., Huang, Z., Harper, M.: Semi-supervised learning for part-of-speech tagging of mandarin transcribed speech. In: 2007 IEEE International Conference on Acoustics, Speech and Signal Processing-ICASSP 2007, vol. 4, pp. IV-137. IEEE (2007)
49. Yarowsky, D.: Unsupervised word sense disambiguation rivaling supervised methods. In: 33rd Annual Meeting of the Association for Computational Linguistics, pp. 189–196 (1995)
50. Zhu, X.J.: Semi-supervised learning literature survey (2005)

Supervised Learning of Keyphrase Extraction Utilizing Prior Summarization

Tingyi Liu and Mizuho Iwaihara(✉)

Graduate School of Information, Production and Systems, Waseda University, Kitakyushu 808-0135, Japan
tyliu@akane.waseda.jp, iwaihara@waseda.jp

Abstract. Keyphrase extraction is the task of selecting a set of phrases that can best represent a given document. Keyphrase extraction is utilized in document indexing and categorization, thus being one of core technologies of digital libraries. Supervised keyphrase extraction based on pretrained language models are advantageous thorough their contextualized text representations. In this paper, we show an adaptation of the pertained language model BERT to keyphrase extraction, called BERT Keyphrase-Rank (BK-Rank), based on a cross-encoder architecture. However, the accuracy of BK-Rank alone is suffering when documents contain a large amount of candidate phrases, especially in long documents. Based on the notion that keyphrases are more likely to occur in representative sentences of the document, we propose a new approach called Keyphrase-Focused BERT Summarization (KFBS), which extracts important sentences as a summary, from which BK-Rank can more easily find keyphrases. Training of KFBS is by distant supervision such that sentences lexically similar to the keyphrase set are chosen as positive samples. Our experimental results show that the combination of KFBS + BK-Rank show superior performance over the compared baseline methods on well-known four benchmark collections, especially on long documents.

Keywords: Keyphrase extraction · Supervised learning · Pretrained language model · Extractive summarization · Document indexing

1 Introduction

Keyphrase extraction is a natural language processing task of automatically selecting a set of representative and characteristic phrases that can best describe a given document. Due to its clarity and practical importance, keyphrase extraction has been a core technology for information retrieval and document classification [1]. For large text collections, keyphrases provide faster and more accurate searches and can be used as concise summaries of documents [2, 18].

For keyphrase extraction, unsupervised methods have played an important role, because of corpus independence and search efficiency. However, compared with supervised methods, unsupervised methods only use statistical information from the target document and the document set. The performance of unsupervised is limited due to the

H.-R. Ke et al. (Eds.): ICADL 2021, LNCS 13133, pp. 157–166, 2021.
https://doi.org/10.1007/978-3-030-91669-5_13

lack of information on the contexts surrounding candidate phrases. Supervised methods can learn contextual information on where keyphrases are likely to occur, but they require training datasets.

In this paper, we discuss supervised keyphrase extraction based on finetuning pretrained language model BERT [6]. Our proposed method consist of two parts. First, Keyphrase-Focused BERT Summarization (KFBS) is applied for prior-summarization, which extracts important sentences that are likely to contain keyphrases. We utilize distant supervision for training of KFBS, such that sentences that contain words lexically similar to reference keyphrases are used as golden summaries for training.

After prior-summarization, part-of-speech (POS) tagging is applied to extract candidate noun phrases. BERT Keyphrase-Rank (BK-Rank) has a cross-encoder architecture which attends over the pair of the extracted summary sentences and a candidate phrase, and scores the candidate phrase. Top-ranked phrases are chosen as keyphrases. Our rigorous experimental evaluations show that our proposed method of KFBS+BK-Rank outperforms the baseline methods in terms of F1@K, by a large margin. The results also show that prior-summarization by KFBS improves the results of BK-Rank alone, especially on long documents.

2 Related Work

KP-Miner [7] is a keyphrase extraction system that considers various types of statistical information beyond the classical method TF-IDF [18]. YAKE [4] considers both statistical and contextual information, and adopts features such as the position and frequency of a term, and the spread of the terms within the document.

TextRank [14], borrowing the idea of PageRank [3], uses part-of-speech (POS) tags to obtain candidates, creates an undirected and unweighted graph in which the candidates are added as nodes and an edge is added between nodes that co-occur within a window of N words. Then the PageRank algorithm is applied. SingleRank [22] is an extension of TextRank which introduces weights on edges by the number of co-occurrences.

Embedding-based methods train low-dimensional distributed representations of phrases and documents for evaluating importance of phrases. EmbedRank [2] extracts candidate phrases from a given document based on POS tags. Then EmbedRank uses two different sentence embedding methods (Sent2vec [17] and Doc2vec [11]) to represent the candidate phrases and the document in the same low-dimensional vector space. Then the candidate phrases are ranked using the normalized cosine similarity between the embeddings of the candidate phrases and the document embedding. SIFRank [20] combines sentence embedding model SIF [1] which is used to explain the relationship between sentence embeddings and the topic of the document, and autoregressive pretrained language model ELMo [19] is used to compute phrase and document embeddings, and achieves the state-of-the-art performance in keyphrase extraction for short documents. For long documents, SIFRank is extended to SIFRank+ [20] by introducing position-biased weighting.

3 Methodology

3.1 Motivations

This section discusses motivations and backgrounds that lead us for designing a new keyphrase extraction method.

Context. Context information is vital in determining whether a phrase is a keyphrase. Local contexts often give clues on whether an important concept is stated or not. Also, phrases that are co-occurring with the main topic of the document can be regarded as representative. EmbedRank [2] utilizes context information through document embeddings, and SIFRank [20] adopts the pretrained language model Elmo [19] for context-aware embedding. Both EmbedRank and SIFRank are unsupervised method. On the contrary, BERT [6] captures deep context information through the multi-head self-attention mechanism. We design a BERT Keyphrase-Ranker, called BK-Rank, where keyphrase extraction is formulated as a phrase ranking problem.

Keyphrase Density. The number of keyphrases annotated by human annotators for a document is around 10–15 in average, as shown in the benchmark document collections in Table 1, which include both short documents, such as abstracts and news articles, and long documents such as scientific papers. This means that the density of keyphrases in long documents is relatively lower than in short documents. Also, long documents contain more diverse phrases that are apart from the main topic of the document. As a consequence, long documents are more difficult in finding keyphrases than short documents.

Considering the above analysis, we propose a new approach that integrates document summarization and keyphrase extraction. Extractive summarization [15] is a task to select sentences from a given target document such that the summary well represents the target document. We adopt the following assumption: Keyphrases are more likely to occur in representative sentences. We remove non-representative sentences from the document before keyphrase extraction, as *prior-summarization*. Our approach has the following expected effects:

1. Prior-summarization can reduce phrases that are remotely related to the topic of the document, while the summary retains local contexts of keyphrases that are utilized for final keyphrase extraction.
2. In a summary, keyphrases are more densely occurring than the original document, so that relations between phrases are more easily captured by the attention mechanism of BK-Rank.
3. Prior-summarization will be especially effective for long documents.

We propose a supervised keyphrase extraction method, based on finetuning pre-trained language models for both prior-summarization and final keyphrase extraction. Our proposed method of KFBS+BK-Rank, illustrated in Fig. 1, consists of the following steps:

1. For a given document, prior-summarization is performed by KFBS, which is trained to extract important sentences that are lexically similar to the list of golden keyphrases, so that the selected important sentences are more likely to contain keyphrases.
2. Candidate phrases are extracted which are noun phrases based on POS tagging from prior-summarization.
3. BK-Rank is finetuned by binary cross-entropy loss on keyphrases and non-keyphrases, and used to score candidate phrases occurring in important sentences selected by KFBS.
4. The top-N phrases ranked by BK-Rank are selected as the keyphrases.

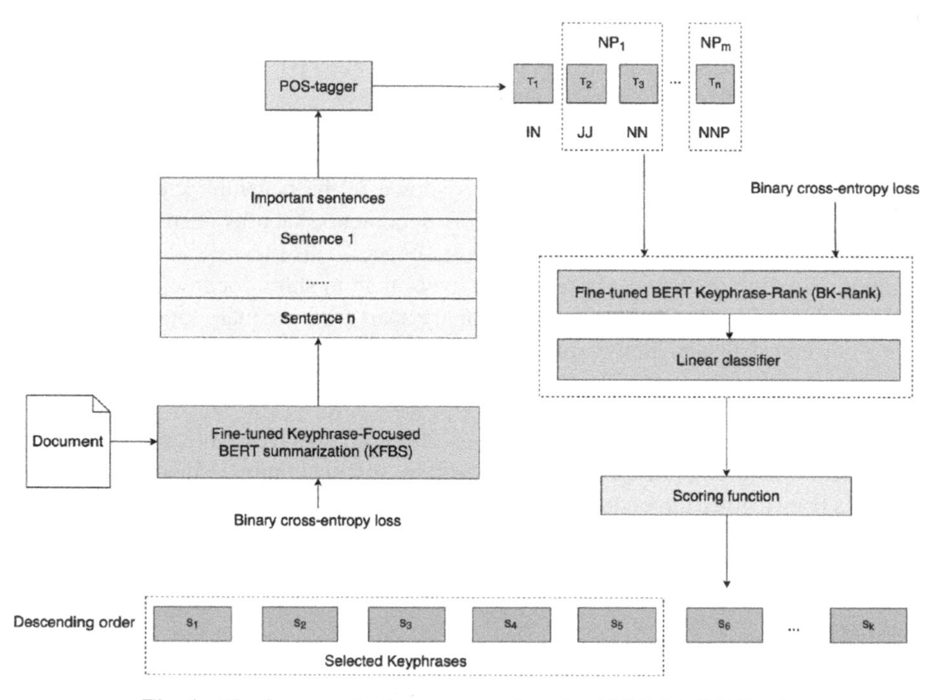

Fig. 1. The framework of our proposed method KFBS + BK-Rank.

3.2 Candidate Phrase Selection

In this stage, we apply **Keyphrase-Focused BERT Summarization (KFBS)** to select important sentences from a document that can represent the document and are more likely to contain keyphrases as the concise summary of this document.

BERTSUM [12] is an extractive summarization method, which changes the input of BERT by adding a CLS-token and a SEP-token at the start and end of each sentence respectively. The output vector at each CLS-token is used as a sentence embedding and entered to the succeeding linear layer, and a fixed number of highly scored sentences are selected as the output summary. When the document exceeds the length limit of 512 tokens of BERT, the leading part of the document is used. In case the given document is already short, prior-summarization is skipped.

To train an extractive summarization model, we need reference summaries. However, since our target task is keyphrase extraction, only reference keyphrases are available as training samples. Therefore, we take the approach of distant supervision such that sentences that contain words or subwords of the reference keyphrases are regarded as quality sentences, and used as positive samples for training the extractive summarization model.

To evaluate overlapping words and subwords between sentences and keyphrases, we utilize the ROUGE-N score, which quantifies the overlap of N-grams. We score the sentences of the target document by the sum of ROUGE-1 + ROUGE-2, and choose the top-ranked sentences as important sentences for training. Binary cross-entropy loss is used for the model to learn the important sentences.

For short documents of length within 200 tokens, KFBS avoids extraction and returns the input document as the final output.

Part-of-speech (POS) Tagging. Keyphrases chosen by humans are often noun phrases that consist of zero or more adjectives followed by one or more nouns (e.g., communication system, supervised learning, word embedding). Thus we utilize part-of-speech (POS) tagging to extract candidate noun phrases as candidate phrases from the prior-summarization performed by KFBS, which are not allowed to end with adjectives, verbs, or adverbs, etc.

3.3 BERT Keyphrase-Rank (BK-Rank)

For final selection of keyphrases from candidate phrases, we construct a BERT model with two inputs: the prior-summarization text and a candidate phrase. We utilize a cross-encoder [9] which computes self-attention between the prior-summarization text and the candidate phrase, to capture relationship between these two parts. Figure 2 shows the configuration of BERT Keyphrase-Rank (BK-Rank). For keyphrase scoring, the classification outcome is whether or not a candidate phrase is a golden keyphrase. So we adopt binary cross-entropy loss for finetuning BK-Rank with a classifier which generates a scalar between 0 and 1. We note that the training documents as well as the target documents receive prior-summarization by KFSB, which needs to be trained before BK-Rank.

4 Experiments

In this section, we report our experimental evaluations of our proposed models, compared with baseline methods, on four commonly used datasets. F1@K is used for evaluating results.

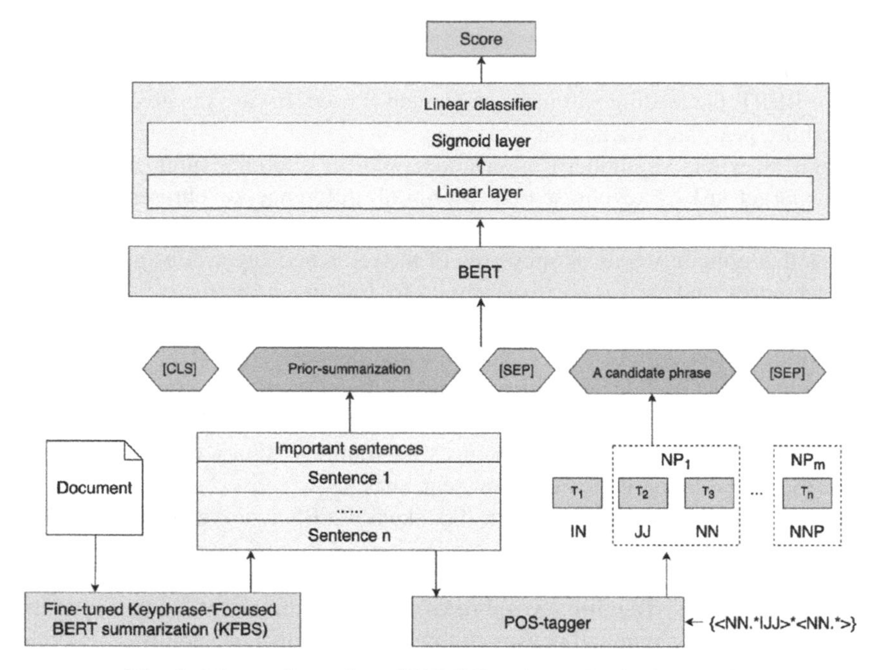

Fig. 2. The configuration of BERT Keyphrase-Rank (BK-Rank).

Table 1. Statistics of four datasets.

Dataset	Documents		Keyphrases			
	Number (test set)	Average tokens	Total	Average	Missing in doc	Missing in candidates
Inspec	500	134.28	4913	9.83	14.46%	38.94%
DUC 2001	123	800.63	1010	8.21	4.11%	10.38%
SemEval 2010	100	7662.42	1467	14.67	15.11%	16.36%
NUS	100	8765.93	1106	11.06	5.68%	11.86%

4.1 Datasets

Table 1 shows the statistics of the four benchmark datasets.

- **Inspec** [8] consists of 2,000 short documents from scientific journal abstracts in English. The training set, validation set, and test set contain 1,000, 500, and 500 documents, respectively.
- **DUC 2001** [22] consists of 308 newspaper articles which are collected from TREC-9, where the documents are organized into 30 topics. The golden keyphrases we used are annotated by X. Wan and J. Xiao. Here we use 145 for training and 123 for test.

- **SemEval 2010** [10] consists of 284 long documents which are scientific papers, 144 documents for training, 100 documents for test and 40 for validation.
- **NUS** [16] consists of 211 long documents which are full scientific conference papers of between 4–12 pages. Here we use 111 for training and 100 for test.

Table 2. Comparison of our method and baseline methods, by F1@K (%).

Method	Inspec		DUC 2001		SemEval 2010		NUS	
	F1@5	F1@10	F1@5	F1@10	F1@5	F1@10	F1@5	F1@10
Baseline method								
TextRank	14.72	15.28	15.17	15.24	3.23	6.55	3.21	6.56
EmbedRank doc2vec	31.51	37.94	24.02	28.12	2.28	3.53	2.35	3.58
EmbedRank sent2vec	29.88	37.09	27.16	31.85	3.31	5.33	3.39	5.42
SIFRank	29.11	38.80	24.27	27.43	3.05	5.43	–	–
SIFRank +	28.49	36.77	30.88	33.37	10.47	12.40	–	–
Proposed method								
BK-Rank	41.99	46.53	42.08	46.88	9.49	13.54	11.46	16.55
KFBS (Top-3) + BK-Rank	38.89	44.22	40.44	44.11	11.12	13.30	**17.60**	**17.70**
KFBS (Top-4) + BK-Rank	38.15	44.31	41.17	45.53	**11.18**	**15.59**	17.41	15.96
KFBS (Top-5) + BK-Rank	**42.01**	**46.62**	**42.16**	**46.93**	10.05	13.46	17.24	16.23

4.2 Baseline Methods

We compare our proposed method with the following baseline methods: **TextRank** [14], **EmbedRank** [2], and **SIFRank/SIFRank+** [20]. SIFRank is an unsupervised method which combines sentence embedding model SIF [1] and pretrained language model ELMo [19] to generate embeddings. For long documents, SIFRank is upgraded to SIFRank + by position-biased weight.

4.3 Experimental Details

In the experiments, we use StandfordCoreNLP [21] to generate POS tags and use AdamW [13] as the optimizer. For training KFBS, which is used to select important sentences, we finetune the model with learning rate in {5e−5, 3e−5, 2e−5, 1e−5}, dropout rate 0.1, batch size 256, and warm-up 5% of the training steps. We finetune

BERT Keyphrase-Rank (BK-Rank) with a batch size of 32, learning rate in {5e−5, 3e−5, 2e−5, 1e−5}, weight decay 0.01, and warm-up 10% of the training data. Then we save the models which achieve the best performances. For the pretrained language models, we use bert-base-uncased model for both BK-Rank and KFBS.

4.4 Performance Comparison

For evaluation, we use the common metrics of F1-score (F1). Table 2 shows the results. KFBS (Top-k) means top-k sentences selected by KFBS are used as important sentences, on which KB-Rank is applied. Due to hardware limitations, SIFRank and SIFRank+ are not obtained on NUS, so we do not report their results.

As shown in Table 2, the performance of KFBS + BK-Rank shows the best results on all the four datasets, both on short documents and long documents, achieving superior performance over the compared baseline methods. When we select top-5 sentences by KFBS, KFBS + BK-Rank achieves the best results on Inspec and DUC 2001 for F1@5 and F1@10. KFBS (Top-4) + BK-Rank achieves the best results on F1@5 and F1@10 on SemEval 2010. On NUS, KFBS (Top-3) + BK-Rank achieves the best results on F1@5 and F1@10.

Prior-summarization by KFBS is improving the results of BK-Rank by 0.02 to 6.17 points. The results show that selecting important sentences before candidate phrase selection by BK-Rank is effective, especially on long document collections of SemEval 2010 and NUS. Prior-summarization by KFBS is effectively removing sentences that are unlikely to contain keyphrases, which also benefits finetuning of BK-Rank. We notice that on Inspec and DUC 2001, KFBS (Top-k) with $k = 5$ is better than $k = 3$ or 4, while on SemEval 2010 and NUS, $k = 5$ is falling behind of $k = 3$ and 4. This can be explained by keyphrase density such that for short documents, keyphrases are relatively evenly occurring in sentences, while for long documents, more selective summarization is advantageous.

5 Conclusion

In this paper, we proposed a supervised method for keyphrase extraction from documents, by combining BERT Keyphrase-Rank (BK-Rank) and Keyphrase-Focused BERT Summarization (KFBS). We introduce KFBS to select important sentences from which candidate phrases are extracted and also used for finetuning BK-Rank. BK-Rank fully exploits contextual text embeddings by the cross-encoder reading a target document and candidate phrase. KFBS is trained by distant supervision to extract important sentences that are likely to contain keyphrases. Our experimental results show that our proposed method has superior performance on this task over the compared baseline methods. Diversity on keyphrases is necessary to avoid the situation that similar keyphrases occupy the result. BK-Rank can be extended to incorporate Maximal Marginal Relevance (MMR) [5] for enhancing diversity.

References

1. Arora, S., Liang, Y., Ma, T.: A simple but tough-to-beat baseline for sentence embeddings. In: ICLR (2016)
2. Bennani-Smires, K., Musat, C.C., Hossmann, A., et al.: Simple unsupervised keyphrase extraction using sentence embeddings. In: Conference on Computational Natural Language Learning (2018)
3. Brin, S., Page, L.: The anatomy of a large-scale hypertextual web search engine. Comput. Netw. ISDN Syst. **30**(1–7), 107–117 (1998)
4. Campos, R., Mangaravite, V., Pasquali, A., Jorge, A.M., Nunes, C., Jatowt, A.: YAKE! Collection-independent automatic keyword extractor. In: Pasi, G., Piwowarski, B., Azzopardi, L., Hanbury, A. (eds.) ECIR 2018. LNCS, vol. 10772, pp. 806–810. Springer, Cham (2018). https://doi.org/10.1007/978-3-319-76941-7_80
5. Carbonell, J., Goldstein, J.: The use of MMR, diversity-based reranking for reordering documents and producing summaries. In: Proceedings of 21st Annual International ACM SIGIR Conference on on Research and Development in Information Retrieval, pp. 335–336 (1998)
6. Devlin, J., Chang, M.W., Lee, K., Toutanova, K.: BERT: pre-training of deep bidirectional transformers for language understanding. In: Proceedings of NAACL-HLT 2019, pp. 4171–4186 (2019)
7. El-Beltagy, S.R., Rafea, A.: KP-miner: participation in SemEval-2. In: Proceedings of. 5th Int. Workshop on Semantic Evaluation, pp. 190–193 (2010)
8. Hulth, A.: Improved automatic keyword extraction given more linguistic knowledge. In: Proceedings of 2003 Conference on Empirical Methods in Natural Language Processing (EMNLP), pp. 216–223 (2003)
9. Humeau, S., Shuster, K., Lachaux, M.A., et al.: Poly-encoders: architectures and pre-training strategies for fast and accurate multi-sentence scoring. In: International Conference on Learning Representations (2019)
10. Kim, S.N., Medelyan, O., Kan, M.Y., et al.: Semeval-2010 task 5: automatic keyphrase extraction from scientific articles. In: Proceedings of 5th International Workshop on Semantic Evaluation, pp. 21–26 (2010)
11. Le, Q., Mikolov, T.: Distributed representations of sentences and documents. In: International Conference on Machine Learning, pp. 1188–1196 (2014)
12. Liu, Y.: Fine-tune BERT for Extractive Summarization. arXiv preprint arXiv:1903.10318 (2019)
13. Loshchilov, I., Hutter, F.: Decoupled weight decay regularization. In: International Conference on Learning Representations (2018)
14. Mihalcea, R., Tarau, P.: TextRank: bringing order into text. In: Proceedings of 2004 Conference on Empirical Methods in Natural Language Processing (EMNLP), pp. 404–411 (2004)
15. Moratanch, N., Chitrakala, S.: A survey on extractive text summarization. In: 2017 International Conference on Computer, Communication and Signal Processing (ICCCSP). IEEE (2017)
16. Nguyen, T.D., Kan, M.-Y.: Keyphrase extraction in scientific publications. In: Goh, D.-L., Cao, T.H., Sølvberg, I.T., Rasmussen, E. (eds.) ICADL 2007. LNCS, vol. 4822, pp. 317–326. Springer, Heidelberg (2007). https://doi.org/10.1007/978-3-540-77094-7_41
17. Pagliardini, M., Gupta, P., Jaggi, M.: Unsupervised learning of sentence embeddings using compositional n-gram features. In: Proceedings of NAACL-HLT, pp. 528–540 (2018)
18. Papagiannopoulou, E., Tsoumakas, G.: A review of keyphrase extraction. Wiley Interdisc. Rev. Data Min. Knowl. Discov. **10**(2) e1339 (2020)

19. Peters, M.E., et al.: Deep contextualized word representations. arXiv preprint arXiv:1802.05365 (2018)
20. Sun, Y., et al.: SIFRank: a new baseline for unsupervised keyphrase extraction based on pre-trained language model. IEEE Access **8**, 10896–10906 (2020)
21. Toutanova, K., et al.: Feature-rich part-of-speech tagging with a cyclic dependency network. In: Proceedings of 2003 Human Language Technology Conference of the North American Chapter of the Association for Computational Linguistics (2003)
22. Wan, X., Xiao, J.: Single document keyphrase extraction using neighborhood knowledge. In: AAAI Conference on Artificial Intelligence (AAAI-08), pp. 855–860 (2008)

Bert-Based Chinese Medical Keyphrase Extraction Model Enhanced with External Features

Liangping Ding[1,2], Zhixiong Zhang[1,2](✉), and Yang Zhao[1,2]

[1] National Science Library, Chinese Academy of Sciences, Beijing 100190, China
{dingliangping,zhangzhx,zhaoyang}@mail.las.ac.cn
[2] Department of Library Information and Archives Management, University of Chinese Academy of Sciences, Beijing 100049, China

Abstract. Keyphrase extraction is a key natural language processing task and has widespread adoption in many information retrieval and text mining applications. In this paper, we construct nine Bert-based Chinese medical keyphrase extraction models enhanced with external features and present a thorough empirical evaluation to explore the impacts of feature types and feature fusion methods. The results show that encoding part-of-speech (POS) feature and lexicon feature generated from descriptive keyphrase metadata into the word embedding space improves the baseline Bert-SoftMax model for 4.82%, meaning that it's beneficial to incorporate features into Chinese medical keyphrase extraction model. Furthermore, the results of the comparative evaluation experiments show that model performance is sensitive to both of feature types and feature fusion methods, so it's advisable to consider these two factors when dealing with feature enhanced tasks. Our study also provides a feasible approach to employ metadata, aiming to help stakeholders of digital libraries to take full advantage of large quantities of metadata resources to boost the development of scholarly knowledge discovery.

Keywords: Digital library · Keyphrase extraction · Feature fusion · Pretrained language model · Scholarly text mining · Metadata

1 Introduction

Keyphrase extraction is related to automatically extract a set of representative phrases from a document that concisely summarize its content [6]. It is a branch of information extraction and lays the foundation for many natural language processing tasks, such as information retrieval [9], text summarization [25], text

The work is supported by the project "Artificial Intelligence (AI) Engine Construction Based on Scientific Literature Knowledge" (Grant No. E0290906) and the project "Key Technology Optimization Integration and System Development of Next Generation Open Knowledge Service Platform" (Grant No. 2021XM45).

H.-R. Ke et al. (Eds.): ICADL 2021, LNCS 13133, pp. 167–176, 2021.
https://doi.org/10.1007/978-3-030-91669-5_14

classification [7], opinion mining [2], document indexing [21]. With the exponential growth of text, it's beneficial to develop a well-performing, robust keyphrase extraction system to help people capture the main points of the text and find relevant information quickly.

In recent years, deep neural networks have achieved significant success in many natural language processing tasks [1,19]. Most of these algorithms are trained end to end, and can automatically learn features from large-scale annotated datasets. However, these data-driven methods typically lack the capability of processing rare or unseen phrases. External features such as lexicon feature and POS feature may provide valuable information for model training [12,13]. While no consensus has been reached about whether external features should be or how to be integrated into deep learning based keyphrase extraction model.

The emergence of pretrained language model provides a new way to incorporate features into deep neural network. Recently, Devlin et al. [4] proposed a new language representation model called Bert whose input representation is comprised by summing token embedding, segment embedding and position embedding. This method proved that features such as position can be embedded into the word embedding space. Inspired by their work, we assumed that external features might also be encoded into the embedding layer, which is also a feasible feature fusion method without introducing additional parameters.

In this paper, our key contributions are summarized as follows:

1. We formulated Chinese medical keyphrase extraction task as a character-level sequence labeling task and constructed nine Bert-based Chinese medical keyphrase extraction models enhanced with external features using different combinations of feature types and feature fusion methods.
2. We conducted experiments to explore the impact of feature types and feature fusion methods, providing insights for feature incorporation researches.
3. We regarded keyphrase metadata field as lexicon resources and integrated lexicon feature into deep learning based scholarly text mining, providing a feasible way to make full use of digital libraries.

2 Related Work

There are various methods exist to extract keyphrases. Traditional keyphrase extraction algorithms usually contain two steps: extracting candidate keyphrases and determining which of these candidate keyphrases are correct. One of the drawbacks of this method is error propagation. To solve this problem, Zhang et al. [22] formulated keyphrase extraction task as a sequence labeling task and constructed a Conditional Random Field (CRF) model to extract keyphrases from Chinese text, providing a unified approach for keyphrase extraction. Sequence labeling formulation has become a prevalent approach to deal with keyphrase extraction task and has been proven to be effective in many works [5,17,24].

In recent years, deep learning based keyphrase extraction models has become dominant in many natural language processing tasks by automatically learning features. While external features still play a significant role in improving model's ability to process unseen cases. Zhang et al. [23] adopted Dynamic Graph Convolutional Network to solve the keyphrase extraction task and concatenated POS embedding and position embedding [20] with word embedding to construct the final word representation. For domain-specific tasks, lexicon features can improve the generalization transferring to the unseen samples [3,8]. Luo et al. [15] introduced various features including stoke feature, lexicon feature into the model for clinical named entity recognition task. While Li et al. [10] introduced lexicon feature and POS feature as additional features into the CRF model and BiLSTM-Att-CRF model and showed that these two neural network models had limited improvements after adding additional features, and the results were even worse than baseline models when added POS feature only.

With respect to the feature fusion methods, usually external features are concatenated to the word embedding or character embedding to incorporate feature information into the model. Lin et al. [12] concatenated character-level representation, word-level representation, and syntactical word representation (i.e. POS tags, dependency roles, word positions head positions) to form a comprehensive word representation. Li et al. [11] built a Chinese clinical named entity recognition model based on pretrained language model BERT [4]. They concatenated radical feature with the final hidden states of BERT to construct the contextual word representation and fed it into BiLSTM-CRF model to further encode the representation enhanced with external features. In addition, since the emergence of BERT, many pretrained language models based on BERT embedded features into the word embedding space without adding additional parameters [14,18]. Nevertheless, the feature types incorporated to the model are limited to position feature and segment feature, not much practices of other feature types to the best of our knowledge.

In this paper, we constructed Bert-based Chinese medical keyphrase extraction models enhanced with external features and chose POS feature and lexicon feature to explore the influence of feature type and explore the impacts of different feature fusion methods such as "concatenated fusion" method and "embedded fusion" method.

3 Methodology

In this section, the method of Bert-based Chinese medical keyphrase extraction model enhanced with external features is described. Figure 1 shows the whole processing flow of our method. The whole process can be split into three steps: text preprocessing, feature processing and model construction. The detailed description of our method is presented in the following sections.

3.1 Data Preprocessing

The data preprocessing process included text preprocessing and feature processing. In the text preprocessing procedure, some preprocessing steps including BIO

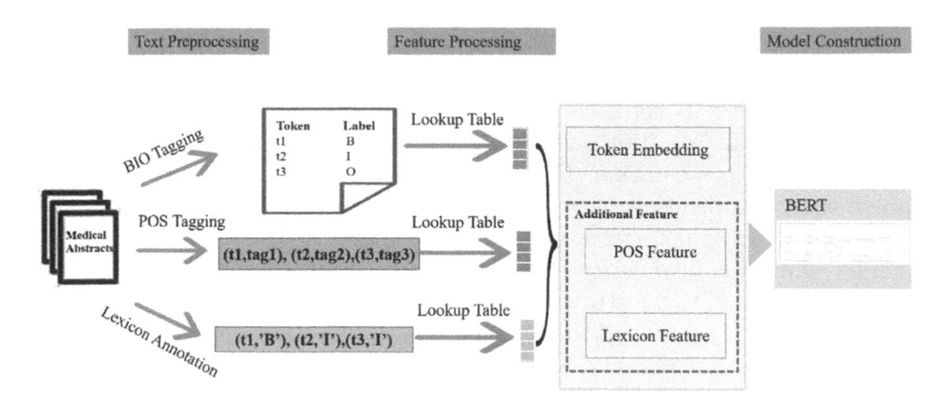

Fig. 1. The processing flow of our method

tagging, POS tagging and lexicon annotation were preformed. We cast keyphrase extraction of scientific medical abstracts as a character-level sequence labeling task. To deal with the phenomenon of mixed Chinese characters and English words, each Chinese character or English word were treated as the basic element of the model input.

The dataset was annotated with BIO tagging scheme [16]. To incorporate POS feature, character-level POS tags are generated by Hanlp[1] segmentation system. By looking up the mapping of the predefined set of POS tags, the POS feature of the input abstract was represented as a sequence of vectors X = (X_1, X_2, $X3$, ..., X_m) ,where m denoted the length of the model input. For the lexicon feature construction, we built our lexicon based on medical keyphrases from Chinese Science Citation Database (CSCD), dictionaries from Sougou, Baidu Baike and some medical knowledge graphs. After removing duplicates and irrelevant words, we constructed a lexicon containing 704,507 medical terms in total. The lexicon feature was encoded in BIO tagging scheme, indicating the position of the token in the matched entry. Finally, a lookup table was used to generate the lexicon embedding and the mapping approach was the same with that of POS feature.

3.2 Model Architecture

According to two types of feature fusion methods, we designed "feature concatenated model" and "feature embedded model" correspondingly. In this section, we describe the model architecture of these two models in details, which have similar neural network architecture, instead using different feature fusion methods. The architectures of the "feature concatenated model" and "feature embedded model" are shown in Fig. 2. In this paper, the pretrained language model Bert was used as the backbone of our model, and a feed-forward neural network was added on top of it and SoftMax classification was performed.

[1] https://github.com/hankcs/HanLP.

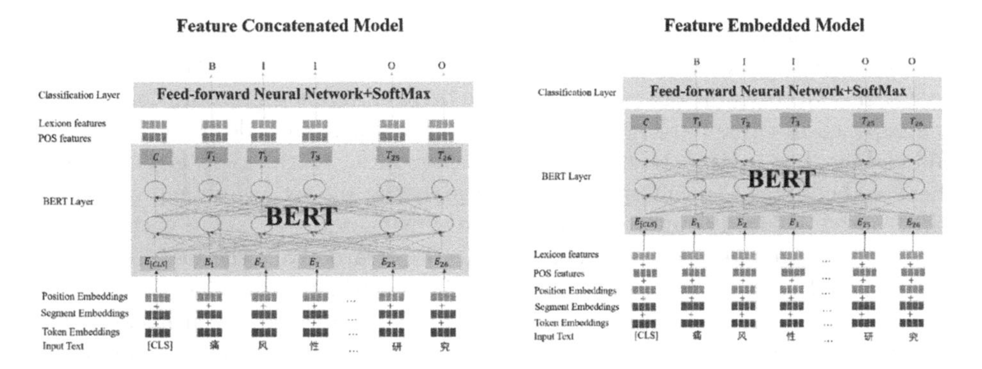

Fig. 2. The architecture of the feature embedded and feature concatenated model

Feature Concatenated Model

Traditionally, external features are concatenated to the word embeddings or character embeddings, acting as additional information to the model training. Pretrained language model Bert is the language representation in essence and it can be concatenated with the feature embeddings. For the feature concatenated model, the extracted feature was represented by a vector and was concatenated with the Bert embedding of each token. And then the concatenated vector for each token was passed on to the feed-forward neural network and followed by a SoftMax classification to output the probability for each category.

Given an abstract $X = \{x_1, x_2, ..., x_m\}$, which is a sequence of m tokens, the first step to concatenate the features is to map them to distributed vectors. Formally, we lookup the POS or lexicon symbols from feature embedding matrix for each token x_i as f_i and $f_i \in R^{d_{feature}}$, where $i \in \{1, 2, ..., m\}$ indicates x_i is the i-th token in X, and $d_{feature}$ is a hyper-parameter indicating the size of feature vector. Denote the Bert embedding vector for each token as e_i and $e_i \in R^{d_{Bert}}$, where d_{Bert} is the size of the last hidden state of Bert. The final contextual embedding of each token fed into the feed-forward neural network is the concatenation of feature vector and Bert embedding with the dimension as $d_{final} = d_{feature} + d_{Bert}$. And the final embedding representation of each token is as follows:

$$B_i = [e_i; f_i] \tag{1}$$

According to the dimension of the final embedding vectors, we constructed the layer of feed-forward neural network with the number of weights as $d_p = d_{final} \times l$, where l indicates the number of categories. So the final embedding representation can be decoded by the feed-forward neural network into probabilities for each label category. Through the feed-forward layer, the decision probability of p_i mapping to the annotated result for each token is expressed as:

$$p_i = SoftMax(W_p B_i) \tag{2}$$

where the weight W_p has the number of parameters as d_p and needs to be learned in the process of training.

Feature Embedded Model
For "feature embedded model", we encoded the feature vector in the embedding layer before feeding into Bert encoder. The original input representation of Bert is composed of three parts: token embedding, segment embedding and position embedding. For a given token, the input embedding to the Bert is the sum of these embeddings. To incorporate features to the model training process, we constructed a 768 dimension feature vectors using the way same with that of "feature concatenated model". As for the feed-forward layer, the feature vectors are embeded to word embedding space without adding additional parameters. So the final embedding of each token fed into feed-forward neural network has the dimension as $d_{Bert}(768)$ instead of $d_{feature} + d_{Bert}$ for "feature concatenated model". The feed-forward neural network has the number of weights as $d_{Bert} \times l$ and the decision probability of p_i for each token is expressed as:

$$p_i = SoftMax(W_p e_i) \tag{3}$$

where W_p is the weight matrix of feed-forward layer and e_i is the final contextual embedding of the token i after Bert encoder.

4 Experiments and Results

4.1 Experimental Design

We used the original data of the publicly available Chinese keyphrase recognition dataset CAKE [5] for the experiments, which is a dataset containing Chinese medical abstracts from Chinese Science Citation Database (CSCD) annotated with BIO tagging scheme [16]. 100,000 abstracts are included in the training set and 3,094 abstracts are included in the test set.

To explore the impacts of different feature fusion methods to the model performance, nine Bert-based keyphrase extraction models were constructed by focusing on two dimensions: (1) feature types and (2) feature fusion methods. And we used Bert-SoftMax model without feature fusion as the model baseline. We used controlled variable method to design two groups of comparative experiments with respect to these two dimensions. We considered two feature types including POS feature and lexicon feature, and the combination of POS feature and lexicon feature was also taken into account. As for the feature fusion methods, we evaluated three feature fusion methods including "embedded fusion", "concatenated fusion" and "embedded and concatenated fusion"[2].

Our neural networks were implemented on Python 3.7 and transformers[3] library, and the POS tagging were preformed by Hanlp. The parameter configurations of our proposed approach are shown in Table 1. In addition, we used

[2] "Embedded and concatenated fusion" is the combination of "embedded fusion" and "concatenated fusion", whose model architecture is also the combination of feature related components of "feature embedded model" and "feature concatenated model".
[3] https://github.com/huggingface/transformers.

CoNLL-2000 Evaluation Script[4] to calculate phrase-level metrics including precision (P), recall (R) and F1-score (F1) to evaluate model performance.

Table 1. Parameter configuration of the proposed approach

Parameters	Values
Batch size	32
Epoch	3
Optimizer	Adam
Learning rate scheduler	Exponential decay
Initial learning rate	1e−4
Max sequence length	512
POS embedding size	20
Lexicon embedding size	20

4.2 Results

In this section, we presented results from our comparative experiments regarding to feature types and feature fusion methods and the evaluation results are shown in Table 2. In particularly, we compared the models using F1-score, which is the weighted average of precision and recall, taking both of them into consideration. We illustrated the results focusing on model performance improvement, which is the improvement of model performance compared to Bert-SoftMax model, adding a feed-forward layer on top of Bert without feature fusion.

Table 2. Experimental results on test set

Feature type	Feature fusion method	P	R	F1	Improvement
Without feature	/	63.54%	66.23%	64.86%	/
POS	Embedded	64.54%	67.72%	**66.09%**	1.23%
	Concatenated	63.62%	65.94%	64.76%	−0.10%
	Embedded and concatenated	63.75%	67.37%	65.51%	0.65%
Lexicon	Embedded	68.18%	69.94%	69.05%	4.19%
	Concatenated	64.14%	67.10%	65.58%	0.72%
	Embedded and concatenated	68.10%	70.23%	**69.15%**	4.29%
POS and lexicon	Embedded	68.87%	70.51%	**69.68%**	4.82%
	Concatenated	63.33%	66.95%	65.09%	0.23%
	Embedded and concatenated	67.12%	70.92%	68.97%	4.11%

[4] https://www.clips.uantwerpen.be/conll2000/chunking/conlleval.txt.

Feature Type Comparative Experiments
In feature type comparative experiments, we analyzed POS feature, lexicon feature, POS and lexical feature together under each of the feature fusion method respectively to explore the contribution of the different feature types to the model performance. We found that incorporating POS feature into the Bert model regardless of the feature fusion methods achieved lower model performance improvement than incorporating lexicon feature. For the "embedded fusion" and "embedded and concatenated fusion" methods, the model performance improvements of POS feature fusion were 1.23% and 0.64% separately. While for lexicon feature fusion, the model performance improvements were significant, which were 4.19% and 4.28% for the above-mentioned two corresponding feature fusion methods. Furthermore, concatenating POS feature to the model achieved even worse result than Bert-SoftMax model without feature fusion.

Feature Fusion Method Comparative Experiments
In feature fusion method comparative experiments, we controlled feature type to explore the impacts of feature fusion method to the model performance. Three feature fusion methods including "embedded fusion", "concatenated fusion" and "embedded and concatenated fusion" were evaluated under each of the feature type. We found that no matter what kind of feature type was incorporated, "concatenated fusion" method achieved the worst model performance improvement among all three feature fusion methods.

To sum up, both of feature type and feature fusion method will influence the model performance. "Embedded fusion" method is a more effective feature fusion method than "concatenated fusion" method. While even using this feature fusion method to incorporate feature, if inappropriate feature type is chose, the model performance improvement is limited. Similarly, lexicon feature fusion is more important for improving the model performance than POS feature fusion. While even integrating lexicon feature into the model, if inappropriate feature fusion method is chosen, the improvement can also be limited.

5 Conclusions

In this paper, we incorporated external features including POS feature and lexicon feature into pretrained language model Bert to deal with Chinese medical keyphrase extraction task. We proposed "embedded fusion" method, which is a feasible feature fusion method to integrate external features by encoding features into the word embedding space. This method is easily to operate compared to "concatenated fusion" method without considering the feature embedding size. Through comparative experiments, we found that both of feature type and feature fusion method are important factors that need to be given thorough consideration. In addition, our study provides an empirical practice to take full advantage of metadata field in digital library, which is an important resource containing hidden knowledge to assist scholarly text mining.

References

1. Alzaidy, R., Caragea, C., Giles, C.L.: Bi-LSTM-CRF sequence labeling for keyphrase extraction from scholarly documents. In: The World Wide Web Conference, pp. 2551–2557 (2019)
2. Berend, G.: Opinion expression mining by exploiting keyphrase extraction (2011)
3. Cai, X., Dong, S., Hu, J.: A deep learning model incorporating part of speech and self-matching attention for named entity recognition of Chinese electronic medical records. BMC Med. Inform. Decis. Mak. **19**(2), 101–109 (2019)
4. Devlin, J., Chang, M.W., Lee, K., Toutanova, K.: BERT: pre-training of deep bidirectional transformers for language understanding. arXiv preprint arXiv:1810.04805 (2018)
5. Ding, L., Zhang, Z., Liu, H., Li, J., Yu, G.: Automatic keyphrase extraction from scientific Chinese medical abstracts based on character-level sequence labeling. J. Data Inf. Sci. **6**(3), 35–57 (2021). https://doi.org/10.2478/jdis-2021-0013
6. Hasan, K.S., Ng, V.: Automatic keyphrase extraction: a survey of the state of the art. In: Proceedings of the 52nd Annual Meeting of the Association for Computational Linguistics (Volume 1: Long Papers), pp. 1262–1273 (2014)
7. Hulth, A., Megyesi, B.: A study on automatically extracted keywords in text categorization. In: Proceedings of the 21st International Conference on Computational Linguistics and 44th Annual Meeting of the Association for Computational Linguistics, pp. 537–544 (2006)
8. Jie, Z., Lu, W.: Dependency-guided LSTM-CRF for named entity recognition. arXiv preprint arXiv:1909.10148 (2019)
9. Jones, S., Staveley, M.S.: Phrasier: a system for interactive document retrieval using keyphrases. In: Proceedings of the 22nd Annual International ACM SIGIR Conference on Research and Development in Information Retrieval, pp. 160–167 (1999)
10. Li, L., Zhao, J., Hou, L., Zhai, Y., Shi, J., Cui, F.: An attention-based deep learning model for clinical named entity recognition of Chinese electronic medical records. BMC Med. Inform. Decis. Mak. **19**(5), 1–11 (2019)
11. Li, X., Zhang, H., Zhou, X.H.: Chinese clinical named entity recognition with variant neural structures based on BERT methods. J. Biomed. Inf. **107**, 103422 (2020)
12. Lin, B.Y., Xu, F.F., Luo, Z., Zhu, K.: Multi-channel BiLSTM-CRF model for emerging named entity recognition in social media. In: Proceedings of the 3rd Workshop on Noisy User-Generated Text, pp. 160–165 (2017)
13. Liu, T., Yao, J.G., Lin, C.Y.: Towards improving neural named entity recognition with gazetteers. In: Proceedings of the 57th Annual Meeting of the Association for Computational Linguistics, pp. 5301–5307 (2019)
14. Liu, W., et al.: K-BERT: enabling language representation with knowledge graph. In: Proceedings of the AAAI Conference on Artificial Intelligence, vol. 34, pp. 2901–2908 (2020)
15. Luo, L., Li, N., Li, S., Yang, Z., Lin, H.: DUTIR at the CCKS-2018 task1: a neural network ensemble approach for Chinese clinical named entity recognition. In: CCKS Tasks, pp. 7–12 (2018)
16. Ramshaw, L.A., Marcus, M.P.: Text chunking using transformation-based learning. In: Armstrong, S., Church, K., Isabelle, P., Manzi, S., Tzoukermann, E., Yarowsky, D. (eds.) Natural Language Processing Using Very Large Corpora. Text, Speech and Language Technology, vol. 11, pp. 157–176. Springer, Dordrecht (1999). https://doi.org/10.1007/978-94-017-2390-9_10

17. Sahrawat, D., et al.: Keyphrase extraction from scholarly articles as sequence labeling using contextualized embeddings. arXiv preprint arXiv:1910.08840 (2019)
18. Sun, Y., et al.: ERNIE: enhanced representation through knowledge integration. arXiv preprint arXiv:1904.09223 (2019)
19. Tang, M., Gandhi, P., Kabir, M.A., Zou, C., Blakey, J., Luo, X.: Progress notes classification and keyword extraction using attention-based deep learning models with BERT. arXiv preprint arXiv:1910.05786 (2019)
20. Vaswani, A., et al.: Attention is all you need. arXiv preprint arXiv:1706.03762 (2017)
21. Wu, Y.F.B., Li, Q., Bot, R.S., Chen, X.: Domain-specific keyphrase extraction. In: Proceedings of the 14th ACM International Conference on Information and Knowledge Management, pp. 283–284 (2005)
22. Zhang, C.: Automatic keyword extraction from documents using conditional random fields. J. Comput. Inf. Syst. **4**(3), 1169–1180 (2008)
23. Zhang, H., Long, D., Xu, G., Xie, P., Huang, F., Wang, J.: Keyphrase extraction with dynamic graph convolutional networks and diversified inference. arXiv preprint arXiv:2010.12828 (2020)
24. Zhang, Q., Wang, Y., Gong, Y., Huang, X.J.: Keyphrase extraction using deep recurrent neural networks on Twitter. In: Proceedings of the 2016 Conference on Empirical Methods in Natural Language Processing, pp. 836–845 (2016)
25. Zhang, Y., Zincir-Heywood, N., Milios, E.: World wide web site summarization. Web Intell. Agent Syst.: Int. J. **2**(1), 39–53 (2004)

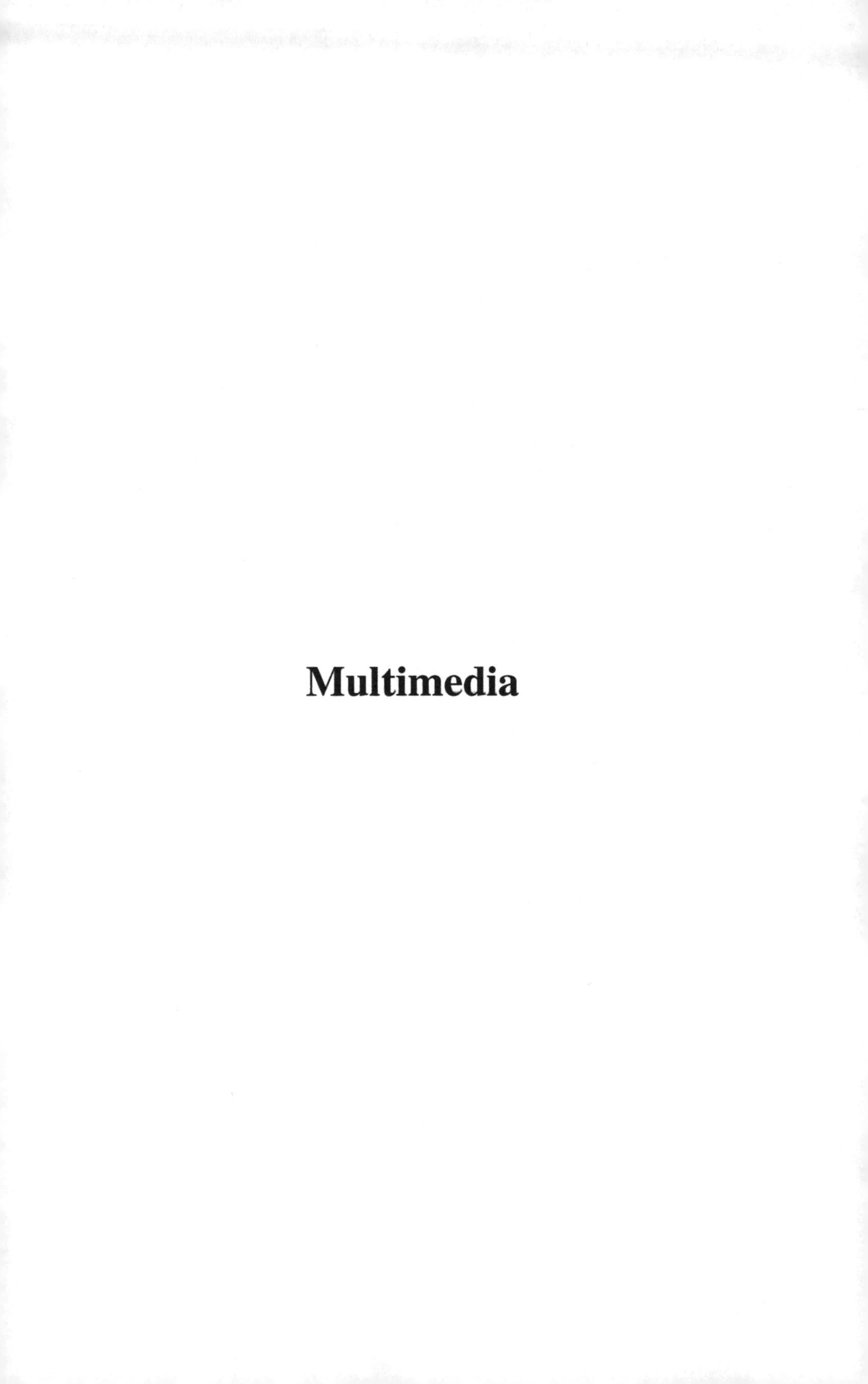

Multimedia

A Novel Approach to Analyze Fashion Digital Library from Humanities

Satoshi Takahashi[1(✉)], Keiko Yamaguchi[2], and Asuka Watanabe[3]

[1] Kanto Gakuin University, Mutsuura, Kanazawa, Yokohama, Kanagawa, Japan
satotaka@kanto-gakuin.ac.jp
[2] Nagoya University, Furocho, Chikusa, Nagoya, Aichi, Japan
[3] Kyoritsu Women's Junior College, 2-6-1 Hitotsubashi, Chiyoda, Tokyo, Japan

Abstract. Fashion styles adopted every day are an important aspect of culture, and style trend analysis helps provide a deeper understanding of our societies and cultures. To analyze everyday fashion trends from the humanities perspective, we need a digital archive that includes images of what people wore in their daily lives over an extended period. In fashion research, building digital fashion image archives has attracted significant attention. However, the existing archives are not suitable for retrieving everyday fashion trends. In addition, to interpret how the trends emerge, we need non-fashion data sources relevant to why and how people choose fashion. In this study, we created a new fashion image archive called Chronicle Archive of Tokyo Street Fashion (CAT STREET) based on a review of the limitations in the existing digital fashion archives. CAT STREET includes images showing the clothing people wore in their daily lives during the period 1970–2017, which contain timestamps and street location annotations. We applied machine learning to CAT STREET and found two types of fashion trend patterns. Then, we demonstrated how magazine archives help us interpret how trend patterns emerge. These empirical analyses show our approach's potential to discover new perspectives to promote an understanding of our societies and cultures through fashion embedded in consumers' daily lives.

Keywords: Fashion trend · Digital fashion archive · Image processing · Machine learning · Deep learning

1 Introduction

Fashion styles adopted in our daily lives are an important aspect of culture. As noted by Lancioni [17], fashion is 'a reflection of a society's goals and aspirations'; people choose fashion styles within their embedded social contexts. Hence, the analysis of everyday fashion trends can provide an in-depth understanding of our societies and cultures. In fashion research, the advantages of digital fashion archives have continued to attract attention in recent times. These digital

This work was supported by the Japan Society for the Promotion of Science (JSPS) KAKENHI [grant number 20K02153].

H.-R. Ke et al. (Eds.): ICADL 2021, LNCS 13133, pp. 179–194, 2021.
https://doi.org/10.1007/978-3-030-91669-5_15

archives have been mainly built based on two purposes. One is enhancing the scholarly values of museum collections by making them public to utilize them digitally. Another motivation is predicting what fashion items will come into style in the short term with images retrieved from the internet for business purposes.

However, these archives are not suitable for analyzing everyday fashion trends from the humanities perspective. To identify everyday fashion trends, we need a new digital archive that includes images of the clothes people wore in their daily lives over an extended period. In addition, we need non-fashion data sources showcasing why and how people choose fashion to interpret the trends.

In this study, we create a new fashion image archive called Chronicle Archive of Tokyo Street Fashion (CAT STREET) showcasing fashion images reflecting everyday women's clothing over a long period from 1970 to 2017 in Tokyo. CAT STREET helps overcome the limitations in the existing digital fashion archives by applying machine learning to identify fashion trend patterns. Then, we demonstrate how magazine archives are suitable in understanding how various trend patterns emerged. The empirical analyses show our approach's potential in identifying new perspectives through fashion trends in daily life, which can promote understanding of societies and cultures.

2 Related Works

2.1 Fashion Trend Analysis

Kroeber [15] analyzed fashion trends manually and measured features of women's full evening toilette, e.g. skirt length and skirt width, which were collected from magazines from 1844 to 1919. After this work, many researchers conducted similar studies [2,21,22,25]. Their work surveyed historical magazines, portraits, and illustrations and focused mainly on formal fashion, rather than clothing worn in everyday life.

The critical issue in the traditional approach is that analyzing fashion trends is labor-intensive. Kroeber [15] is among the representative works in the early stage of quantitative analysis of fashion trends. This type of quantitative analysis requires a considerable amount of human resources to select appropriate images, classify images, and measure features. This labor-intensive approach has long been applied in this field. Furthermore, manually managing large modern fashion image archives is cumbersome. To solve this issue, we utilize machine learning in this study. Machine learning is a computer algorithm that learns procedures based on sample data, e.g. human task results, and imitates the procedures. In recent years, machine learning has contributed to the development of digital humanities information processing [13,18,37]. In the field of fashion, modern fashion styles are complex and diverse. Applying machine learning to fashion image archives may be expected to be of benefit in quantifying fashion trends more precisely and efficiently.

2.2 Digital Fashion Archives

In recent years, digital fashion archives have been proactively built in the fields of museology and computer science. However, the construction of these archives was prompted by different research objectives. Many museums and research institutions have digitized their fashion collections [6,8,11,20,23,33–35] not only to conserve these historically valuable collections but also to enhance the scholarly values by making them available to the public. Vogue digitized their magazines from 1892 to date, the Japanese Fashion Archive collected representative Japanese fashion items from each era, and the Boston Museum of Fine Arts archived historical fashion items from all over the world.

On the other hand, in computer science, the motivation to build digital fashion archives is generally short-sighted and primarily to predict the next fashion trends that can be used by vendors to plan production or by online stores to improve recommendation engines. Several studies have created their own fashion image databases to fit the fashion business issues they focused on [1,14,19,29,32,38]. Table 1 presents the best-known public databases from fashion studies in computer science.

Unfortunately, the fashion databases proposed in previous studies have several limitations in terms of the approach to everyday fashion trends. First, the level of detail of location annotation in existing databases is insufficient to analyze everyday fashion trends. People belong to a social community, and some social communities have their own distinct fashion styles and their territory. For instance, 'Shibuya Fashion' is a fashion style for young ladies that originated from a famous fashion mall in Shibuya [12]. Young ladies dressed in 'Shibuya Fashion' frequent Shibuya, one of the most famous fashion-conscious streets in Japan. Hence, we need fashion image data with location annotations at the street level to focus on what people wear in their daily lives.

Second, most databases consist of recent fashion images from the last decade because they were obtained from the internet. The periods covered by these databases might not be sufficiently long to determine fashion trends over longer periods. By examining how fashion changes over extended periods, sociologists and anthropologists have found that fashion has decadal-to-centennial trends and cyclic patterns [15]. An example is the hemline index, which is a well-known hypothesis that describes the cyclic pattern in which skirt lengths decrease when economic conditions improve [7]. This pattern was determined based on observations of skirt lengths in the 1920s and 30s.

Furthermore, we need data on how and what consumers wear to express their fashion styles. However, fashion photographs on the internet are one of two types: one displays clothes that consumers themselves choose to wear, and the other consists of photographs taken by professional photographers to promote a clothing line. As previous studies built databases by collecting fashion images from the internet, existing databases do not reflect only the fashion styles chosen by consumers. This is the third limitation of existing fashion data archives.

Taken together, no existing database has everyday fashion images that span over a long period with both timestamp and location information. In this study,

Table 1. Popular fashion databases used in computer science.

Database name	Num. of images	Geographical information	Time stamp	Fashion style tag
Fashionista [38]	158,235			
Hipster Wars [14]	1,893			✓
DeepFashion [19]	800,000			✓
Fashion 144k [29]	144,169	City unit		✓
FashionStyle14 [32]	13,126			✓
When Was That Made? [36]	100,000		1900–2009 Decade unit	
Fashion Culture Database [1]	76,532,219	City unit	2000–2015 Date unit	
CAT STREET (Our Database)	14,688	Street unit	1970–2017 Date unit	

we define everyday fashion trends as trends of fashion styles that consumers adopt in their daily lives. Everyday fashion exists on the streets [12,26], changes over time, and trends and cyclic patterns can be observed over extended periods. Hence, we create a new fashion database, CAT STREET, is an attempt to solve these limitations and analyze everyday fashion trends.

3 CAT STREET: Chronicle Archive of Tokyo Street-Fashion

We created CAT STREET via the following steps. We collected street-fashion photographs once or twice a month in fashion-conscious streets such as Harajuku and Shibuya in Tokyo from 1980 to date. In addition, we used fashion photographs from a third-party organization taken in the 1970s at monthly intervals in the same fashion-conscious streets. Next, by using images from the two data sources, we constructed a primary image database with timestamps from 1970 to 2017 for each image. The photographs from the third-party organization did not have location information; hence, we could only annotate the fashion images taken since 1980 with street tags.

Fashion styles conventionally differ between men and women. To focus on women's fashion trends in CAT STREET, two researchers manually categorized the images by the subjects' gender, reciprocally validating one another's categorizations, and we selected only images of women from the primary image database. Some images from the 1970s were in monochrome; therefore, we grayscaled all images to align the color tone across all images over the entire period.

Street-fashion photographs generally contain a large amount of noise, which hinders the precise detection of combinations of clothes worn by the subjects of photographs. To remove the noise, we performed image pre-processing as the final step in the following manner. We identified human bodies in the photographs

Year	1970s	1980s	1990s	2000s	2010s
Total No.	870	1,955	1,014	1,324	9,525
Harajuku	0	788	414	568	2,933
Shibuya	0	632	394	465	1,368
Others	0	328	203	291	5,224
Unlabeled	870	207	3	0	0

Fig. 1. Data overview of CAT STREET. Harajuku and Shibuya are the most famous streets in Japan and Others include images taken in other fashion-conscious streets such as Ginza and Daikanyama.

using OpenPose [4], an machine learning algorithm for object recognition, and removed as much of the background image as possible. Subsequently, we trimmed the subject's head to focus on clothing items based on the head position, which was detected using OpenPose. Figure 1 shows an overview of the data contained in CAT STREET. The total number of images in the database was 14,688.

At the end of the database creation process, we checked whether CAT STREET met the requirements for a database to capture everyday fashion trends. First, CAT STREET comprises photographs captured on fashion-conscious streets in Tokyo. It reflects the fashion styles women wore in their real lives and does not include commercial fashion images produced for business purposes. Second, CAT STREET has necessary and sufficient annotations to track fashion trends: monthly timestamps from 1970 to 2017 and street-level location tags. Images in the 1970s do not have street tags; however, they were taken in the same streets as the photographs taken with street tags since 1980. Hence, we could use all the images in CAT STREET to analyze the overall trends of everyday fashion in fashion-conscious streets as a representative case in Japan.

4 Retrieving Everyday Fashion Trends

In this section, we estimated a fashion style clustering model to identify the fashion styles adopted in fashion images. Then, we applied the fashion clustering model to CAT STREET and retrieved everyday fashion trends indicating the extent to which people adopted the fashion styles in each year. Finally, we verified the robustness of clustering model's result.

4.1 Fashion Style Clustering Model

To build a fashion style clustering model, we selected FashionStyle14 [32] as the training dataset. It consists of fourteen fashion style classes and their tags, as shown in the first row of Fig. 2. Each class consists of approximately 1,000 images, and the database consists of a total of 13,126 images. The fashion styles of FashionStyle14 were selected by an expert as being representative of modern fashion trends in 2017. By applying the fashion clustering model to CAT STREET, we measured the share of each modern style in each year and regarded it as the style's trend.

We trained four deep learning network structures as options for our fashion clustering model: InceptionResNetV2 [31], Xception [5], ResNet50 [10], and VGG19 [30]. We set weights trained on ImageNet [27] as the initial weights and fine-tuned them on FashionStyle14, which we gray-scaled to align the color tone of CAT STREET, using the stochastic gradient descent algorithm at a learning rate of 10^4. For fine-tuning, we applied k-fold cross-validation with k set as five.

The F1-scores are presented in Table 2. InceptionResNetV2 yielded the highest F1-scores among the deep learning network structures. Its accuracy was 0.787, which is higher than the benchmark accuracy of 0.72 established by ResNet50 trained on FashionStyle14 in the study by Takagi et al. [32]. Therefore, we concluded that InceptionResNetV2 could classify gray-scaled color images to fashion styles and adopted the deep learning network structure InceptionResNetV2 as the fashion style clustering model in this study. We applied this fashion style clustering model to the images in CAT STREET, and Fig. 2 shows sample images classified into each fashion style.

4.2 Verification

The fashion style clustering model consisted of five models as we performed five-fold cross-validations when the deep learning network structure was trained, and each model estimated the style prevalence share for each image. To verify the clustering model's robustness in terms of reproducing style shares, we evaluated the time-series correlations among the five models. Most fashion styles had high time-series correlation coefficients of over 0.8, and the unbiased standard errors were small. Some fashion styles, such as those designated Dressy and Feminine styles, exhibited low correlations because these styles originally had low style shares. These results indicate that our fashion style clustering model is a robust instrument for reproducing the time-series patterns of style shares. There were no

Table 2. F1-scores of fine-tuned network architectures. The highest scores are underlined and in boldface.

	InceptionResNetV2	Xception	ResNet50	VGG19
Macro avg.	**<u>0.787</u>**	0.782	0.747	0.690
Weighted avg.	**<u>0.786</u>**	0.781	0.747	0.689

Fig. 2. Overview of FashionStyle14 [32] and sample images in CAT STREET classified into different fashion styles using the fashion style clustering model. 'Conserv.' is an abbreviation for Conservative. The A&F is originally labeled as Ethnic [32]. We changed this original label to 'A&F' because this category contains multiple fashion styles such as Asian, African, south-American, and Folklore styles.

images for 1997 and 2009; we replaced the corresponding zeros with the averages of the adjacent values for the analysis in the next section.

5 Analyzing Fashion Trend Patterns

There are two representative fashion-conscious streets in Tokyo: Harajuku and Shibuya. Geographically, Harajuku and Shibuya are very close and only a single transit station away from each other; however, they have different cultures. In particular, the everyday fashion trends in these streets are famously compared to each other. Some qualitative studies pointed out triggers that form the cultural and fashion modes in each street using an observational method [9,12]. However, they simply reported the triggers with respect to the street and style and did not compare the functioning of these triggers on the mode formations in everyday fashion trends from a macro perspective. This section sheds light on this research gap and interprets how the fashion trends emerged with non-fashion data sources.

First, we compared two everyday fashion trends in fashion-conscious streets with CAT STREET to investigate how fashion styles 'boom' or suddenly increase in prevalence, and classified the resultant patterns. Fourteen styles were classified into two groups: a group comprising styles that emerged on both streets simultaneously and a group with differing timings for trends observed in the streets. We selected two fashion styles from each group as examples, including A&F and Retro from the first group and Fairy and Gal from the second group; Figs. 3(a), 4(a), 5(a), and 6(a) show their average style shares, respectively.

As a non-fashion data source, we focused our attention on magazines because fashion brands have built good partnerships with the magazine industry for a long time. Fashion brands regard magazines as essential media to build a bridge between themselves and consumers, and magazines play a role as a dispatcher of fashion in the market. We quantified the 'magazine' trends to capture how magazines or media sent out information to consumers. For this purpose, we used the digital magazine archive of Oya Soichi Library [24]. Oya Soichi Library has archived Japanese magazines since the late 1980s and built the digital archive. The digital archive houses about 1,500 magazines, and one can search for article headlines of about 4.5 million articles.

The archive covers a range of age groups and mass-circulation magazines that people can easily acquire at small bookstores, convenience stores, and kiosks. By searching for headlines including fashion style words and specific topics from articles with tags related to fashion, we could quantify the 'magazine' trends that indicate how many articles dealt with the styles and specific topics to spread the information to a mass audience (Figs. 3(b), 4(b), 5(b), 7, and 8). The 'magazine' trends labeled as Articles in the figures were normalized to the range of 0 to 1 for comparison.

Finally, to infer why and how the everyday fashion trends emerged, we compared the fashion style shares extracted from CAT STREET to the 'magazine' trends in the Oya Soichi Library database in chronological order.

5.1 Simultaneous Emergence of Fashion Styles

We selected two styles, A&F and Retro, as examples in the group of styles that simultaneously emerged on both streets. The A&F style is inspired by native costumes [3]. Figure 3(a) shows the A&F style's upward trend in the late 1990s in Harajuku. Simultaneously, the A&F style gradually became accepted in Shibuya in the late 1990s and reached the same share level in the mid-2000s. The Retro style, another example in the first group, is an abbreviation of retrospective style [3,39]. According to this style's definition, an overall downward trend is plausible in both streets, as shown in Fig. 4(a). Fashion revival is one of the relevant fashion trend phenomena of the Retro style, and there are a wide variety of substyles representing fashion revivals under the Retro style, such as the 60s look, 70s look, and 80s look. In Fig. 4(a), slight peaks can be observed in the early 1980s and late 1990s, suggesting that the 80s look, which was in vogue in the early 1980s, was revived in the late 1990s.

To quantify how magazines spread information about these styles, we plotted the 'magazine' trends by searching for articles with headlines that included terms related to the style names. Figures 3(b) and 4(b) indicate two relationship patterns between style shares and 'magazine' trends; Fig. 3(b) shows a synchronization phenomenon between style shares and magazine trends, while Fig. 4(b) shows that magazine trends follow share trends.

We interpreted the first pattern as 'mixed'; the shape of 'magazine' trends roughly matches that of style share trends, and this pattern is observed in the A&F style. The 'mixed' patterns in the trends suggest that they include two

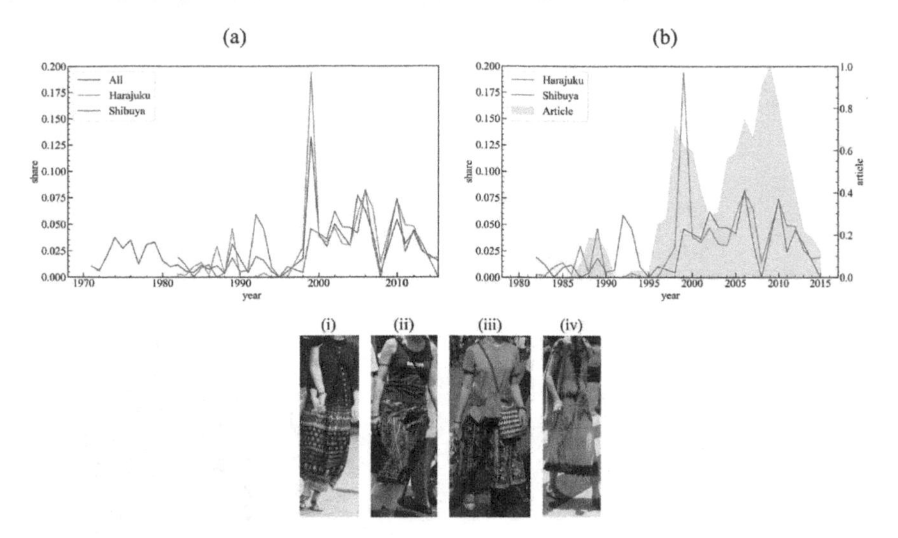

Fig. 3. A&F style's simultaneous emergence in both streets. (a) Average A&F style share. (b) Number of articles about the A&F style. Four images taken in Harajuku (i, ii) and Shibuya (iii, iv) in the late 1990s.

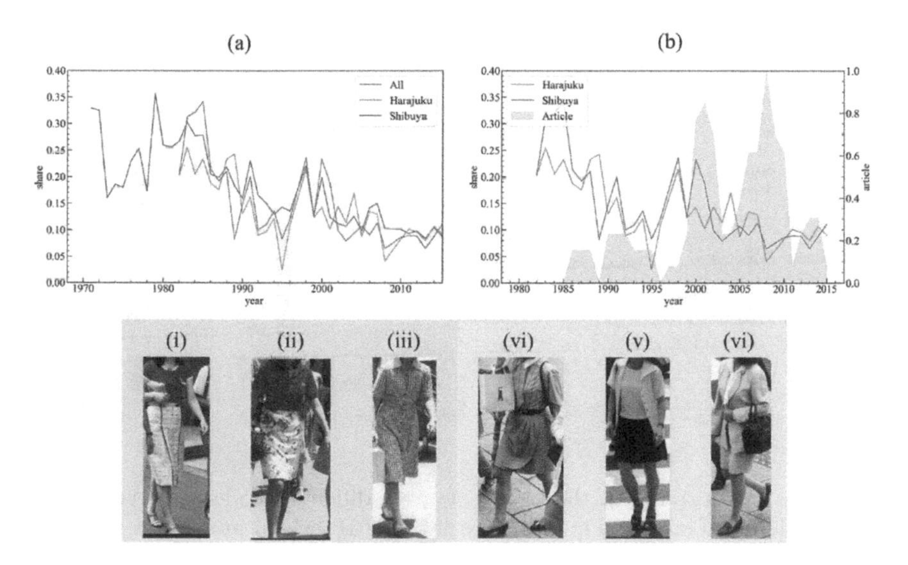

Fig. 4. Retro style's simultaneous emergence in both streets. (a) Average Retro style share. (b) Number of articles about the 80s look. The three images on the left (i–iii) are examples of the Retro style taken in the early 1980s, and the three images on the right (iv–vi) are examples of the Retro style that came into fashion in the late 1990s.

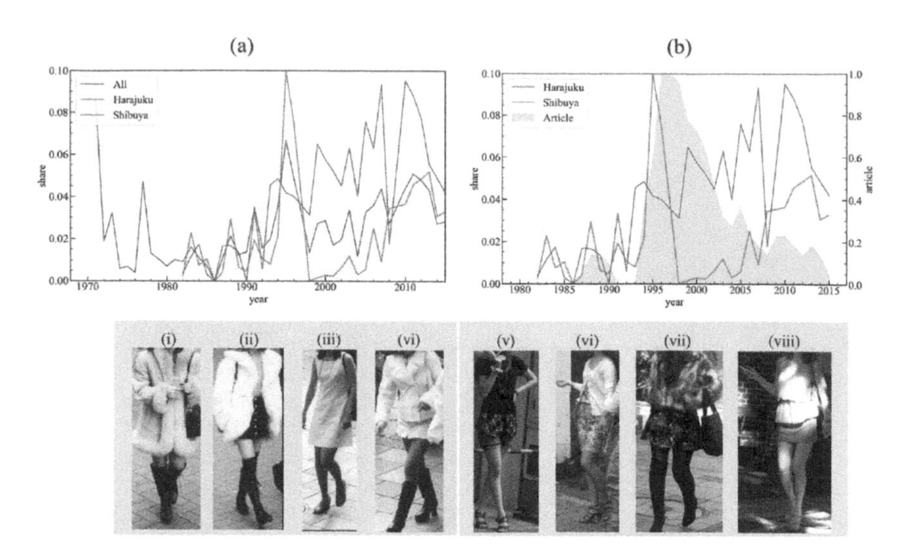

Fig. 5. Gal style's trends in the streets showing different modes. (a) Average Gal style share. (b) Number of articles about the Gal style. The four images on the left were taken in Harajuku (i, ii) and Shibuya (iii, iv) in 1995–1996, and the four images on the right were taken in Harajuku (v, vi) and Shibuya (vii, viii) in the late 2000s.

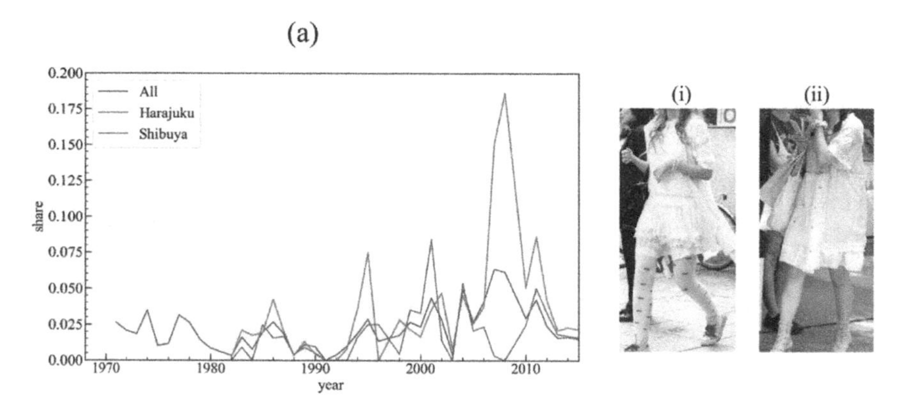

Fig. 6. Fairy style's trends in the streets showing different modes. (a) Average Fairy style share. The two images (i and ii) were taken in Harajuku in the late 2000s.

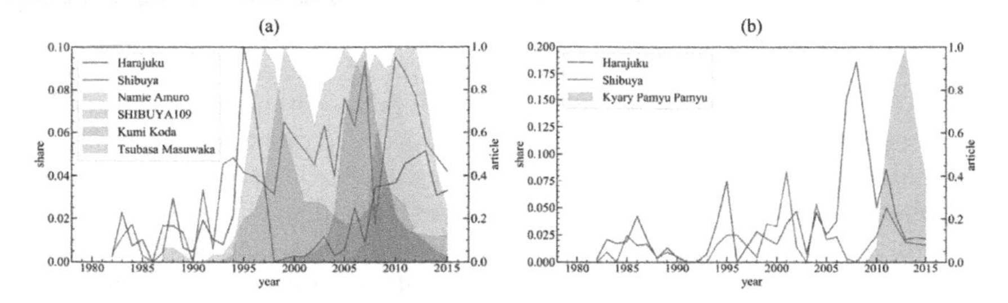

Fig. 7. (a) Gal style's share and 'magazine' trends for its icons. (b) Fairy style's share and 'magazine' trends for its icon.

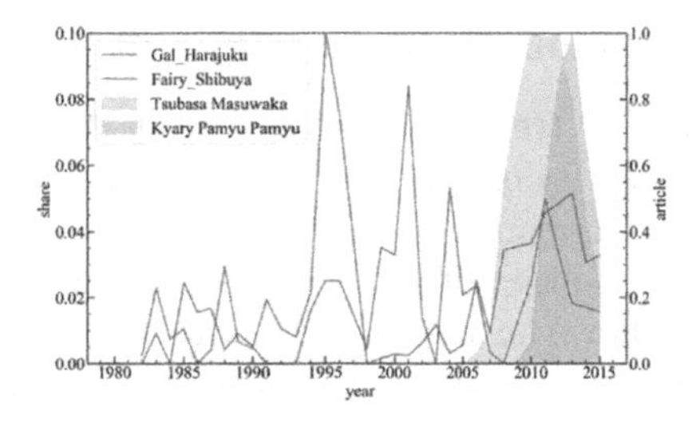

Fig. 8. Gal and Fairy style shares and 'magazine' trends for their icons.

types of phenomena between consumers and magazines. One is that the articles create new trends, spread them, and lead consumers to follow them. The other is that consumers create new style trends by themselves, and articles catch up on the new trends. These two types of phenomena could occur simultaneously or alternately. On the other hand, the second pattern is a 'follow-up'; the articles dealt with the modes that were already in vogue on the streets and contributed to keeping their momentum for some time by spreading the information, and this pattern is observed in the Retro style during the 2000s.

In this analysis, we found two relationship patterns between style shares and 'magazine' trends: 'mixed' and 'follow-up' patterns. We also inferred what types of interactions occur between consumers and magazines in each pattern. However, we could not perform an in-depth analysis on how to distinguish between the two types of consumer-magazine interactions in the case of the 'mixed' pattern because 'magazine' trends generated by searching for article headlines including terms related to the style names did not always reflect the contents of articles accurately. We also found it difficult to uncover why the article peak in the Retro style during the late 2000s was seemingly irrelevant to the Retro style share. To approach these unsolved research questions as future work, we must perform text mining on article contents and headlines, and categorize the articles into the creative type, which creates and spreads new trends, or the reporting type, which refers to presently existing new consumer trends after they develop; e.g. 'ten trends coming in autumn, checker-board pattern, check pattern, military, big tops, shoes & boots, foot coordination, fur, A&F vs. Nordic, beautiful romper, best of the season' (translated from Japanese by the authors) is categorized as a creative-type article title, whereas 'A&F style is a hot topic this spring. We will introduce you to some of the Japanese items that are getting a lot of attention. Crepe, goldfish prints, and more...' (translated from Japanese by the authors) is categorized as a reporting-type article title. By measuring the number of articles about the fashion styles and identifying the contents of articles to reflect the types of consumer-magazine relationships, we expect to decompose relationship types in the 'mixed' pattern, interpret seemingly irrelevant relationships, and clarify the role of each relationship type in the trend formation process.

5.2 Emergence of Fashion Styles at Different Types in the Streets

For the case where fashion trends are observed at different times in the two streets, we focused on two styles: the Gal style, which can be characterized as an exaggeration of the American teenage party style, and the Fairy style, which involves the fashion coordination of frilly dresses reminiscent of fairies [39].

Figures 5 and 6 show how the acceptances of these two styles changed in each street. The Gal style came into fashion around 1995 simultaneously in Harajuku and Shibuya. The style remained in Shibuya, whereas it lost its popularity quickly in Harajuku but re-emerged in the late 2000s. The Fairy style emerged in the late 2000s in Harajuku only and lost its momentum in the early 2010s. As with the first case, we searched 'magazine' trends for these style names and compared them to the style shares. For the Gal style, we found that the style and 'magazine' trends correlated with each other in the mid-80s and the mid-90s. However, magazines gradually lost interest in the style after the mid-00s, and the relationship between the style and 'magazine' trend disappeared accordingly. For the Fairy style, on the other hand, no article in the digital magazine archive included the style name in the headlines. This is because 'Fairy' is a kind of jargon among people interested in the style and hence not used in magazines.

What accounts for this difference between the first and the second case? We assumed that the critical factor determining people's choice in fashion, i.e. 'why people adopt a given style,' is for some purpose; some styles have specific features, such as colors, silhouettes, and patterns, that people consume as a fashion or which represent a certain social identity and have some characteristic features. The Gal and Fairy styles are in the latter group and represent a 'way of life' for some people [9,16].

Icons representing 'way of life,' such as celebrities, have substantial power to influence people to behave in a certain manner; for instance, celebrities can influence people to purchase products that they use or promote. Choosing or adopting fashion styles is no exception, and some articles identified fashion icons for the Gal and the Fairy styles [9,28,39]. We attempted to explain why the style trends in the streets showed different modes from the perspective of fashion icons and the media type that the icons utilized.

Figures 7 and 8 show the relationship between the style shares and 'magazine' trends for style icons. Previous works introduced the following Gal style icons:

Namie Amuro for the mid-90s, SHIBUYA 109 for the late-00s, Ayumi Hamasaki for the early-00s, and Kumi Koda and Tsubasa Masuwaka for the late-00s [9,28].

The first icon who boosted the Gal style was Namie Amuro, who debuted nationwide as a pop singer in 1992. Girls yearned to imitate her fashion style, and magazines focused heavily on her as a style icon in the mid-90s. Simultaneously, as the 'magazine' trend for the style shows in Fig. 7 (a), many tabloid magazines were interested in this phenomenon and spread it as a new 'way of life' among youth consumers. The first icon boosted the style in both streets because she appeared nation-wide. However, the second icon that sustained the Gal style's upward trend is regionally specific: an iconic fashion mall named SHIBUYA 109 (pronounced Ichi-maru-kyū) in Shibuya. Many tenants in the mall sold Gal style products, which is why the Gal style is also known as 'maru-kyū fashion' [39]. This street-specific image created by the second icon might have influenced the third and fourth icons, Ayumi Hamasaki and Kumi Koda, who were nationwide pop singers in the early-00s and the mid-00s, because the style share increased during that time only in Shibuya.

The icons prompted the last increase in the prevalence share of the Gal style in Shibuya, and the spike in the Fairy style's share in Harajuku exhibited the same characteristics. Both Tsubasa Masuwaka for the Gal style in Shibuya and 'Kyary Pamyu Pamyu' for the Fairy style in Harajuku were active as exclusive reader models in street-based magazines at the beginning, and they created the booms in each street [28]. However, the street-based magazines they belonged to were not included in the digital magazine archive in Oya Soichi Library; hence, the 'magazine' trends for their names missed their activities as exclusive reader models when the style shares showed an upward trend (Figs. 7(b) and 8). Around 2010, they debuted as nationwide pop stars and frequently appeared in mass-circulation magazines and on television. Additionally, Tsubasa Masuwaka proposed a new style named 'Shibu-Hara' fashion, a combination of the styles in both Shibuya and Harajuku, and also referred to interactions with 'Kyary Pamyu Pamyu' on her social networking account. The Gal style's second peak in Harajuku and the Fairy style's spike in Shibuya in the early 2010s aligned with these icons' nationwide activities; this indicates the styles' cross-interactions driven by the icons (Figs. 7(b) and 8).

5.3 Discussion

The findings in Sect. 5.1 prompted a new research question of determining the importance of the different roles that media play in fashion trends: the role of the 'mixed' pattern for creating and reporting new trends and the 'follow-up' effect to keep their momentum. To theorize how fashion trends are generated in more detail, our findings indicate that quantifying the trends from multiple digital archives is essential to test the research questions about the effect of the media's role, which has not been discussed thoroughly.

In previous studies, researchers analyzed fashion style trends at a street level independently. However, to analyze the recent, more complicated trend patterns, the findings in Sect. 5.2 suggests that it would be thought-provoking to focus on what the icons represent for people's social identities, rather than the style itself, and how they lead the trends. An example is the analysis of the types of media that icons use and the reach of that media. The viewpoints gained from our findings can be useful in tackling unsolved research questions such as how styles interact and how new styles are generated from this process. If we can determine people's social identities from the aims of their adopted fashions using the digital archives, researchers can retrieve beneficial information to delve into the research questions and expand fashion trend theories.

6 Conclusions

In this study, we reviewed the issues in the existing digital fashion archives for the analysis of everyday fashion trends. As one of the solutions to these issues, we built CAT STREET, which comprises fashion images illustrating what people wore in their daily lives in the period 1970–2017, along with street-level geographical information. We demonstrated that machine learning retrieved how fashion styles emerged on the street using CAT STREET and the magazine archives helped us interpret these phenomena. These empirical analyses showed our approach's potential to find new perspectives to promote the understanding of our societies and cultures through fashion embedded in the daily lives of consumers.

Our work is not without limitations. We used the fashion style categories of FashionStyle14 [32], which was defined by fashion experts and is considered to represent modern fashion. However, the definition does not cover all contemporary fashion styles and their substyles in a mutually exclusive and collectively exhaustive manner. Defining fashion styles is a complicated task because some fashion styles emerge from consumers, and others are defined by suppliers. Hence, we must further refine the definition of fashion styles to capture everyday fashion trends more accurately. Furthermore, prior to building CAT STREET, only printed photos were available for the period 1970–2009. Consequently, the numbers of images for these decades are not equally distributed because only those images from printed photos that have already undergone digitization are currently present in the database. The remainder of the printed photos will be digitized and their corresponding images added to the database in future work.

References

1. Abe, K., et al.: Fashion culture database: construction of database for world-wide fashion analysis. In: Proceedings of the 2018 15th International Conference on Control, Automation, Robotics and Vision (ICARCV), pp. 1721–1726. IEEE, November 2018. https://doi.org/10.1109/ICARCV.2018.8581148
2. Belleau, B.D.: Cyclical fashion movement: women's day dresses: 1860–1980. Cloth. Text. Res. J. **5**(2), 15–20 (1987). https://doi.org/10.1177/0887302X8700500203
3. Bunka Gakuen Bunka Publishing Bureau, Bunka Gakuen University Textbook Department (eds.): Fashion Dictionary. Bunka Gakuen Bunka Publishing Bureau, Tokyo, Japan (1993)

4. Cao, Z., Hidalgo, G., Simon, T., Wei, S.E., Sheikh, Y.: OpenPose: realtime multi-person 2D pose estimation using part affinity fields. IEEE Trans. Pattern Anal. Mach. Intell. **43**(1), 172–186 (2019)
5. Chollet, F.: Xception: deep learning with depthwise separable convolutions. In: Proceedings of the 2017 IEEE Conference on Computer Vision and Pattern Recognition (CVPR), pp. 1251–1258. IEEE, July 2017. https://doi.org/10.1109/CVPR.2017.195
6. Condé Nast: Vogue Archive. https://archive.vogue.com/. Accessed 11 July 2021
7. Dhanorkar, S.: 8 unusual indicators to gauge economic health around the world. The Economic Times (2015)
8. Europeana: Europeana, Fashion. https://www.europeana.eu/en/collections/topic/55-fashion. Accessed 11 July 2021
9. Hasegawa, S.: The History of Two Decades for Gals and "Guys" 1990's–2010's. Akishobo Inc., Tokyo (2015)
10. He, K., Zhang, X., Ren, S., Sun, J.: Deep residual learning for image recognition. In: Proceedings of the 2016 IEEE Conference on Computer Vision and Pattern Recognition (CVPR), pp. 770–778. IEEE, June 2016. https://doi.org/10.1109/CVPR.2016.90
11. japanese.fashion.archive: Japanese Fashion Archive. http://japanesefashionarchive.com/about/. Accessed 11 July 2021
12. Kawamura, Y.: Japanese teens as producers of street fashion. Curr. Sociol. **54**(5), 784–801 (2006). https://doi.org/10.1177/0011392106066816
13. Kestemont, M., De Pauw, G., van Nie, R., Daelemans, W.: Lemmatization for variation-rich languages using deep learning. Digit. Sch. Humanit. **32**(4), 797–815 (2017). https://doi.org/10.1093/llc/fqw034
14. Kiapour, M.H., Yamaguchi, K., Berg, A.C., Berg, T.L.: Hipster wars: discovering elements of fashion styles. In: Fleet, D., Pajdla, T., Schiele, B., Tuytelaars, T. (eds.) ECCV 2014. LNCS, vol. 8689, pp. 472–488. Springer, Cham (2014). https://doi.org/10.1007/978-3-319-10590-1_31
15. Kroeber, A.L.: On the principle of order in civilization as exemplified by changes of fashion. Am. Anthropol. **21**(3), 235–263 (1919)
16. Kyary Pamyu Pamyu: Oh! My God!! Harajuku Girl. POPLAR Publishing Co. Ltd., Tokyo (2011)
17. Lancioni, R.A.: A brief note on the history of fashion. J. Acad. Mark. Sci. **1**, 128–131 (1973). https://doi.org/10.1007/BF02722015
18. Lang, S., Ommer, B.: Attesting similarity: supporting the organization and study of art image collections with computer vision. Digit. Sch. Humanit. **33**(4), 845–856 (2018). https://doi.org/10.1093/llc/fqy006
19. Liu, Z., Luo, P., Qiu, S., Wang, X., Tang, X.: DeepFashion: powering robust clothes recognition and retrieval with rich annotations. In: Proceedings of the 2016 IEEE Conference on Computer Vision and Pattern Recognition (CVPR), pp. 1096–1104. IEEE, June 2016
20. Los Angeles County Museum of Art: LACMA, Fashion, 1900–2000. https://collections.lacma.org/node/589000. Accessed 11 July 2021
21. Lowe, E.D.: Quantitative analysis of fashion change: a critical review. Home Econ. Res. J. **21**(3), 280–306 (1993). https://doi.org/10.1177/0046777493213004
22. Lowe, E.D., Lowe, J.W.: Velocity of the fashion process in women's formal evening dress, 1789–1980. Cloth. Text. Res. J. **9**(1), 50–58 (1990). https://doi.org/10.1177/0887302X9000900107

23. Museum of Fine Arts, Boston: Boston Museum of Fine Arts. https://collections.mfa.org/collections/419373/tfafashionable-dress; jsessionid=05B040E674A761851F36C9A1C70DBB7F/objects. Accessed 11 July 2021
24. OYA-bunko: Web OYA-bunko. https://www.oya-bunko.com/. Accessed 11 July 2021
25. Robenstine, C., Kelley, E.: Relating fashion change to social change: a methodological approach. Home Econ. Res. J. **10**(1), 78–87 (1981). https://doi.org/10.1177/1077727X8101000110
26. Rocamora, A., O'Neill, A.: Fashioning the street: images of the street in the fashion media. In: Shinkle, E. (ed.) Fashion as Photograph Viewing and Reviewing Images of Fashion, chap. 13, pp. 185–199. I.B.Tauris, London (2008). https://doi.org/10.5040/9780755696420.ch-013
27. Russakovsky, O., et al.: ImageNet large scale visual recognition challenge. Int. J. Comput. Vis. **115**(3), 211–252 (2015). https://doi.org/10.1007/s11263-015-0816-y
28. Shinmura, K., Okamoto, A.: 2012 the frontline of the never-ending "reader models" boom. Weekly Playboy **23**, 89–100 (2012)
29. Simo-Serra, E., Fidler, S., Moreno-Noguer, F., Urtasun, R.: Neuroaesthetics in fashion: modeling the perception of fashionability. In: Proceedings of the 2015 IEEE Conference on Computer Vision and Pattern Recognition (CVPR), pp. 869–877. IEEE, June 2015
30. Simonyan, K., Zisserman, A.: Very deep convolutional networks for large-scale image recognition. In: Bengio, Y., LeCun, Y. (eds.) Proceedings of the 3rd International Conference on Learning Representations (ICLR), May 2015
31. Szegedy, C., Ioffe, S., Vanhoucke, V., Alemi, A.: Inception-v4, inception-ResNet and the impact of residual connections on learning. In: Proceedings of the Thirty-First AAAI Conference on Artificial Intelligence, pp. 4278–4284. AAAI Press, February 2017
32. Takagi, M., Simo-Serra, E., Iizuka, S., Ishikawa, H.: What makes a style: experimental analysis of fashion prediction. In: Proceedings of the 2017 IEEE International Conference on Computer Vision Workshops (ICCVW), pp. 2247–2253. IEEE, October 2017. https://doi.org/10.1109/ICCVW.2017.263
33. The Metropolitan Museum of Art: THE MET 150, Libraries and Research Centers, ART, Thomas J. Watson Library Digital Collections, Costume Institute Collections. https://www.metmuseum.org/art/libraries-and-research-centers/watson-digital-collections/costume-institute-collections. Accessed 11 July 2021
34. The Museum at FIT: The Museum at FIT, Collections. https://www.fitnyc.edu//museum/collections/. Accessed 11 July 2021
35. University of North Texas: UNT Digital Library, Texas Fashion Collection. https://digital.library.unt.edu/explore/collections/TXFC/. Accessed 11 July 2021
36. Vittayakorn, S., Berg, A.C., Berg, T.L.: When was that made? In: Proceedings of the 2017 IEEE Winter Conference on Applications of Computer Vision (WACV), pp. 715–724. IEEE, March 2017. https://doi.org/10.1109/WACV.2017.85
37. Wevers, M., Smits, T.: The visual digital turn: using neural networks to study historical images. Digit. Sch. Humanit. **35**(1), 194–207 (2020). https://doi.org/10.1093/llc/fqy085
38. Yamaguchi, K., Kiapour, M.H., Ortiz, L.E., Berg, T.L.: Parsing clothing in fashion photographs. In: Proceedings of the 2012 IEEE Conference on Computer Vision and Pattern Recognition (CVPR), pp. 3570–3577. IEEE, June 2012
39. Yoshimura, S.: The Fashion Logos. Senken Shinbun Co., Ltd., Tokyo (2019)

The Effect of Library Virtual Tour on Library Image Construction: Study on Perpustakaan BPK RI

Zaidan Abdurrahman Qois(✉) and Luki Wijayanti(✉)

Library and Information Department, Universitas Indonesia, Kota Depok, Indonesia
{Zaidan.abdurrahman,Luki_w}@ui.ac.id

Abstract. Image is very important for every institution, including libraries. The library image greatly influences the behavior of library users, especially those related to their actions towards the library, where a bad image about the library is still developing and is trusted by the public, such as warehouse to store books, unfriendly librarians, and others. However, the development of information technology has been used by various industries to improve the image of their institutions, such as through social media and virtual tour technology, which have also been implemented in Perpustakaan BPK RI. Based on that, this study aims to evaluate the effect of a library virtual tour on the library image, focusing on Perpustakaan BPK RI which has implemented virtual tour technology as a means of their promotion. The study using quantitative approach to collect quantitative data with survey from 26 March until 3 April 2021 and using inferential statistics to determine the relationship and effect between library virtual tour and library image. The results show that there is a significant relationship and effect between library virtual tour and library image, besides it also shows that Perpustakaan BPK RI has a good image to the users and its Virtual Tour has a good quality based on user assessments.

Keywords: Library virtual tour · Library image · Perpustakaan BPK RI

1 Introduction

The current reality shows that the image of an object plays a very important role in influencing user behavior, especially in shaping the user experience. Among them, in the decision-making process related to the visit or use of the object; in the process of comparing satisfaction and perceived quality of an object; revisiting; and the process of disseminating information and object recommendations to friends and family [1].

Today, the use of increasingly sophisticated information and communication technology is a priority that must be owned and understood by every organization, both in the fields of education, economy, sports, and so on. One of them is Virtual Reality (VR) technology which has begun to be implemented and has become an inseparable part of every industry. This is as explained by Kim & Hall (2019), VR is currently

H.-R. Ke et al. (Eds.): ICADL 2021, LNCS 13133, pp. 195–209, 2021.
https://doi.org/10.1007/978-3-030-91669-5_16

the most important topic in contemporary information management given its increasing application in every different industry, including in tourism [2].

In the world of libraries, Perpustakaan Badan Pemeriksa Keuangan Republik Indonesia (BPK RI) is one of the libraries that has started and has implemented VR technology to create virtual tours of their library, namely the Perpustakaan BPK RI Virtual Tour (https://www.iheritage.id/public/bpk/). Perpustakaan BPK RI Virtual Tour is a service innovation created by Perpustakaan BPK RI since August 2020 as an effort to overcome the Covid-19 pandemic which causes users to be unable to come to the library, as well as a form of promotion and support for the services provided by Perpustakaan BPK RI. However, evaluation of library virtual tours has not been done, especially in side of users.

Therefore, this study seeks to evaluate the Perpustakaan BPK RI Virtual Tour by seeing whether there is an effect given by the Perpustakaan BPK RI Virtual Tour on the Image Construction of the Perpustakaan BPK RI. More specifically, this study answers the following research questions: (1) How the user's assessments of the Perpustakaan BPK RI Virtual Tour? and (2) How is the effect of the Perpustakaan BPK RI Virtual Tour on the Image Construction of the Perpustakaan BPK RI?

2 Theoretical Background

2.1 Virtual Tour

VR is defined as the use of a computer-generated 3D environment or also known as a "Virtual Environment", which allows navigation and interaction with it, and produces real-time simulations of one or more of the five consecutive user [3]. The history of VR begins in the late 1950s and early 1960s. At that time, it began with the emergence of the first VR simulator in 1970, then in 1980 the development of video games became a driving factor for the progress of VR, then in the late 1990s, the development of the internet finally brought VR further and more sophisticated.

In the field of tourism, Guttentag (2010) points out that VR has been used in various key areas of tourism, such as management and planning, preservation of a national heritage, marketing, accessibility, education, entertainment, and information provision. VR technology even removes distance barriers for potential tourists to gain information and understanding about tourist destinations before finally deciding to make a visit.

Likewise in the field of libraries, VR technology has begun to be applied and developed in libraries. Based on a survey on library tours conducted by Academic ARL Libraries, it is known that library virtual tours are the most popular tours in the library together with library guided tours and independent tours [4]. This is also in line with the 3 (three) main progressive developments in the use of tours or tours as library learning (instructional) media, namely physical tours (walking tours), virtual web tours (web tours), and virtual reality tours (virtual reality tours) [5].

This study refers to the research of Lee, et al. (2020) regarding the dimensions of the virtual tour quality to take measurements of the Perpustakaan BPK RI Virtual Tour, that are (1) Content Quality, which refers to the quality of information provided through the virtual tour, which includes content accuracy, completeness, and presentation format or content presentation format [6]. (2) System Quality refers to a system that is able to provide the characteristics of mobile devices (mobile devices) and web browsing services (web browsing devices) so that they can be used by users [7], also includes reliability, convenience of access, response time, and system flexibility. (3) Vividness or clarity refers to the representation or method used by the virtual environment in presenting information to each of the user's senses. The level of vividness or clarity can be increased by enriching the depth and breadth of a system, in this case related to the quality of the presentation or presentation of information, such as the use of multimedia that includes video, audio, and animation from a virtual environment.

2.2 Perpustakaan BPK RI Virtual Tour

The Perpustakaan BPK RI Virtual Tour was created in August 2020. To access and utilize the Perpustakaan BPK RI Virtual Tour, users can visit the https://www.iheritage.id/publik/bpk/ (Fig. 1).

Fig. 1. The Beginning of the Perpustakaan BPK RI Virtual Tour

Generally, the Perpustakaan BPK RI Virtual Tour will invite users to take a virtual tour of the Perpustakaan BPK RI, in other words, users will be navigated to visit every room and section contained in Perpustakaan BPK RI. In addition to inviting users to take a virtual tour of the library, the Perpustakaan BPK RI Virtual Tour also provides various features or menus that also increase knowledge and understanding, and are able to meet the information needs of users. Here are some features or menus of the Perpustakaan BPK RI Virtual Tour, like (1) Information Point, (2) Navigation, (3) Search System (OPAC).

2.3 Library Image

Image is a set of impressions or images in the user's mind of an object [8]. In relation to tourism, the image of a tourist destination is closely related to the subjective interpretation made by an individual and influences tourist behavior (Agapito, et al., 2013). This shows that the library image can also be defined as a view or interpretation of the user about the library, which also affects the behavior of the user towards the library.

Based on the Image Construction process model by Gartner which has also been studied by McFee, et al. (2019), Kim, et al. (2017), and Agapito, et al. (2013), It is known that there are 3 (three) main dimensions in measuring Image Construction, that are (1) Cognitive Image, which refers to beliefs and knowledge about a place or destination, in which it is related to various components of a place or destination, such as attractions to behold, the environment (such as weather and cleanliness), and experiences that influence memory as the basis of cognitive structures in destination images. (2) Affective Image dimension, refers to the emotions and feelings felt by users when using various features found in a place or destination. (3) Conative Image, refers to the behavior or actions and intentions that will be carried out by users or visitors in the future (such as the intention to visit a destination, or also including comments related to the destination).

Based on a review of the literature that has been carried out regarding the relationship and effect between virtual tours on the formation of destination image, it was found that there is a positive relationship and effect between virtual tours and destination Image Construction. Like Lee, et al. (2020) found that Content Quality, System Quality, and Vividness positively affect attitudes and impressions of presence (telepresence), which also positively affects users' behavioral intentions to visit destinations. McFee, et al. (2019), found that virtual tours have a positive effect on Cognitive Image, Affective Image, and Overall Image, which also have a positive effect on tourist's intention to visit [9]. Kim, et al. (2017), found that information quality in social media positively affects Cognitive Image, Affective Image, and Conative Image [10]. However, most of the research is only conducted with an orientation in the tourism sector. In the field of libraries, research on virtual tours also only focuses on the process of creating and implementing library virtual tours, and there has been no study on evaluating library virtual tours from a user perspective. Thus, this study also contributes to the development of library virtual tour literature and library imagery, because it investigates the relationship and influence of library virtual tours on library imagery, as well as a form of evaluation of library virtual tours from a user perspective.

3 Research Methodology

3.1 Research Models and Hypotheses

This research is a bivariate study, where there are 2 (two) variables in the study, that are library virtual tour as the independent variable in the study, and library image as the dependent variable in the study. Therefore, the research model is formed as follows: (Fig. 2)

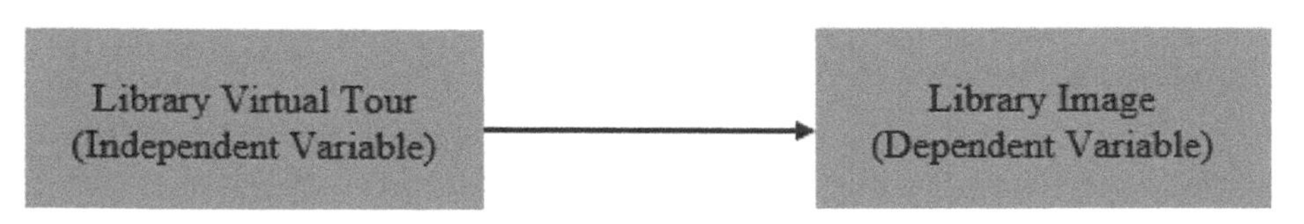

Fig. 2. Research model

Based on previous literature Lee, et al. (2020), McFee, et al. (2019), and Kim, et al. (2017), which explains that virtual tours affect the Image Construction of a destination, based on that, the hypotheses formulated in the study:

H1: Library Virtual Tour Has an Effect on Library Image.

3.2 Data Collection

This study collects quantitative data with surveys and literature studies to support the findings. Respondents are required to use a virtual library tour from the Perpustakaan BPK RI on its official website (https://www.iheritage.id/publik/bpk/), then answer the questions provided in the questionnaire.

Respondents collected amounted to 385 respondents. 30.91% (n = 119) of respondents were male, and 69.09% (n = 266) of respondents were female. In addition, the majority of respondents were in the age group <21 years and 21–25 years (n = 372, 96.63%). After evaluating the portrait of the use of the Perpustakaan BPK RI, it was found that 90.39% (n = 348) of respondents had never made a direct visit to the Perpustakaan BPK RI. The following are the demographic characteristics of the sample which are summarized in Table 1.

Table 1. Demographic characteristics of respondents

Characteristics	Jumlah	%
Gender		
• Male	119	30,91%
• Female	269	69,09%
Age		
• <21tahun	167	43,38%
• 21–25 tahun	205	53,25%
• 26 – 30 tahun	7	1,82%
• 31 – 35 tahun	1	0,78%
• 41 – 45 tahun	3	0,52%
• >45 tahun	2	0,26%
Profession		
• Not yet working	23	5,97%
• Student	311	80,78%
• BUMN/BUMD Employee	1	0,26%
• Private Employee	31	8,05%
• PNS/TNI/Polri	3	0,78%

(continued)

Table 1. (*continued*)

Characteristics	Jumlah	%
• Entrepreneur	4	1,04%
• Etc.	12	3,12%
Direct Visit to Perpustakaan BPK RI in the Last 3 Years		
• <3 times	31	8,05%
• 3 – 6 times	2	0,52%
• 7 – 10 times	2	0,52%
• >10 times	2	0,52%
• Never	348	90,39%
Utilization of Perpustakaan BPK RI Virtual Tour in the Last 5 Months		
• Never	195	50,65%
• 1 – 5 times	187	48,57%
• 6 – 10 times	2	0,52%
• >10 times	1	0,26%

Source: Researcher Processed Data, 2021

3.3 Data Measurement

Data obtained from 385 respondents from the distribution of the questionnaire from March 26 to April 3, 2021 were transformed into interval data with MSI, then descriptive analysis of the research variables and inferential statistical analysis consisted of the Pearson Correlation Test and Simple Linear Regression Test, to investigate the relationship and influence of the library virtual tour variable on the library image variable.

4 Data Analysis

4.1 Descriptive Analysis of Research Variables

Because this study also uses a Likert Scale from 1–5 to see and determine the effect of library virtual tours on the formation of library images, the mean of respondents' answers will be calculated to see the tendency of respondents' assessment of each indicator in the questionnaire. After that the mean value will be categorized into a certain class, because the class division consists of 5 (five) classes, then the interval of each class is as follows: (Table 2).

The following is a descriptive analysis of the research variables, which consist of "Library Virtual Tour" and "Library Image":

4.1.1 "Library Virtual Tour" Variable

The variable "Library Virtual Tour" consists of the content quality dimension, system quality dimension, and the vividness dimension. The following are the results of

Table 2. Scale range of all research variables

Mean	Library Virtual Tour	Library Image
$1,00 < x \leq 1,80$	Very Low	Very Low
$1,81 < x \leq 2,60$	Low	Low
$2,61 < x \leq 3,40$	Moderate	Moderate
$3,41 < x \leq 4,20$	High	High
$4,21 < x \leq 5,00$	Very High	Very High

descriptive analysis by calculating the mean (mean) of each measurement item in each dimension: (Table 3).

Table 3. Descriptive statistics "Library Virtual Tour" variable

No	Indicators	Mean	Category
Content Quality			
1	Perpustakaan BPK RI Virtual Tour gave me an overview (location, room, layout, etc.) about the Perpustakaan BPK RI	4,87	Very High
2	Perpustakaan BPK RI Virtual Tour provided the information I needed (relevant and updated) about the Perpustakaan BPK RI	4,88	Very High
3	Perpustakaan BPK RI Virtual Tour provided the information I needed (relevant and updated) about the Perpustakaan BPK RI Collection	4,07	High
4	Perpustakaan BPK RI Virtual Tour provides the information I need (relevant and updated) regarding Perpustakaan BPK RI Services	4,36	Very High
5	Perpustakaan BPK RI Virtual Tour provides easy-to-understand information about the Perpustakaan BPK RI Collection	4,11	High
6	Perpustakaan BPK RI Virtual Tour provides easy-to-understand information about Perpustakaan BPK RI Services	4,36	Very High
7	Perpustakaan BPK RI Virtual Tour provides complete information about Perpustakaan BPK RI	4,14	High

(*continued*)

Table 3. (*continued*)

No	Indicators	Mean	Category
Mean		**4,40**	**Very High**
System Quality			
1	Perpustakaan BPK RI Virtual Tour has easy navigation to move	4,29	Very High
2	Perpustakaan BPK RI Virtual Tour has easy navigation to use	4,09	High
3	Perpustakaan BPK RI Virtual Tour has easy navigation to move	4,19	High
4	Perpustakaan BPK RI Virtual Tour has a good interaction response (such as when touching or clicking a button that displays information in the form of images and videos can be displayed correctly and quickly)	4,09	High
Mean		**4,17**	**High**
Vividness			
1	Perpustakaan BPK RI Virtual Tour provides clean and clear visual images (no noise or blurry and grainy images)	4,09	High
2	Perpustakaan BPK RI Virtual Tour provides clean and clear video visuals (no noise or blurry and grainy images)	4,65	Very High
3	Perpustakaan BPK RI Virtual Tour presents visual images that look real	4,46	Very High
4	Perpustakaan BPK RI Virtual Tour presents video visuals that look real (like the real BPK RI Library)	4,09	High
5	Perpustakaan BPK RI Virtual Tour presents detailed visual images	4,32	Very High
6	Perpustakaan BPK RI Virtual Tour presents detailed video visuals	4,09	High
Mean		**4,29**	**Very High**
Mean Value of Library Virtual Tour Variables		**4,28**	**Very High**

Source: Researcher Processed Data, 2021

4.1.2 "Library Image" Variable

"Library Image" consists of the cognitive image dimension, the affective image dimension, and the conative image dimension. The following are the results of descriptive analysis of the "Library Image" variable: (Table 4).

Table 4. Descriptive statistics "Library Image" variable

No	Indicators	Mean	Category
Cognitive Image			
1	Perpustakaan BPK RI Virtual Tour gave me the knowledge that the Perpustakaan BPK RI provides reliable collections and services for users	4,36	Very High
2	Perpustakaan BPK RI Virtual Tour gave me the knowledge that the Perpustakaan BPK RI Library has a clean and tidy environment	5,00	Very High
3	Perpustakaan BPK RI Virtual Tour gave me the knowledge that Perpustakaan BPK RI provides historical tours about BPK RI in an interesting way	4,41	Very High
Mean		**4,62**	**Very High**
Affective Image			
1	Perpustakaan BPK RI Virtual Tour makes me feel that the Perpustakaan BPK RI will provide convenience for users when physically visit the library	4,75	Very High
2	Perpustakaan BPK RI Virtual Tour makes me feel that the Perpustakaan BPK RI will provide convenience for users when physically visit the library	4,75	Very High
3	I feel happy when I use the Perpustakaan BPK RI Virtual Tour	4,75	Very High
Mean		**4,75**	**Very High**
Conative Image			
1	After using the Perpustakaan BPK RI Virtual Tour, I would like to know more information about the Perpustakaan BPK RI	4,58	Very High
2	After using the Perpustakaan BPK RI Virtual Tour, I was interested in visiting the Perpustakaan BPK RI	4,09	High

(continued)

Table 4. (*continued*)

No	Indicators	Mean	Category
3	After using the Perpustakaan BPK RI Virtual Tour, I want to spread information and positive things about the Perpustakaan BPK RI Virtual Tour	4,46	Very High
4	After using the Perpustakaan BPK RI Virtual Tour, I am willing to recommend the Perpustakaan BPK RI to others	4,09	High
5	After using the Perpustakaan BPK RI Virtual Tour, I am willing to recommend library virtual tours to others	4,58	Very High
Mean		**4,36**	**Very High**
Mean Valeu of Library Image Variables		**4,58**	**Very High**

Source: Researcher Processed Data, 2021

4.2 Pearson Correlation Test

The Pearson Correlation test was conducted to determine whether there was a relationship and how strong the relationship was between the research variables. In the Correlation Test, the research variables are stated to be related if the significance value is <0.05. To determine the strength of the relationship between variables, the degree of relationship is used [11], as follows: (Table 5).

Table 5. Degree of Pearson Correlation

Value of Pearson Correlation	Correlation
0,00–0,20	No Correlation
0,21–0,40	Weak
0,41–0,60	Medium
0,61–0,80	Strong
0,81–1,00	Very Strong

The following are the results of the Pearson Correlation Test of research variables, which show that there is a “Strong” correlation or relationship with a positive direction between the research variables: (Table 6).

Table 6. Correlation test results of "Library Virtual Tour" against variable "Image Library"

		Library Virtual Tour	Library Image
Library Virtual Tour	Pearson Correlation	1	.778**
	Sig. (2-tailed)		.000
	N	385	385
Library Image	Pearson Correlation	.778**	1
	Sig. (2-tailed)	.000	
	N	385	385

Source: Researcher Processed Data, 2021

4.3 Simple Linear Regression Test

The Simple linear regression analysis is used to measure the influence of one independent variable on the dependent variable [12]. This analysis is used to answer the hypothesis in the study, that are "Library Virtual Tour has an effect on Library Image". Furthermore, the data obtained will be tested for hypotheses in the form of a t test which will show the effect of the independent variable on the dependent variable. In conducting the t-test, there are several steps that need to be carried out, namely (1) making a hypothesis, (2) setting testing rules, (3) comparing t-count and t-tables, and (4) making decisions [13].

Table 7. Results of the "Library Virtual Tour" coefficient of determination calculation against "Image Library"

Model Summary				
Model	R	R Square	Adjusted R Square	Std. Error of the Estimate
1	.778[a]	.605	.604	4.68573

a. Predictors: (Constant), Library Virtual Tour
Source: Researcher Processed Data, 2021

The Table 7 shows the effects of "Library Virtual Tour" on "Library Image", where it is known that the R Square value is 0.605, which indicates that "Library Virtual Tour" affects "Library Image" by 60.5%.

The Table 8 shows the acceptance of the hypothesis in this study, that are the acceptance of Ha. Library Virtual Tour has an effect on Library Image. It can be seen by comparing the value of t-count with t-table. The calculated t value obtained is 24.232, this value is higher than the t table ($n = 385$, $df = 383$) which has a value of 1.966. In addition, the level of influence of "Library Virtual Tour" on "Library Image" has a positive direction of influence with a value of 0.539, which shows that every 1% addition of the value of the "Library Virtual Tour" variable, it will affect the value of the "Library Image" variable of 0.539.

Table 8. "Library Virtual Tour" coefficient test against "Library Image"

Coefficients[a]						
Model		Unstandardized Coefficients		Standardized Coefficients	t	Sig
		B	Std. Error	Beta		
1	(Constant)	10.493	1.644		6.384	.000
	Library Virtual Tour	.539	.022	.778	24.232	.000

a. Dependent Variable: Library Image
Source: Researcher Processed Data, 2021

5 Advanced Discussion and Analysis

5.1 Virtual Tour and Media Social Are Effective Media to Attract Young Generation to Visit and Use Information Institutions (Libraries, Archives, Museums and Galleries)

The survey on "Characteristics of Respondents" it is shows that the majority of respondents are aged <21 years and 21–25 years who are students. These results are in line with research conducted by Dateportal:digital Indonesia 2021 [14], which shows that there are 202.6 million active internet users in Indonesia, of which 170 million are active users of social media with the majority of users being aged 13–34 years (Kemp, 2021). In addition, based on descriptive analysis, it is known that the conative image indicator regarding the intentions or desires of users after using library virtual tours gets a score with the "Very High" category, of course indicating that virtual tours are able to influence the younger generation to be interested in visiting, utilizing, and recommending libraries. These results also certainly become a strong basis for information institutions (Libraries, Archives, Museums, and Galleries) to start using, implementing, and improving information technology in carrying out business activities and promoting themselves to the wider community.

5.2 Perpustakaan BPK RI Virtual Tour has a Very Good Quality and the Image of the Perpustakaan BPK RI has a Very Good Image Based on User Rating

The Descriptive statistical analysis of the "Library Virtual Tour" variable got an average value in the "Very High" category, this shows that the Perpustakaan BPK RI Virtual Tour has very good quality, seen from each measurement indicator in the three virtual tour quality dimensions (Content Quality gets "Very High", System Quality scored "High", and Vividness also received "Very High"), including indicators regarding the quality of virtual tour content which is very good because it is able to provide an overview and collection information about the Perpustakaan BPK RI, has navigation that is easy to move, and presents visual images and videos that look real.

Likewise, the results of the descriptive analysis of the variable "Library Image" which get an average value in the "Very High" category, this also shows that the formation of the image of the Perpustakaan BPK RI gets very good results, seen from each measurement indicator in the three dimensions of the library image. (Cognitive Image, Affective Image, and Conative Image) get an average value in the "Very High" category based on user ratings, including cognitive image indicators that let users know that the Perpustakaan BPK RI has a clean and tidy environment, affective images that make users feel happy when using virtual tours and feel that the Perpustakaan BPK RI will provide comfort when they visit and use the library directly, as well as a conative image indicator that shows that after using the virtual tour, users are interested in visiting the Perpustakaan BPK RI even they are willing to recommend library virtual tours to others. Thus, it can be concluded that the Perpustakaan BPK RI Virtual Tour is able to form a very good user image (cognitive, affective, and conative) regarding the BPK RI Library or in other words, shows that the Perpustakaan BPK RI has a very good image in terms of users.

5.3 There is a Strong Relationship with Positive Direction Between Library Virtual Tour with Library Image

The Pearson Correlation test shows that there is a strong relationship with a positive direction between "Library Virtual Tour" and "Library Image". This can be seen based on the significance value <0.05 which indicates that there is a relationship between the two variables, besides the Pearson Correlation value of "Library Virtual Tour" with "Library Image" is 0.778 which is included in the "Strong" category based on the degree of relationship (Raharjo, 2017) with a positive relationship direction.

5.4 Library Virtual Tour Affects Library Image

The Simple Linear Regression Test shows that there is an effect of "Library Virtual Tour" on "Library Image". This can be seen from the t-count value of 24.232 which is greater than the t-table of 1.966, where this result also indicates that the hypothesis Ha. The Library Virtual Tour has an effect on the library image is accepted, the results of the Correlation Test and Simple Linear Regression Test are also in accordance with the research of Lee, et al. (2020), McFee, et al. (2019), and Kim, et al. (2017), where the results of their study found that there is a positive relationship and influence of virtual tours on images from the user perspective, both cognitive images, affective images, and conative images. This also indicates that library virtual tours are not only able to shape and influence user responses or assumptions about the library, but are also able to shape and influence user behavior related to their future intentions and actions regarding the library.

5.5 Constraints/Barriers in Utilizing the Perpustakaan BPK RI Virtual Tour

Based on the results of observations and utilization of the Perpustakaan BPK RI Virtual Tour, found a technical error (error) from the Perpustakaan BPK RI Virtual Tour. This can be seen from the information points provided that are not very clear and errors (can be seen in the information point of the "Journal" section and the "Accounting and Auditing" shelf, which provides information that is not so clear and even inaccessible to users) and also navigation that is not easy to access, not clickable and does not direct the user to the next section (can be seen in the navigation to want to go to the 2nd floor stairs). This fact becomes the basis for the Perpustakaan BPK RI to conduct a review of the virtual tour that has been made, as well as carry out continuous improvement and development. In addition, by looking at the fact that there is a relationship and effects between the "Library Virtual Tour" on the "Library Image", it also becomes a strong basis that improvements and enhancements to the virtual tour are very necessary, because it is able to shape the user's image of the library, and also influences user behavior. against the library.

The fact shows that there is a significant relationship and effects between "Library Virtual Tour" and "Library Image", which of course becomes a strong basis for information institutions (Libraries, Archives, Museums, and Galleries) to start increasing the application of information technology in adapting to current developments. This fact also shows that virtual tours can not only be a means of promotion, but can also be a medium of information, recreation and entertainment for the community, and this is very important to be implemented and improved in information institutions, especially in museums and galleries. By starting to implement and improve information technology, especially virtual tours, it is able to become a progressive step for information institutions to continue to survive, compete, and show seriousness to the wider community, that information institutions are not only able to meet primary needs such as information, but also strive to fulfill their recreational and entertainment needs of the community.

In addition, various facts found from research also show that by implementing a virtual tour into the world of libraries, of course, being able to encourage the development of the concept of digital libraries in the future. A digital library [15] is defined as "a collection that focuses on digital objects, including text, video, and audio, which is also related to access and retrieval methods, and has functions for selecting, organizing, and maintaining collections" (Witten, Bainbridge, & Nichols., 2010), with the implementation of virtual tours in libraries, of course it will encourage the development of digital libraries, which are no longer limited to library websites that provide virtual collections and services, but also allow users to navigate and interact virtual (such as utilizing collections and virtual services) in real-time, so that the current pandemic (Covid-19) is no longer an obstacle for libraries to continue to provide services excellence to the community, especially the potential community of libraries.

6 Conclusions and Suggestions

This research contributes in 3 (three) ways. First, this study shows that the Perpustakaan BPK RI Virtual Tour has very good quality and the Perpustakaan BPK RI has a very good image based on user assessments, by adapting the virtual tour evaluation model and Image Construction model, so that it certainly helps the Perpustakaan BPK RI in evaluating the Perpustakaan BPK RI Virtual Tour in user perspective.

Second, the research shows that there is a significant relationship and influence between the Perpustakaan BPK RI Virtual Tour and the Perpustakaan BPK RI image. It can be seen based on the results of the analysis of the Pearson Correlation Test and the Simple Linear Regression Test, which of course proves that the research hypothesis can be accepted, that are "Library Virtual Tour" has effects on "Library Image". In addition, this study also suggests that information institutions (Libraries, Archives, Museums, and Galleries) need to implement and improve the application of information technology and social media in carrying out their business activities, especially virtual tour technology, which can not only be a means of promotion for institutions, but can also be an effective information medium that is able to meet and satisfy the information, recreation, and entertainment needs of users.

Finally, this research also contributes to the development of literature related to virtual tours, especially in the field of libraries and library Image Construction. In addition, based on the literature review that has been carried out in the study, this research is the first study in evaluating library virtual tours from a user perspective that provides empirical evidence to support research ideas. Therefore, this research is expected to be a reference for further research in a wider scope.

References

1. Agapito, D., Valle, P., Mendes, J.: The cognitive-affective-conative model of destination image: a confirmatory analysis. J. Travel Tour. Market. 571–481 (2013)
2. Kim, M., Hall, M.: A hedonic motivation model in virtual reality tourism: comparing visitors. Int. J. Inf. Manag. 236–249 (2019)
3. Guttentag, D.: Virtual reality: applications and implications for tourism. Tour. Manag. 637–651 (2010)
4. Oling, L., Mach, M.: Tours trends in academic ARL libraries. Coll. Res. Libr. 13–23 (2002)
5. Xiao, D.: Experiencing the library in a panorama virtual reality environment. Library Hi Tech, pp. 177–184 (2000)
6. Lee, M., Lee, S., Jeong, M., Oh, H.: Quality of virtual reality and its impacts on behavioral intention. Int. J. Hosp. Manag. 1–9 (2020)
7. Chen, L.: The quality of mobile shopping and its impacts on purchase intention and performance. Int. J. Manag. Inf. Technol. (MIT) 23–32 (2013)
8. Astuti, P.: Membangun citra perpustakaan perguruan tinggi. JIPI (Jurnal Ilmu Perpustakaan dan Informasi) 206–225 (2016)
9. McFee, A., Mayrhofer, T., Baratova, A., Neuhofer, B., Rainoldi, M., Egger, R.: The effects of virtual reality on destination image construction. Inf. Commun. Technol. Tour. 107–119 (2019)
10. Kim, S., Lee, K., Shin, S., Yang, S.: Effects of tourism information quality in social media on destination image construction: the case of Sina Weibo. Inf. Manag. 687–702 (2017)
11. Raharjo, S.: https://www.youtube.com/watch?v=jq6N3waOQPU&t=496s. Accessed 6 March 2021
12. Raharjo, S.: https://www.spssindonesia.com/2014/02/analisis-korelasi-dengan-spss.html. Accessed 6 March 2021
13. Siregar, S.: Metode penelitian kuantitatif: dilengkapi dengan perbandingan perhitungan manual & SPSS. Kencana Prenamedia Group, Jakarta (2013)
14. Kemp, S.: https://datareportal.com/reports/digital-2021-indonesia. Accessed 6 April 2021
15. Witten, I., Bainbridge, D., Nichols, D.: How to Build a Digital Library. Morgan Kaufmann Publishers, USA (2010)

Exploring the Research Utility of Fan-Created Data in the Japanese Visual Media Domain

Senan Kiryakos and Magnus Pfeffer(✉)

Stuttgart Media University, Stuttgart, Nobelstraße 10, 70569 Stuttgart, Germany
{kiryakos,pfeffer}@hdm-stuttgart.de

Abstract. Researchers wishing to study the Japanese visual media domain do not currently have access to a large set of descriptive data for analysis. To remedy this, the Japanese Visual Media Graph project (JVMG) seeks to build a knowledge graph consisting of descriptive metadata sourced from various online fan communities, described using RDF ontologies. To better understand how this informal, crowdsourced data can be both described using a formal ontologies, and made useful to researchers, this paper presents a summary of the properties of community data from a number of fan sites, and discusses the impact it has on the creation of a unified dataset and on possible research use cases. We find that the data sources are of high quality and coverage. They complement each other well and our central ontology should enable these connections between related resources across communities. Certain niche research topics are enabled by the use of community created data alone, but others encourage the incorporation of additional authoritative sources.

Keywords: Ontologies · Knowledge graphs · Visual media · Fan communities · Data integration

1 Introduction

Japanese visual media, such as animation, manga, and video games, is a topic of interest for researchers of a variety of disciplines. As global interest in Japanese visual media has grown commercially, so too has academic interest [1]. This interest ranges from studying the domain broadly [2], to research on fandom interaction with the domain online [3]. This latter group includes studies on the utility of user-generated content [4], and intersects with more general studies on fandom and participatory culture [5,6]. Those wishing to conduct data-driven research on topics such as these are faced with the fact that there is no central database on the domain and authoritative data, e.g. from libraries, is limited in detail and coverage. On the other hand there are a lot of fan communities that

This work has been funded by Deutsche Forschungsgemeinschaft (German Research Foundation) with a grant from the funding program e-Research Technologies.

H.-R. Ke et al. (Eds.): ICADL 2021, LNCS 13133, pp. 210–218, 2021.
https://doi.org/10.1007/978-3-030-91669-5_17

create and curate a significant amount of unique and granular information on the domain of Japanese visual media. But these communities are focused on different aspects of the domain and while data across these communities often deals with the same resources, the fact is that these draw from different data sources, use distinct data models and vocabularies and describe the resources at different granularity levels. This results in a level of heterogeneity that can be an obstacle for researchers wishing to analyze the domain broadly, or who desire as much data as possible. The Japanese Visual Media Graph project (JVMG)[1] seeks to address these issues through the creation of a central knowledge graph which combines data from multiple community providers and a unified, RDF-based ontology.

Foundational to the creation of this knowledge graph is the collection and analysis of descriptive data from a number of online fan communities. This not only serves as a necessary step in the technical development, but also functions as a type of limited domain analysis, resulting in a unified ontology that better reflects the ways that different communities understand and describe their data. Additionally, this allows for fundamental aspects and limitations of the domain to be better communicated to researchers, such as levels of data quality and authority, descriptive granularity of various resource types and entities, and domain coverage, to better guide and inform their prospective studies.

In early phases of the project we identified and analyzed over 70 online community databases and selected several based on quality and quantity of data, diversity in coverage and languages, and unique site-specific information. In instances where the data was not already openly available, agreements on data exchange were made, followed by the collection and processing of data for each community separately. In order to preserve as much of the semantic richness and diversity of the sources, a Resource Description Framework (RDF) based ontology consisting of a class structure and vocabulary is created for each dataset. The processing consists primarily of a transformation from the source formats, typically SQL tables, to an RDF serialization based on the respective ontologies. While more labor intensive than simply mapping the heterogeneous community data to a central ontology, this provides several advantages[7]. Most importantly, it means that the semantic richness of each source dataset is maintained, as the meanings and constraints of individual properties reflect the original data, while also allowing for easier alterations or updates based on changes made to the original source data.

In this paper, we will provide an overview of the individual communities from which data was collected in order to show how a more thorough understanding of the domain based will both impact our unified ontology, and impact the types of research and analysis that can be performed using this data.

2 Data Structures and Vocabularies of Community Data

Community data that the JVMG has already or seeks to incorporate into its knowledge graph can be organized into three general categories. First,

[1] https://jvmg.iuk.hdm-stuttgart.de/.

communities that cover the Japanese visual media domain broadly, or contain data for the domain as part of a larger subset. These include the Media Arts Database[2], Wikidata[3] and AnimeClick[4]. Second are sources that cover a particular medium within the domain, such as anime or video games. Examples of this type are AniDB[5], and the Visual Novel Database[6], for anime and visual novels, respectively. The last group are communities that focus on a single aspect of the domain, but without a medium restriction, such as the Anime Characters Database[7].

The communities we have collaborated with all offer large amounts of data for different resource types, a varying but often high amount of descriptive granularity, and structured, though informal, data models. To illustrate these aspects, this section will feature a representative of each group and briefly discuss important features of each, such as the coverage and scope, primary entities and relationships, and descriptive properties. A table summarizing the data quantity of these three communities is shown in Table 1.

Table 1. Approximate quantities of sample community data.

Community	Works	Characters	Tags	Producers
AnimeClick	120,000	102,000	5,000	67,000
Visual Novel Database	28,000	91,000	5,400	31,000
Anime Characters Database	11,000	101,000	5,000	4,900

2.1 The Visual Novel Database

The Visual Novel Database (VNDB) is an online database for the visual novel genre of video games. This community focuses solely on visual novels (VNs), but features an extremely high granularity for most aspects of the genre, such as creative works, characters, and contributors. Interestingly, this granularity extends to the carriers of VNs (i.e. physical items for sale, or digital downloads) rather than only the content, which is the primary focus of most other communities. Both the content and carrier are represented by distinct entity types, Visual Novel and Release, with each having its own set of distinct properties and values. Content entities include those such as Characters and descriptive Trait and Tags, along with entities representing contributor roles such as Staff and Producer.[8]

[2] https://mediaarts-db.bunka.go.jp/.
[3] https://www.wikidata.org/.
[4] https://www.animeclick.it/.
[5] https://anidb.net/.
[6] https://vndb.org/.
[7] https://www.animecharactersdatabase.com/.
[8] A description of the dataset, along with the RDF ontology created by the authors for the VNDB is available at https://doi.org/10.5281/zenodo.5506936.

Though the scope of VNDB is limited to a single medium, it is by far the most thorough dataset for that medium. Both creative works and their contents are described in extreme detail, and connections between relevant entities have been established. Relationships between entities are plentiful, with all related characters, releases, contributors, and other VNs being connected via a single umbrella VN entity. The granular description of contents is perhaps the greatest strength of VNDB, as the Tag and Trait hierarchies are robust, logically connected, and occasionally informally defined through links to external resources, primarily Wikipedia. Other communities may apply a trait such as "criminal" to a character, whereas VNDB features the more specific "blackmails" trait, which is a part of the "Engages In – Crime – Blackmail" hierarchy, is defined via a scope note, contains a link to the Wikipedia article on Blackmail, and provides "extortion" as an alias. This granularity extends to Tags attributed to VNs, where the parent group "Theme" extends to "Drama–Health Issues–Psychological Problems–Eating Disorder". Though this hierarchy is not a traditional subject authority file, building formal vocabularies of descriptive tags for niche mediums with user-generated content has previously been undertaken [4]. While the focus on a single medium is a clear limitation for a database covering Japanese visual media broadly, and descriptive data regarding content is arguably subjective, the opportunity to extend VNDB's Tag and Trait hierarchy to a wide range of applicable Japanese visual media, along with its fairly thorough coverage of an entire medium, are important reasons for its inclusion in a unified community dataset.

2.2 The Anime Characters Database

The Anime Characters Database (ACDB) is a community database dealing primarily with anime characters, though characters sharing an 'anime aesthetic' from other mediums, such as original art, video games, and manga, are also included. ACDB refers to itself as a 'visual search engine', and indexes characters according to a specific set of traits, such as hair and eye color, age, and type of clothing worn. This results in a uniform set of available traits for characters, and less ambiguity due to the limited numbers of descriptive traits available, but also results in a limited level of granularity compared to what a character may receive on VNDB. For example, characters can have their general hair length identified, but not a specific cut or style name. In addition to the central `Character` entity are `People` entities that represent voice artists, `Work` entities which are the source material for characters, and the descriptive `Character Tag` and `Series Tag` entities describing characters and works respectively.[9]

Though descriptive granularity is limited when compared to VNDB, ACDB contains data for a variety of mediums, and this has significant implications for the types of relationships between entities in the dataset. While both VNDB and ACDB contain relationships between creative works and characters, ACDB

[9] A description of the dataset, along with the RDF ontology created by the authors for the ACDB is available at https://doi.org/10.5281/zenodo.5508699.

covering a variety of mediums beyond VNs means that descriptive data applied to a character is applied to many more instances of that character than those in VNDB. Similarly, while both contain recursive relationships between creative works, these relationships in ACDB extend beyond related VNs found in VNDB. One unique outcome of this is that ACDB can connect `Work` entities representing specific medium instances to a `Work` entity representing a broad series or franchise, something not possible in VNDB, as it by nature does not feature entities representing multimedia franchises. While the descriptive properties are still useful and do contribute unique data, relationships between entities are the most significant contribution of ACDB to a unified community dataset.

2.3 AnimeClick

AnimeClick is an Italian language fan site for Japanese anime, manga, and live action drama. The site is a general fandom wiki, and covers these mediums broadly, without a particular focus. Data are represented by four primary entities - `Animation Work`, `Comic Work`, `Character` and `Staff`. The `Animation Work` entity includes various animated formats, such as TV series, films, or original video animations, while `Comic Works` are primarily manga. The granularity of the data varies between entity types, with work entities being fairly detailed, and information on characters being more limited. Relationships between relevant entities are present, including adaptations of `Animation` and `Comic Works` to one another, and `Characters` to the `Staff` that voiced them. While some descriptive data is of limited utility due to it being in Italian, there is also a significant amount of useful language-agnostic data, such as episode counts, completion statuses, and connections between related resources.[10]

AnimeClick's dataset can be seen as a type of broad middle ground between the types of datasets represented by VNDB and ACDB. While VNs are largely absent from AnimeClick's dataset, it does describe multiple mediums, similar to ACDB. Descriptive granularity for characters is quite low, while works are described in greater detail. AnimeClick also features the parent-child relationship found in ACDB, connecting individual entities to a high-level franchise. Unlike ACDB, relationships are also present between members of a given franchise, with their relationship type defined, indicating sequels, prequels, derivative works, etc. Descriptive data for mediums beyond those covered in VNDB, provided by additional relationships between entities than those provided by ACDB, are the primary contributions that AnimeClick provides to a unified community dataset. Additionally, the ability to incorporate Italian data that is able to contribute usable and relatable data to a largely English dataset encourages the inclusion of other international communities, both for any additional data they may provide, and to expand the audience of the JVMG database.

[10] A description of the dataset, along with the RDF ontology created by the authors for AnimeClick is available at https://doi.org/10.5281/zenodo.5508683.

3 Discussion and Impact of the Data Analysis

After having summarized each community's data in Sect. 2, this section explains the impact that the analysis of the data has had, and continues to have, on the ongoing development of the JVMG project.

3.1 Impact on the JVMG Ontology

The intention of the JVMG is to present a unified view on the domain of Japanese visual media and as such a unified domain model is needed. The analysis of the models derived from the fan databases has been very helpful in determining key aspects of the unified domain model. In particular, analyzing VNDB revealed that, while limited to only describing VNs and their contents, the granularity was unmatched by other sources, even when compared to other sources, such as Wikidata or the Media Arts Database. The benefits of this for researchers interested in VNs is clear, but immediate advantages for those interested in Japanese visual media more broadly, or simply other mediums, are less so. However, the phenomenon of multimedia franchises or 'media-mix' in Japan is extremely common, meaning a lot of VNDB's data, particular for characters and producers, can be used in conjunction with other datasets, once relationships between them have been established. For example, the granular trait data based on a character's appearance in a VN may be able to be applied to the same character's appearance in a manga, anime, or live action adaptation.

The need to facilitate and establish relationships between communities were also the takeaways of both the ACDB and AnimeClick datasets. Though descriptive granularity from these communities is limited when compared to VNDB, their inclusion of more media types and their ability to contribute unique data means that the linking of data from communities is greatly expanding the amount of information available to JVMG database users, and not simply providing redundant data from multiple sources. The media-mix mentioned previously is again important here; opportunities for connections between related works, creators, and characters from VNDB, ACDB, and AnimeClick are plentiful, so long as relationships can be identified and established. The importance of these relationships has informed the development of the JVMG RDF ontology, affecting both its properties and classes. First, the vocabulary needs properties that are able to link related entities. While this type of property is common in many ontologies, such as the `relation` or `isVersionOf` properties from the DCMI Terms vocabulary[11], the granularity of community data means that relationship types are often defined, e.g. sequels, prequels, or adaptations. To maintain the semantic richness of the source data, the granularity of relationship properties in our ontology should be equal to community data that the ontology is describing. While relational properties can link related resources while maintaining entity separation, combining related data into a single entity is beneficial for the visual browsing of multi-community data via a single access point. Research

[11] https://www.dublincore.org/specifications/dublin-core/dcmi-terms/.

has shown that this is a preferred method for accessing a group of related media by various audiences [8,9], and its use for Japanese visual media can be seen online in Wikipedia articles that describe multimedia franchises rather than single medium instantiations. The ontology should therefore also provide classes able to represent a similar set of related data from varying mediums and entities. This has been the subject of past research [10], and Sect. 2 touched upon how AnimeClick and ACDB have versions of this, but an extension of this concept and its modeling in the JVMG dataset should be implemented. The merging of data from multiple sources into franchise entities can also create problems relating to redundant or conflicting data stemming from communities using different primary source data, differing transliteration styles for Japanese titles or names, etc. Providing a way to mitigate these issues by creating properties allowing for the labeling of specific data as ground truth, where it has been able to be accurately identified, is also a feature that the ontology should support.

3.2 Impact on Researchers

The impact that community data has on research avenues was determined in part by three primary researchers that are a part of the internal JVMG team, and whose areas of interest include digital humanities and media studies. While additional external researchers have been collaborative partners as the project has progressed, with more planned in the future, feedback from these internal researchers during ongoing development was helpful in guiding various factors of the database; some of their findings and concerns are discussed here.

One example research question focussed on the relationship between character archetypes and the audience. Here, the VNDB data proved extremely valuable, as this medium often deals with romance as a theme, with the player taking the role of the protagonist. Due to the high granularity of the data, the researcher was able to identify and define distinct archetypes through analyzing co-occurring traits, and hypothesize the reasons for the frequency of certain archetypes and the effects they may have on the audience. While the preliminary analysis used only the VNDB dataset, it can be expanded on other media using the connections between the data sources.

Another researcher conducting more broad data-driven research found that despite the accumulation of multiple datasets providing a significant quantity of granular data, issues still arose when attempting to definitively address their various hypotheses. Research into topics such as the existence or lack of temporal changes in the relative frequency of shared character traits, and comparing networks of creators of large multimedia franchises, were impeded by either a limited level of data granularity, or by missing or conflicting data. These critiques suggest that the knowledge graph needs both additional granular data, as well as some amount of authoritative data and quality control.

As we are constantly seeking to incorporate additional data into the JVMG, the former of these concerns will hopefully be addressed as the project progresses, but the amount of extreme granular data for Japanese visual media is often limited, and there may simply be research questions that desire a subset of data that is not readily available. With regards to data quality concerns, we have already begun to address this through the inclusion of data from the Media Arts Database (MADB). This source is a Japanese government funded database that includes descriptive metadata for anime, manga, and video games. Unlike the fan community data, MADB's data is, depending on the medium, sourced directly from producers, publishers, and library catalogues. While this will not result in a lot of additional granular data, specifically for the contents of creative works, it will be able to act as an authoritative source for important fields such as titles and names, addressing some data conflicts between different communities and acting as the source of ground truth mentioned in Sect. 3.1.

4 Conclusion and Future Work

In this paper, we have examined the current landscape of descriptive data for the Japanese visual media domain from the perspective of fan communities online, and the role that this data has had in informing the JVMG. We have identified three categories of fan communities and presented a representative example for each. We have shown that the data models can be connected to merge the information that is present in the individual datasets into a larger model. Also, while some data might be redundant, given the diverse interests of the individual communities, a significant part of the data augments the other sources and allows for the creation of a richer, more granular description of the included entities. This better understanding of the domain informed key aspects of the JVMG, including how the central ontology should enable as many meaningful connections between community data as possible, and how we can better meet the needs of prospective researchers by continuing to incorporate more granular and more authoritative data into the knowledge graph.

Plans for future work are to address the issues revealed during the data analysis and impact assessments, i.e. concerns with data quality/authority, and the ability to establish relationships between community data. A separate analysis on data quality within the communities is currently underway, and the previously mentioned incorporation of the Media Arts Database dataset will directly address some authority concerns. The identification of related data available to connect is an ongoing process, and we are continuing to explore ways in which this can be done. Matching fields such as titles and creators has been successful, but we are currently exploring other ways to identify eligible data, such as via characters or matching Wikipedia links. After additional data has been connected and data quality further addressed, we hope to work with additional outside researchers and collect more feedback to further expand and improve the knowledge graph.

References

1. Fennell, D., Liberato, A.S.Q., Hayden, B., Fujino, Y.: Consuming anime. Telev. New Media **14**(2), 440–456 (2013). https://doi.org/10.1177/1527476412436986

2. Berndt, J.: Anime in academia: representative object, media form, and Japanese studies. Arts **7**(4), 56–69 (2018). https://doi.org/10.3390/arts7040056
3. Lee, J.H., Shim, Y., Jett, J.: Analyzing user requests for anime recommendations. In: Proceedings of the 15th ACM/IEEE-CS Joint Conference on Digital Libraries (JCDL 2015), pp. 269–270. Association for Computing Machinery, New York (2015). https://doi.org/10.1145/2756406.2756969
4. Windleharth, T.W., Jett, J., Shmalz, M., Lee, J.H.: Full steam ahead: a conceptual analysis of user-supplied tags on steam. Cat. Classif. Q. **54**(7), 418–441 (2016). https://doi.org/10.1080/01639374.2016.1190951
5. Mittell, J.: Sites of participation: wiki fandom and the case of lostpedia. Transform. Works Cult. **3** (2019). https://doi.org/10.3983/twc.2009.0118
6. Popova, M.: Fan studies, citation practices, and fannish knowledge production. Transform. Works Cult. **33** (2020). https://doi.org/10.3983/twc.2020.1861
7. Kiryakos, S., Pfeffer, M.: The benefits of RDF and external ontologies for heterogeneous data: a case study using the Japanese visual media graph. In: Information between Data and Knowledge: Proceedings of the 16th International Symposium of Information Science (ISI 2021), pp. 308–320. Glückstadt: Verlag Werner Hülsbusch Regensburg (2021). https://doi.org/10.5283/epub.44950
8. Lee, J.H., Clarke, R.I., Rossi, S.: A qualitative investigation of users' discovery, access, and organization of video games as information objects. J. Inf. Sci. **42**(6), 833–850 (2016). https://doi.org/10.1177/0165551515618594
9. Tallerås, K., Dahl, J.H.B., Pharo, N.: User conceptualizations of derivative relationships in the bibliographic universe. J. Doc. **74**(4), 894–916 (2018). https://doi.org/10.1108/JD-10-2017-0139
10. Kiryakos, S., Sugimoto, S.: Building a bibliographic hierarchy for manga through the aggregation of institutional and hobbyist descriptions. J. Doc. **75**(2), 287–313 (2019). https://doi.org/10.1108/JD-06-2018-0089

Interactive Curation of Semantic Representations in Digital Libraries

Tim Repke[1(✉)] and Ralf Krestel[2,3]

[1] Hasso Plattner Institute, University of Potsdam, Potsdam, Germany
tim.repke@hpi.uni-potsdam.de
[2] Kiel University, Kiel, Germany
[3] ZBW Leibnitz Information Centre for Economics, Kiel, Germany
r.krestel@zbw.eu

Abstract. Digital libraries often contain many heterogeneous documents and cover a variety of topics. Computer generated virtual maps of such collections can help to get an overview and explore the data. The position of each document from the corpus on this virtual two-dimensional map is determined by its semantic similarity to the other documents. However, the computed layout of the data may not adhere to the expectation of domain experts. To this end, we propose a novel approach that enables users to interactively curate the layout of the data. By dragging only a few documents on the canvas, the user can adjust the computed layout to better reflect the expected interpretation of the underlying data. We demonstrate the effectiveness and robustness of our approach using a series of real world datasets.

Keywords: Artificial intelligence for digital libraries · Dimensionality reduction · Data visualisation · Interactive machine learning

1 Introduction

Many digital libraries contain large numbers of heterogeneous documents covering a variety of topics. In order to get an overview and explore these collections, suitable semantic representations are needed to allow intuitive visualisations later. Deep learning methods provide one way to embed documents into semantic spaces, where each document can be represented by a high-dimensional vector. Vectors of semantically similar documents hereby reside closer to one another in this space. In the context of digital libraries, there is a plethora of related work on learning and applying representations for documents. High-dimensional vector representations are utilised in explainable models to interactively gather insights into a digital library [36,37], for visual search interfaces [25,34,38], as well as for interactive clustering or classification [11,12]. Visualisations, such as overview maps of an entire digital library, are powerful tools to explore the data [15]. On such a map of a digital library, which we call the *document landscape*, each document is represented by a point on a two-dimensional canvas,

H.-R. Ke et al. (Eds.): ICADL 2021, LNCS 13133, pp. 219–229, 2021.
https://doi.org/10.1007/978-3-030-91669-5_18

such that semantically similar documents are near one another. The layout of the points on the document landscape is typically done by a dimensionality reduction algorithm that projects the high-dimensional representations into a two-dimensional space, while preserving their pairwise cosine similarities as best as possible. This process has several drawbacks: First, the dimensionality reduction inevitably looses information and usually only favours local similarities. Second, the underlying data may be interpreted in different ways depending on the use-case and fine-tuning the embedding model in an unsupervised setting is also not possible. Once domain experts can explore the layout using an appropriate interactive visualisation, they may be able to suggest edits by dragging documents to different locations to better fit their mental model based on their valuable background knowledge. For example, assuming the documents mostly consist of business reports, a financial expert may want to group documents by industry sectors, whereas an environmental expert may prioritise geographical and technological aspects. The ability to interactively curate the visualisation of embedded documents is easier than, for example, developing a special document embedding model for fiction novels that learns explicitly designed aspects [30]. Therefore, we propose to use an algorithm, that enables users to manipulate the layout with only a few drag-and-drop edits of data-points. The algorithm should update the layout accordingly, while preserving the overall arrangement where possible. This reduces the manual effort to create usable maps of a dataset for a specific use-case, as single edits are augmented. Such an algorithm needs to take the intent behind a user's edit into account. We define an *edit* to be the action of dragging a single point on the map to a new location. There are three fundamental intents, namely *Separate*, *Merge*, and *Arrange*.

In this paper, we propose *ediMAP*, which is based on UMAP [23] to augment the curation of two-dimensional maps of data. The feedback provided by the user by updating the position of a few data points on the map is used to update the underlying similarity graphs and thus the two-dimensional layout of the data. We demonstrate the effectiveness of our approach using several real-world datasets.

2 Related Work

Visualisations are valuable tools to explore digital libraries and gain insights in an intuitive manner. In this paper, we focus on the computer assisted interactive curation of map-like visualisations, which are two-dimensional semantic layouts of a digital library. This form of visualisation has been used on text corpora in different domains, for example in medicine [37], for climate change research [6], and patents [16]. Pang et al. [27] found that transferring concepts and analogies from geographic maps to these artificial maps helps users to get a better overview of their digital library. Depending on the use-case, users may want to be able to interactively manipulate the layout of the data. We identify three ways to incorporate user feedback into the layout process. First, by preconditioning the layout process. Therefore, dimensionality reduction algorithms either use (partially) user annotated data [3,26], manually annotated pairs of very similar or dissimilar items from the digital library [24], or the map is initialised by placing a few items on the empty canvas [31]. Second, by interactive model parametrisation, which allows users to update the layout by changing model parameters [32] or composing a mixture of multiple models [14]. Third, by directly editing existing layouts, where users can manipulate the position of points in an existing two-dimensional layout by dragging.

In this paper, we focus on the third way of incorporating user feedback. Endert et al. [10] discussed interaction patterns for semantic landscapes. In their work, they also proposed a framework of updating a force-based layout of the data. Spathis et al. [33] use a very similar framework. They however use a neural network to first replicate a reference layout provided by an arbitrary dimensionality reduction algorithm. Edits made by a user are then used to update the model. Both these approaches are limited to handle only very small datasets. Contrary to directly editing the landscape, Yuan et al. [18] proposed a dimensionality reduction algorithm that can be influenced by combinations of quality metrics. However, their goal is to optimise visualisations of multi-variate data reduced to more than two dimensions. In either of these setups, a fundamental requirement is the interpretability of the resulting visualisation as discussed by Ding et al. [9]. Furthermore, Lespinats et al. [22] raised the question, whether it is even possible to find faithful two-dimensional mappings of the originally high-dimensional data. Bian et al. [5] avoid this issue by updating the input data itself. Each edit done by a user in the two-dimensional visualisation produced by multi-dimensional scaling of a pre-trained BERT model [8] is propagated back to update the model's last layer.

In our work, we acknowledge the fact, that a layout cannot preserve both, the global and local similarities of all points. Especially in embedding models of textual corpora, there are many ambiguities and overlapping word senses [4] as well as semantic and syntactic subspaces [28]. Others argue that the concept of proximity in a high-dimensional space may not be qualitatively meaningful [1] or that the distribution of distances in this space is closely related, among other characteristics, to the discriminability of the data [17]. Therefore it is essential to identify different interpretations and aspects of the data a user intents to prioritise by their edits. Kobak et al. [19] have demonstrated that dimensionality reduction algorithms can be strongly influenced by their initialisation. User edits could be incorporated in the initialisation to influence the overall result. This approach would be limited to very coarse-grained changes to the overall layout. Most of the recent state-of-the-art algorithms for dimensionality reduction are based on a graph representation of similarity neighbourhoods: either by directly embedding graphs using a neural model [35], manifold approximation [23], or as most recently proposed by minimising distortion functions [2]. We utilise the representation of the data as simplicial sets in UMAP [23] by updating the simplices based on the similarity neighbourhoods affected by user edits.

3 Computer-Assisted Curation of Document Landscapes

The layout of a document landscape of a digital library is typically computed by reducing the dimensionality of high-dimensional semantic representations of all documents from that library. We assume this set of high-dimensional vector representations $x_i \in X \subset \mathbb{R}^n$ to be given. Dimensionality reduction algorithms use a similarity metric, defined by a distance measure $d(x_i, x_j)$ between high-dimensional points, to compute a projection that faithfully preserves pairwise similarities in the layout. This projection $P : \mathbb{R}^n \mapsto \mathbb{R}^2$ maps each item x_i to its two-dimensional counterpart $y_i \in Y$. For the scope of this work, we assume that such an *initial layout* already exists. Given the *initial layout* of the digital library, a user curates the map by dragging points from their *source location* y_k to their *target location* $\mathring{y}_k$. The proposed algorithm assists this curation process by using these changes and computing an *updated layout* of the digital library. Thereby, the updated layout should fulfil the following set of objectives.

1. A manipulated point should be positioned at or near the target location in the updated layout.
2. The position of points in the proximity of the target location shall not differ significantly from the initial layout.
3. Points in the proximity of the source location may also move to a different location if necessary.
4. Changes to the general layout of the map shall be minimal to preserve the user's mental map of the data. This can also be referred to as stability or robustness.

Note, that algorithms for assisting the curation process may not be able to accommodate all these objectives, as the actually intended edit may contradict some of these objectives.

Our proposed algorithm assisting the curation of document landscapes is based on the popular UMAP algorithm [23]. The core principle of UMAP is to use a network of similar items from the dataset to create the two-dimensional layout of the data. The similarities are represented as a weighted network of items that are neighbours in the high-dimensional space, so-called fuzzy simplicial sets. A spectral embedding of this network is used to initialise the layout, which is then fine-tuned with a force-directed layout algorithm.

We utilise this concept of manifold approximation and force-directed graph drawing in *ediMAP*. Hereby, the edits to the document landscape suggested by a user are used to update the similarity network. As shown by related work, the starting point of any (re-)layout process has a significant impact on the resulting layout [19,29]. By continuing to update the initial layout and only partially changing the underlying similarities, we preserve the mental map of the data as much as possible. Since we assume that only the layout of a digital library is provided and not a fitted UMAP model, we first need to construct the normalised similarity graph. The similarities are based on the distance measure $d(x_i, x_j)$ between high-dimensional vectors, each representing a respective document from the library. Let $\mathcal{N}^k_{x_i}$ be the set of k nearest neighbours of document x_i. For each $x_j \in \mathcal{N}^k_{x_i}$, we add an edge (x_i, x_j) to the similarity graph. The edge weights are defined as $\exp(-\max\{0, d(x_i, x_j) - \rho\}/\sigma)$, where ρ is the distance to the closest neighbour to x_i and σ the distance to the k-th closest neighbour to x_i. Note, that UMAP is actually defining σ to be the smoothed k nearest neighbour distance.

This similarity graph is based on the high-dimensional vector representations, but should also reflect the proximity between documents in the initial layout, as most dimensionality reduction algorithms commonly aim to preserve local similarity neighbourhoods.

As a user moves a document x_i from its source location y_i in the initial layout to its target location $\mathring{y}_i$, we update the similarity graph as follows. First, we determine the k nearest neighbours $\mathcal{N}_{\mathring{y}_i}^k$ of the manipulated document at the target location $\mathring{y}_i$. As before, we add edges for each neighbour to the similarity graph and weigh them by their normalised distance. This time, however, we use the average of the normalised distances in the high-dimensional and two-dimensional space. Using only either one space to determine the similarity weight would either neglect the actual similarity in the semantic representation or what the user actually sees while curating the document landscape. Furthermore, we update the edge weights of the original neighbourhood of the manipulated document x_i in the similarity graph as follows. Edges, if they exist, connecting x_i to the k nearest neighbours $x_j \in \mathcal{N}_{y_i}^k$ at the source location y_i, are reduced by the factor $\xi \in (0, 1)$. All edges, apart from the aforementioned, connecting any $x_j \in \mathcal{N}_{y_i}^k$ are reduced by the factor of ξ^2. This reduction of edge weights limits the otherwise counteracting forces in the update phase of the layout. Furthermore, it could be exposed in a user interface for curating document landscapes as a user defined parameter to influence, how much the documents in the source neighbourhood should be moved along with the document that was edited.

Finally, we revise the layout of the document landscape using a force-directed layout algorithm based on the updated similarity graph. For each node that was affected by updating the similarity graph, we iteratively update the location of the respective document's position y_i on the landscape over several epochs. In each epoch of the layout optimisation, the location is updated to $\tilde{y}_i$ using

$$\tilde{y}_i = y_i - \eta \sum_{(y_i, y_j, w_{i,j}) \in \mathcal{G}} w_{i,j} \frac{(y_i - y_j)}{||y_i - y_j||},$$

where η is the learning rate, which decays with each iteration. Typically, force directed layout algorithms require an additional repelling force. However, since we only manipulate the locations of documents affected by the update of the similarity graph and use all the remaining that are connected to these documents in the graph as fixed references, we only need the attracting force defined above.

4 Evaluation

In this section, we apply our model for interactive dimensionality reduction to several real world datasets. We simulate user interactions to measure how well our model can fulfil the expectations and objectives we defined earlier across several different setups. The resulting maps of the datasets are quantitatively and qualitatively evaluated in a series of experiments.

For our experiments, we use six datasets with different characteristics: real world datasets with multivariate, image, and text data, as well as an artificial dataset. This includes the well-known *MNIST*[1] dataset of written digits [21] and the *MNIST-1D*[2] variant, which is derived from the original data but harder to separate [13]. We also use *FashionMNIST*[3], which contains greyscale images of fashion articles like shoes and sweaters across ten categories [39]. Aside from image data, we also use the multivariate *Seeds*[4] dataset [7]. It is comprised of measurements of wheat kernels and thus provides intuitively interpretable dimensions. Furthermore, we evaluate our approach on textual data using the *20-Newsgroups*[5] dataset [20]. Real-world datasets often contain overlapping or ambiguous latent aspects, which makes them difficult to use for evaluation. Thus we generate the artificial *Blobs* dataset to control the latent aspects within the high-dimensional space.

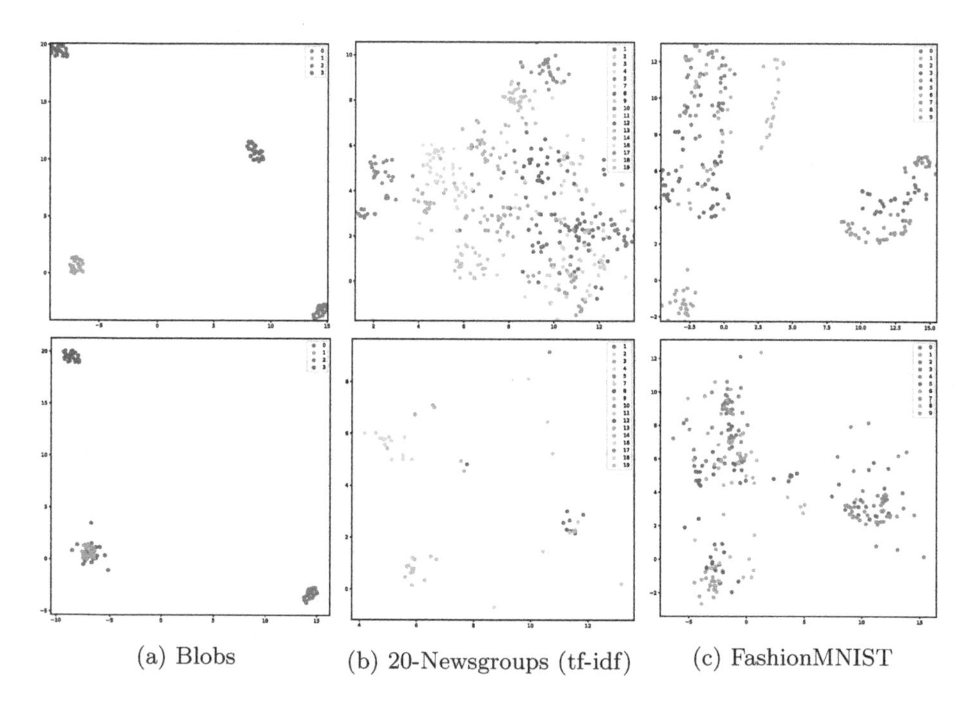

(a) Blobs (b) 20-Newsgroups (tf-idf) (c) FashionMNIST

Fig. 1. Scatterplots of dataset layouts before (top) and after (bottom) curation.

We simulate user edits by moving one or more points with a specific label towards the centroid of points with another label. In this way, we mimic the user intent of merging two clusters of points. This procedure is repeated for different numbers of manipulated points and different sets of labels. In particular,

[1] http://yann.lecun.com/exdb/mnist/.
[2] https://github.com/greydanus/mnist1d.
[3] https://github.com/zalandoresearch/fashion-mnist.
[4] https://archive.ics.uci.edu/ml/datasets/seeds.
[5] http://qwone.com/~jason/20Newsgroups/.

we simulate the following sets of edits: For the *20-Newsgroups* dataset, we simulate edits with the intent to create a document landscape of four clusters of messages. These four clusters are based on the respective subtopics of *computer* (`comp.*`), *recreational* (`rec.*`), *science* (`sci.*`), and *talk* (`talk.*`). Similar to the *20-Newsgroups* dataset, we simulate a curation of merging clusters for the *FashionMNIST* dataset. Of the originally more fine-grained labels, the intent is to form groups of articles, in particular *footwear* (sneaker, boot, sandal), tops (tshirt, pullover, shirt, coat), and others (trouser, dress, bag). Lastly, for the *Blobs* dataset, we simulate the intended merging of two clusters.

Note, that in an actual interface for curating document landscape, the data may not be annotated as described here. We only use these datasets for a clear definition of semantically similar groups of items in a dataset. Figure 1 shows the three datasets before and after applying edits and adapting the layout with *ediMAP*. The artificially generated *Blobs* dataset has a clear separation between differently labelled points in the initial layout. *ediMAP* is able to perfectly achieve the goal of merging the orange and blue clusters without altering the location of any other points. When moving many points from different source locations, as done for the *20-Newsgroups* and *FashionMNIST* datasets, the individual intents are hard to distinguish and many points overlap in the resulting layout. To circumvent this issue, we split the edits across smaller batches and do repeated partial updates. In these settings, adding a repulsion factor as in traditional force directed layout algorithms may resolve this issue.

We also evaluate our approach quantitatively and compare the results to *iSP*, a neural network based approach that learns to replicate the initial layout and can be retrained after edits [33]. We improved the performance by adding another term to the objective function of *iSP* in order to stabilise the layouts.

Table 1. Displacement measures after updating the layout with *iSP* (left) and *ediMAP* (right) based on simulated edits of 10% of points.

	Blobs		News		Seeds		MNIST		MNIST1D		F-MNIST	
TOTAL	0.41	0.08	0.50	0.01	0.45	0.13	0.43	0.06	0.57	0.06	0.36	0.07
TARGET	0.39	0.09	0.55	0.00	0.51	0.25	0.45	0.08	0.59	0.08	0.33	0.10
DIST	0.43	0.39	0.60	0.26	0.58	0.22	0.49	0.17	0.62	0.13	0.39	0.24
DTT	−0.39	−0.03	−0.27	−0.06	−0.23	−0.04	−0.29	−0.01	−0.12	−0.02	−0.31	−0.02

Corresponding to the previously defined objectives for the updated layout, we determine (1) the distance of the manipulated points between the target location and their location in the updated layout (DIST); (2) the displacement of points between the initial layout and the updated layout of points near the target location (TARGET); (3) the difference between all pairwise distances of all points in the source and target cluster, based on the label information, before and after updating the layout (DTT); and (4) the total displacement of all points between the two layouts (TOTAL). Smaller numbers are better, DTT should be negative in most cases. However, since the overall purpose of our approach is to update the layout, some movement has to necessarily occur. Thus, we exclude all points from the calculations that were explicitly edited. All results are averaged over multiple runs and normalised by the relevant total number of points and the size of the respective landscape. We move randomly selected points that have the same label towards the centroid of points that have a different label and repeat this process for multiple pairs of labels. All values are normalised to reduce the effect of different sizes of the initial layouts and the varying sizes of the datasets. We list the results for all metrics and configurations after updating the layout in Table 1. Here we can see, that the layouts updated using *ediMAP* generally cause less displacement to the overall layout, as shown by the TOTAL and TARGET metrics. Although *iSP* uses a mask to minimise the effect it has on the general layout, there is still more movement overall. Both algorithms show a similar performance in the DTT and DIST metrics, however *ediMAP* requires less points to be edited to minise these numbers.

In conclusion, we were able to demonstrate that our proposed *ediMAP* algorithm provides useful assistance for curating a document landscape. In judging the performance of *ediMAP* or any other curation assistance it is important to note, that identifying the intent of a user's edit is almost impossible. Using only the feedback of dragging a document to a new target location can be interpreted in many different ways. With *ediMAP*, we focused on one aspect, the intent of merging clusters of documents a user determined to be similar and were able to show its effectiveness to achieve that goal.

5 Conclusion

In this paper, we presented an approach for the interactive curation of semantic representations in digital libraries. Given a two-dimensional projection, which we call the document landscape, our algorithm is able to assist the curation process based on only a few suggested edits by a user. We described, how our *ediMAP* algorithm uses a similarity graph to update existing layouts of document landscapes. Additionally, we improved a neural network based model from related work to use as a baseline. In the evaluation on several real-world datasets, we were able to demonstrate the effectiveness of our approach.

However, as discussed in the evaluation, identifying the user's edit intent is important, yet very challenging. Conditioning models on specific intents to generate several suggested updated document landscapes for a given edit will be part of future work, along with an actual user interface. Intuitive and semantically meaningful visualisations of digital libraries heavily rely on good high-dimensional semantic representations of the documents. Utilising the user feedback given during the curation process could be propagated back to a representation model to improve the model itself, not just the 2-dimensional visualisation.

References

1. Aggarwal, C.C., Hinneburg, A., Keim, D.A.: On the surprising behavior of distance metrics in high dimensional space. In: Van den Bussche, J., Vianu, V. (eds.) ICDT 2001. LNCS, vol. 1973, pp. 420–434. Springer, Heidelberg (2001). https://doi.org/10.1007/3-540-44503-X_27
2. Agrawal, A., Ali, A., Boyd, S.P.: Minimum-distortion embedding (2021). arXiv:2103.02559
3. An, S., Hong, S., Sun, J.: Viva: semi-supervised visualization via variational autoencoders. In: Plant, C., Wang, H., Cuzzocrea, A., Zaniolo, C., Wu, X. (eds.) Proceedings of the International Conference on Data Mining (ICDM), pp. 22–31. IEEE (2020). https://doi.org/10.1109/ICDM50108.2020.00011
4. Arora, S., Li, Y., Liang, Y., Ma, T., Risteski, A.: Linear algebraic structure of word senses, with applications to polysemy. Trans. Assoc. Comput. Linguist. (TACL) **6**, 483–495 (2018)
5. Bian, Y., North, C.: DeepSI: interactive deep learning for semantic interaction. In: Proceedings of the International Conference on Intelligent User Interfaces (IUI), pp. 197–207. ACM Press, Geneva, Switzerland (2021)
6. Callaghan, M., Minx, J., Forster, P.: A topography of climate change research. Nat. Clim. Change **10**, 118–123 (2020). https://doi.org/10.1038/s41558-019-0684-5
7. Charytanowicz, M., Niewczas, J., Kulczycki, P., Kowalski, P.A., Łukasik, S., Żak, S.: Complete gradient clustering algorithm for features analysis of x-ray images. In: Piętka, E., Kawa, J. (eds.) Information Technologies in Biomedicine. Advances in Intelligent and Soft Computing, vol. 69, pp. 15–24. Springer, Berlin, Heidelberg (2010). https://doi.org/10.1007/978-3-642-13105-9_2
8. Devlin, J., Chang, M., Lee, K., Toutanova, K.: BERT: pre-training of deep bidirectional transformers for language understanding. In: Proceedings of the Annual Conference of the North American Chapter of the Association for Computational Linguistics (NAACL), pp. 4171–4186. ACL (2019). https://doi.org/10.18653/v1/n19-1423
9. Ding, J., Condon, A., Shah, S.P.: Interpretable dimensionality reduction of single cell transcriptome data with deep generative models. Nat. Commun. **9**(1), 1–13 (2018)
10. Endert, A., Fiaux, P., North, C.: Semantic interaction for visual text analytics. In: Proceedings of the SIGCHI conference on Human Factors in Computing Systems (CHI), pp. 473–482. ACM Press (2012)
11. Ghosal, T., Raj, A., Ekbal, A., Saha, S., Bhattacharyya, P.: A deep multimodal investigation to determine the appropriateness of scholarly submissions. In: Proceedings of the Joint Conference on Digital Libraries (JCDL), pp. 227–236. IEEE (2019). https://doi.org/10.1109/JCDL.2019.00039
12. Ghosal, T., Sonam, R., Ekbal, A., Saha, S., Bhattacharyya, P.: Is the paper within scope? Are you fishing in the right pond? In: Proceedings of the Joint Conference on Digital Libraries (JCDL), pp. 237–240. IEEE (2019). https://doi.org/10.1109/JCDL.2019.00040
13. Greydanus, S.: Scaling *down* deep learning. CoRR abs/2011.14439 (2020). arXiv:2011.14439
14. Hilasaca, G.M.H., Paulovich, F.V.: User-guided dimensionality reduction ensembles. In: Proceedings of the International Conference on Information Visualisation (IV), pp. 228–233. IEEE (2019)

15. Hogräfer, M., Heitzler, M., Schulz, H.J.: The state of the art in map-like visualization. In: Computer Graphics Forum, vol. 39, pp. 647–674. Wiley Online Library (2020)
16. Hoo, C.S.: Impacts of patent information on clustering in derwent innovation's themescape map. World Pat. Inf. **63**, 102001 (2020). https://doi.org/10.1016/j.wpi.2020.102001
17. Houle, M.E.: Dimensionality, discriminability, density and distance distributions. In: Proceedings of the International Conference on Data Mining (ICDM), pp. 468–473. IEEE (2013). https://doi.org/10.1109/ICDMW.2013.139
18. Johansson, S., Johansson, J.: Interactive dimensionality reduction through user-defined combinations of quality metrics. Trans. Vis. Comput. Graph. (TVCG) **15**(6), 993–1000 (2009)
19. Kobak, D., Linderman, G.C.: Initialization is critical for preserving global data structure in both t-SNE and UMAP. Nat. Biotechnol. 1–2 (2021)
20. Lang, K.: Newsweeder: learning to filter netnews. In: Proceedings of the International Conference on Machine Learning (ICML), pp. 331–339. Morgan Kaufmann (1995). https://doi.org/10.1016/b978-1-55860-377-6.50048-7
21. LeCun, Y., Bottou, L., Bengio, Y., Haffner, P.: Gradient-based learning applied to document recognition. Proc. IEEE **86**(11), 2278–2324 (1998)
22. Lespinats, S., Aupetit, M.: CheckViz: sanity check and topological clues for linear and non-linear mappings. Comput. Graph. Forum **30**(1), 113–125 (2011)
23. McInnes, L., Healy, J., Saul, N., Großberger, L.: UMAP: uniform manifold approximation and projection. J. Open Source Softw. **3**(29), 861 (2018)
24. Meng, M., Wei, J., Wang, J., Ma, Q., Wang, X.: Adaptive semi-supervised dimensionality reduction based on pairwise constraints weighting and graph optimizing. Int. J. Mach. Learn. Cybern. **8**(3), 793–805 (2015). https://doi.org/10.1007/s13042-015-0380-3
25. Mesbah, S., Fragkeskos, K., Lofi, C., Bozzon, A., Houben, G.-J.: Facet embeddings for explorative analytics in digital libraries. In: Kamps, J., Tsakonas, G., Manolopoulos, Y., Iliadis, L., Karydis, I. (eds.) TPDL 2017. LNCS, vol. 10450, pp. 86–99. Springer, Cham (2017). https://doi.org/10.1007/978-3-319-67008-9_8
26. Mikalsen, K.Ø., Soguero-Ruíz, C., Bianchi, F.M., Jenssen, R.: Noisy multi-label semi-supervised dimensionality reduction. Pattern Recogn. **90**, 257–270 (2019). https://doi.org/10.1016/j.patcog.2019.01.033
27. Pang, P.C.I., Biuk-Aghai, R.P., Yang, M., Pang, B.: Creating realistic map-like visualisations: results from user studies. J. Vis. Lang. Comput. (JVLC) **43**, 60–70 (2017)
28. Reif, E., et al.: Visualizing and measuring the geometry of BERT. In: Proceedings of the Conference on Neural Information Processing Systems (NIPS), pp. 8592–8600. NIPS Foundation Inc., San Diego, USA (2019). https://proceedings.neurips.cc/paper/2019/hash/159c1ffe5b61b41b3c4d8f4c2150f6c4-Abstract.html
29. Repke, T., Krestel, R.: Robust visualisation of dynamic text collections: measuring and comparing dimensionality reduction algorithms. In: Proceedings of the Conference for Human Information Interaction and Retrieval (CHIIR) (2020)
30. Risch, J., Garda, S., Krestel, R.: Book recommendation beyond the usual suspects. In: Dobreva, M., Hinze, A., Žumer, M. (eds.) ICADL 2018. LNCS, vol. 11279, pp. 227–239. Springer, Cham (2018). https://doi.org/10.1007/978-3-030-04257-8_24
31. Saket, B., Endert, A., Rhyne, T.: Demonstrational interaction for data visualization. IEEE Comput. Graph. Appl. **39**(3), 67–72 (2019)

32. Shi, X., Yu, P.S.: Dimensionality reduction on heterogeneous feature space. In: Proceedings of the International Conference on Data Mining (ICDM), pp. 635–644. IEEE (2012). https://doi.org/10.1109/ICDM.2012.30
33. Spathis, D., Passalis, N., Tefas, A.: Interactive dimensionality reduction using similarity projections. Knowl.-Based Syst. **165**, 77–91 (2019)
34. Vahdati, S., Fathalla, S., Auer, S., Lange, C., Vidal, M.-E.: Semantic representation of scientific publications. In: Doucet, A., Isaac, A., Golub, K., Aalberg, T., Jatowt, A. (eds.) TPDL 2019. LNCS, vol. 11799, pp. 375–379. Springer, Cham (2019). https://doi.org/10.1007/978-3-030-30760-8_37
35. Wang, Y., Jin, Z., Wang, Q., Cui, W., Ma, T., Qu, H.: DeepDrawing: a deep learning approach to graph drawing. Trans. Vis. Comput. Graph. (TVCG) **26**(1), 676–686 (2020)
36. Wawrzinek, J., Balke, W.-T.: Measuring the semantic world – how to map meaning to high-dimensional entity clusters in PubMed? In: Dobreva, M., Hinze, A., Žumer, M. (eds.) ICADL 2018. LNCS, vol. 11279, pp. 15–27. Springer, Cham (2018). https://doi.org/10.1007/978-3-030-04257-8_2
37. Wawrzinek, J., Hussaini, S.A.R., Wiehr, O., Pinto, J.M.G., Balke, W.: Explainable word-embeddings for medical digital libraries - a context-aware approach. In: Proceedings of the Joint Conference on Digital Libraries (JCDL), pp. 299–308. ACM Press (2020). https://doi.org/10.1145/3383583.3398522
38. Wohlmuth, C., Correia, N.: User interface for interactive scientific publications: a design case study. In: Doucet, A., Isaac, A., Golub, K., Aalberg, T., Jatowt, A. (eds.) TPDL 2019. LNCS, vol. 11799, pp. 215–223. Springer, Cham (2019). https://doi.org/10.1007/978-3-030-30760-8_19
39. Xiao, H., Rasul, K., Vollgraf, R.: Fashion-mnist: a novel image dataset for benchmarking machine learning algorithms. CoRR abs/1708.07747 (2017). arXiv:1708.07747

Text Classification and Matching

Automating the Choice Between Single or Dual Annotation for Classifier Training

Satoshi Fukuda[1(✉)], Emi Ishita[2], Yoichi Tomiura[2], and Douglas W. Oard[3]

[1] Chuo University, Tokyo 112-8551, Japan
sfukuda277@g.chuo-u.ac.jp
[2] Kyushu University, Fukuoka 819-0395, Japan
[3] University of Maryland, College Park, MD 20742, USA

Abstract. Many emerging digital library applications rely on automated classifiers that are trained using manually assigned labels. Accurately labeling training data for text classification requires either highly trained coders or multiple annotations, either of which can be costly. Previous studies have shown that there is a quality-quantity trade-off for this labeling process, and the optimal balance between quality and quantity varies depending on the annotation task. In this paper, we present a method that learns to choose between higher-quality annotation that results from dual annotation and higher-quantity annotation that results from the use of a single annotator per item. We demonstrate the effectiveness of this approach through an experiment in which a binary classifier is constructed for assigning human value categories to sentences in newspaper editorials.

Keywords: Text classification · Efficient annotation · Multi-armed bandits

1 Introduction

As a result of open data promotion, various types of data have become available through data repositories and data archives. New types of research, for example, digital humanities or computational social science, are greatly benefiting from them. Many studies that use digital data and text have been conducted using machines. By contrast, qualitative research is still useful because it can provide deeper knowledges than the insights. However, qualitative research requires human effort; hence, the amount of text and data that can be analyzed is limited. We are interested in introducing machines to qualitative research methods while maintaining the quality of thought and improving efficiency. This would enable us to support digital humanities and computational social sciences. Moreover, it would increase the importance of digital archives and data repositories.

Supervised machine learning is widely used for text classification in tasks that range from academic studies of collaboration behavior to commercial tracking of brand mentions on social media. To construct good classifiers, substantial quantities of annotated training data are typically required. Human coders are typically used to create such training data, but this choice results in limits to the amount of annotation that can be

H.-R. Ke et al. (Eds.): ICADL 2021, LNCS 13133, pp. 233–248, 2021.
https://doi.org/10.1007/978-3-030-91669-5_19

performed for cost reasons. Therefore, we are interested in techniques to create the most useful annotated training data for a given annotation cost.

For manual annotation, Ishita et al. explored two approaches, each assuming that two coders were available [11]. One approach had coders working alone to annotate the largest possible number of different texts for a given level of annotation effort, and the other had coders working together to create the best possible annotation for each text, albeit at a higher cost per annotation. Ishita et al. compared the classifiers trained using each approach by plotting learning curves to investigate how well the classifiers performed as the number of annotations increased for six annotation tasks, and found three broad categories of patterns: (1) rapidly generating large quantities of annotations was sometimes a good approach, particularly when positive examples were rare, (2) generating higher quality annotations by having two coders annotate each text independently and then discussing their disagreements was also sometimes a good approach, particularly when the task was difficult, and (3) starting out with single annotation but switching to dual annotation at some point in the process was sometimes the best overall approach. Clearly, it may also be the case that one coder is better than another, and because of learning and fatigue effects, such differences may change over time. Therefore, in this paper, we regard the choice between a single annotator and dual annotation, and for single annotation the choice of the coder, as an optimization problem, and present a method that consistently makes near-optimal choices over a range of tasks.

If the outcome of each choice could be known with certainty, it would be best to select the pair of a specific coder and a specific text that yields the greatest possible improvement in the performance of the classifier after the next annotation. Clearly, we do not know which label would actually be assigned in each case, but despite this, we can compute the expected score of this improvement using the results of past annotations. Two types of annotations are possible when two coders are available: labeling an unannotated text to increase the quantity of training data or improving an existing label as a result of adjudication (i.e., discussion and joint decision) between coders after obtaining a second annotation for a text that has already been annotated once. If the labeling of unannotated text by a coder improves the performance of the classifier, then the expectation that this coding method will be further improved in terms of classifier performance in the next annotation will be higher. If, by contrast, the expectation of a second annotation that results from adjudication improves the performance of the classifier as the annotation progresses and its expected score is higher than that of a new annotation action, a higher performance classifier will be constructed when an adjudication action is selected for the next annotation.

The sequential choice of the coding method to obtain a higher-performance classifier is modeled in the framework of a multi-armed bandit problem, which is a problem setting for reinforcement learning. Reinforcement learning is a learning approach that seeks to learn to act in a manner that maximizes a reward. In the multi-armed bandit problem, the actions are choices (called "arms") whose past probability of success can be observed, and the goal is to develop a policy for sequentially selecting the best choice at each point in time to maximize the total reward over a finite time.[1] For our goal of affordably building the most useful training data, the size of the reward is the quality of the resulting classifier, and the choices are between single or dual annotation. For comparability with prior results, we tested our approach using the datasets constructed by Ishita et al. [11], and found that our approach consistently made near-optimal choices. From these results, we conclude that a similar multi-armed bandit approach to choosing between single or dual annotation could be useful for a broad range of applications in which text classifiers must be trained in a cost-effective manner.

2 Related Work

In this section, we describe prior work on efficiently constructing training data and multi-armed bandits.

2.1 Efficient Construction of Training Data

Several information retrieval evaluation organizations (e.g., TREC and NTCIR) have created large-scale annotated datasets using pooling [12, 17]. In pooling, the annotation task is partitioned by topic, with a single coder judging all texts found above some rank cutoff for that topic using any system that contributed to the creation of the pool. This approach models a case in which the appropriate annotation is an individual opinion, as is typically the case for relevance judgements in a search task. Although this is an efficient approach, in contrast to personal opinions about search topics, our interest is in the annotation of phenomena on which people can agree.

In pooling, machines work alone to select items for annotation. Other approaches are possible, including active learning in which humans and machines work together to determine which documents should be annotated. For example, Cai et al. [4] proposed an active learning method to automatically select the text to be labeled next from an unlabeled dataset using the already annotated items. In this method, the non-annotated text furthest from the already annotated items is chosen to be labeled next. Somewhat closer in spirit to our proposed approach, Culotta and McCallum [7] proposed a method called expected model change maximization to determine the text to be labeled next, which is the text that is expected to most change the discriminative model after the new annotation is added to the labeled dataset, comparing the discriminative model generated from the already-labeled dataset. The distinction between their study and our approach is that they proposed choosing the item to be annotated, whereas we propose choosing the coder who will perform the annotation. Both are important problems, and in future work, the two should be explored together. However, for the initial study reported in this paper, considerable simplification results from our choice to focus simply on the coder selection problem.

All the aforementioned approaches focus on a single annotator. However, we are interested in annotation tasks in which the phenomenon to be annotated is defined by

[1] The name comes from a colloquial reference to slot machines used by gamblers called "one-armed bandits." In the imagined multi-armed bandit scenario, the gambler seeks to pull the arm that would yield the greatest profit.

consensus rather than an individual opinion. To evaluate the quality of the annotations between two coders, κ [5, 6], *S* [3], and π [15] coefficients have been widely used. For example, Artstein and Poesio [1] and Fort et al. [8] verified the reliability of the agreement between two coders using various coefficients. When using coefficients to measure the reliability of annotations between coders, it is necessary to conduct the annotation on some fixed collection that both coders annotate. We use such a collection in our experiments, but our goal is to determine optimal choices that can limit dual annotation to those cases in which it is actually adding value.

In recent years, the use of many annotators has been widely studied in the context of crowdsourcing [10], where it is possible to request annotations from a broad range of coders. A number of frameworks for obtaining useful results from crowdworkers who exhibit varying levels of expertise, ability, and diligence have been studied. For example, Nguyen et al. [13] proposed a method to measure the quality of annotations obtained from crowdworkers using a dataset labeled by domain experts using active learning. Welinder et al. [18] proposed a method for estimating true labels by modeling the worker's answering process. Zhang et al. [19] proposed a framework to manage the quality of coders by clustering coders based on their annotation history. However, crowdworkers normally operate independently, whereas we are interested in settings in which disagreeing coders can meet to resolve their disagreements.

2.2 Multi-armed Bandit Problem

Many algorithms have been proposed for the bandit problem: ε-greedy algorithm, UCB (Upper Confidence Bound) [2], and Thompson sampling [16] are typical algorithms. The ε-greedy algorithm is a method in which an arm is randomly selected with a probability of ε ("exploration") and the arm with the highest expected score for the reward at that time is selected with a probability of $1 - \varepsilon$ ("exploitation"). By considering the balance between "exploration" and "exploitation" for the arm selection, the risk of continuing to select one arm can be mitigated. UCB is a method for calculating the score of each arm using the average of the observed reward and the value calculated based on the number of times the arm is selected. In this method, the lower the number of times the arm is selected, the higher the score. Thompson sampling is a method for selecting an arm based on the probability that the arm can obtain the maximum expected score. In the method, the posterior distribution of the expected score of the reward is modeled in the framework of Bayesian statistics. In a general bandit problem, rewards follow the Bernoulli distribution; that is rewards were obtained or not obtained. By contrast, in the construction of training data, elaborate rewards for the annotation action can be measured by calculating the differences of the performance between the current classifier and previously constructed the classifier. We therefore apply the ε-greedy algorithm, which is the simplest of the bandit algorithms, which can flexibly incorporate the above element.

The above bandit strategies assume that the reward acquisition probability of each arm does not change. However, in reality, the reward distribution of the arm can change over time (e.g., changes in the click rate for the advertisements caused by changes in ad design and trends). The bandit problem corresponding to such a temporal change of the reward distribution is called a non-stationary multi-armed bandit problem. Discounted

UCB [9] and discounted Thompson sampling [14] are the algorithms used to apply UCB and Thompson sampling to this problem, respectively, and they introduce the concept of the discount. In the annotation for the construction of the training data, past and present scenarios may have changed the nature of the labeled dataset and coders. Therefore, by introducing a decay function when evaluating the arm, we expect that an arm with a high expected score can be selected effectively while the influence of past observation results is diminished.

3 Selection Algorithm

3.1 Overview

We are interested in binary annotation tasks in which some phenomenon (in our case, a consensus definition of topical relevance) is to be annotated as present or absent. Our goal is to maximize some single-valued measure of classification effectiveness, such as F1. Assuming, without loss of generality, that there are two coders, and assuming, as a computational simplification, that the order of texts to be annotated is fixed, there are four choices to be made about which coder annotates which text:

- Arm 1: Select coder 1 and the first text in the unannotated set.
- Arm 2: Select coder 2 and the first text in the unannotated set.
- Arm 3: Select coder 2 and the first text in the labeled text set annotated by coder 1, and decide the final annotation after discussion.
- Arm 4: Select coder 1 and the first text in the labeled text set annotated by coder 2, and decide the final annotation after discussion.

For the first two arms, the selected unannotated text is annotated by coder 1 or 2. For the last two arms, coder 1 or 2 annotates the selected text; the two coders discuss which annotation is correct, if their annotations are different; and then the final annotation is decided.

For the calculation of the expected score for each arm, we approximate the average of the increase or decrease of the F1 value. Specifically, we calculate the difference between the F1 value of the classifier based on the labeled dataset constructed by a certain arm at time t and the F1 value of the classifier based on the dataset constructed at time $t-1$. Using the difference between the F1 values, we consider that a highly effective annotation action reflects the intuition that it will continue to improve the performance of the classifier. Additionally, when the improvement range of the arm that has been selected decreases or deteriorates as the annotation progresses, this may suggest that the nature of coders or labeled datasets is changing, and this leads to an opportunity to select another arm that matches the current annotation environment and improves the annotation. When reflecting the change of the nature of coders from the viewpoint of the time series, it can be helpful to use only the results in a certain section from the present to the past n times or gradually reduce the influence of past results when calculating the expected score of each arm. If only the results of a certain interval from the present are used, the expected score for the arm that was not selected even once during that period cannot be calculated. Therefore, we adopt a strategy to reduce the influence of past results and use a decay function.

The formula for the expected score for each arm incorporating these elements is as follows:

$$Score\_Arm_i = \frac{\sum_{t=1}^{T} \Delta F_i(t) \times e^{-\alpha(T-t)}}{\sum_{t=1}^{T} c_i(t) \times e^{-\alpha(T-t)}} \quad (1)$$

$$\Delta F_i(t) = \begin{cases} F_t - F_{t-1} \ (\textit{if Arm}_i \textit{ is operated at time } t) \\ 0 \quad\quad (\textit{otherwise}) \end{cases} \quad (2)$$

$$c_i(t) = \begin{cases} 1 \ (\textit{if Arm}_i \textit{ is operated at time } t) \\ 0 \quad\quad (\textit{otherwise}) \end{cases},$$

where T is the total number of annotation actions and F_t is the F1 value of the classifier constructed using the labeled dataset in time t. If all labels in the dataset are negative or the precision at time t cannot be computed, we set $F_t = 0$. We use $e^{-\alpha(T-t)}$ as the decay function, where α is the weighting parameter used for adjusting the damping speed. The weight for the R-th past result from the present is $e^{-\alpha R}$. If we set α so that this weight becomes W, then

$$\alpha = -\frac{1}{R} \log W.$$

For example, if the weight for the past 100th result from the present is 0.1, then α is $-1/100 \times \log 0.1$.

3.2 Algorithm

In this section, we describe a method for sequentially selecting a coder and text based on the ε-greedy algorithm. Table 1 contains a summary of the notation used in our method. Note that our method needs to set C, D, F^{th}, N, R, W and ε in advance. Additionally, D_1, D_2, and A are initially set to the empty set ∅. In our method, if Arm 1 is operated, the algorithm removes the selected text from D and adds the text with the label annotated by coder 1 to D_1. If Arm 2 is operated, the algorithm removes the selected text from D and adds the text with the label annotated by coder 2 to D_2. If Arm 3 is operated, the algorithm removes the selected text from D_1 and adds that text with the label by the adjudication to A. If Arm 4 is operated, the algorithm removes the selected text from D_2 and adds that text with the label by the adjudication to A.

Our method consists of three steps. We describe the details of each step below.

(Step 1) Initial setting for the classifier

As shown in Eq. (1), the expected score of each arm is calculated based on the F1 value of the classifier constructed from the labeled dataset. Therefore, before executing the ε-greedy algorithm, it is necessary to construct a scenario in which the F1 value can be calculated from the labeled dataset. Additionally, a classifier with an extremely low F1 value can be said to have poor dataset quality, which may adversely affect the

Table 1. Notation.

	Description
D	Ordered set of texts to be annotated
D_1	Set of pairs of text and its annotation by coder 1
D_2	Set of pairs of text and its annotation by coder 2
A	Set of pairs of text and its annotation by the two coders' adjudication
C	Number of annotations that can be executed with the given budget
F^{th}	Threshold for F1 value of classifier constructed by the labeled dataset
N	Parameter for selecting each arm $N/4$ times to evaluate evenly
ε	Parameter for determining "exploitation" or "exploration"

calculation of the expected score. Furthermore, from the findings in [11], the F1 value tends to increase as the amount of data increases for the initial annotation cost. Therefore, in step 1, the following process is executed.

First, Arm 1 and Arm 2 are selected alternately when D is not empty, and Arm 3 and Arm 4 are selected alternately when D is empty. After the arm is operated, C is decremented by 1. If there is at least one positive example and negative example in the labeled dataset and the F1 value of the classifier constructed using the labeled dataset is greater than or equal to F^{th}, the algorithm proceeds through step 2 and sets F_0 to the F1 value of the classifier constructed using the labeled dataset constructed. Note that this process is not regarded as an annotation action by the arm. If C is 0, our method is terminated.

(Step 2) Initial setting of the expected score for each arm
In this step, the initial score for $Score\_Arm_i$ ($1 \leq i \leq 4$) is obtained. Each arm is essentially selected in the order Arm 1, Arm 2, Arm 3, and Arm 4. If there is no text to execute for the selected arm, the operation on the arm is skipped and N is decremented by 1, and then the next arm is selected (this process is repeated until a viable arm is selected). After the arm operation is complete, the F1 value of the classifier constructed using the labeled dataset is assigned to F_t, N is decremented by 1, and C is decremented by 1. If N is 0, the algorithm proceeds to step 3. If C is 0, our method is terminated.
(Step 3) Arm selection based on the ε-greedy algorithm
This step selects the arm based on the ε-greedy algorithm. First, r is randomly selected from [0, 1). If r is less than ε, the algorithm switches to "exploration," and if r is greater than or equal to ε, the algorithm switches to "exploitation." In "exploration" and "exploitation," the following processes are conducted.

- **Exploration**: The algorithm randomly selects one arm from the four types of arms with equal probability. If no text is available for annotation by the selected arm, the other arm is randomly selected again. This process is performed repeatedly until a viable arm is selected.

- **Exploitation**: The algorithm calculates $Score\_Arm_i$ ($1 \leq i \leq 4$) using Eq. (1), and then selects the arm with the highest score for $Score\_Arm_i$. If no text is available for annotation by the selected arm, another arm with the next highest score for $Score\_Arm_i$ is selected (this process is performed repeatedly until a viable arm is selected).

After "exploration" or "exploitation," F_t is calculated. Then, C is decremented by 1, and the end condition in step 3 (whether C is 0) is confirmed.

4 Experiment

4.1 Experimental Settings

Our method can be applied to any binary classification task, but for our experiments, we chose to use the annotation task in [11]. This task was set originally to support social science research on the role of human values in the nuclear power debate, and was part of a multistep process that was ultimately intended to provide automated assistance for topically focused investigations of the role of human values in public debates. In this study, 120 Japanese newspaper editorials from the Mainichi Shimbun newspaper that substantively addressed the nuclear power debate in Japan following the Fukushima Daiichi disaster were divided into six subsets of 20, and each sentence in each document was annotated independently by two coders (both of whom were native Japanese speakers) for its relevance to eight predefined human value categories. Table 2 shows the definitions of the human value categories. For example, if a sentence described a judgement or evaluation based on results that included future prospects or macro/long-term perspectives, the sentence was considered to be positive for the human value "Consequence." Note that this annotation process allows multiple human value categories to be assigned to a sentence because more than one human value may be expressed in a sentence. The coders then decided the adjudicated annotation of a batch of 20 documents by discussion in one sitting. They repeated these processes six times because there were 120 documents. The resulting dataset that was created for each human value category by this annotation process contained three types of annotation for each sentence: coder A's annotation labels, coder B's annotation labels, and the adjudicated annotation labels that were ultimately arrived at through discussion between the coders.

In our experiment, annotated sentences in rounds 2 to 5 were used as training data and annotated sentences in round 6 were used as evaluation data. Sentences in round 1 were not used to minimize annotator learning effects. Additionally, six types of categories with more than 50 positive examples in the adjudicated training data for eight types of human value categories were chosen. Table 3 shows the details of the dataset for each human value. Each training data cell shows the number of sentences annotated by coder A/coder B/adjudication. Each evaluation data cell shows the number of sentences annotated as a result of the adjudication; that is, we regarded the adjudication in the evaluation data as the gold standard. The sizes of training and evaluation data were 2,188 and 570, respectively. In our method, 2,188 ordered sentences were set in D, described in Table 1, and 540 sentences were reserved for the calculation of the actual F1 values of

Table 2. Definitions of the human value categories.

Human Value	Definition
Consequence	Values on judgement or evaluation based on results including future prospects (e.g. outcomes, objectives) or macro/long-term perspectives
Safety	Values of safety and security; the state of being free from danger, injury, threat or fear; measures to prevent accidents and hazards
Human Welfare	Values on fulfilling benefits common to human beings and related to society as a whole; Clear benefits to the public
Intention	Values on emotion or feelings including impression, attitude, empathy, prudence, and sincerity; The quality of being honest and integrity; adherence to moral principles
Social Order	Values on social structure, including rules, norms, common sense and expectations as well as social responsibility; Institutional, legal, and political decisions involving governments and states
Wealth	Values on pursuing any economic goals, such as money, material possessions, resources, and profit including business activities
Freedom	Value of individual freedom and choices; the state of being unconstrained; freedom from interference or influence by others;
Personal Welfare	Values on personal needs, growth, and self-actualization

the classifier constructed from the labeled dataset. Additionally, the annotation cost for annotating 2,188 sentences by adjudication was 4,376. To show the full range of results, we set C, described in Table 1, to 4,376.

Because our task was order dependent, we repeated the entire process 1,000 times, with 1,000 random shuffles of the sentences in the training data document-by-document. Each time a sentence was annotated, we retrained a support vector machine (SVM), implemented in TinySVM[2] using a linear kernel. We used Juman version 7.01[3] to tokenize and perform morphological analysis for Japanese, and we used all the resulting words as features for the classifier after removing period and comma characters. We used whether the word appeared or not as each feature's weight.

We plotted learning curves for the evaluation data by placing the annotation cost on the horizontal axis and the F1 value on the vertical axis. Note that the horizontal axis represents the number of annotations, not the number of sentences, because one sentence may have two annotations. The learning curves were for the following:

- **Adjudicated (baseline)**: we used the adjudicated data for training, and counted this as two annotations in each iteration.
- **Alternating (baseline)**: we alternated between using coder A's and coder B's individual annotations for training until all sentences were annotated, and then replaced single annotations with adjudicated annotations in the same order.

[2] http://chasen.org/~taku/software/TinySVM/
[3] http://nlp.ist.i.kyoto-u.ac.jp/EN/index.php?JUMAN.

Table 3. Distribution of the dataset for each human value category.

	Training data		Evaluation data	
	Positive	Negative	Positive	Negative
Consequence	667/758/959	1521/1430/1229	270	270
Safety	586/663/719	1602/1525/1469	249	291
Human Welfare	196/196/224	1992/1992/1964	73	467
Intention	167/152/205	2021/2036/1983	66	474
Social Order	1445/1473/1570	743/715/618	431	109
Wealth	263/251/289	1925/1937/1899	118	422

- **Selection**: we sequentially selected the next sentence and coder using the ε-greedy algorithm.

We plotted the above learning curves for each human value category in Table 3[4].

We determined the parameters F^{th}, N, R, W, and ε used in our method using the following approach. We used the datasets for Consequence, Safety, and Human Welfare as tuning data, and repeatedly performed the following: we observed the transition of F1 values in the three types of dataset and manually tuned each parameter based on the results. We fixed N at 40 because step 2 played a part in obtaining the initial value for each arm used in step 3, and it was not necessary to incur a large cost in step 2 because the score of each arm was updated in step 3. Additionally, we fixed ε at 0.1, which has been commonly used in various tasks that used the ε-greedy algorithm. Finally, we manually determined the parameters with good F1 value transitions for the three types of dataset. The parameters we determined were $\left(F^{th}, N, R, W, \varepsilon\right) = (0.3, 40, 200, 0.1, 0.1)$. We applied this combination of parameters to the other three types of dataset (Intention, Social Order, and Wealth).

4.2 Experimental Results

Figure 1 shows how F1 improved based on the macro average over 1,000 random simulations as the number of annotations increased for each human value. First, we observe the results when the parameters described in Sect. 4.1 were tuned on the same categories (Consequence, Safety, Human Welfare). For Consequence (Fig. 1(a)), the Selection curve shows F1 values close to those of the Adjudicated curve, which achieved higher values than the Alternating curve for the early annotation cost. For Safety (Fig. 1(b)) and Human Welfare (Fig. 1(c)), the Selection curve shows F1 values close to those of the Alternating curve, which achieved higher values than the Adjudicated curve for the

[4] Note the difference in annotation strategy between constructing annotated data in [11] and applying the multi armed bandit (MAB)-problem method: in [11], coders assigned several labels to each sentence in one sitting; however for the MAB-problem method, one label is assigned or not for each sentence.

early annotation cost, and then later shows almost the same F1 values as the Adjudicated curve, which achieved higher values than the Alternating curve, because the annotation costs where the transitions of the F1 value occurred when using the Adjudicated and Alternating curves intersected.

Next, we investigate the results when the optimum parameters determined using the above categories were applied to the other categories (Intention, Social Order, Wealth). For Intention (Fig. 1(d)), the Selection curve shows F1 values close to those of the Alternating curve, which achieved higher values than the Adjudicated curve for the early annotation cost. For Social Order (Fig. 1(e)), the Selection curve shows F1 values close to those of the Adjudicated curve, which achieve higher values than the Alternating curve for the early annotation cost. For Wealth (Fig. 1(f)), the Selection curve shows almost the same F1 values as the Alternating curve, which achieved higher values than the Adjudicated curve for the early annotation cost (annotation cost interval 1–600), and then shows F1 values between the Adjudicated and Alternating curves. These results indicate that our method generally yielded good results that were between the Adjudicated and Alternating curves. Additionally, to further improve the learning curves of Selection for categories such as Wealth, we would need to automatically estimate optimum parameters according to task and work scenario.

4.3 Discussion

In this section, we verify the effectiveness of Fth, $e^{-\alpha(T-t)}$, and difference between the F1 values ($\Delta F_i(t)$) in Eq. (2) in Selection to investigate how these elements added to or changed the conventional ε-greedy algorithm's effect on the F1 value's curve. For these verifications, we used the datasets of Consequence and Human Welfare, which were used for parameter tuning. For F^{th}, we set $F^{th} = 0$ to verify whether it is better to move to step 2 with a value above a certain level for the F1 value of the classifier constructed from the training data created in step 1. Note that all parameters other than F^{th} used the values described in Sect. 4.1. For $e^{-\alpha(T-t)}$, we compared the cases with and without this weight in Eq. (1). Note that all parameters other than R and W used the values described in Sect. 4.1. For $\Delta F_i(t)$, we compared the case where the score follows the difference between F1 values with the case where the score follows the Bernoulli distribution used in the conventional bandit algorithm. In this verification, we set $\Delta F_i(t)$ to 1 if the F1 value improved from the previous classifier, and 0 otherwise. For this case, we used the parameter values described in Sect. 4.1.

Figure 2 shows the transitions of the F1 values for the above cases. Selection (Fth = 0) shows the curve when F^{th} is set to 0 in Selection. Selection (without weight) shows the curve when $e^{-\alpha(T-t)}$ is not used in Eq. (1) in Selection. Selection (binary) shows the curve when binary evaluation is used instead of the difference between the F1 values ($\Delta F_i(t)$) in Eq. (2) in Selection. First, we compare the results for the settings of Fth. For Consequence (Fig. 2(a)), the transitions of the F1 values when $F^{th} = 0$ and $F^{th} = 0.3$ were almost the same. However, for Human Welfare (Fig. 2(b)), the results for $F^{th} = 0.3$ demonstrated higher performance than the results for $F^{th} = 0$. From these results, we conclude that to construct a high-performance classifier consistently in Selection, we need to construct a classifier that has a certain F1 value before moving to step 2 in Selection. Next, we compare the results with and without $e^{-\alpha(T-t)}$. In Human Welfare

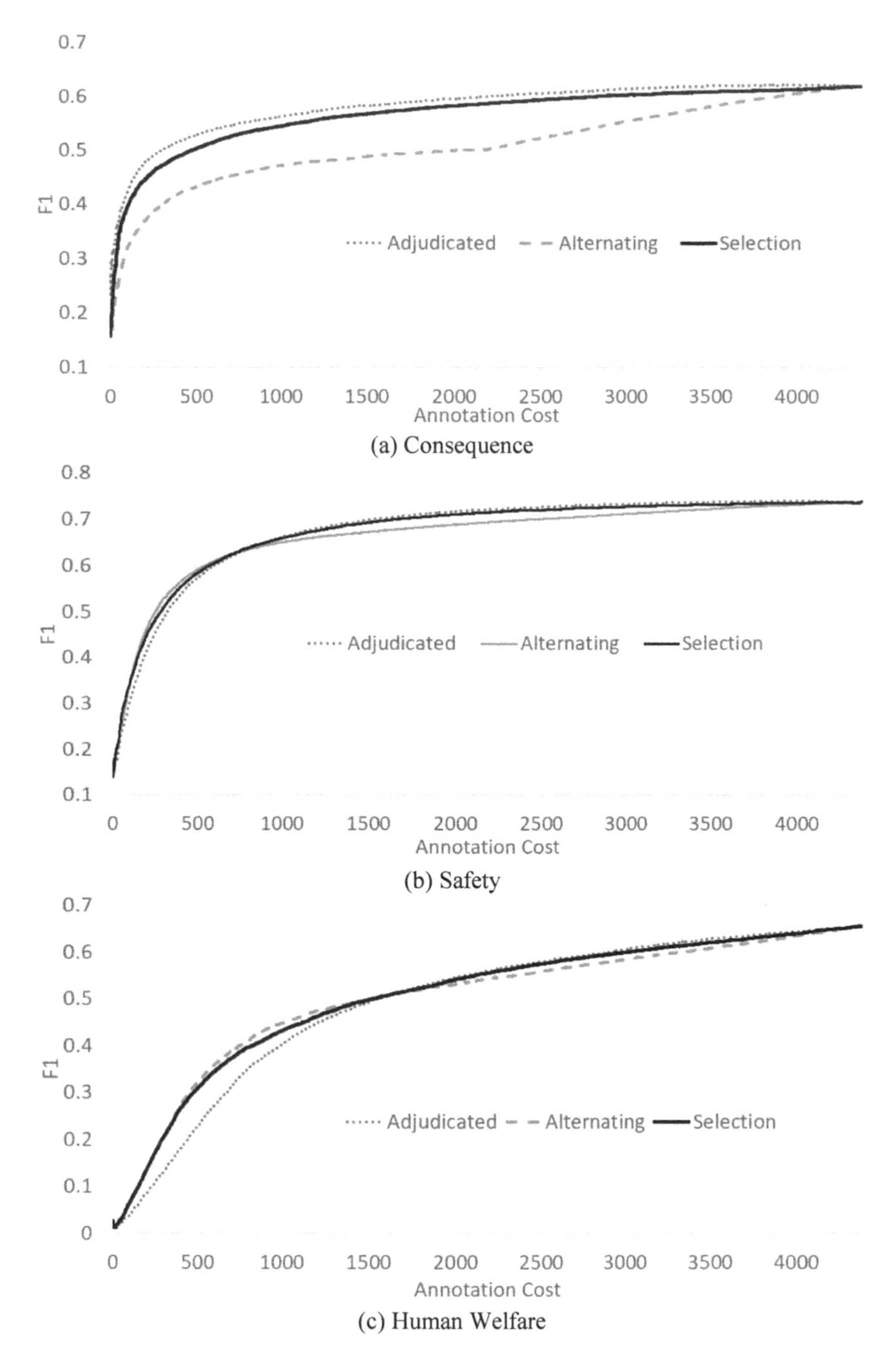

(a) Consequence

(b) Safety

(c) Human Welfare

Fig. 1. Learning curves for each human value.

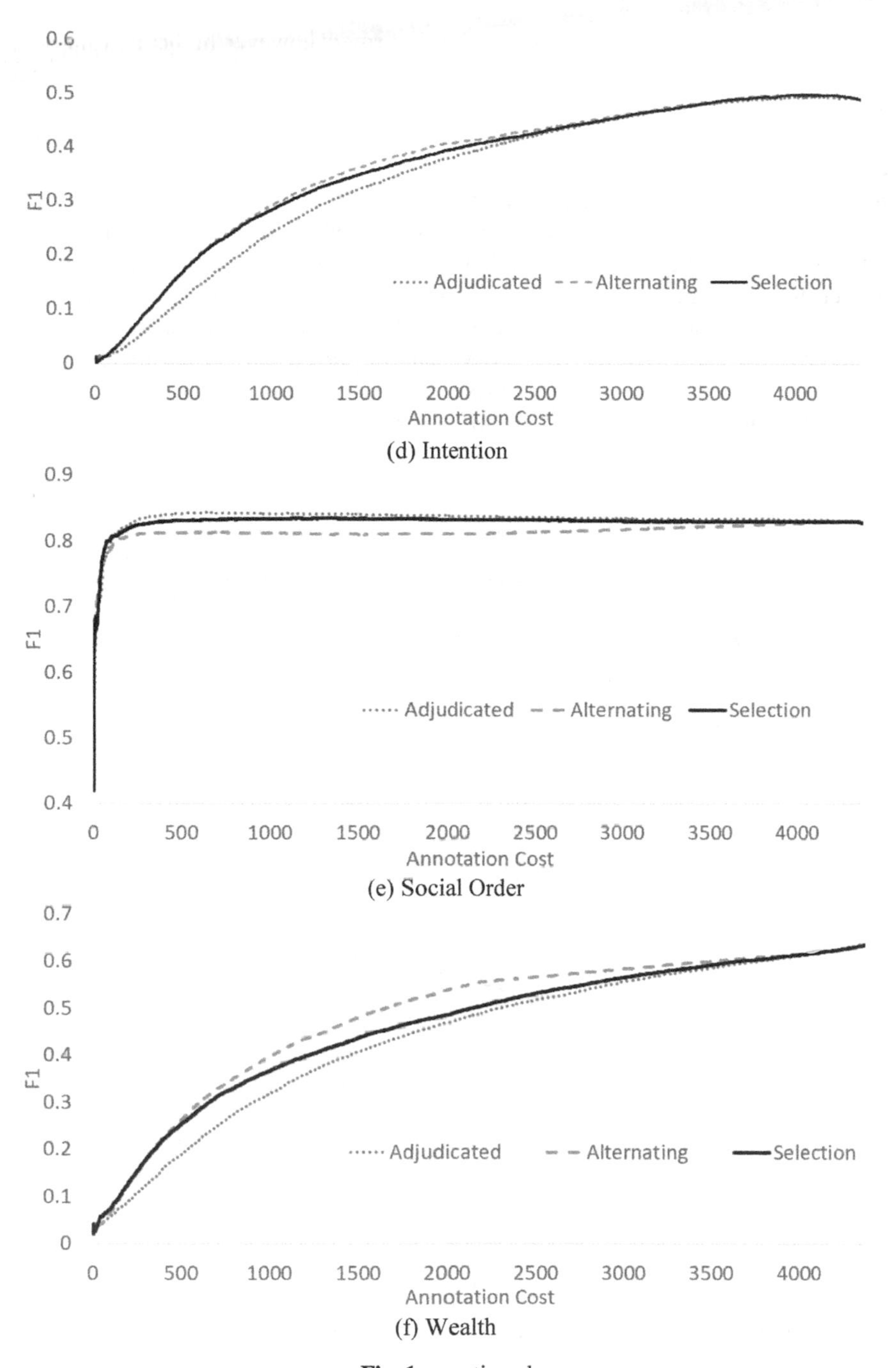

(d) Intention

(e) Social Order

(f) Wealth

Fig. 1. continued

(Fig. 2 (b)), the effect of using weights was not confirmed; however, higher F1 values were confirmed when the weight was used for Consequence (Fig. 2(a)). These results suggest that the probabilities of the improved performance of some arms may differ between past and current work, depending on the task. Therefore, we can possibly calculate the expected score of each arm more precisely by reducing the influence of past work. Finally, we compare the results when using the difference between F1 values and when using the binary values 1 (F1 value improved) or 0 (otherwise) in Eq. (2). For Human Welfare (Fig. 2(b)), the F1 value was relatively good when the binary values were used; however, the F1 values were very low, close to those of Alternating, in Consequence (Fig. 2(a)). These results suggest that the binary indicator used in the conventional bandit algorithms to indicate whether the F1 value had improved did not accurately measure the expected score of the arm for some tasks, and it was better to use the elaborated score by evaluating how much the F1 value had improved or deteriorated.

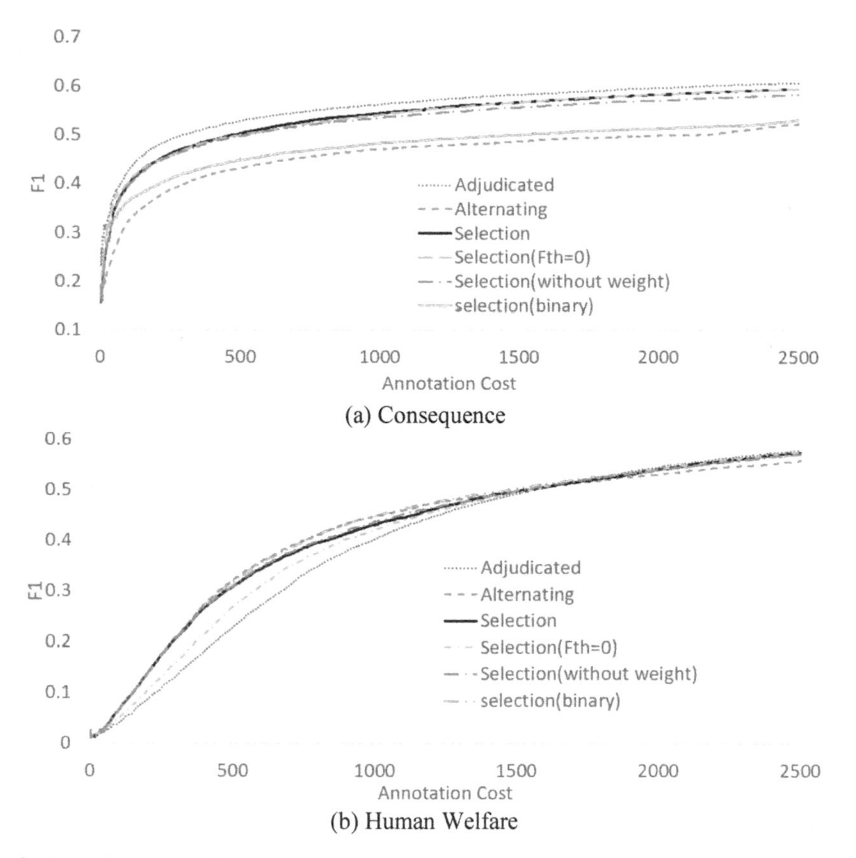

(a) Consequence

(b) Human Welfare

Fig. 2. Learning curves for changes in various settings in the annotation cost interval 1–2,500.

5 Conclusion

We have demonstrated that a high-performance classifier can be consistently and affordably constructed for a range of annotation tasks using a multi-armed bandit algorithm to select the coder(s) when creating a labeled training set. The key idea behind our method is to automatically select the coder that the annotation history would lead us to expect would be the most likely to increase the classifier's effectiveness, measured in our experiments using F1. Over six annotation tasks, we showed that this approach tends to show the better of two reasonable baselines. In the future, we plan to apply this technique to other digital library applications, notably including topic classification, we plan to integrate active learning approaches so that we select not just the best choice of coder but also the best choice of document to be coded, we plan to explore approaches to further improve efficiency using asynchronous batch updates, and we plan to explore approaches to efficiently tune model parameters to specific application settings.

Acknowledgements. This work was supported by JSPS KAKENHI Grant Number JP18H03495.

References

1. Artstein, R., Poesio, M.: Inter-coder agreement for computational linguistics. Comput. Linguist. **34**(4), 555–596 (2008)
2. Auer, P., CesaBianchi, N., Fischer, P.: Finitetime analysis of the multiarmed bandit problem. Mach. Learn. **47**(23), 235–256 (2002)
3. Bennett, E.M., Alpert, R., Goldstein, A.C.: Communications through limited response questioning. Public Opin. Q. **18**(3), 303–308 (1954)
4. Cai, W., Zhang, Y., Zhou, J.: Maximizing expected model change for active learning in regression. In: Proceedings of the ICDM, pp. 51–60 (2013)
5. Carletta, J.: Assessing agreement on classification tasks: the kappa statistic. Comput. Linguist. **22**(2), 249–254 (1996)
6. Cohen, J.: Weighted kappa: nominal scale agreement with provision for scaled disagreement or partial credit. Psychol. Bull. **70**(4), 213–220 (1968)
7. Culotta, A., McCallum, A.: Reducing labeling effort for structured prediction tasks. In: Proceedings of the AAAI, pp. 746–751 (2005)
8. Fort, K., François, C., Galibert, O., Ghribi, M.: Analyzing the impact of prevalence on the evaluation of a manual annotation campaign. In: Proceedings of the LREC, pp. 1474–1480 (2012)
9. Garivier, A., Moulines, E.: On upper-confidence bound policies for switching bandit problems. In: Kivinen, J., Szepesvári, C., Ukkonen, E., Zeugmann, T. (eds.) Algorithmic Learning Theory. ALT 2011. LNCS, vol. 6925, pp. 174–188. Springer, Berlin, Heidelberg (2011). https://doi.org/10.1007/978-3-642-24412-4_16
10. Howe, J.: Crowdsourcing: Why the Power of the Crowd is Driving the Future of Business. Crown Publishing Group, New York (2008)
11. Ishita, E., Fukuda, S., Oga, T., Tomiura, Y., Oard, D.W., Fleischmann, K.R.: Cost-effective learning for classifying human values. In: Proceedings of the iConference (2020)
12. Kuriyama, K., Kando, N., Nozue, T., Eguchi, K.: Pooling for a large-scale test collection: an analysis of the search results from the first NTCIR workshop. Inf. Retr. **5**(1), 41–59 (2002)

13. Nguyen, A.T., Wallace, B.C., Lease, M.: Combining crowd and expert labels using decision theoretic active learning. In: Proceedings of the HCOMP, pp. 120–129 (2015)
14. Raj, V. and Kalyani, S.: Taming nonstationary bandits: A bayesian approach. arXiv preprint arXiv:1707.09727 (2017)
15. Scott, W.: Reliability of content analysis: the case of nominal scale coding. Public Opin. Q. **19**, 321–325 (1955)
16. Thompson, W.R.: On the likelihood that one unknown probability exceeds another in view of the evidence of two samples. Biometrika **25**(3–4), 285–294 (1933)
17. Voorhees, E.M., Harman, D.K.: TREC: Experiment and Evaluation in Information Retrieval. The MIT Press, Cambridge (2005)
18. Welinder, P., Branson, S., Belongie, S., Perona, P.: The multidimensional wisdom of crowds. In: Proceedings of the NIPS, pp. 2424–2432 (2010)
19. Zhang, Y., Cui, L., Huang, J., Miao, C.: CrowdMerge: achieving optimal crowdsourcing quality management by sequent merger. In: Proceedings of the ICCSE, pp. 1–8 (2018)

Integrating Semantic-Space Finetuning and Self-Training for Semi-Supervised Multi-label Text Classification

Zhewei Xu(✉) and Mizuho Iwaihara

Graduate School of Information, Production and Systems, Waseda University, Kitakyushu 808-0135, Japan
xuzhewei@toki.waseda.jp, iwaihara@waseda.jp

Abstract. To meet the challenge of lack of labeled data in document classification tasks, semi-supervised learning has been studied, in which unlabeled samples are also utilized for training. Self-training is one of the iconic strategies for semi-supervised learning, in which a classifier trains itself by its own predictions. However, self-training has been mostly applied to multi-class classification, and rarely applied to the multi-label scenario. In this paper, we propose a self-training-based approach for semi-supervised multi-label document classification, in which semantic-space finetuning is introduced and integrated into the self-training process. Newly discovered credible predictions are used not only for classifier finetuning, but also for semantic-space finetuning, which further benefit label propagation for exploring more credible predictions. Experimental results confirm the effectiveness of the proposed approach and show a satisfactory improvement over the baseline methods.

Keywords: Semi-supervised learning · Multi-label classification · Self-training · Semantic-space finetuning · Label propagation

1 Introduction

Text classification occupies an important role in natural language processing and information retrieval. It is widely used in the fields like document management, news categorization and sentiment analysis, being one of the key technologies for digital libraries. However, the shortage of labeled training documents has been always a challenge that cannot be ignored, for which semi-supervised learning has been studied.

Self-training is one of the symbolic semi-supervised learning approaches, which has been proved effective through plenty of researches [11, 19, 25]. Although self-training has achieved a great success, most of the existing methods only apply self-training on multi-class classification, in which each sample falls into just one category. However, in real document collections, each document is often labeled with multiple categories, as depicted in Fig. 1. The task of assigning a set of category labels is called *multi-label classification.*

H.-R. Ke et al. (Eds.): ICADL 2021, LNCS 13133, pp. 249–263, 2021.
https://doi.org/10.1007/978-3-030-91669-5_20

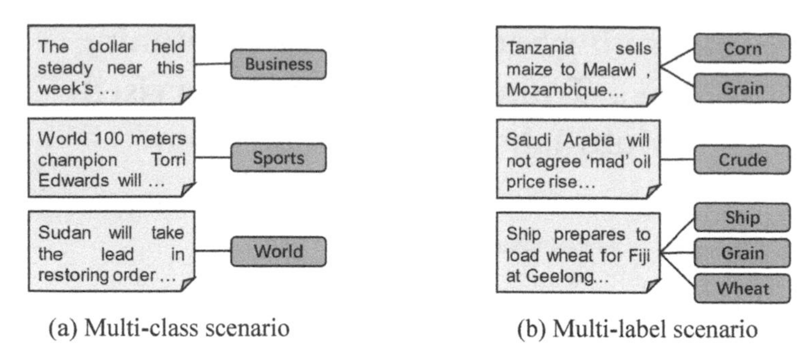

(a) Multi-class scenario (b) Multi-label scenario

Fig. 1. Comparison of multi-class and multi-label scenario

It is natural to consider extending self-training to the multi-label scenario. However, there is a tough nut in judging credible predictions. Unlike the case of multi-class classification in which labels are mutually exclusive, each sample in the multi-label scenario can belong to multiple categories, causing difficulties in judging whether a prediction is credible enough to be used in the next training loop. There also exists a greater risk of introducing noise into the system, thereby damaging the self-training process.

In this paper, we firstly propose **Sem**antic **L**abel **P**ropagation **A**lgorithm (SemLPA), which extends the label propagation algorithm (LPA) [24], propagating confident labels over the semantic-space of documents. SemLPA is integrated into the self-training loop, in which the semantic-space is iteratively finetuned, using document pairs constructed by high-confidence predictions introduced in Sect. 4.3. Then, we propose a method named **D**ynamic **S**elf-**T**raining integrating semantic-space for semi-supervised **M**ulti-**L**abel text classification (ML-DST), extending self-training to the multi-label scenario, where the introduction of noise is reduced as much as possible through the cooperation of the multi-label classifier ML-BERT and SemLPA. Experiments on three datasets from different domains validate the superiority of our proposed approach.

The rest of the paper is organized as follows. Section 2 reviews the related work in the fields of self-training, label propagation and semi-supervised multi-label learning. Section 3 formulates the target task and introduces the self-training structure. In Sect. 4, we describe the models of SemLPA and ML-BERT, and introduce ML-DST. Section 5 shows evaluation results and analysis. Finally, Sect. 6 summarizes our work and draws conclusion.

2 Related Work

Self-training. The original idea of self-training can be traced back to 1965 [5]. Since then, it is constantly being extended and improved. Until now, excellent works based on self-training are still emerging. One of the notable works is [8], in which pseudo labels were introduced to self-training and achieved great improvement. Following [8], deeper studies on pseudo labels appeared [13, 26]. In the meantime, the common issue of data imbalance in the self-training process was noticed and reducing the impact of data imbalance for self-training [18, 25] have been discussed. Furthermore, self-training has been extended to tougher tasks such as few-shot learning [9, 13] and hierarchical classification [12], demonstrating its great potential in the semi-supervised scenario.

Label Propagation Algorithm (LPA) [24] is a classic semi-supervised learning method designed for the transductive setup. There has been research on label propagation for multi-label data as early as 2006 [6]. After that, Zhang et al. [23] innovatively introduced non-metric distance to label propagation, and Wang et al. [16] proposed dynamic label propagation in which label information is involved in the label propagation process. With the rise of deep learning, label propagation has also been integrated with deep learning methods according to recent researches [4, 10].

Semi-supervised Multi-label Classification. There have been several trials for multi-label classification under the semi-supervised scenario. Kong et al. [7] proposed TRAM, which transforms the classification problem into an optimization problem of estimating the label composition for each unlabeled instance. Zhan et al. [22] and Xing et al. [20] applied the co-training structure for semi-supervised multi-label classification, in which the label information of unlabeled samples is alternately communicated between two classifiers. Wang et al. [17] jointly explored the feature distribution and the label relation, where the predictions on unlabeled data are also leveraged. The above methods select reliable predictions by ranking composite scores, which is usually an average score among all the categories, lacking examination on each specific category.

3 Preliminaries

3.1 Problem Formulation

Given a collection of n documents $X = \{x_1, x_2, \ldots, x_l, x_{l+1}, \ldots, x_n\}$ and a collection of m categories $C = \{c_1, c_2, \ldots, c_m\}$, we assume that each document x_i is associated with a vector of labels $\boldsymbol{y}_i = [y_{i1}, y_{i2}, \ldots, y_{im}]$, $y_{ij} \in \{0, 1\}$, where $y_{ij} = 1$ indicates the i-th document belongs to c_j, otherwise $y_{ij} = 0$. In the given document collection X, the labels of the first l documents $x_1, x_2, \ldots, x_l$ are known to us. We denote this labeled document set as X_{label}. The labels of the remaining $(l - n)$ documents $x_{l+1}, x_{l+2}, \ldots, x_n$ are unknown to us. We denote this unlabeled document set as $X_{unlabel}$. The target is to predict the labels of the unlabeled document set $X_{unlabel}$ by utilizing all the information provided by the document collection X.

3.2 Self-training

Self-training is a process that a classifier trains itself by its own predictions repeatedly. Specifically, the classifier is firstly trained on the labeled data only. Then the classifier is used to predict the labels for the unlabeled data. The credibility of each prediction is then estimated and those credible predictions will be picked out, assigned pseudo labels and then, participate in the next round of classifier training.

4 Proposed Approach

In this paper, we propose a new method named *dynamic self-training integrating semantic-space for semi-supervised multi-label text classification* (ML-DST) which adapts self-training to multi-label text classification. Figure 2 shows the overview of our proposed approach. At the beginning, since the high-confidence set is empty, the semantic label propagation algorithm (SemLPA) and the multi-label classifier (ML-BERT) finetuning are executed on the given labeled set separately. After that, with the aim to lift the credibility of predictions, comprehensive prediction scores are produced with respect to the results of SemLPA and ML-BERT. Finally, high-confidence predictions are picked out to update the high-confidence set, which will be combined with the labeled set and participate in the following training iteration.

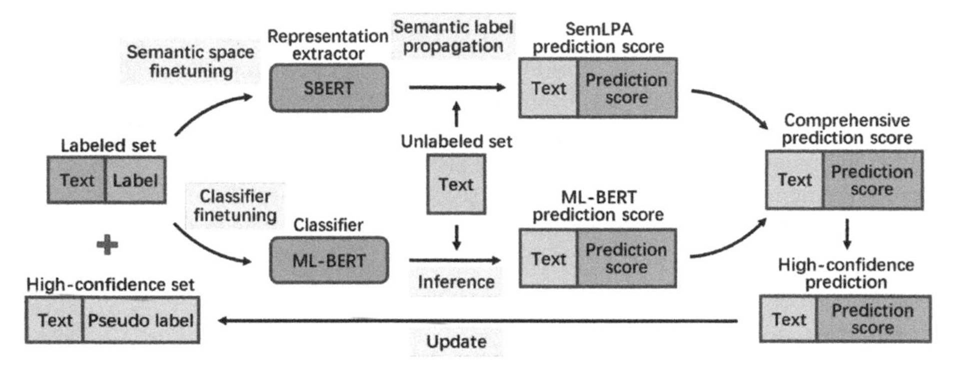

Fig. 2. The overview of the proposed approach ML-DST

4.1 Semantic Label Propagation Algorithm (SemLPA)

In the classical label propagation algorithm, labels are propagated to neighboring samples, where the distance measure between samples has a significant influence on the performance of LPA. We need to consider an appropriate distance measure for multi-label document classification. We also point out that LPA assumes a predefined distance measure on document samples, lacking opportunities to integrate credible predictions discovered by a separate method.

Sentence-BERT (SBERT) [14] is a pretrained language model adopting Siamese network structure to produce fixed-length document representations, which enables to efficiently compute the distance between two documents and find neighbors of a document. We utilize SBERT to construct a semantic-space of documents, where the distance between two documents x_u and x_v is measured by the cosine distance of two document representations $\boldsymbol{u}$ and $\boldsymbol{v}$:

$$D(x_u, x_v) = 1 - \cos(\boldsymbol{u}, \boldsymbol{v}) \tag{1}$$

Semantic-space Finetuning. We introduce the notion of *semantic-space finetuning*, such that the *semantic-space*, defined by the distance measure on documents, is improved to better cluster similarly labeled documents, through finetuning document representations. We also propose the cycle of repeating semantic-space finetuning and label propagation, to progressively find new credible predictions and improve the semantic-space. Semantic-space finetuning also has the effect of finetuning SBERT to the target corpus.

For a pair of documents (x_u, x_v), we intend to finetune the semantic-space to make cosine similarity of the two text representations $\boldsymbol{u}$ and $\boldsymbol{v}$ positively correlated with that of label vectors $\boldsymbol{y}_u$ and $\boldsymbol{y}_v$:

$$\min \left|\cos(\boldsymbol{u}, \boldsymbol{v}) - \cos(\boldsymbol{y}_u, \boldsymbol{y}_v)\right| \quad (2)$$

Therefore, the loss function of semantic-space finetuning is defined as:

$$L_{SSF} = \sum\nolimits_{(x_u, x_v) \in S_{pair}} \left|\cos(\boldsymbol{u}, \boldsymbol{v}) - \cos(\boldsymbol{y}_u, \boldsymbol{y}_v)\right|, \quad (3)$$

where S_{pair} is a set of document pairs. To avoid high computational cost, we randomly sample N positive pairs and N negative pairs from labeled documents for each category to construct document pair set S_{pair}. A positive pair of the i-th category is such that both documents are positive on the i-th category, while the labels on the other categories are arbitrary. Similarly, a negative pair of the i-th category is such that two documents have opposite labels on the i-th category without regard to the other categories.

Label Propagation. For label propagation, a transition matrix $P \in R^{n \times n}$ is constructed from the document distances $D(x_u, x_v)$:

$$P_{uv} = \frac{D(x_u, x_v)}{\sum_{x_w \in X} D(x_u, x_w)} \quad (4)$$

Given a labeled set X_{label} and a label matrix of labeled data Y_{label}, our proposed algorithm SemLPA shown in Algorithm 1 predicts a label matrix $Y_{unlabel}$ for all the unlabeled data. It should be noted that in the label propagation, positive and negative labels exist in the form of 1 and -1 respectively, so that it can be distinguished whether a spread is from a positive or a negative label. We name the values indicating the polarity of labels the prediction scores. After label propagation, prediction scores are finally rescaled to the range of [0, 1] through min-max normalization.

Algorithm 1: **Semantic label propagation algorithm (SemLPA)**

Input: $X_{label} = \{x_1, x_2, \dots, x_l\}$, $Y_{label} = \{y_1, y_2, \dots, y_l\}$,
Output: $Y_{unlabel} = \{y_{l+1}, y_{l+2}, \dots, y_n\}$

Process:
1: Construct document pair set S_{pair} using X_{label}
2: Finetune semantic-space on S_{pair} by loss L_{SSF}
3: Calculate document distances and the transition matrix P
4: $Y^0 = [Y^0_{label}; \mathbf{0}]$
5: Repeat V times:
6: $Y^{t+1} = P * Y^t$
7: $Y^{t+1}_{label} = Y^0_{label}$
8: Output $Y_{unlabel}$

4.2 Classifier Finetuning

In the self-training process, the classifier uses its own prediction to train itself iteratively. Intuitively, the better the performance of the initial classifier, the higher the improvement in self-training. In this paper, we adopt BERT [3], a context-aware pretrained language model, to construct a multi-label classifier, called ML-BERT.

As Fig. 3 shows, we construct the classifier by adding a fully connected layer on top of the BERT model. The fully connected layer produces logit vectors $\boldsymbol{s_i} = [s_{i1}, s_{i2}, \dots, s_{im}]$, after which the element-wise sigmoid function is applied to produce prediction score $\boldsymbol{ps_{ij}} = \left[ps_{i1}, ps_{i2}, \dots, ps_{im}\right]$, $0 \le ps_{ij} \le 1$, where ps_{ij} represents the likelihood that the i-th document belongs to the j-th category c_j.

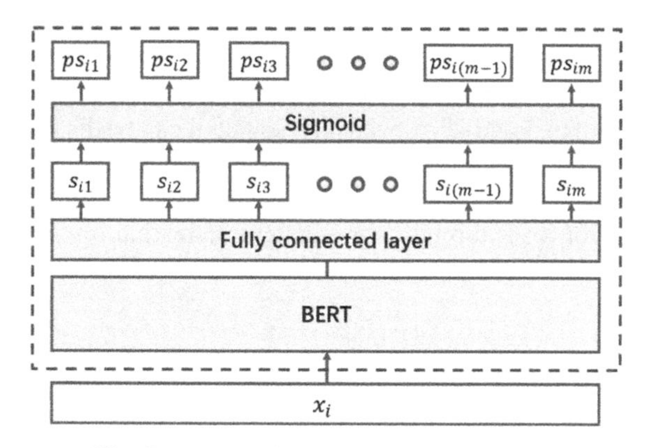

Fig. 3. The multi-label classifier ML-BERT

Weighted Binary Cross Entropy Loss (WBCE). Binary cross entropy loss is commonly used for multi-label classification. However, pseudo labels in self-training cannot always correctly indicate the polarity of the labels. Therefore, we modify the binary cross entropy loss to reduce the negative effects from mislabeling through assigning weights to high-confidence predictions according to their certainties. The *weighted binary cross entropy loss* is represented as below:

$$L_{WBCE}\left(PS, Y_{label}, \hat{Y}_{conf}\right) = \sum_{x_i \in X_{label}} L_{BCE}(\boldsymbol{ps}_i, \boldsymbol{y}_i) + \sum_{x_i \in X_{conf}} w_i L_{BCE}(\boldsymbol{ps}_i, \hat{\boldsymbol{y}}_i) \quad (5)$$

$$L_{BCE}(\boldsymbol{ps}_i, \boldsymbol{y}_i) = -\frac{1}{m}\sum_{j=1}^{m}\left(y_{ij}\log ps_{ij} + (1 - y_{ij})\log(1 - ps_{ij})\right)$$

$$w_i = 1 - \max_{j=1,2,\ldots,m} H(\hat{y}_{ij})$$

$$H(\hat{y}_{ij}) = -\hat{y}_{ij}\log\hat{y}_{ij} - (1 - \hat{y}_{ij})\log(1 - \hat{y}_{ij})$$

Here, PS is the prediction score matrix of the input data, Y_{label} is the label matrix of the labeled data. Pseudo label matrix $\hat{Y}_{conf}$ for high-confidence data will be introduced in Sect. 4.3. Symbols $\boldsymbol{ps}_i, \boldsymbol{y}_i$ and $\hat{\boldsymbol{y}}_i$ represent the i-th row of PS, Y_{label} and $\hat{Y}_{conf}$, respectively. We evaluate the certainty of the prediction by information entropy $H(\cdot)$, aiming to assign larger weights to more convinced predictions and smaller weights to the predictions with low certainty. During the calculation of $L_{BCE}(\boldsymbol{ps}_i, \hat{\boldsymbol{y}}_i)$, we regard 0.5 as the dividing line and set pseudo labels larger than 0.5 to 1 and pseudo labels less than or equal to 0.5 to 0.

Weighted Multi-label Categorical Cross Entropy Loss (WMCE). Based on the categorical cross entropy loss function, Su [15] newly proposed multi-label categorical cross entropy loss (MCE) for the multi-label scenario, which is derived from the idea of label ranking. Different from the other label ranking-based losses, MCE is able to determine the polarity of labels, making it possible to integrate into the self-training framework. Similarly to WBCE, we assign weights to high-confidence predictions and introduce the *weighted multi-label categorical cross entropy loss*, as follows:

$$L_{WMCE}\left(S, Y_{label}, \hat{Y}_{conf}\right) = \sum_{x_i \in X_{label}} L_{MCE}(\boldsymbol{s}_i, \boldsymbol{y}_i) + \sum_{x_i \in X_{conf}} w_i L_{MCE}(\boldsymbol{s}_i, \hat{\boldsymbol{y}}_i) \quad (6)$$

$$L_{MCE}(\boldsymbol{s}_i, \boldsymbol{y}_i) = \log\left(1 + \sum_{j \in \Omega_i^{neg}} e^{s_{ij}}\right) + \log\left(1 + \sum_{j \in \Omega_i^{pos}} e^{-s_{ij}}\right)$$

$$w_i = 1 - \max_{j=1,2,\ldots,m} H(\hat{y}_{ij})$$

$$H(\hat{y}_{ij}) = -\hat{y}_{ij}\log\hat{y}_{ij} - (1 - \hat{y}_{ij})\log(1 - \hat{y}_{ij})$$

Here, S is the logit matrix of the input data. Symbol $\boldsymbol{s}_i$ represents the i-th row of S. Ω_i^{pos} and Ω_i^{neg} refer to the index sets of the positive and negative labels of the i-th sample. During the calculation of $L_{MCE}(\boldsymbol{s}_i, \hat{\boldsymbol{y}}_i)$, we regard pseudo labels larger than 0.5 as positive while less than or equal to 0.5 as negative.

4.3 High-Confidence Prediction Selection

We denote the prediction scores of the i-th unlabeled document predicted by SemLPA and ML-BERT, as $\boldsymbol{ps}_i^{lpa} = \left[ps_{i1}^{lpa}, ps_{i2}^{lpa}, \ldots, ps_{im}^{lpa}\right]$ and $\boldsymbol{ps}_i^{clf} = \left[ps_{i1}^{clf}, ps_{i2}^{clf}, \ldots, ps_{im}^{clf}\right]$, respectively, and obtain the *comprehensive prediction score* by element-wise averaging:

$$\boldsymbol{ps}_i = \frac{1}{2}\left(\boldsymbol{ps}_i^{lpa} + \boldsymbol{ps}_i^{clf}\right) \tag{7}$$

For multi-label classification, one document may belong to multiple categories. It is necessary to check that predictions on all the categories are reliable. Any mislabeling on even one category may introduce considerable noise to the training process, which may lead self-training into a vicious circle. Therefore, in self-training, we use only *high-confidence predictions* such that the prediction score vector $\boldsymbol{ps}_i = \left[ps_{i1}, ps_{i2}, \ldots, ps_{im}\right]$ satisfies the following boundaries:

$$\forall j \in \{1, 2, \ldots, m\}, \qquad ps_{ij} > \theta_+ \vee ps_{ij} < \theta_- \tag{8}$$

Here, θ_+ and θ_- are predefined thresholds. Each of the high-confidence predictions associates with a pseudo label vector, which is generally the vector of comprehensive prediction score. We call the set of the unlabeled documents having high-confidence predictions as *high-confidence set* X_{conf}.

It is necessary to point out that self-training is an iterative process and high-confidence predictions change after each loop. When new high-confidence predictions are derived, we update X_{conf} by merging them with the existing high-confidence set X_{conf}. For the case that one document already exists in the previous X_{conf}, we take the element-wise average of pseudo labels as its new pseudo label.

Our proposed approach is summarized in Algorithm 2.

5 Experiments

5.1 Datasets

We use three multi-label text datasets from different domains to evaluate the performance of our proposed approach.

Reuters. Reuters-21578 dataset collects the documents published on the Reuters newswire. We employ the ModApte split [2] of the Reuters-21578 text collection which contains a total of 10 topics. The first 8,000 samples are used in the experiment.

Algorithm 2: Dynamic self-training integrating semantic-space for semi-supervised multi-label text classification (ML-DST)

Input: $X_{label} = \{x_1, x_2, \ldots, x_l\}$, $Y_{label} = \{y_1, y_2, \ldots, y_l\}$, $X_{unlabel} = \{x_{l+1}, x_{l+2}, \ldots, x_n\}$
Output: $Y_{unlabel} = \{y_{l+1}, y_{l+2}, \ldots, y_n\}$

Process:
1: $X_{conf} \leftarrow \emptyset$
2: Repeat T times:
3: Finetune ML-BERT using $X_{label} \cup X_{conf}$
4: Semantic-space finetuning and label propagation (SemLPA)
5: Produce comprehensive prediction scores by ML-BERT and SemLPA
6: Select high-confidence predictions and update X_{conf}
7: Infer $Y_{unlabel}$ using the final ML-BERT

Arxiv Academic Paper Dataset (AAPD). AAPD dataset published in [21] is a collection of academic paper abstracts. We reconstruct the dataset by retaining the top 20 subjects and 20,000 samples. Each sample belongs to at least one of the 20 categories.

Blurb Genre Collection (BGC). BGC [1] consists of advertising descriptions of books, which are called blurbs. Like the processing for AAPD, we retain the top 20 categories and 20,000 samples for experiment.

5.2 Baselines

In this paper, the performance of our proposed approach is compared against two semi-supervised methods: LPA [24] and TRAM [7]. For both LPA and TRAM, we produce document embeddings through SBERT as input feature vectors. To present the gain from pretrained language models, ML-BERT alone is also evaluated.

LPA. Labels are propagated from labeled data to unlabeled data according to the distance between documents. Compared with documents that are far away, labels are more likely to be propagated between documents that are close to each other.

TRAM. The multi-label classification is formulated as an optimization problem of estimating label concept compositions. Then, a closed-form solution is derived to this optimization problem and an effective algorithm is used to assign labels to the unlabeled instances.

ML-BERT. A dense layer and a sigmoid layer are added on top of the pretrained BERT model, forming a BERT-based multi-label classifier, which is then finetuned with the labeled data.

5.3 Implementation Details

Composition of Labeled and Unlabeled Document Sets. Randomly taking a small part of samples as labeled data may cause a situation that some categories have no positive samples, making training intractable. Therefore, we randomly extract s positive samples on each category to compose the labeled document set X_{label}. Suppose there are m categories, then the labeled document set X_{label} contains a total of $s \cdot m$ samples, where each sample has $p \geq 1$ positive labels and $(m - p)$ negative labels. The remaining documents are naturally assigned to the unlabeled document set $X_{unlabel}$. Due to the construction, there may exist duplication in X_{label} and each category may have more than s positive samples in X_{label}.

For Reuters, s is ranging on [8, 16, 24, 40, 80]. For both AAPD and BGC, s is ranging on [10, 20, 30, 50, 100]. By this configuration, the labeled samples account for exactly $1\%, 2\%, 3\%, 5\%, 10\%$ of the size of the dataset. For classifier finetuning, X_{label} is further divided into the training set and the validation set by the ratio of 4:1.

Hyperparameters. The adopted BERT and SBERT models are "bert-base-uncased" and "bert-base-nli-stsb-mean-tokens" respectively. The learning rate of BERT finetuning is 5×10^{-5} and the batch size is 24. For semantic-space finetuning, we use 1,000,000 document pairs to finetune the semantic-space for one epoch in the first loop. For the following loops, the size of S_{pair} is 200,000. In SemLPA, labels are propagated for a total of 400 times. θ_+ and θ_- are fixed to 0.8 and 0.2, respectively, in all the experiments. The maximum iteration T is set to 5 in our experiments. We run five times and take the average of the evaluation results. All the experiments are executed on a hardware with an NVIDIA RTX3090 GPU.

5.4 Results

For multi-label classification, a single metric is difficult to comprehensively evaluate the performance of a model. Here, not only the overall accuracy but also the performance on each class should be considered. Therefore, we adopt macro-F1, micro-F1 and hamming loss to evaluate the model performance.

Macro-F1 and micro-F1 are calculated based on true positives, false negatives, and false positives, while hamming loss globally evaluates the proportion of misclassified sample-label pairs, such that a relevant label is missed or an irrelevant label is predicted.

Tables 1, 2 and 3 show the results on Reuters, AAPD and BGC datasets, respectively.

According to the results, our proposed ML-DST outperforms the other models in most cases. On the Reuters dataset, ML-DST with WBCE loss function is outperforming the other models in all the cases. On the other two datasets, two ML-DST models with different loss functions are comparable, where WMCE tends to achieve higher F1 scores, while WBCE is more advantageous on hamming loss.

The difference in the principle of the loss function is one of the factors that causes the different results. WBCE treats the predictions of each category equally without considering its polarity, which weakens the role of minority polarity, while WMCE considers the pairwise ranking between positive and negative labels, thereby increasing

Table 1. Evaluation results on Reuters, by marco-F1, micro-F1, and hamming loss

Label rate	1%	2%	3%	5%	10%
Macro-F1					
LPA	0.625	0.761	0.777	0.862	0.818
TRAM	0.570	0.698	0.702	0.722	0.735
ML-BERT (BCE)	0.652	0.826	0.870	0.895	0.913
ML-BERT (MCE)	0.672	0.840	0.886	0.891	0.906
ML-DST (WBCE)	**0.813**	**0.870**	**0.895**	**0.902**	**0.917**
ML-DST (WMCE)	0.777	0.858	0.887	0.892	0.914
Micro-F1					
LPA	0.744	0.858	0.869	0.918	0.896
TRAM	0.776	0.827	0.835	0.843	0.853
ML-BERT (BCE)	0.697	0.884	0.919	0.937	0.955
ML-BERT (MCE)	0.759	0.858	0.920	0.927	0.948
ML-DST (WBCE)	**0.872**	**0.908**	**0.935**	**0.940**	**0.956**
ML-DST (WMCE)	0.819	0.883	0.926	0.932	0.952
Hamming loss					
LPA	0.062	0.031	0.029	0.018	0.023
TRAM	0.047	0.037	0.035	0.033	0.031
ML-BERT (BCE)	0.051	0.024	0.017	**0.013**	0.010
ML-BERT (MCE)	0.045	0.029	0.017	0.015	0.011
ML-DST (WBCE)	**0.027**	**0.020**	**0.014**	**0.013**	**0.009**
ML-DST (WMCE)	0.038	0.025	0.016	0.015	0.010

Table 2. Evaluation results on AAPD, by marco-F1, micro-F1, and hamming loss

Label rate	1%	2%	3%	5%	10%
Macro-F1					
LPA	0.416	0.525	0.557	0.598	0.636
TRAM	0.485	0.561	0.592	0.613	0.638
ML-BERT (BCE)	0.366	0.536	0.578	0.642	0.661
ML-BERT (MCE)	0.457	0.598	0.618	0.659	0.669
ML-DST (WBCE)	0.485	0.596	0.623	0.658	0.680
ML-DST (WMCE)	**0.495**	**0.605**	**0.631**	**0.664**	**0.681**
Micro-F1					
LPA	0.453	0.568	0.589	0.618	0.666
TRAM	0.536	0.601	0.620	0.640	0.666
ML-BERT (BCE)	0.444	0.594	0.621	0.666	0.691
ML-BERT (MCE)	0.520	0.628	0.647	0.678	0.688
ML-DST (WBCE)	0.546	0.637	0.649	0.679	**0.704**
ML-DST (WMCE)	**0.554**	**0.639**	**0.657**	**0.684**	0.701
Hamming loss					
LPA	0.110	0.085	0.084	0.079	0.066
TRAM	0.097	0.082	0.077	0.074	0.069
ML-BERT (BCE)	0.081	0.069	0.068	0.063	**0.059**
ML-BERT (MCE)	0.079	0.068	0.067	0.065	0.062
ML-DST (WBCE)	**0.075**	**0.067**	**0.066**	**0.062**	**0.059**
ML-DST (WMCE)	0.077	0.068	**0.066**	0.063	0.061

Table 3. Evaluation results on BGC, by marco-F1, micro-F1, and hamming loss

Label rate	1%	2%	3%	5%	10%
Macro-F1					
LPA	0.378	0.514	0.580	0.630	0.713
TRAM	0.503	0.603	0.645	0.676	0.697
ML-BERT (BCE)	0.393	0.610	0.638	0.696	0.723
ML-BERT (MCE)	0.504	0.636	0.669	0.695	0.721
ML-DST (WBCE)	0.564	0.667	0.674	0.706	0.736
ML-DST (WMCE)	**0.566**	**0.668**	**0.692**	**0.718**	**0.743**
Micro-F1					
LPA	0.532	0.619	0.666	0.698	0.770
TRAM	0.595	0.677	0.706	0.739	0.754
ML-BERT (BCE)	0.598	0.708	0.727	0.761	0.782
ML-BERT (MCE)	0.653	0.723	0.742	0.758	0.776
ML-DST (WBCE)	0.668	**0.733**	0.740	0.759	0.785
ML-DST (WMCE)	**0.677**	0.731	**0.753**	**0.770**	**0.790**
Hamming loss					
LPA	0.119	0.096	0.083	0.074	0.056
TRAM	0.105	0.078	0.075	0.066	0.062
ML-BERT (BCE)	0.079	0.064	0.061	**0.057**	**0.052**
ML-BERT (MCE)	0.075	0.064	**0.060**	0.058	0.054
ML-DST (WBCE)	**0.074**	**0.063**	0.063	0.059	0.053
ML-DST (WMCE)	**0.074**	0.066	**0.060**	**0.057**	**0.052**

the impact of the minority polarity. In the datasets of AAPD and BGC, positive labels are minorities and the polarity imbalance is particularly significant. In such cases, WMCE is likely to obtain more true positive predictions, contributing to higher F1 scores, while WBCE tends to produce more true negative predictions, decreasing the hamming loss.

We also notice that when only labeled samples are used for training, BERT-based models are generally better than TRAM and LPA, except for several cases when the label rate is extremely small, which reveals the weakness of BERT-based models without self-training on small-scale training sets.

5.5 Exploring the Impact of the Semantic-Space on SemLPA

Semantic-space finetuning updates the semantic-space and distance between documents, which affects the label propagation performance. We further perform experiments on how varying percentage of labeled samples affect semantic-space finetuning and the result of label propagation.

The following tests are applied on the BGC dataset. We firstly construct three document pair sets using 1%, 10% and 50% labeled samples, respectively. Each of the document pair set consists of 1,000,000 document pairs. Then, three new semantic-spaces are separately generated by semantic-space finetuning, using the prepared three document pair sets. After that, SemLPA is applied on these three semantic-spaces separately, and also on the semantic-space without finetuning. Figure 4 depicts the results of the comparison, in which the x-axis represents the proportion of labeled samples in

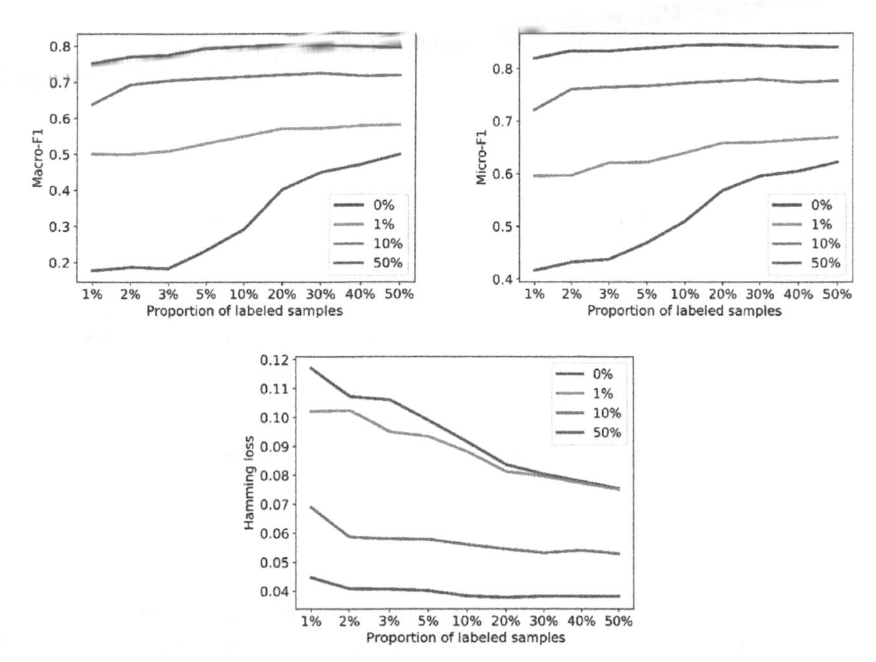

Fig. 4. The influence of semantic-space finetuning on SemLPA, where 0%, 1%, 10%, and 50% labeled samples are used for finetuning.

SemLPA. In the cases of the same proportion of labeled samples, we use the same labeled data and unlabeled data in the experiment.

According to the results, semantic-space finetuning significantly improves the performance of SemLPA. With the increase of labeled data used for semantic-space finetuning, SemLPA results are obviously improved.

5.6 Comparing the High-Confidence Set Derived from Different Models

To verify the combination of the results of SemLPA and ML-BERT shown in (7) is improving the quality of confidence data and reducing introduction of noise, we examine the evaluation results on high-confidence sets which are respectively derived from 1) the result of SemLPA, 2) the result of ML-BERT and 3) the result of the combination of SemLPA and ML-BERT. Here, the Reuters dataset is used and the results of the first iteration in different label rates are shown in Table 4. According to the evaluation results, the quality of the confidence data has been improved to varying degrees by the combination, validating the approach of combining two results for high-confidence selection.

Table 4. Evaluation results on high-confidence sets of Reuters

Label rate	1%	2%	3%	5%	10%
Macro-F1					
SemLPA	0.722	0.771	0.917	0.886	0.929
ML-BERT	0.569	0.901	0.955	0.942	0.950
SemLPA and ML-BERT	**0.730**	**0.951**	**0.977**	**0.971**	**0.968**
Micro-F1					
SemLPA	0.866	0.851	0.923	0.916	0.941
ML-BERT	0.845	0.961	0.986	0.970	0.975
SemLPA and ML-BERT	**0.931**	**0.976**	**0.987**	**0.980**	**0.983**
Hamming losss					
SemLPA	0.036	0.033	0.018	0.018	0.012
ML-BERT	0.030	0.008	**0.003**	0.006	0.005
SemLPA and ML-BERT	**0.023**	**0.005**	**0.003**	**0.004**	**0.004**

6 Conclusion

In this paper, we proposed Dynamic Self-Training integrating semantic-space for Multi-Label document classification (ML-DST), which incorporates self-training into the multi-label scenario. We firstly proposed a semantic label propagation method named SemLPA, which can be embedded in our self-training framework, so that label propagation can be performed iteratively as credible predictions are newly discovered. SemLPA assists the multi-label classifier ML-BERT to find high-quality predictions, which in turn improve the performance of SemLPA and ML-BERT.

Our experimental results show that ML-DST yields superior performance over the representative baselines, which confirms the effectiveness of our proposed method. Further analysis on SemLPA confirms the necessity of semantic-space finetuning for label propagation. The role of SemLPA in finding credible predictions has also been proven through examining intermediate results.

References

1. Aly, R., Remus, S., Biemann, C.: Hierarchical multi-label classification of text with capsule networks. In: ACL: Student Research Workshop (2019)
2. Apte, C., Damerau, F., Weiss, S.M.: Towards language independent automated learning of text categorization models. In: Croft, B.W., van Rijsbergen, C.J. (eds.) SIGIR 1994. Springer, London (1994). https://doi.org/10.1007/978-1-4471-2099-5_3
3. Devlin, J., et al.: BERT: pre-training of deep bidirectional transformers for language understanding. In: NAACL-HLT (2019)
4. Iscen, A., et al.: Label propagation for deep semi-supervised learning. In: CVPR (2019)
5. Scudder, H.J.: Probability of error of some adaptive pattern-recognition machines. IEEE Trans. Inf. Theory **11**(3), 363–371 (1965)
6. Kang, F., Jin, R., Sukthankar, R.: correlated label propagation with application to multi-label learning. In: CVPR (2006)
7. Kong, X., Ng, M.K., Zhou, Z.H.: Transductive multilabel learning via label set propagation. IEEE Trans. Knowl. Data Eng. **25**(3), 704–719 (2011)

8. Lee, D.: Pseudo-label: the simple and efficient semi-supervised learning method for deep neural networks. In: Workshop on Challenges in Representation Learning, ICML, vol. 3, no. 2 (2013)
9. Li, X., et al.: Learning to self-train for semi-supervised few-shot classification. In: NeurIPS (2019)
10. Liu, Y., et al.: Learning to propagate labels: transductive propagation network for few-shot learning. In: ICLR (2019)
11. Meng, Y., et al.: Weakly-supervised neural text classification. In: CIKM (2018)
12. Meng, Y., et al.: Weakly-supervised hierarchical text classification. In: AAAI (2019)
13. Mukherjee, S., Ahmed, A.: Uncertainty-aware self-training for few-shot text classification. In: NeurIPS (2020)
14. Reimers, N., Gurevych, I.: Sentence-bert: sentence embeddings using siamese bert-networks. In: EMNLP-IJCNLP (2019)
15. Su, J.: Blog post. https://www.spaces.ac.cn/archives/7359. Accessed 13 July 2021
16. Wang, B., Tu, Z., Tsotsos, J.K.: Dynamic label propagation for semi-supervised multi-class multi-label classification. In: ICCV (2013)
17. Wang, L., et al.: Dual relation semi-supervised multi-label learning. In: AAAI (2020)
18. Wei, C., et al.: CReST: a class-rebalancing self-training framework for imbalanced semi-supervised learning. In: CVPR (2021)
19. Xie, Q., et al.: Self-training with noisy student improves imagenet classification. In: CVPR (2020)
20. Xing, Y., et al.: Multi-label co-training. In: IJCAI (2018)
21. Yang, P., et al.: SGM: sequence generation model for multi-label classification. In: COLING (2018)
22. Zhan, W., Zhang, M.L.: Inductive semi-supervised multi-label learning with co-training. In: SIGKDD (2017)
23. Zhang, Y., Zhou, Z.: Non-metric label propagation. In: IJCAI (2009)
24. Zhu, X., Ghahramani, Z.: learning from labeled and unlabeled data with label propagation. Technical report CMU-CALD-02–107, Carnegie Mellon University (2002)
25. Zou, Y., Yu, Z., Vijaya Kumar, B.V.K., Wang, J.: Unsupervised domain adaptation for semantic segmentation via class-balanced self-training. In: Ferrari, V., Hebert, M., Sminchisescu, C., Weiss, Y. (eds.) ECCV 2018. LNCS, vol. 11207, pp. 297–313. Springer, Cham (2018). https://doi.org/10.1007/978-3-030-01219-9_18
26. Zou, Y., et al.: Confidence regularized self-training. In: ICCV (2019)

Named Entity Recognition Architecture Combining Contextual and Global Features

Tran Thi Hong Hanh[1,2,4], Antoine Doucet[2], Nicolas Sidere[2], Jose G. Moreno[3], and Senja Pollak[4(✉)]

[1] University of Science and Technology in Hanoi, Hanoi, Vietnam
[2] L3i Laboratory, University of La Rochelle, La Rochelle, France
[3] IRIT, University of Toulouse, Toulouse, France
[4] Jozef Stefan Institute, Ljubljana, Slovenia
senja.pollak@ijs.si

Abstract. Named entity recognition (NER) is an information extraction technique that aims to locate and classify named entities (e.g., organizations, locations, ...) within a document into predefined categories. Correctly identifying these phrases plays a significant role in simplifying information access. However, it remains a difficult task because named entities (NEs) have multiple forms and they are context dependent. While the context can be represented by contextual features, the global relations are often misrepresented by those models. In this paper, we propose the combination of contextual features from XLNet and global features from Graph Convolution Network (GCN) to enhance NER performance. Experiments over a widely-used dataset, CoNLL 2003, show the benefits of our strategy, with results competitive with the state of the art (SOTA).

Keywords: NER · XLNet · GCN · Contextual embeddings · Global embeddings

1 Introduction

The proliferation of large digital libraries has spurred interest in efficient and effective solutions to manage the collections of digital contents (documents, images, videos, etc.) which are available, but not always easy to find. As an alternative to better handle information in digital libraries, named entity recognition (NER) was introduced.

NER is an information extraction technique that aims to locate named entities (NEs) in text and classify them into predefined categories. Correctly identifying entities plays an important role in natural language understanding and numerous applications such as entity linking, question answering, or machine translation, to mention a few.

H.-R. Ke et al. (Eds.): ICADL 2021, LNCS 13133, pp. 264–276, 2021.
https://doi.org/10.1007/978-3-030-91669-5_21

A crucial component that contributes to the recent success of NER progress is how meaningful information can be captured from original data via the word embeddings, which can be divided into two major types: global features and contextual features (in the scope of this paper, "features" and "embeddings" are interchangeable terms).

- **Global features** [30] capture latent syntactic and semantic similarities. They are first constructed from a global vocabulary (or dictionary) of unique words in the documents. Then, similar representations are learnt based on how frequently the words appear close to each other. The problem of such features is that the words' meaning in varied contexts is often ignored. That means, given a word, its embedding always stays the same in whichever sentence it occurs. Due to this characteristic, we can also define global features as "*static*". Some examples are word2vec [4], GloVe [30], and FastText [13].
- **Contextual features** [6] capture word semantics in context to address the polysemous and context-dependent nature of words. By passing the entire sentence to the pretrained model, we assign each word a representation based on its context, then capture the uses of words across different contexts. Thus, given a word, the contextual features are "*dynamically*" generated instead of being static as the global one. Some examples are ELMo [31], BERT [6], and XLNet [40].

In terms of global features, there exist several tokens that are always parts of an entity. The most obvious cases, as an example in the CoNLL 2003 dataset, are the names of countries include U.S. (377 mentions), Germany (143 mentions), Australia (136 mentions), to mention a few. However, it is not true for all tokens in an entity. The token may or may not be part of an entity (e.g., "Jobs said" vs. "Jobs are hard to find") and may belong to different entity types depending on the context (e.g., "Washington" can be classified as a person or a location). Meanwhile, the contextual features are based on neighboring tokens, as well as the token itself. They aim to represent word semantics in context to solve the problem of using global features, so as to improve the prediction performance (e.g., "Jobs" in "Jobs said" and "Jobs are hard to find" will have different representations).

In this paper, we present a joint architecture to enhance the NER performance simultaneously with static and dynamic embeddings[1]. Extensive experiments on CoNLL 2003 dataset suggest that our strategy surpasses the systems with standalone feature representation. The main contributions of this paper are:

- We introduce a new architecture that combines the contextual features from XLNet and the global features from GCN to enhance NER performance.
- We demonstrate that our model outperforms the systems using only contextual or global features alone and has a competitive result compared with SOTAs on CoNLL 2003 dataset.

[1] Link to the code: github.com/honghanhh/ner-combining-contextual-and-global-features.

This paper is organised as follows: Sect. 2 presents the related work, which leads to our approach's descriptions in Sect. 3 and the corresponding experimental details in Sect. 4. The results are reported in Sect. 5, before we conclude and present future works in Sect. 6.

2 Related Work

2.1 Named Entity Recognition

The term "named entity" (NE) first appeared in the 6th Message Understanding Conference (MUC-6) [9] to define the recognition of the information units. Regarding the surveys on NER techniques [18,29,39], we can broadly divide them into four categories: Rule-based, unsupervised learning, feature-based supervised learning, and deep learning based approaches.

Rule-Based Approaches. Rule-based NER is the most traditional technique that does not require annotated data as it relies on manually-crafted rules well-designed by the domain experts (e.g., LTG [26], NetOwl [14]). Despite good performance when the lexicon is exhaustive, such systems often achieve high precision and low recall due to the limitation on domain-specific rules and incomplete dictionaries.

Unsupervised Learning. Another approach that also needs no annotated data is unsupervised learning, typically NE clustering [5]. The key idea is to extract NEs from the clustered groups based on context similarity. The lexical resources, lexical patterns, and statistics are computed on a large corpus and then applied to infer mentions of NEs. Several works proposed the unsupervised systems to extract NEs in diverse domains [8,28].

Feature-Based Supervised Learning. Given annotated data, features are carefully designed so that the model can learn to recognize similar patterns from unseen data. Several statistical methods have been proposed, notably Markov models, Conditional Random Fields (CRFs), and Support Vector Machines (SVMs). Among them, CRF-based NER has been widely applied to identify entities from texts in various domains [21,33,34]. However, these approaches depend heavily on hand-crafted features and domain-specific resources, which results in the difficulty to adapt to new tasks or to transfer to new domains.

Deep Learning. Neural networks offer non-linear transformation so that the models can learn complex features and discover useful representations as well as underlying factors. Neural architectures for NER often make use of either Recurrent Neural Networks (RNNs) or Convolution Neural Networks (CNNs) in conjunction with CRFs [3] to extract information automatically. With further researches on contextual features, RNNs plus LSTM units and CRFs have been proposed [15] to improve the performance. Moreover, the conjunction of bidirectional LSTMs, CNNs, and CRFs [25] is introduced to exploit both word- and character-level representations. The combination of Transformer-based models, LSTMs, and CRFs [20] is also applied to extract knowledge from raw texts and empower the NER performance.

2.2 Embeddings

A key factor that contributes to the success of NER is how we capture meaningful information from original data via word representations, especially global features and contextual features.

Global Features. Global features are context-free word representations that can capture meaningful semantic and syntactic information. It can be represented at different levels such as word-level features [19], lookup features [10], document and corpus features [12]. Recently, the global sentence-level representation [42] has been proposed to capture global features more precisely and it outperforms various sequence labeling tasks. Furthermore, the Graph Neural Network [41] is getting more attention to not only have rich relational structure but also preserve global structure information of a graph in graph embeddings.

Contextual Features. Contextual features are context-aware word representations that can capture word semantics under diverse linguistic contexts. That is, a word can be represented differently and dynamically under particular circumstances. The contextual embeddings are often pretrained on large-scale unlabelled corpora and can be divided into 2 types: unsupervised approaches [16,17] and supervised approaches [36].

The contextual embeddings succeed in exploring and exploiting the polysemous and context-dependent nature of words, thereby moving beyond global word features and contributing significant improvements in NER. In contrast, the global features are still less-represented.

3 Methodology

In this section, we explain how we extract global as well as contextual features and how to combine them. For global features, we take advantage of GCN [2,35] to better capture the correlation between NEs and the global semantic information in text, and to avoid the loss of detailed information. For contextual features, we apply XLNet [40], a Transformer-XL pretrained language model that exhibits excellent performance for language tasks by learning from bi-directional context. The details are explained in the following subsections.

3.1 GCN as Global Embeddings

Graph Convolutional Network (GCN) aims to learn a function of signals/features on a graph $G = (V, E)$ with V as Vertices and E as Edges. Given N as number of nodes, D as number of input features, and F as the number of output features per node, GCN takes 2 inputs: (1) An $N \times D$ feature matrix X as feature description; (2) An adjacency matrix A as representative description of the graph; Finally, it returns as output Z, an $N \times F$ feature matrix [7].

Every neural network layer can then be written in the form of a non-linear function:

$$H^{(l+1)} = f(H^{(l)}, A) \tag{1}$$

where $H^{(0)} = X$, $H^{(L)} = Z$, L being the number of layers.

In our specific task, we capture the global features by feeding feature matrix X and adjacent matrix A into a graph using two-layer spectral convolutions in GCN. Raw texts are first transformed into word embeddings using GloVe. Then, universal dependencies are employed so that the input embeddings are converted into graph embeddings where words become nodes and dependencies become edges. After that, two-layer GCN is applied to the generated matrix of nodes feature vectors X and the adjacent matrix A to extract meaningful global features.

Mathematically, given a specific graph-based neural network model $f(X, A)$, spectral GCN follows the layer-wise propagation rule:

$$H^{(l+1)} = \sigma(\tilde{D}^{\frac{-1}{2}} \tilde{A} \tilde{D}^{\frac{-1}{2}} H^{(l)} W^{(l)}) \tag{2}$$

where A is the adjacency matrix, X is the matrix of node feature vectors (given sequence x), D is the degree matrix, $f(\cdot)$ is the neural network like differentiable function, $\tilde{A} = A + I_N$ is the adjacency matrix of the undirected graph G with added self-connections, I_N is the identity matrix of N nodes, $\tilde{D}_i = \sum_j \tilde{A}_{ij}$, $W^{(l)}$ is the layer-specific trainable weight matrix, $\sigma(\cdot)$ is the activation function, and $H^{(l)} \in \mathbb{R}^{(N \times D)}$ is the matrix of activation in the lth layer (representation of the lth layer), $H^{(0)} = X$.

After calculating the normalized adjacency matrix $\tilde{D}^{\frac{-1}{2}} \tilde{A} \tilde{D}^{\frac{-1}{2}}$ in the preprocessing step, the forward model can be expressed as:

$$Z = f(X, A) = softmax(\tilde{A} ReLU(\tilde{A} X W^0) W^1) \tag{3}$$

where $W^{(0)} \in R^{C \times H}$ is the input-to-hidden weight matrix for a hidden layer with H feature maps and $W^{(1)} \in R^{H \times F}$ is the hidden-to-output weight matrix.

$W^{(0)}$ and $W^{(1)}$ are trained using gradient descent. The weights before feeding into Linear layer with Softmax activation function are taken as global features to feed into our combined model. We keep the prediction results of GCN after feeding weights to the last Linear layer to compare the performance and prediction qualities with our proposed architecture's results.

3.2 XLNet as Contextual Embeddings

XLNet is an autoregressive pretraining method based on a novel generalized permutation language modeling objective. Employing Transformer-XL as the backbone model, XLNet exhibits excellent performance for language tasks involving long context by learning from bi-directional context and avoiding the disadvantages in the autoencoding language model.

The contextual features are captured from the sequence using permutation language modeling objective and two-stream self-attention architecture, integrating relative positional encoding scheme and the segment recurrence mechanism from Transformer-XL [40]. Given a sequence x of length T, the permutation language modeling objective can be defined as:

$$\max_{\theta} \mathbb{E}_{\mathbf{z}\sim\mathcal{Z}_T}\left[\sum_{t=1}^{T}\log p_\theta\left(x_{z_t} \mid \mathbf{x}_{\mathbf{z}_{<t}}\right)\right] \tag{4}$$

where $\mathcal{Z}_T$ is the set of all possible permutations of the index sequence of length T $[1, 2, ..., T]$, z_t is the tth element of a permutation $\mathbf{z} \in \mathcal{Z}_T$, $\mathbf{z} < t$ is the first $(t-1)^{th}$ elements of a permutation $\mathbf{z} \in \mathcal{Z}_T$, and p_θ is the likelihood. θ is the parameter shared across all factorization orders during training so x_t is able to see all $x_i \neq x_t$ possible elements in the sequence.

We also use two-stream self-attention to remove the ambiguity in target predictions. For each self-attention layer $m = 1, ..., M$, the two streams of representation are updated schematically with a shared set of parameters:

$$\begin{aligned} g_{z_t}^{(m)} &\leftarrow \mathit{Attention}\left(\mathrm{Q} = g_{z_t}^{(m-1)}, \mathrm{KV} = \mathbf{h}_{\mathbf{z}<t}^{(m-1)}; \theta\right) \\ h_{z_t}^{(m)} &\leftarrow \mathit{Attention}\left(\mathrm{Q} = h_{z_t}^{(m-1)}, \mathrm{KV} = \mathbf{h}_{\mathbf{z}\leq t}^{(m-1)}; \theta\right) \end{aligned} \tag{5}$$

where $g_{z_t}^{(m)}$ is the query stream that uses z_t but cannot see x_{z_t}, $h_{z_t}^{(m)}$ is the content stream that uses both z_t and x_{z_t}, and K, Q, V are the key, query, value, respectively.

To avoid slow convergence, the objective is customized to maximize the log-likelihood of the target sub-sequence conditioned on the non-target sub-sequence as in Eq. 6.

$$\max_{\theta} \mathbb{E}_{\mathbf{z}\sim\mathcal{Z}_T}\left[\log p_\theta\left(x_{\mathbf{z}_{>c}} \mid \mathbf{x}_{\mathbf{z}_{\leq c}}\right)\right] = \mathbb{E}_{\mathbf{z}\sim\mathcal{Z}_T}\left[\sum_{t=c,the+1}^{|z|}\log p_\theta\left(x_{z_t} \mid \mathbf{x}_{\mathbf{z}<t}\right)\right] \tag{6}$$

where $\mathbf{z}_{>c}$ is the target sub-sequence, $\mathbf{z}_{\leq c}$ is the non-target one, and c is the cutting point.

Furthermore, we make use of the relative positional encoding scheme and the segment recurrence mechanism from Transformer-XL. While the position encoding ensures the reflection in the positional information of text sequences, the attention mask is applied so the texts are given different attention during the creation of input embedding. Given 2 segments $\mathbf{x} = s_{1:T}$ and $\mathbf{x} = s_{T:2T}$ from a long sequence s, $\mathbf{z}$ and z referring to the permutations of [1 ... T] and [T + 1 ... 2T], we process the first segment, and then cache the obtained content representations $\mathbf{h}^{(m)}$ for each layer m. After that, we update the attention for the next segment x with memory, which can be expressed as in Eq. 7.

$$h_{z_t}^{(m)} \leftarrow \mathit{Attention}\left(\mathrm{Q} = h_{z_t}^{(m-1)}, \mathrm{KV} = \left[\tilde{\mathbf{h}}^{(m-1)}, \mathbf{h}_{\mathbf{z}\leq t}^{(m-1)}\right]; \theta\right) \tag{7}$$

Similar to global features, we capture the weights before feeding to the last Linear layer and use it as contextual embeddings of our combined model. For the purpose of comparison, we also keep the prediction results of XLNet after feeding weights to the last Linear layer.

3.3 Joint Architecture

Given global and contextual features from GCN and XLNet, respectively, we concatenate and feed them into a Linear layer, which is simplest way to show the most evident impact of these features to the NER task. The proposed approach is presented in Fig. 1.

4 Experimental Setup

In this section, we describe the dataset, the evaluation metrics, as well as present our implementations and experimental configurations on XLNet, GCN, and the joint models in detail.

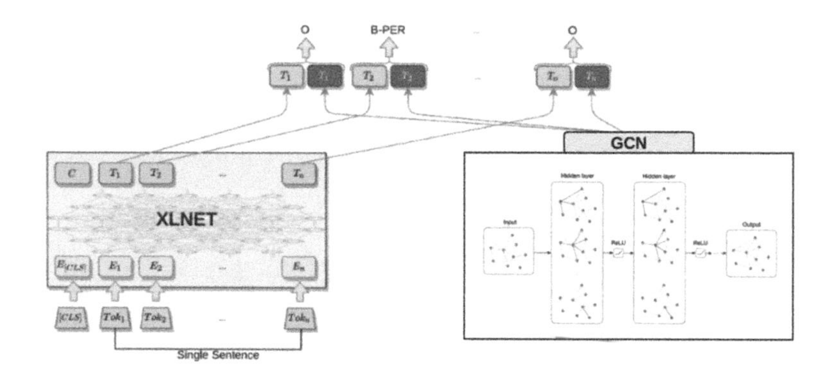

Fig. 1. Visualization of the global architecture of our proposed approach.

4.1 Dataset

We opted for the CoNLL 2003 [38], one of the widely-adopted benchmark datasets for NER tasks. The English version is collected from the Reuters Corpus with news stories between August 1996 and August 1997. The dataset concentrates on 4 types of NEs: persons (PER), locations (LOC), organizations (ORG), and miscellaneous (MISC).

4.2 Implementation Details

Global Embeddings with GCN. The sentences are annotated with universal dependencies from spaCy to create a graph of relations where words become nodes and dependencies become edges. The dataset is then converted into 124 nodes and 44 edges with the training corpus size of approximately 2 billion words, the vocabulary size of 222,496, and the dependency context vocabulary size of 1,253,524. Next, the graph embeddings are fed into 2 Graph Convolution layers with a Dropout of 0.5 after each layer to avoid overfitting. The global features are captured before the last Linear layer. We perform batch gradient descent using the whole dataset for every training iteration, which is a feasible option as long as the dataset fits in memory. We take advantage of TensorFlow for efficient GPU-based implementation of Eq. 2 using sparse-dense matrix multiplications.

Contextual Embeddings with XLNet. We have investigated on diverse embeddings such as FastText [27][2], Flair [1][3], Stanza [32][4] and XLNet [40][5] pretrained embeddings. Preliminary results suggest that XLNet (XLNet-Base, Cased) outperforms others, therefore, is chosen for our final implementation. The word embedding of size 768 with 12 layers were used for XLNet. Each layer consists of 3 sublayers: XLNet Relative Attention, XLNet Feed Forward, and Dropout layer. The XLNet Relative Attention is a two-stream self-attention mechanism as mentioned in Eq. 7. A Normalization layer with element-wise affine and a Dropout layer are employed around this sub-layer. Meanwhile, XLNet Feed Forward is a fully connected feed-forward network, whose outputs are also of dimension 768, the same as the outputs of the embedding layers. Like the previous sublayers, the Feed Forward layer is surrounded by a Normalization layer and a Dropout layer, however, another 2 Linear layers are added between them. Then, an additional Dropout layer is counted. It is notable that we only take the rate of 0.1 for every Dropout layer inside our model, from sublayers to inside sublayers. After 12 XLNet layers, another Dropout layer is added before the last Linear layer. We capture the intermediary output before the last Linear layer as the contextual features.

Proposed Model. Additional steps were taken to maintain alignments between input tokens and their corresponding labels as well as to match corresponding representations from global features to contextual features in the same sentence. First, we define an attention mask in XLNet as a sequence of 1s and 0s, with 1s for the first sub-word as the whole word embedding after tokenization and 0s for all padding sub-words. Then, in GCN features, we map the corresponding word representation at the position that the XLNet attention mark returns 1s and pad 0 otherwise. Therefore, each sentence has the same vector dimension in both global and contextual embeddings, which simplifies the concatenation.

In our implementation, we used a GPU 2070 Super and a TitanX GPU with 56 CPUs, 128 GB RAM. The hyperparameters were 300 as embedding size, 16 as batch size, 5e-5 as learning rate, 0.5 as dropout rate, 4 for number of epochs.

4.3 Metrics

We choose "relaxed" micro averaged F_1-score, which regards a prediction as the correct one as long as a part of NE is correctly identified. This evaluation metric has been used in several related publications, journals, and papers on NER [11,15,25,37].

[2] https://fasttext.cc/.
[3] https://github.com/flairNLP/flair.
[4] https://github.com/stanfordnlp/stanza.
[5] https://github.com/zihangdai/xlnet.

5 Results

We conducted multiple experiments to investigate the impact of global and contextual features on NER. Specifically, we implemented the architecture with only global features, only contextual features, and then the proposed joint architecture combining both features.

As shown in Table 1a, the proposed model achieves **93.82%** in F_1-score, which outperforms the two variants using global or contextual features alone. In terms of recognition of specific entity types, the details are provided in Table 1b, showing that PER is the category where the best results are achieved, while the lowest results are with the MISC, that is, the category of all NEs that do not belong to any of the three other predefined categories. Note that using only training data and publicly available word embeddings (GloVe), our proposed model has competitive results without the need of adding any extra complex encoder-decoder layers.

Table 1. Evaluation on the prediction results of our proposed model.

(a) Results of the proposed joint architecture compared to only contextual or only global features.

Embeddings	F_1 scores
Global features	88.63
Contextual features	93.28
Global + contextual features	**93.82**

(b) Performance evaluation per entity type.

Entity types	Precision	Recall	F_1-score
LOC	94.15	93.53	93.83
MISC	81.33	81.89	81.62
ORG	88.97	92.29	90.60
PER	96.67	97.09	96.88

Furthermore, the benefit of the joint architecture is illustrated in Fig. 2. While contextual features (XLNet), which is used in the majority of recent SOTA approaches, misclassifies the entity, the prediction from GCN and the combined model correctly tags "MACEDONIA" as the name of a location, confirming our hypothesis on the effect of global features.

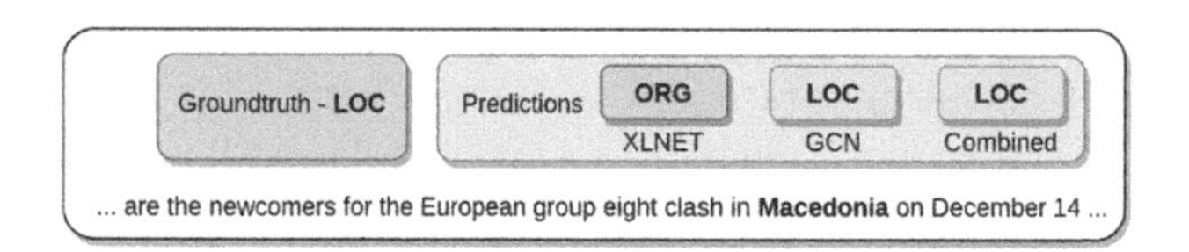

Fig. 2. XLNet, GCN, and the combined model's prediction on CoNLL 2003's example.

In Fig. 3, we compare our results with reported SOTA results on the same dataset from 2017 up to now. It can be observed that our results are competitive compared with SOTA approaches as the difference is by a small margin (the current benchmark is 94.3% F_1-score, compared to 93.82% achieved by our approach). Moreover, we notice that NER performance can be boosted with external knowledge (i.e. leveraging pretrained embeddings), as proven in our approach as well as in top benchmarks [22–24]. More importantly, complex decoder layers (CRF, Semi-CRF, ...) do not always lead to better performance in comparison with softmax classification when we take advantage of contextualized language model embeddings.

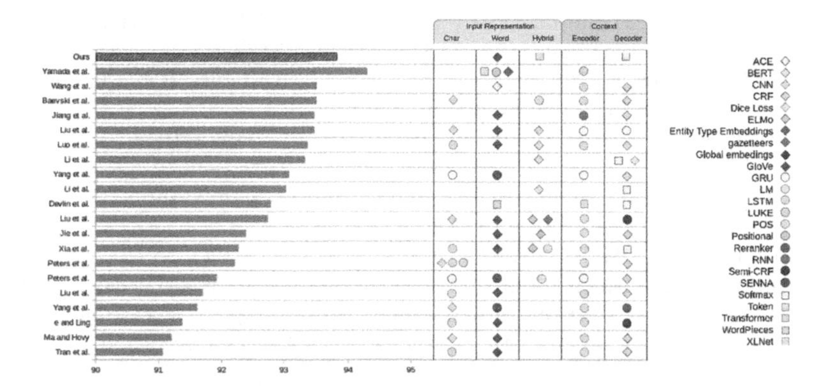

Fig. 3. Comparison of our proposal against SOTA techniques on the CoNLL 2003 dataset in terms of F_1-score. Values were taken from original papers and sorted by descending order.

6 Conclusion and Future Work

We propose a novel hierarchical neural model for NER that uses both global features captured via graph representation and contextual features at the sentence level via XLNet pretrained model. The combination of global and contextual embeddings is proven to have a significant effect on the performance of NER tasks. Empirical studies on the CoNLL 2003 English dataset suggest that our approach outperforms systems using only global or contextual features, and is competitive with SOTA methods. Given the promising results in English, our future work will consist of adapting the method to other languages, as well to a cross-lingual experimental setting. In addition, we will consider further developing the method by also incorporating background knowledge from knowledge graphs and ontologies.

Acknowledgements. This work has been supported by the European Union's Horizon 2020 research and innovation program under grants 770299 (NewsEye) and 825153 (EMBEDDIA). The work of S. P. has also received financial support from the Slovenian Research Agency for research core funding for the Knowledge Technologies programme (No. P2-0103) and the project CANDAS (No. J6-2581).

References

1. Akbik, A., Bergmann, T., Blythe, D., Rasul, K., Schweter, S., Vollgraf, R.: FLAIR: an easy-to-use framework for state-of-the-art NLP. In: Proceedings of the 2019 Conference of the North American Chapter of the Association for Computational Linguistics (Demonstrations), pp. 54–59 (2019)
2. Cetoli, A., Bragaglia, S., O'Harney, A., Sloan, M.: Graph convolutional networks for named entity recognition. In: Proceedings of the 16th International Workshop on Treebanks and Linguistic Theories, pp. 37–45 (2017)
3. Chiu, J.P., Nichols, E.: Named entity recognition with bidirectional LSTM-CNNs. Trans. Assoc. Comput. Linguist. **4**, 357–370 (2016)
4. Church, K.W.: Word2vec. Nat. Lang. Eng. **23**(1), 155–162 (2017)
5. Collins, M., Singer, Y.: Unsupervised models for named entity classification. In: 1999 Joint SIGDAT Conference on Empirical Methods in Natural Language Processing and Very Large Corpora (1999)
6. Devlin, J., Chang, M.W., Lee, K., Toutanova, K.: BERT: pre-training of deep bidirectional transformers for language understanding. In: NAACL-HLT (1) (2019)
7. Duvenaud, D.K., et al.: Convolutional networks on graphs for learning molecular fingerprints. In: Advances in Neural Information Processing Systems, pp. 2224–2232 (2015)
8. Etzioni, O., et al.: Unsupervised named-entity extraction from the web: an experimental study. Artif. Intell. **165**(1), 91–134 (2005)
9. Grishman, R., Sundheim, B.M.: Message understanding conference-6: a brief history. In: COLING 1996 Volume 1: The 16th International Conference on Computational Linguistics (1996)
10. Hoffart, J., et al.: Robust disambiguation of named entities in text. In: Proceedings of the 2011 Conference on Empirical Methods in Natural Language Processing, pp. 782–792 (2011)
11. Huang, Z., Xu, W., Yu, K.: Bidirectional LSTM-CRF models for sequence tagging. arXiv:1508.01991 (2015)
12. Ji, Z., Sun, A., Cong, G., Han, J.: Joint recognition and linking of fine-grained locations from tweets. In: Proceedings of the 25th International Conference on World Wide Web, pp. 1271–1281 (2016)
13. Joulin, A., Grave, E., Bojanowski, P., Douze, M., Jégou, H., Mikolov, T.: FastText.zip: compressing text classification models. arXiv preprint arXiv:1612.03651 (2016)
14. Krupka, G., IsoQuest, K.: Description of the NEROWL extractor system as used for MUC-7. In: Proceedings of the 7th Message Understanding Conference, Virginia, pp. 21–28 (2005)
15. Lample, G., Ballesteros, M., Subramanian, S., Kawakami, K., Dyer, C.: Neural architectures for named entity recognition. In: Proceedings of the 2016 Conference of the North American Chapter of the Association for Computational Linguistics: Human Language Technologies, pp. 260–270 (2016)
16. Lample, G., Conneau, A.: Cross-lingual language model pretraining. arXiv:1901.07291 (2019)
17. Lan, Z., Chen, M., Goodman, S., Gimpel, K., Sharma, P., Soricut, R.: ALBERT: a lite BERT for self-supervised learning of language representations. In: International Conference on Learning Representations (2019)
18. Li, J., Sun, A., Han, J., Li, C.: A survey on deep learning for named entity recognition. IEEE Trans. Knowl. Data Eng. (2020)

19. Liao, W., Veeramachaneni, S.: A simple semi-supervised algorithm for named entity recognition. In: Proceedings of the NAACL HLT 2009 Workshop on Semi-supervised Learning for Natural Language Processing, pp. 58–65 (2009)
20. Liu, L., Shang, J., Ren, X., Xu, F.F., Gui, H., Peng, J., Han, J.: Empower sequence labeling with task-aware neural language model. In: 32nd AAAI Conference on Artificial Intelligence, AAAI 2018, pp. 5253–5260. AAAI Press (2018)
21. Liu, S., Sun, Y., Li, B., Wang, W., Zhao, X.: HAMNER: headword amplified multi-span distantly supervised method for domain specific named entity recognition. In: AAAI, pp. 8401–8408 (2020)
22. Liu, T., Yao, J.G., Lin, C.Y.: Towards improving neural named entity recognition with gazetteers. In: Proceedings of the 57th Annual Meeting of the Association for Computational Linguistics, pp. 5301–5307 (2019)
23. Liu, Y., Meng, F., Zhang, J., Xu, J., Chen, Y., Zhou, J.: GCDT: a global context enhanced deep transition architecture for sequence labeling. In: Proceedings of the 57th Annual Meeting of the Association for Computational Linguistics, pp. 2431–2441 (2019)
24. Luo, Y., Xiao, F., Zhao, H.: Hierarchical contextualized representation for named entity recognition. In: AAAI, pp. 8441–8448 (2020)
25. Ma, X., Hovy, E.H.: End-to-end sequence labeling via bi-directional LSTM-CNNs-CRF. In: ACL (1) (2016)
26. Mikheev, A., Moens, M., Grover, C.: Named entity recognition without gazetteers. In: Ninth Conference of the European Chapter of the Association for Computational Linguistics (1999)
27. Mikolov, T., Grave, E., Bojanowski, P., Puhrsch, C., Joulin, A.: Advances in pre-training distributed word representations. In: Proceedings of the International Conference on Language Resources and Evaluation (LREC 2018) (2018)
28. Nadeau, D., Turney, P.D., Matwin, S.: Unsupervised named-entity recognition: generating gazetteers and resolving ambiguity. In: Lamontagne, L., Marchand, M. (eds.) AI 2006. LNCS (LNAI), vol. 4013, pp. 266–277. Springer, Heidelberg (2006). https://doi.org/10.1007/11766247_23
29. Palshikar, G.K.: Techniques for named entity recognition: a survey. In: Bioinformatics: Concepts, Methodologies, Tools, and Applications, pp. 400–426. IGI Global (2013)
30. Pennington, J., Socher, R., Manning, C.D.: GloVe: global vectors for word representation. In: Proceedings of the 2014 Conference on Empirical Methods in Natural Language Processing (EMNLP), pp. 1532–1543 (2014)
31. Peters, M.E., et al.: Deep contextualized word representations. In: Proceedings of NAACL-HLT, pp. 2227–2237 (2018)
32. Qi, P., Zhang, Y., Zhang, Y., Bolton, J., Manning, C.D.: Stanza: a python natural language processing toolkit for many human languages. arXiv preprint arXiv:2003.07082 (2020)
33. Ritter, A., Clark, S., Etzioni, O., et al.: Named entity recognition in tweets: an experimental study. In: Proceedings of the 2011 Conference on Empirical Methods in Natural Language Processing, pp. 1524–1534 (2011)
34. Rocktäschel, T., Weidlich, M., Leser, U.: ChemSpot: a hybrid system for chemical named entity recognition. Bioinformatics **28**(12), 1633–1640 (2012)
35. Seti, X., Wumaier, A., Yibulayin, T., Paerhati, D., Wang, L., Saimaiti, A.: Named entity recognition in sports field based on a character-level graph convolutional network. Information **11**(1), 30 (2020)

36. Subramanian, S., Trischler, A., Bengio, Y., Pal, C.J.: Learning general purpose distributed sentence representations via large scale multi-task learning. In: International Conference on Learning Representations (2018)
37. Takeuchi, K., Collier, N.: Use of support vector machines in extended named entity recognition. In: COLING-02: The 6th Conference on Natural Language Learning 2002 (CoNLL-2002) (2002)
38. Tjong Kim Sang, E.F., De Meulder, F.: Introduction to the CoNLL-2003 shared task: language-independent named entity recognition. In: Daelemans, W., Osborne, M. (eds.) Proceedings of CoNLL-2003, Edmonton, Canada, pp. 142–147 (2003)
39. Yadav, V., Bethard, S.: A survey on recent advances in named entity recognition from deep learning models. In: Proceedings of the 27th International Conference on Computational Linguistics, pp. 2145–2158 (2018)
40. Yang, Z., Dai, Z., Yang, Y., Carbonell, J., Salakhutdinov, R.R., Le, Q.V.: XLNet: generalized autoregressive pretraining for language understanding. In: Advances in Neural Information Processing Systems, pp. 5753–5763 (2019)
41. Yao, L., Mao, C., Luo, Y.: Graph convolutional networks for text classification. In: Proceedings of the AAAI Conference on Artificial Intelligence, vol. 33, pp. 7370–7377 (2019)
42. Zhang, Y., Liu, Q., Song, L.: Sentence-state LSTM for text representation. In: Proceedings of the 56th Annual Meeting of the Association for Computational Linguistics (Volume 1: Long Papers), pp. 317–327 (2018)

A Consistency Analysis of Different NLP Approaches for Reviewer-Manuscript Matchmaking

Nishith Kotak[1(✉)], Anil K. Roy[1], Sourish Dasgupta[2], and Tirthankar Ghosal[3]

[1] DA-IICT, Gandhinagar, India
{201921004,anil_roy}@daiict.ac.in
[2] RAx Labs Inc., Delaware, USA
sourish@raxter.io
[3] Institute of Formal and Applied Mathematics, Charles University, Prague, Czech Republic
ghosal@ufal.mff.cuni.cz

Abstract. Selecting a potential reviewer to review a manuscript, submitted at a conference is a crucial task for the quality of a peer-review process that ultimately determines the success and impact of any conference. The approach adopted to find the potential reviewer needs to be consistent with its decision of allocation. In this work, we propose a framework for evaluating the reliability of different NLP approaches that are implemented for the match-making process. We bring various algorithmic approaches from different paradigms and an existing system Erie, implemented in IEEE INFOCOM conference, on a common platform to study their consistency of predicting the set of the potential reviewers, for a given manuscript. The consistency analysis has been performed over an actual multi-track conference organized in 2019. We conclude that Contextual Neural Topic Modeling (CNTM) with a balanced combinatorial optimization technique showed better consistency, among all the approaches we choose to study.

Keywords: Reviewer-manuscript matching · Semantics analysis · Consistency analysis

1 Introduction

The peer-review process in a conference is the cornerstone in the current academic and research field which is majorly regarded as an important part of scholarly communications. The selection of an expert reviewer plays a crucial role in the peer-review process. A reviewer, while reviewing, needs to focus on a) technical quality of the work b) reproducibility of the work c) impact of paper over the community, and d) extent of the work to be original and novel. For this, the reviewer assigned to the manuscript must be an expert in the domain of the submitted manuscript.

H.-R. Ke et al. (Eds.): ICADL 2021, LNCS 13133, pp. 277–287, 2021.
https://doi.org/10.1007/978-3-030-91669-5_22

A framework is required to be developed that scrutinizes all the allocations of the expert reviewers to the submitted manuscripts. *This work is not an attempt to propose a better reviewer-manuscript match-making system but rather to propose a framework for evaluating the reliability of match-making algorithms. This framework is agnostic to any conference, of whether the actual (semi)-manual allocation is perfect or not.*

Certain attempts have been made to develop automated systems like TPMS [9], GRAPE [10], SubSift [11], Erie [20] to find a perfect match. The authors [28] have generalized the range of approaches for matching a reviewer with the manuscript. The authors in [12,14,16,17,24] have considered keywords as a matching parameter. The authors in [4,15,18,26] have used Latent Dirichlet Allocation (LDA) approach while in [27], apart from LDA, authors also considered the concept of freshness for understanding the change in the research interest of a reviewer with time. Even the bibliography-based matching was been proposed by the authors in [21]. The authors in [22] worked on expertise, authority and diversity parameters while the authors in [23] considered a set of references and pedagogical facets. Hiepar-MLC approach [31] used a two-level bidirectional GRU with an attention mechanism to capture word-sentence-document information. To the best of our knowledge, any kind of consistency analysis of the implemented approaches in the context of reviewer-manuscript matching has not been performed yet.

By consistency, we here show that, if the approach agrees with a certain set of reviewers by providing a higher similarity score, then it should provide a significantly lower similarity score to the other set of reviewers, proving the system to be less ambivalent. A detailed explanation of consistency is given in Sect. 2. We attempt to bring different paradigms together to perform the analysis over the actual dataset provided by the conference organized in 2019. Over the analysis we performed, Contextual Neural Topic Modeling (CNTM) approach provided us with more stable and reliable results giving a new direction to explore CNTM in a more further detailed version that can be used in developing a reviewer-manuscript match-making system.

2 Problem Formulation

The reviewer-manuscript match-making process is accomplished majorly by imposing two constraints: a) workload constraint and b) review coverage constraint. Workload constraint is the maximum number of manuscripts that can be allocated to an individual reviewer to review, while review coverage constraint deals with the number of reviews required per manuscript to fulfill the peer-review process.

Let's consider $\mathcal{R} = \{r^{(i)}\}_{i=1}^{n}$ be the set of n-reviewers, $\mathcal{M} = \{m^{(j)}\}_{j=1}^{m}$ be the set of m-manuscripts submitted to review. Let $[\Pi]^n$ denote the profiles of n reviewers defined as $[\Pi]^n = (\pi^{(1)}, \pi^{(2)}, \dots \pi^{(n)})$. Here, profile of reviewers represents the expertise of reviewers. The process of formulation of profiles is mentioned in Sect. 4.1. We define sigma (σ_{rt}) as the match-making similarity function applied over the reviewer's and manuscript profile, to obtain the similarity score matrix in-between the reviewers and manuscripts, using any match-making representational technique (let's say rt). A similarity tensor S can be obtained as:

$$\mathrm{S} = \sigma_{rt}[\Pi, \mathcal{M}]$$

$S_{ij} \in [0,1]^{nxm}$ be the similarity matrix between the reviewer and manuscript. Higher the similarity score, more inclined the reviewer's expertise to the manuscript's theme. Let $\{\mathcal{R}^{(ar)}\}$ be the set of K-allocated reviewers to a particular manuscript and $\{\mathcal{R}^{(nar)}\}$ be the set of non-allocated reviewers. Here, $\{\mathcal{R}^{(nar)}\} = \mathcal{R} - \{\mathcal{R}^{(ar)}\}$.

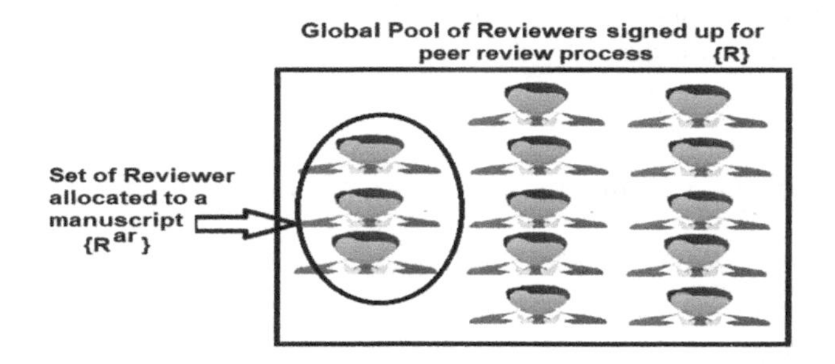

Fig. 1. Example of consistency for an algorithmic approach, selecting a set of reviewers out of the global pool of reviewers who signed up for the review process

It is necessary to determine the consistency of the approach adopted to calculate the similarity. By consistency, we mean the agreement of any match-making algorithmic approach to a certain set of reviewers by providing a higher similarity score, while it should disagree with the remaining set of the reviewers by providing a significantly lower similarity score. We define a term, here, a degree of consistency, denoted as Δ, that shows the consistency in the decision of predicting the reviewers by a particular algorithm. Figure 1 shows the set of reviewers predicted by any match-making algorithm to review a particular manuscript out of the global pool of the reviewers who actually signed up for the review process. The degree of consistency can be defined as, the absolute difference in the average similarity score of the predicted reviewers and the average similarity score of the remaining set of reviewers.

$$AS_{ar} = \left[\frac{\sum_{i=1}^{m} S_{ir_k}}{m}\right], r_k \in \{\mathcal{R}^{(ar)}\}, 0 \leq k \leq K$$

$$AS_{nar} = \left[\frac{\sum_{i=1}^{m} S_{ir_k}}{m}\right], r_k \in \{\mathcal{R}^{(nar)}\}, 0 \leq k \leq n-K$$

$$\Delta = abs\left(AS_{ar} - AS_{nar}\right) \tag{1}$$

Here, AS_{ar} is the average similarity score of allocated reviewers, while AS_{nar} is the average similarity score of non-allocated reviewers. Δ represents the degree of consistency. More the value of Δ, more consistent the algorithm is, with its decision of predicting the reviewers.

3 Conference Dataset Description

The Technical Program Committee Chair of the "MultiTrack Conf"[1] conference provided us with the complete data of a) all submitted manuscripts, b) the full list of reviewers with their affiliations (which we call Global pool), c) track-wise list of reviewers (which we call Track pool), and d) manuscripts allocated to a set of reviewers (which we call Original allocation). "MultiTrack Conf" was an engineering domain multi-track conference organized in 2019. Table 1 gives a summary of the conference data.

Table 1. "MultiTrack Conf" conference dataset details

Parameter	Value
Conference name	"MultiTrack Conf"
Number of tracks	15+
Number of submitted manuscripts	600+
Number of accepted manuscripts	200+
Number of signed up reviewers	500+
Average number of papers per reviewer	3.93
Average number of reviews per paper	3.68
Avg. no. of words (Title + Abstract)	109

4 Methodologies Implemented and Result Analysis

This section includes various representation approaches that have been used for the match-making process. This section also focuses on the experimental setup and the evaluation method that has been undertaken to evaluate the consistency of different approaches.

4.1 Experimental Setup

The first step is to create profiles of the manuscripts and the reviewers. From the reviewers' names and affiliation, the publications title and the publication years are extracted using Orcid [1]. Using the publication details, the DOI number is

[1] Due to the data privacy and confidentiality conditions, the original conference's name is not revealed.

extracted using Crossref [3]. Finally, the abstracts of the papers are extracted using Semantic Scholar [2]. Publication details of some reviewers are not available in Orcid, hence their abstracts are extracted by web scraping.

In order to build the reviewer's profile, we hypothesize that the past 5 years or recent 20 papers (which we empirically derive from the publication frequency of reviewers in our dataset) are an indicator of the research domain of operation/interests of the reviewer. Hence, for each reviewer, publications of last 5 years or recent 20 publications, whichever was earlier, are profiled. The title and abstract of the publications collectively formed the reviewer's profile. The title and abstract provided by the conference, are used to build the manuscript's profile. From the generalized structural property of the research papers, it is evident that the title and abstract reflect the core theme of the entire paper.

Before applying any match-making algorithm, a pre-processing task involving the removal of English stopwords and research stopwords was carried out. The research stopwords like *author*, *efficiency*, *proposal*, *study*, etc. are the set of words which are frequently repeated in the publications. They generally, do not convey sufficient information as a standalone entity.

Algorithm 1: Match-Making Algorithm

Input : Reviewer Workload list $[\mu]^n$, Manuscript Coverage list $[\Lambda]^m$
Similarity Matrix S^{nxm}, Maximum papers per reviewer μ
Reviews required per manuscript λ, Number of Manuscripts m
Output: Allocation Dictionary $\mathcal{A}^m$
Algorithm:
Initialize $\mathcal{A}^m$: empty lists
Cost matrix S'(i,j) := max(S) - S(i,j); $\forall i \in [1,n], \forall j \in [1,m]$
while sum($[\Lambda]^m$) $< \lambda * m$ **do**
  $\mathcal{P}_{ij} \leftarrow$ Hungarian Assignment(S')
  for each $\mathcal{P}_{ij} = (i,j)$; $i \in [n], j \in [m]$ **do**
    S'(i,j) $\leftarrow$ DISALLOWED for pair(r^i, m^j) in $\mathcal{P}_{ij}$
    $\mathcal{A}^{(j)}.append(r^{(i)})$
    $[\mu]^i$+=1
    $[\Lambda]^j$+=1
    if $[\mu]^i == \mu$ **do**
      delete i^{th} row from cost matrix S'
    if $[\Lambda]^j == \lambda$ **do**
      delete j^{th} column from cost matrix S'
  end for
end while

The match-making process between the reviewer and manuscript is mentioned in Algorithm 1. During each allocation, the reviewer workload and manuscript coverage constraints are taken into consideration. The balanced optimized Hungarian approach [19] is adopted for the assignment process. The constraint pair (μ, λ) is taken as (6, 3). The similarity matrix S between reviewer and manuscript can be obtained using different approaches mentioned in Sect. 4.3.

Table 2. Description of algorithms implemented with corresponding approaches adopted in Reviewer Matching Problem (No. of latent topics for topic modeling approaches is set to 20. All hyper-parameters are set to default.)

Representation technique	Algorithm	Method description	Approach implemented
Statistical approach based Keywords Extraction Methods	TextRank	This approach works on the typical pagerank algorithm giving more importance to words having more adjacency	The set of the keyphrases are extracted using these approaches, from reviewer's profile and manuscripts separately. The set of extracted keyphrases are matched using the **n-gram based scoring approach**. This scoring approach gave a higher similarity score to the reviewer manuscript pair which has more number of matching continuous words
	RAKE	This approach takes into account the frequency of the words with its co-appearance to generate a ranked list of keywords	
	YAKE	This algorithm deals with the statistical features of the words and identify the most significant keywords based on words co-occurrences in different sentences	
Probabilistic topic modeling approach	LDA	Any document can be considered to be representing certain theme of topic. The set of the vocabularies are representational for any particular theme. LDA is a probabilistic approach that considers each document to have a certain theme, which on training, clusters the documents into the latent topics	Trained LDA using reviewer and manuscript profiles to generate Reviewer-Topic and Manuscript-Topic probabilistic distribution matrix. **Cosine similarity** is applied over the matrices to generate reviewer-manuscript similarity matrix
Transformer based embedding	Universal Sentence Encoder (USE)	This transfer learning-based technique generates 512-dimensional generic encoded vectors that are efficient enough to retain the information within the sentence while discarding the noise	The sentences of each of the documents from reviewer's profile and manuscript's profile are encoded using transformer based embedding approach. Thereafter, the **cosine similarity** is applied between these embedded vectors to generate the reviewer-manuscript similarity matrix
	Sentence-BERT (SBERT)	A modified version of BERT that derives the semantically relevant and meaningful 768-dimensional sentence embeddings which further can be utilized directly to compute the similarity between the sentences	
Transformer based topic modelling	CNTM using training-testing approach	CNTM establishes the coherency and semantic relations among the words that are present in the document. The coherency between the topic-word can be increased by considering the contextual embeddings, which can be obtained from the pre-trained BERT model	The reviewer's profile is used to train the CNTM model that generated the reviewer-topic distribution matrix. The manuscript's profile is tested using the trained model to generate the manuscript-topic distribution matrix. The **cosine similarity** between the matrices was performed to obtain the similarity score between the reviewer and manuscript

(continued)

Table 2. (*continued*)

Representation technique	Algorithm	Method description	Approach implemented
	CNTM using inference based approach	Same as CNTM using training-testing approach	CNTM model is trained over combined profiles of manuscripts and reviewers to generate a topic distribution matrix. Then by using inferencing, manuscript-topic and reviewer-topic distribution matrices were extracted, over which **cosine similarity** is applied to generate similarity between reviewer and manuscript
	CNTM using word embeddings	Same as CNTM using training-testing approach	768-dimensional embedding vector is generated corresponding to each vocabulary provided during the training process. The **cosine similarity** between the embedding of the vocabs of reviewer's profile and manuscript's profile is computed to generate the similarity matrix between the reviewer and manuscript
	CNTM using topic vector	Same as CNTM using training-testing approach	The CNTM model provided the weightage to each of the latent topic vectors for a document. Here, only top-4 contributing topics are considered. For these top-4 topic vectors, the representative set of 20 words are extracted over which **Jaccard similarity** is applied to produce similarity between the reviewer and the manuscript
	CNTM using concepts of vectors	Same as CNTM using training-testing approach	The CNTM model generated the topic-word distribution vector. Here, the word vector for each topic (top-20 words for each topic) is formed and **Jaccard similarity** is applied to produce the similarity between the reviewer and the manuscript
	BERTopic	It considers class based TF-IDF (c-TF-IDF) to create clusters which helps in extracting the interpretable and interconnective topics with reference to the words	The combined set of documents from the reviewer's profile and the manuscript's profile were used to train the BERTopic model. This clustered the class based words into latent topics and generated the document-topic probabilistic matrices. **Cosine similarity** between these matrices is applied and the manuscript-reviewer similarity matrix is generated
Existing system	Erie	Erie was investigated over three different approaches that involves LDA, TF-IDF Vectorization and Latent Semantic Indexing (LSI). Based on the nature of scattering of similarity scores, LSI was found to be a better choice.	Trained the LSI model using the combined profiles of reviewer and manuscript to generate the reviewer-topic and manuscript-topic distribution matrices. **Cosine similarity** is applied over the matrices to generate reviewer-manuscript similarity matrix

4.2 Evaluation Method

To evaluate the approaches, top-3 reviewers are assigned to each of the submitted manuscripts using Algorithm 1. This allocation is compared with the original allocation done by the track chair. Three other modes of allocation are also done to study the consistency of approaches. These modes include the allocation to the reviewers among the global pool, which we call here as global pool allocation. The allocation was also studied by restricting the reviewers to the track they have selected, which we call as track-based reviewer allocation. The third mode of allocation includes the allocation of a particular manuscript among the set of reviewers who were actually not being allocated that particular manuscript to review, which we call it as global pool minus original allocation. Let AS_{oa} be the average similarity score of original allocated reviewers, while AS_{goa} be the average similarity score of global pool minus original allocated reviewers. The Eq. 1 for the degree of consistency can now be moulded as:

$$\Delta = abs\,(AS_{oa} - AS_{goa}) \tag{2}$$

4.3 Methodologies Description and Implementation

This subsection includes discussion of representational paradigms of queries (manuscripts) and targets (reviewers), along with their implementation to obtain the similarity between the reviewer and manuscript. We have implemented various approaches that includes Statistical approach based Keyword extraction methods like TextRank [25], RAKE [30] and YAKE [7], probabilistic topic modeling approach like Latent Dirichlet Allocation (LDA) [6], neural topic modeling approaches like Contextual Neural Topic Modeling (CNTM) [5] and BERTopic [13], Transformer based embedding approaches like Universal Sentence Encoder (USE) [8] and Sentence BERT (SBERT) [29]. We also observed the consistency over the existing system Erie [20] implemented in IEEE INFOCOM conference. Table 2 shows the description of approaches with their implementations, to obtain the similarity between the reviewers and manuscripts.

4.4 Result Analysis

Using the approaches mentioned in Sect. 4.3, the similarity between the reviewer and manuscript has been obtained. Now, to calculate the consistency of each of these approaches, the degree of consistency (Δ) is been computed using Eq. 2. Figure 2 is the comparison graph showing the consistency of different approaches and existing system Erie over the original allocation and the three other modes of allocation of reviewers. With the perspective of the degree of consistency (Δ), it can be seen from the Fig. 2 that CNTM using word embedding variant shows better consistency among other approaches.

Keywords are the important facets of any paper. Authors tend to provide very specific yet peripheral keywords (e.g. Adam optimizer) or the broader category of keywords (e.g. Artificial intelligence). It may not serve a good idea to rely

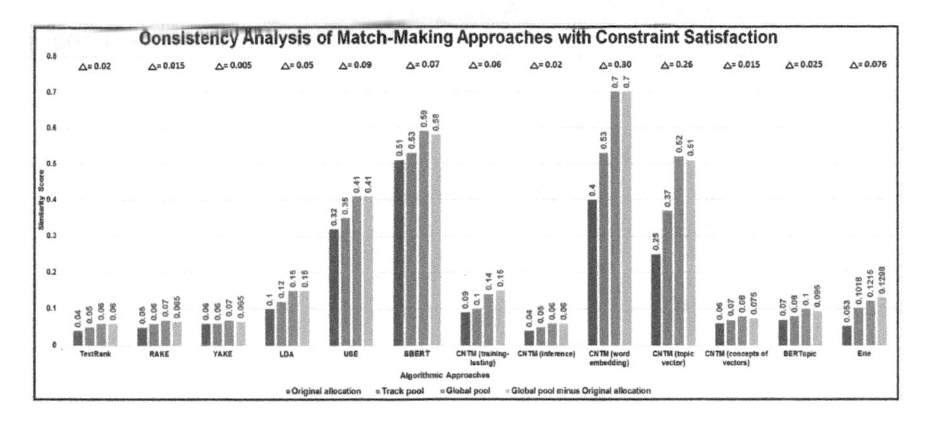

Fig. 2. Comparison chart of consistency in terms of similarity score with the delta differential (Δ) as a measure of consistency

only on author-tagged keywords. Hence, a technique like keyword extraction is adopted to extract the core concepts of the paper. It can be observed that these approaches showed consistency with low delta differential value. Keywords based matching doesn't consider semantically relevant concepts like *plagiarism* and *copy* as similar ones. So, we decided to introduce transformer-based contextual embedding in the representation to study their consistency. They showed higher similarity agreements but have lower delta differential component.

A publication is a collection of (latent) topics representing certain themes. So, we analyzed a topic modeling approach like LDA to study consistency. This approach clustered topics based on the representing words, but the issue of semantic relevance still persists. So, we decided to introduce and test the contextual embedding over the topic modeling approach like CNTM, where the semantically relevant words were classified in the same topic cluster. For instance, *biological cell* and *electrolytic cell*, despite having common word *cell*, would fall in different clusters representing biological/medical topic and in electronics domain respectively. Variants of CNTM are also applied to study their consistency. The consistency analysis is also performed over the existing reviewer assignment system, Erie implemented in the IEEE INFOCOM conference. As seen in Fig. 2, the CNTM model using word embeddings proved to have better consistency than any other approaches that we have considered in this study.

5 Conclusion and Future Work

We bring various algorithmic approaches from different paradigms and an existing system Eire, on a common platform, to study a framework of consistency, in evaluating match-making approaches. From the analysis performed, it can be established that the reviewer-manuscript match-making system based on Contextual Neural Topic Modelling (CNTM) using Word Embedding approach may result in a better match, as it directly considers SBERT embeddings used in the model. In the future, we plan to develop a match-making system considering Conflict of Interests (COIs), with sentiment analysis performed over the reviews provided by the reviewers. This will help in identifying the detailed quality reviews. We would like to extend the study of consistency over the full text of publications. We plan to develop a match-making system that may reduce the burden over the TPCs and thus promising a better quality of peer-review process in the conference.

References

1. Orcid. https://orcid.org/. Accessed 14 July 2021
2. Semantic scholar | AI-powered research tool. https://www.semanticscholar.org/. Accessed 14 July 2021
3. You are crossref - crossref. https://www.crossref.org/. Accessed 14 July 2021
4. An automated conflict of interest based greedy approach for conference paper assignment system. J. Informetr. **14**(2), 101022 (2020). https://doi.org/10.1016/j.joi.2020.101022. https://www.sciencedirect.com/science/article/pii/S1751157719301373
5. Bianchi, F., Terragni, S., Hovy, D.: Pre-training is a hot topic: contextualized document embeddings improve topic coherence. arXiv preprint arXiv:2004.03974 (2020)
6. Blei, D.M., Ng, A.Y., Jordan, M.I.: Latent Dirichlet allocation. J. Mach. Learn. Res. **3**(null), 993–1022 (2003)
7. Campos, R., Mangaravite, V., Pasquali, A., Jorge, A., Nunes, C., Jatowt, A.: YAKE! Keyword extraction from single documents using multiple local features. Inf. Sci. **509**, 257–289 (2020). https://doi.org/10.1016/j.ins.2019.09.013
8. Cer, D., et al.: Universal sentence encoder (2018)
9. Charlin, L., Zemel, R.: The Toronto paper matching system: an automated paper-reviewer assignment system (2013)
10. Di Mauro, N., Basile, T., Ferilli, S.: GRAPE: an expert review assignment component for scientific conference management systems, pp. 789–798, June 2005. https://doi.org/10.1007/11504894_109
11. Flach, P.A., et al.: Novel tools to streamline the conference review process: experiences from SIGKDD'09. SIGKDD Explor. Newsl. **11**(2), 63–67 (2010). https://doi.org/10.1145/1809400.1809413
12. Goldsmith, J., Sloan, R.: The AI conference paper assignment problem. In: AAAI Workshop - Technical Report, January 2007
13. Grootendorst, M.: BERTopic: leveraging BERT and c-TF-IDF to create easily interpretable topics (2020). https://doi.org/10.5281/zenodo.4430182
14. Hartvigsen, D., Wei, J., Czuchlewski, R.: The conference paper-reviewer assignment problem*. Decis. Sci. **30**, 865–876 (2007). https://doi.org/10.1111/j.1540-5915.1999.tb00910.x
15. Jin, J., Geng, Q., Zhao, Q., Zhang, L.: Integrating the trend of research interest for reviewer assignment. In: Proceedings of the 26th International Conference on World Wide Web Companion. WWW 2017 Companion, International World Wide Web Conferences Steering Committee, Republic and Canton of Geneva, CHE, pp. 1233–1241 (2017). https://doi.org/10.1145/3041021.3053053
16. Kalmukov, Y.: Architecture of a conference management system providing advanced paper assignment features. Int. J. Comput. Appl. **34**, 51–59 (2011). https://doi.org/10.5120/4083-5888

17. Kalmukov, Y.: Describing papers and reviewers' competences by taxonomy of keywords. Comput. Sci. Inf. Syst. **9**, 763–789 (2012). https://doi.org/10.2298/CSIS110906012K
18. Kou, N.M., Leong Hou, U., Mamoulis, N., Gong, Z.: Weighted coverage based reviewer assignment. In: Proceedings of the 2015 ACM SIGMOD International Conference on Management of Data, SIGMOD 2015, New York, NY, USA, pp. 2031–2046. Association for Computing Machinery (2015). https://doi.org/10.1145/2723372.2723727
19. Kuhn, H.W.: The Hungarian method for the assignment problem. Naval Res. Logist. Q. **2**(1–2), 83–97 (1955)
20. Li, B., Hou, Y.T.: The new automated IEEE INFOCOM review assignment system. IEEE Netw. **30**(5), 18–24 (2016). https://doi.org/10.1109/MNET.2016.7579022
21. Li, X., Watanabe, T.: Automatic paper-to-reviewer assignment, based on the matching degree of the reviewers. Procedia Comput. Sci. **22**, 633–642 (2013). https://doi.org/10.1016/j.procs.2013.09.144. https://www.sciencedirect.com/science/article/pii/S187705091300937X. 17th International Conference in Knowledge Based and Intelligent Information and Engineering Systems - KES 2013
22. Liu, X., Suel, T., Memon, N.: A robust model for paper reviewer assignment. In: Proceedings of the 8th ACM Conference on Recommender Systems, RecSys 2014, New York, NY, USA, pp. 25–32. Association for Computing Machinery (2014). https://doi.org/10.1145/2645710.2645749
23. Medakene, A.N., Bouanane, K., Eddoud, M.A.: A new approach for computing the matching degree in the paper-to-reviewer assignment problem. In: 2019 International Conference on Theoretical and Applicative Aspects of Computer Science (ICTAACS), vol. 1, pp. 1–8 (2019). https://doi.org/10.1109/ICTAACS48474.2019.8988127
24. Merelo-Guervós, J.J., Castillo-Valdivieso, P., et al.: Conference paper assignment using a combined greedy/evolutionary algorithm. In: Yao, X. (ed.) PPSN 2004. LNCS, vol. 3242, pp. 602–611. Springer, Heidelberg (2004). https://doi.org/10.1007/978-3-540-30217-9_61
25. Mihalcea, R., Tarau, P.: TextRank: bringing order into text. In: EMNLP (2004)
26. Nguyen, J.: Knowledge aggregation in people recommender systems: matching skills to tasks (2019)
27. Peng, H., Hu, H., Wang, K., Wang, X.: Time-aware and topic-based reviewer assignment. In: Bao, Z., Trajcevski, G., Chang, L., Hua, W. (eds.) DASFAA 2017. LNCS, vol. 10179, pp. 145–157. Springer, Cham (2017). https://doi.org/10.1007/978-3-319-55705-2_11
28. Price, S., Flach, P.A.: Computational support for academic peer review: a perspective from artificial intelligence. Commun. ACM **60**, 70–79 (2017)
29. Reimers, N., Gurevych, I.: Sentence-BERT: sentence embeddings using siamese BERT-networks (2019)
30. Rose, S., Engel, D., Cramer, N., Cowley, W.: Automatic keyword extraction from individual documents. Text Min. Appl. Theory **1**, 1–20 (2010). https://doi.org/10.1002/9780470689646.ch1
31. Zhang, D., Zhao, S., Duan, Z., Chen, J., Zhang, Y., Tang, J.: A multi-label classification method using a hierarchical and transparent representation for paper-reviewer recommendation. ACM Trans. Inf. Syst. **38**(1), February 2020. https://doi.org/10.1145/3361719

Data Infrastructure for Digital Libraries

When Expertise Gone Missing: Uncovering the Loss of Prolific Contributors in Wikipedia

Paramita Das[1(✉)], Bhanu Prakash Reddy Guda[2], Debajit Chakraborty[1], Soumya Sarkar[3], and Animesh Mukherjee[1]

[1] IIT Kharagpur, Kharagpur 721302, India
paramita.das@iitkgp.ac.in
[2] Adobe Research, Bangalore 560087, India
[3] TU Darmstadt, 64289 Darmstadt, Germany

Abstract. Success of planetary-scale online collaborative platforms such as Wikipedia is hinged on active and continued participation of its voluntary contributors. The phenomenal success of Wikipedia as a valued multilingual source of information is a testament to the possibilities of collective intelligence. Specifically, the sustained and prudent contributions by the experienced prolific editors play a crucial role to operate the platform smoothly for decades. However, it has been brought to light that growth of Wikipedia is stagnating in terms of the number of editors that faces steady decline over time. This decreasing productivity and ever increasing attrition rate in both newcomer and experienced editors is a major concern for not only the future of this platform but also for several industry-scale information retrieval systems such as *Siri, Alexa* which depend on Wikipedia as knowledge store. In this paper, we have studied the ongoing crisis in which experienced and prolific editors withdraw. We performed extensive analysis of the editor activities and their language usage to identify features that can forecast *prolific Wikipedians*, who are at risk of ceasing voluntary services. To the best of our knowledge, this is the first work which proposes a scalable prediction pipeline, towards detecting the *prolific Wikipedians*, who might be at a risk of retiring from the platform and, thereby, can potentially enable moderators to launch appropriate incentive mechanisms to retain such 'would-be missing' valued *Wikipedians*.

Keywords: Wikipedia · Missing editor · Prolific contributor · Platform moderation

1 Introduction

Wikipedia has emerged as an immensely popular digital encyclopedia, over the last decade. This is primarily attributed to the "be bold" policy, which permits anyone to contribute on almost all wikipages. Encouragement toward global collaboration generates hundreds of millions of views monthly which resembles

H.-R. Ke et al. (Eds.): ICADL 2021, LNCS 13133, pp. 291–307, 2021.
https://doi.org/10.1007/978-3-030-91669-5_23

Wikipedia as a reliable source of information, irrespective of socio-economic backgrounds and cultures [26,39]. The workhorse behind this success story is a large pool of voluntary editors. These group of people maintain Wikipedia pages behind the scenes which includes creating new pages, changing contents, making sure that the fact provided is appropriate etc. It is imperative for the survival of this platform that this resource is nurtured so that they can operate with due diligence.

In these lines there have been several works, which attempt to understand diverse roles editors play in the community [33,42,45]. These studies revealed valuable insights about editors' motivation, group behaviour and productivity, thus further reinforcing our knowledge about community health. A complementary direction of exploration includes developing personalised recommendation system to identify appropriate pages and mentors for a newbie Wikipedia editor so that seamless onboarding to the platform can be achieved [8,43]. However notwithstanding former efforts, there has been a steady attrition of Wikipedia editors [1]. One possible reason could be personal disillusionment from the project. However, some notable reasons are *increased bureaucracy* and *incivility* [12,25]. In particular, to maintain the high quality standards of encyclopedic content, Wikipedia community has gradually become impermeable for the editors, resulting in unreasonable abuse of power such as blocking of users, deletion of good faith edits, and unexplained conflict arbitration [23]. Out-flux of contributors has been a chronic problem for other knowledge sharing platforms as well such as Yahoo! Answers, Baidu Knows etc. [14,15,22].

Loss of Wikipedians is a Loss of Wikipedia: Editors, or Wikipedians [35,40] are the soul of Wikipedia; without their active participation, Wikipedia will eventually become stagnant. As mentioned earlier, the loss of editors and simultaneously *editor retention* [17] is a Wikipedia-wide problem.

Related Wikiprojects: Several wikiprojects [4], hosted by Wikipedia aim to construct a stronger bonding among Wikipedians and also figuring out several measures [2,3] to establish the objective of editor retention. There exists corresponding wikiproject [5] dedicated to find the cause of editor depletion and early identification of editors at risk of forsaking Wikipedia. Often expert editors withdraw themselves because of the discontent with Wikipedia's policies and widespread norms resulting them into semi-retired or permanent-retired often not disclosing the prime reasons behind the retirement. As an example, the Wikipedia project (WP:MISS)[1] maintains a list of experienced Wikipedians who have made no edits to any article after a fixed time point in their life and are defined as *missing Wikipedians.* All of these editors contributed at least 1000 edits in their lifetime and can be considered as *prolific editors.* Even after being inactive, many of them are still featured among top editors[2]. As of June 2020, this list includes 1226 such editors whom we refer to henceforth as *missing Wikipedians* or *missing editors* interchangeably. In contrast, the pool of editors who are continuing to contribute as active participants till date, will be referred as *active editors* hereafter. Recently, a wikiproject named as *Community Health*

[1] https://en.wikipedia.org/wiki/Wikipedia:Missing_Wikipedians.
[2] https://tinyurl.com/3ewzb468.

Metrics: Understanding Editor Drop-off[3] has been initiated to understand the editor drop-off across multiple language versions of Wikipedia. The project aims to understand various dynamics in editors' life-cycle with a special attention to the veteran editors of the community who are at the risk of leaving Wikipedia. Our work is in line with the objectives of these projects - identifying the decline of prolific editors whose contributions are measured significant in the community. We observed that majority of editors faced some form of impediment from the Wikipedia community that curbed their interest in further editing. Finally, these missing editors have left the platform and there is no assurance of their return[4]. Most of these editors do not reveal the explicit reason for leaving Wikipedia. Some editors have pointed out typical instances of disagreement, bureaucracy on their personal user talk pages that influenced them for announcing retirement. Among many example reasons that we observed on editors' user pages, a typical example is the following:

*"I have left Wikipedia. I do not see it as acceptable to have advertisements, whether they be for brand identity or for a product, on Wikipedia." –**Im*****v***

The above text shows an editor's disagreement with the community for compromising NPOV policy[5] of Wikipedia as a prime cause of his/her withdrawal from Wikipedia.

Editor Attributes as a Cue? Our hypothesis is that the early signals of retirement of a missing editor could be hidden in his/her last trail of *editing activities*. In other words, the editing activity patterns of the missing editors would have difference with the existing active editors. Further, the typical sentiments, emotions expressed on user talk pages should strike different attitude of missing editors compared to the active group of editors. Due to the challenges and hindrance of the platform, missing editors might follow different profile of quality as compared to active editors. Based on this hypothesis, in this paper, we develop a framework for the discovery of key editors who have stopped content generation in Wikipedia. For this purpose, we first identify 1146 editors who have no edits in the calendar year of 2020 and denote them as missing editors. We next select a set of 2569 editors whose editing activity (i) are ongoing and (ii) match the overall editing activity level of the missing editors. This second set is denoted as active editors. We compare the two sets of editors on a longitudinal scale in the attempt to identify various properties related to their activity patterns and language usage that could significantly differentiate them. Our main research questions are as follows.

RQ1: To what extent activity/linguistic/quality features of editors can help us discern missing from active editors?

RQ2: Using the discriminative editor activities can we predict currently prolific editors who have a possibility to leave the platform in the near future?

[3] https://tinyurl.com/3e6jj6zz.
[4] https://tinyurl.com/35n624a6.
[5] https://en.wikipedia.org/wiki/Wikipedia:Neutral_point_of_view.

Our Contributions: We make the following novel contributions in this paper.

- We curate a first ever dataset of missing editors, a comparable dataset of active editors along with all the associated metadata that can appropriately characterise the editors from each dataset (see Sect. 3).
- First we put forward a number of features describing the editors (activity and behaviour) which portray significant differences between the active and the missing editors (see Sect. 4).
- Next we use SOTA machine learning approaches to predict the currently prolific editors who are at the risk of leaving the platform in near future. Our best models achieve an overall accuracy of 82% in the prediction task (see Sect. 5).
- We perform rigorous ablation studies to provide further insights into our results. We discuss various nuanced observations that get manifested in the course of our study. An intriguing finding is that some very simple factors like how often an editor's edits are reverted or how often an editor is assigned administrative tasks could be monitored by the moderators to determine whether an editor is about to leave the platform (see Sect. 5).

To the best of our knowledge this is the first work which proposes an automatic approach to predict, early on, the editors who have a propensity to leave the platform soon. We believe that the moderators can use these attributes to launch suitable *platform governance* [18] measures to appropriately incentivize these editors and, thereby, retain them on the platform. Further the observations and insights that we obtain from our results can he useful in designing the specific incentive strategies [6].

2 Related Work

The reason behind the popularity of Wikipedia is often attributed to the hundreds of thousands of volunteers from all around the world, but several tens of thousands of Wikipedians and their collaboration are very crucial for the generation and maintenance of healthy and informative content [24,37,41]. Researchers are trying to overcome different challenges disseminated by open source of knowledge and information especially the quality and trust issues [10,36] of this gigantic platform.

With the constant effort of regulation and maintenance of encyclopedia content, it appears that Wikipedia is at the peak of its popularity; however, experts noted that it is at the danger of sharp decline of its active editors [20,44]. The revert of edits is highly effective in controlling vandalism [19], sometimes it becomes harsh to the editors, especially the newly joining ones [21]. The topic of volunteer retention [25] is one of the basic concerns of all peer-production systems, tied to their preliminary survival. Hence, researchers of Wikipedia shed light on the *retention* of *new* contributors [11,27,31]. However, works on early identification and retention of *prolific editors* at risk of retiring are scarce. Therefore, in this work we set out to develop an automatic framework that can accurately identify editing activity signatures that point to the early signals resulting in the abdication of an editor.

3 Dataset Description

Wikipedians are the prime elements of our experiment. In order to curate two sets of editors – missing and active, we follow a two step approach.

List of Missing Editors: First, we prepare a list of editors who are mentioned as *missing* by Wikipedia. We extracted the list of missing Wikipedians[6] and their last date of contribution using XTools[7]. Wikipedia has mentioned 1226 such editors in the list at the time of our dataset collection (June, 2020). From this list we remove those that we found still contributing in 2020 despite being declared as missing by Wikipedia. This results in 1146 unique editors who further have not edited in the calendar year 2020.

List of Active Editors: Next, we collect a set of 2569 editors who are still active, i.e., are still editing different Wikipedia articles. In order to build a comparable set of active editors, we performed the following.

- First, we collect the list of top 60 pages (top 20 from each namespace – 'Main', 'Wikipedia' and 'Talk') that have been most frequently edited by each missing editor. This leads us to an initial set of 68,760 Wikipedia pages.
- Next we look into the 100 latest revisions of each of the 68,760 pages and the editors who contributed to the latest revisions are extracted from the revision history of the corresponding pages. Further, for every missing editor, only top 10 active editors are selected based on the frequency of their edits on the above mentioned top edited 60 pages. Formally, for a missing editor M, the editors $A_1, A_2, ..., A_{10}$ are assumed to be the most active if they have the highest edit counts in the last 100 revisions of the top edited pages $P_1, P_2, P_3, ..., P_{60}$ by the editor M. These editors are placed into the pool of active editors. For every missing editor we perform this exercise leading us to a set of 5213 unique active editors. We confirm that none of these editors are listed in the missing Wikipedians page.
- As a final step, we compute the average number of edits per day for the active editors and the missing editors. We next compute the mean of the average edits per day for the missing editors (say m). Next we compute the L2-norm of m with the average number of edits per day of each of the active editors. We consider those active editors for whom the L2-norm is within one standard deviation of m. This results in 2569 active editors. The difference in the distributions of the average edits per day for the missing and active editors so chosen is not statistically significant (Mann Whitney U test, $p = 0.14$) thus making them comparable.

4 Feature Space Design

In this section we describe our feature space which we develop to differentiate the active from the missing editors. The features can be categorised into three broad classes–

[6] https://en.wikipedia.org/wiki/Wikipedia:Missing_Wikipedians.

[7] https://www.mediawiki.org/wiki/XTools/API.

Activity features, Quality features, Linguistic features.

Resources for the work is available at: https://github.com/paramita08/Missing-Active-Wikipedians.

4.1 Activity Features

The features we propose in this category describe the activity patterns of editors. The choice of these features is motivated by the fact that these would reflect different aspects of editor's interactions with the community over time. These features, we believe should be useful to capture the temporal change in the behavior of the editors who have a possibility to leave the platform. We extract the details of the latest 50 non-automated edits of every user using the *User API* of XTools[8] and identify the following activity features.

Edits in Different Namespaces: Among different namespaces[9] mentioned, we consider the editor contribution in 4 namespaces - article pages i.e., namespace 0; article talk pages, i.e., namespace 1; administrative main pages - Wikipedia, i.e., namespace 4 and the administrative talk pages - Wikipedia talk, i.e., namespace 5. Majority of contribution of editors is restricted in these respective namespaces [30] and we denote the four namespaces as F1, F2, F3 and F4 following the order as mentioned above. Our hypothesis is that the frequency of edits in the first two namespaces (i.e., 0, 1) will be affected drastically for the missing editors in their latest series of contributions. We have included these two namespaces to observe if the missing editors exhibit a decline in participation in Wikipedia's policy decisions. As shown in Fig. 1(a), the average count of edits in all the namespaces except namespace 5 is higher in case of active editors. This trend implies a slow decline in contribution from missing editors possibly pointing to their absence in near future.

Major and Minor Edits: A *minor*[10] edit is the one denoting minor changes such as typographical errors, formatting errors, reversion of definite vandalism etc. that the editors believe need no further reviews. In contrast, any contribution that changes the article content and needs to be reviewed for the acceptability to the community of editors is denoted as *major* edit. We hypothesize that an increase in the number of minor edits as compared to major edits can bear an early signature of their disengagement, finally abdicating the platform. Hence, we included the count of edits in both categories - major (F5) and minor (F6) of the latest revisions in our feature set. The cumulative edit count of major and minor edits in Fig. 2 show that for major edits, active editors exhibit a steady state in the count over time but the decline is extremely unstable for the missing editors and almost always below that of the active editors. On the other hand, close to their retirement, the missing editors seem to be engaging themselves in an increasingly more number of minor edits as compared to the active editors.

[8] https://www.mediawiki.org/wiki/XTools/API/User#Non-automated_edits.
[9] https://en.wikipedia.org/wiki/Wikipedia:Namespace.
[10] https://en.wikipedia.org/wiki/Help:Minor_edit.

Length of the Edits: The average length of the edits is captured by this feature in terms of the bytes added or deleted per edit. Further, we classify this feature into four categories - (i) addition in major edits (F7), (ii) deletion in major edits (F8), (iii) addition in minor edits (F9), (iv) deletion in minor edits (F10). In each of the cases, we compute the average bytes added (or deleted) in case of a major (or minor) edit over the latest contribution of the editors. Once again the hypothesis is that the length of the edits done by the missing editors shall diminish over time. An interesting observation that we have here is that the missing editors seem to engage more in deleting minor edits than adding minor edits (see Fig. 1(b)).

Span (in Months): We compute the time span (denoted as F11) in months taken by each editor to complete the latest revisions of contribution. The span is expected to be larger in case of missing editors compared to the active editors indicating a possible loss of overall interest in the platform. We find that the average span in case missing editors is 2.04 months while for active editors this is 0.82 months.

ORES Score of the Edits: ORES[11], the web-service API by Wikimedia Foundation is used in automating several wiki management tasks such as assigning scores to individual edits of editors and predicting the edits to be damaging or good-faith. We compute the average ORES score (represented as F12) for each of the editors in our dataset over their contribution span. In addition, we consider the count of the good faith (denoted as F13) vs the damaging (denoted as F14) edits for every editor in our dataset. On average we did not see any difference in the quality of edits done by the two classes of editors. Further both the classes have a very small number of damaging edit contributions.

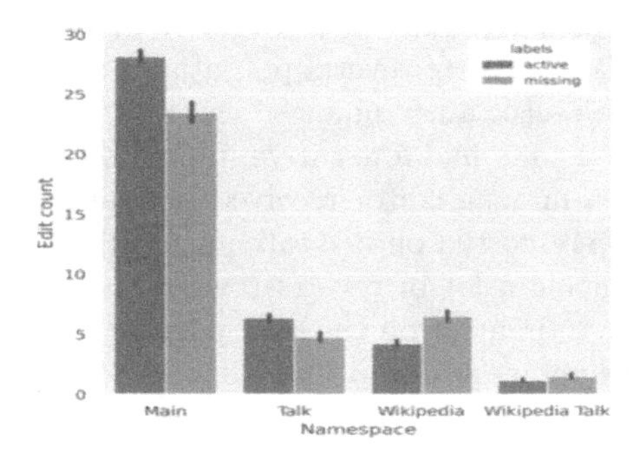

(a) The average count of edits in different namespaces of active and missing editors.

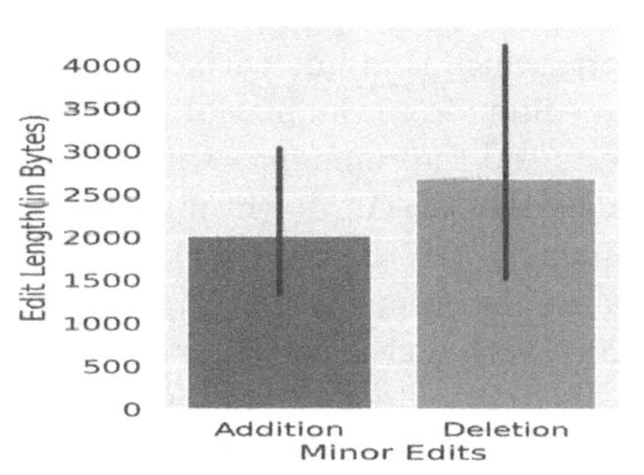

(b) Average length of edits(in bytes) in minor edits (addition and deletion) by missing editors.

Fig. 1. Plots showing different feature values for the two classes of editors.

[11] https://www.mediawiki.org/wiki/ORES.

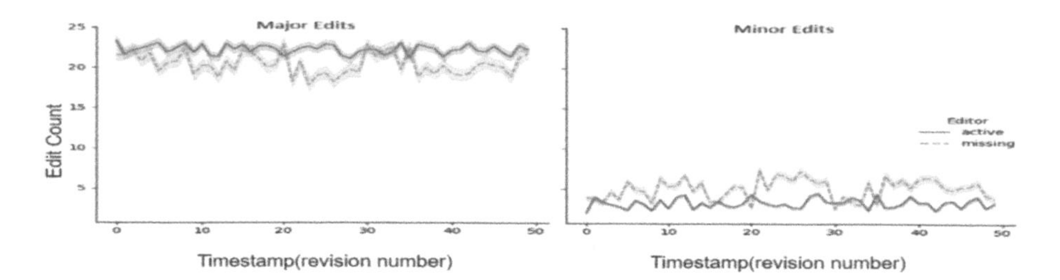

Fig. 2. Temporal trends showing major and minor edit counts individually over the latest revisions for both the sets of editors. The cumulative count at every revision exhibits that missing editors contributed more minor edits than the active editors and vice-versa.

4.2 Quality Features

Although the content of any Wikipedia article is factual, its quality heavily depends on the dexterity of the editors. We formulate the following quality features that could be potential indicators of the future churn.

Count of Reverts: Reverting is the way of reversing a prior edit or undoing the effects of one or more edits, which typically results the article to be restored to a version that existed sometime previously. Reverts are discouraging to any editor, irrespective of their experience [21,23] and can potentially impede an editor toward making future contributions to the platform. To investigate this, we scrape the number of reverted edits of every individual editor in their respective top 50 most edited pages in the main namespace only. This is since almost all editors make their largest chunk of contributions to the main namespace. From Table 1 we observe that missing editors have experienced a larger number of reverts which is in line with our hypothesis that more reverts potentially disengage editors from the platform. Further we revisit a few nuanced cases in which reverts and the comments with the reverted edits had a negative impact on the missing editors. As shown in the Table 2, a missing editor received several foul comments with the reverted edits which were posted on her talk page publicly. Finally, she stopped editing after experiencing a lot of reverts toward the end of 2004 with a clear message declaring her retirement on the user page in 2005. We also look into the talk pages of active editors and find that they had experienced less sensitive comments for their reverted edits and an example scenario is depicted in the second row of the Table 2.

Admin Score: Wikipedia maintains a pool of editors, known as admins, who are responsible to perform various administrative tasks and the tasks are pivotal in monitoring various quality issues of Wikipedia. These set of people act voluntarily and every wikiproject tries to find the admins from among the prolific and experienced editors. XTools provides the *admin score*[12], intended to find how suitable an user is for serving as an admin. The more the admin-score, the

[12] https://xtools.readthedocs.io/en/3.1.6/tools/adminscore.html.

Table 1. Mean (μ) and std. deviation (σ) of the quality features for missing and active editors.

Quality	Missing (μ, σ)	Active (μ, σ)
Avg. revert count per page	(0.031, 0.041)	(0.026, 0.031)
Admin score	(734.38, 165.38)	(823.86, 174.93)

higher is the chance of the editor to become an admin. We observe in Table 1 that active editors on average have a higher admin score and therefore higher chances of being an admin as compared to the missing editors.

Table 2. Example comments on editors' talk page showing comments received with reverted edits. The comments received by an active editor is less harsh as compared to a missing editor.

Comments received by	Comments
A missing editor	*If that happens I will, with considerable regret, withdraw from Wikipedia altogether. An encyclopedia that can't or won't defend itself against Stalinist and LaRouche wreckers will never succeed and doesn't deserve to.*—**A****, 01:05, 16 Nov 2004 (UTC)**
An active editor	*Greetings. I noticed you had undid my edits on H**** C**** and J**** K****, specifically the ones....I've since reverted them. But I'm opening discussion here, in case there's disagreement. Looking forward to your thoughts.*—**Ga****, 14:06, 23 Dec 2015 (UTC)**

4.3 Linguistic Features

As we have discussed earlier a number of editors have shared their basic information, Wikipedia-related activities, awards and badges earned etc. and sometimes their grievances[13] about the platform on their user pages that finally hold them back in continuing further contributions. We assume that these pages to be the profile of the editors and leverage the user generated text in characterizing the two sets of editors. We extract the text from the HTML version of the latest editor page and perform pre-processing to remove various links and HTML tags. We next extract various features from this pre-processed text.

POS Tags: We compute the POS tags of the text in the editor pages using the NLTK parser [28]. Prior to tagging, we tokenized the sentences and words common to both the groups of editors (missing and active) are removed from the corpus. We also remove the stop words and the non-English words. We observe

[13] https://en.wikipedia.org/wiki/User:Bcrowell.

several frequent POS tag categories such as "JJ", "NN", "NNS", "VBD", "VBP" that exhibit significant differences between the missing and the active editors.

Empath Categories: Empath is a text analysis tool [16] which can identify psycholinguistic signs hidden in a text in the form of pre-validated lexical categories. For our purpose, we compute the normalised value for each predefined category for each sentence. We then average out this value for all the sentences present in the profile of an editor. We observe that a number of lexical categories (21 in number), i.e., "Internet", "Noise", "Trust", "Reading", "Violence", "Negative Emotion", "Positive Emotion" etc. show the difference in distributions for two classes of editors is statistically significant as per the Mann-Whitney U test. However the absolute differences between the two classes in terms of the values of the various lexical categories is negligibly small. However the absolute differences between the two classes in terms of the values of the various lexical categories is negligibly small.

Sentence Vector: We generate the sentence vector using the Universal Sentence Encoder [13] which outputs a 512 dimensional vector representation of the text. We compute the sentence vector for every editor which can be thought of as a summary of the language usage of the users on the platform. To visualise how different the sentence vectors from the two classes are, we plot them using t-SNE [29]. We consider the first two principal components of the vectors for visualisation. Figure 4a shows that there is no clear separation between the sentence vector drawn from the two classes. However, there are at least two small but very distinct pockets of vectors from a particular class clustering together expressing opposing sentiments toward the platform. We handpick a few examples from the talk pages of editors belonging to each of these two clusters. These are shown in Fig. 3. The difference in sentiments toward the platform is apparent in the selected sentences from the two clusters. While missing editors seem to express grudges against the platform, the active editors mostly narrate their positive experiences with the platform.

4.4 Feature Correlation

To examine the relationship among the different features described above, we compute the Pearson's correlation coefficients [9] between the all-pair features. Figure 4b shows the correlation coefficient matrix for the 16 features (except the linguistic features). We observe that some features are positively correlated and some features are negatively correlated (e.g., major edits and minor edits) with each other. However almost all the coefficients are in the range -0.05 to 0.05 (i.e., ~ 0 correlation) except for the pairs (addition in major edits, deletion in major edits) and (addition in minor edits, deletion in minor edits), which belong to the same category. We do not include the linguistic features in this study as they are characteristically very different from these features.

I know there are good people out there who understand what's really going on. Alas, you are not in charge. –V****

I've no interest in working on a system where people break down your work without discussing it on appropriate page. –A*****

Many articles seem to fluctuate around a kind of equilibrium, with a whole bunch of people editing, but mostly just undoing each other's edits. I'm no longer actively participating much as an editor. –B****

Thank you kindly and I hope you like this Tea as a token of my appreciation for your non aggressive and educated reply! –M****

Congratulations, W****! The article you nominated, has been promoted to FA status, recognizing it as one of the best articles on Wikipedia. –I***

I often find that just as satisfying-these experiences have only made it clearer to me that devoting myself to Wikipedia, was the right thing to do, and I plan on sticking around for a while yet. –W***

Fig. 3. A set of sample sentences, narrated by some anonymous editors belonging to the two prominent sentence clusters (see the t-SNE plot in Fig. 4a). In the cluster of the missing editors, we find sentences expressing grudges against the platform (red sentences). In the cluster of active editors, we find positive sentences about the platform (blue sentences). (Color figure online)

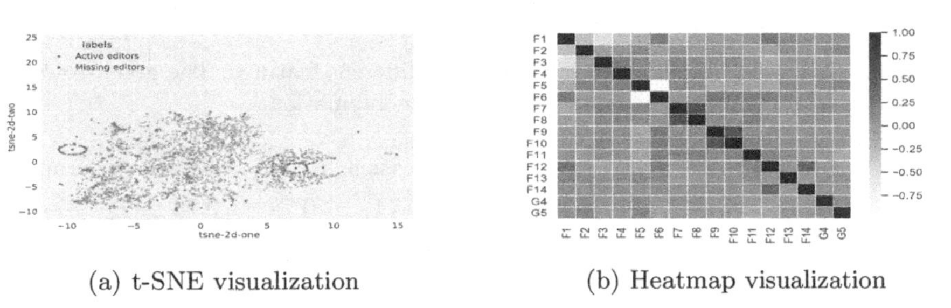

(a) t-SNE visualization

(b) Heatmap visualization

Fig. 4. (a) t-SNE showing sentence vectors expressed by two sets of editors – missing (in green) and active (in red) and (b) Heatmap showing the correlation between the various features from the activity and quality categories. (Color figure online)

5 Classification of Editors

In this section we use the mentioned features to automatically classify the missing editors from the active ones. However, in order to make the model outcomes interpretable we incrementally include the features to better understand their effect on the overall performance. We name the feature groups as follows-

- Activity features as G1
- POS tags and Empath as G2
- Sentence vector as G3
- Admin score as G4
- Count of reverts as G5

Classification Model: Corresponding to each editor in our dataset, we construct a normalised feature vector. We experiment with multiple classifiers such as Random Forest, Logistic Regression SVMs, XGBoost, AdaBoost. In this task, a binary classifier is built to predict if the given editor is a missing or active one.

The train-test split for our experiments is set to 80:20. We use accuracy along with weighted precision, recall and F1-score as the evaluation measures.

Results: Among the different features, we observe that the 14 activity features, reverts and admin score seem to be performing much better than the linguistic features (see Table 3). The G1 feature group, i.e., the activity features seem to be the strongest individual discriminator resulting in an accuracy of 75% (F1-score 74%). Together with G4, this accuracy goes to 78% (F1-score 79%). Among the linguistic features, the sentence vector seems to be the most effective in conjunction with activity and the admin score features. Together this feature group G1, G3 and G4 attains an accuracy of 82% (F1-score 81%). The feature group G1, G4 and G5 also attains a similar accuracy of 82% (F1-score 81%). Larger groups of features do not seem to improve performance. These results further corroborate that missing editors hardly leave any special linguistic cues before they abdicate the platform.

Table 3. Prediction outcomes: combination of different features. The green rows indicate the best result using the respective feature combination.

Features	Classifier	Precision	Recall	F-score	Accuracy
G1	XGBoost	0.74	0.74	0.74	0.75
G2	XGBoost	0.63	0.68	0.64	0.68
G3	AdaBoost	0.63	0.67	0.63	0.67
G1 ⊕ G2	Random Forest	0.77	0.78	0.76	0.78
G1 ⊕ G3	XGBoost	0.75	0.76	0.75	0.76
G1 ⊕ G4	AdaBoost	0.78	0.79	0.79	0.78
G1 ⊕ G5	AdaBoost	0.78	0.79	0.79	0.77
G1 ⊕ G2 ⊕ G4	XGBoost	0.77	0.78	0.77	0.78
G1 ⊕ G3 ⊕ G4	XGBoost	0.81	0.82	0.81	0.82
G1 ⊕ G4 ⊕ G5	AdaBoost	0.81	0.82	0.81	0.82
G1 ⊕ G3 ⊕ G5	XGBoost	0.78	0.79	0.78	0.79
G1 ⊕ G2 ⊕ G4 ⊕ G5	Random Forest	0.82	0.82	0.82	0.81
G1 ⊕ G3 ⊕ G4 ⊕ G5	XGBoost	0.80	0.81	0.80	0.81

Feature Importance Analysis: In this section we investigate the importance of the individual features in this feature group that consists of only activity and quality features (16 in all). To this purpose, we use three different ways to compute the importance.

Feature Importance Function: We use the *Gini impurity* bases feature importance which is a standard technique in tree based ensemble methods [34] to compute the feature importance values. We observe the features that dominate the top of the rank list are - reverts, ORES score, admin score, deletion in major edits, edits in namespace 0 and span.

Permutation Importance Function: We compute feature importance using *permutation importance* [7] and observe almost similar importance order as that of the previous function. The features that come at the top 6 places of the importance list are - reverts, ORES score, admin score, span, edits in namespace 0 and edits in namespace 4.

LIME: LIME [38] is an explainability tool meant to generate explanations regarding the workings of a ML model. We use all the instances in the test data for generating the explanations. We note the top five features returned by LIME that are responsible for the prediction outcomes for each instance. We observe that across all the instances the set of features {reverts, admin score, edits in namespace 0, edits in namespace 1, ORES damaging *true*, edits in namespace 4} present the best explanations.

Overall, we observe that the quality features are the most discriminatory among all. This points to the fact that they constitute the best signals for identifying whether an editor is about to leave the platform. Platform moderators can deploy simple monitoring schemes to understand these signals early on and take preventive steps to minimise the loss of such prolific contributors.

6 Discussion

The primary objective of this work has been to bring forth the issue of the growing depletion of editors, especially the experienced editors in Wikipedia. One of the most important findings of our work is that missing editors can be differentiated from the existing active ones by a set of simple and interpretable characteristics of editors extracted from their interaction with the platform.

Dissatisfaction of Prolific Editors: Further to test the accuracy of our model, we have observed a number of specific cases where the classifiers are confident in predicting the missing editors. We revisited their talk pages and found that in many such cases the editors had to face bullying[14] from other editors. Those bullying instances are usually mentioned as a part of the article talk pages. Sometimes the user has also been found to express his/her grievances on their personal user page. Figure 5 shows few example cases of toxic arguments and explicit reasons for leaving the platform that the editor had to face before they abdicated. Although a number of editors have stopped their contribution silently, the external causes of discontent with the platform that trigger the churn of prolific editors need to be considered carefully.

Enabling Platform Moderation: We collected a total of 8 false positive cases (mentioned as active but predicted as missing) in which our model is highly confident (confidence probability >0.8). We found that out of these sample cases, the last edit for 4 editors were done latest in 2017. We believe that all these editors should be potentially included in the list of missing editors but somehow got overlooked by the Wikipedia community. In fact, one out of these 4 cases

[14] https://en.wikipedia.org/wiki/Wikipedia:WikiBullying.

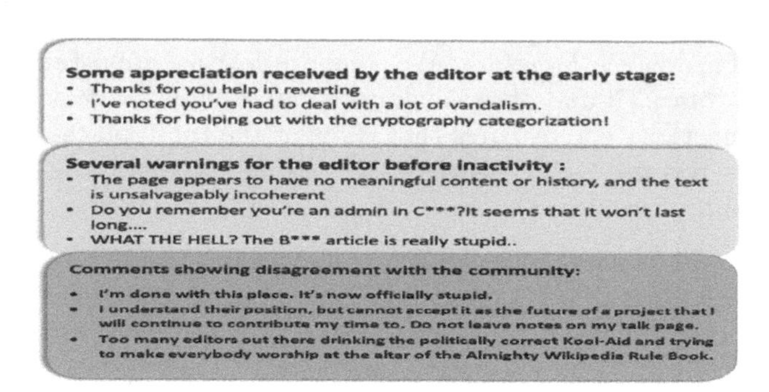

Fig. 5. Examples showing talk page arguments of a missing editor. The text in green colour box shows some appreciation received by the editor at early stage. In later phase, a number of toxic comments (shown in red colour box) alleged editing activity. A few examples from his user page (shown in blue colour box) indicates editor's dissatisfaction toward the community. (Color figure online)

have been actually identified as a missing editor in the most recent edition of the missing editor list, i.e., June 2021. From an application point of view, our system can monitor the current editor pool and flag any missing editor which the community may overlook. Suitable policies may be adopted to retain/reinstate such editors.

Design Implication: Our idea was to keep the model as much interpretable as possible so that it is easy to understand whether the editing activity and the language usage across the platform are useful for classification. Further, our dataset is unbalanced and the total number of data points is limited to only 3715. Hence, in our work, we have followed simple feature based classification model. Moreover, we proposed a novel task in which the feature set design is such that they can be calculated in all category of Wikipedia articles. The existing literature tried to divide editors either with respect to a certain number of Wikiprojects or based on a set of Wikipedia articles. Therefore suitable baselines are scarce. Thus here we consider feature ablations as baselines as is normative in many NLP applications [32] where suitable baselines are not available.

7 Conclusion

In this paper we take a deep dive into the emerging issue of editor depletion in Wikipedia. Our investigation shows that longitudinal activity traces as well as linguistic clues in their profile, i.e., user talk pages provide necessary signals to ascertain whether a prolific contributor will be quitting the Wikipedia platform in near future. We identify a number of important features that summarise the level of activity and behaviour of an editor and can be used to investigate the casual connection for leaving the platform. For instance, if two missing editors have almost equivalent qualities and extremely similar editing activity patterns and have retired almost at the same time, then it might well be the case that both of them left the platform for very similar reasons. In future, we plan to use these features to identify the causes of a group of editor leaving Wikipedia. This would allow for a better streamlining of the governance measures based on the nature of the cause of disengagement. Upon acceptance, we plan to place all our data and code in the public domain to facilitate further research in this area.

References

1. The future of Wikipedia: Wikipeaks? https://www.economist.com/international/2014/03/04/wikipeaks. Accessed 04 Mar 2014
2. Wikipedia: Editor activation. https://tinyurl.com/2njfds4y
3. Wikipedia: Surviving new editor. https://tinyurl.com/4cjdyfs4
4. Wikipedia: Teahouse. https://tinyurl.com/jzf346
5. Wikipedia: Wikiproject editor retention. https://tinyurl.com/acmyrp8s
6. Aaltonen, A., Lanzara, G.F.: Building governance capability in online social production: insights from Wikipedia. Organ. Stud. **36**(12), 1649–1673 (2015)
7. Altmann, A., Tolosi, L., Sander, O., Lengauer, T.: Permutation importance: a corrected feature importance measure. Bioinformatics **26**(10), 1340–1347 (2010)
8. Balali, S., Steinmacher, I., Annamalai, U., Sarma, A., Gerosa, M.A.: Newcomers' barriers... is that all? An analysis of mentors' and newcomers' barriers in OSS projects. Comput. Support. Cooper. Work (CSCW) **27**(3), 679–714 (2018)
9. Benesty, J., Chen, J., Huang, Y., Cohen, I.: Noise Reduction in Speech Processing, vol. 2. Springer, Heidelberg (2009). https://doi.org/10.1007/978-3-642-00296-0
10. Chen, X., Iwaihara, M.: Weakly-supervised neural categorization of Wikipedia articles. In: Jatowt, A., Maeda, A., Syn, S.Y. (eds.) ICADL 2019. LNCS, vol. 11853, pp. 16–22. Springer, Cham (2019). https://doi.org/10.1007/978-3-030-34058-2_2
11. Choi, B., Alexander, K., Kraut, R.E., Levine, J.M.: Socialization tactics in Wikipedia and their effects. In: Proceedings of the 2010 ACM Conference on Computer Supported Cooperative Work, pp. 107–116 (2010)
12. Ciampaglia, G.L., Taraborelli, D.: MoodBar: increasing new user retention in Wikipedia through lightweight socialization. In: Proceedings of the 18th ACM Conference on Computer Supported Cooperative Work and Social Computing, pp. 734–742 (2015)
13. Conneau, A., Kiela, D., Schwenk, H., Barrault, L., Bordes, A.: Supervised learning of universal sentence representations from natural language inference data. arXiv preprint arXiv:1705.02364 (2017)
14. Danescu-Niculescu-Mizil, C., West, R., Jurafsky, D., Leskovec, J., Potts, C.: No country for old members: user lifecycle and linguistic change in online communities. In: Proceedings of the 22nd International Conference on World Wide Web, pp. 307–318 (2013)
15. Dror, G., Pelleg, D., Rokhlenko, O., Szpektor, I.: Churn prediction in new users of Yahoo! answers. In: Proceedings of the 21st International Conference on World Wide Web, pp. 829–834 (2012)
16. Fast, E., Chen, B., Bernstein, M.S.: Empath: understanding topic signals in large-scale text. In: Proceedings of the 2016 CHI Conference on Human Factors in Computing Systems, pp. 4647–4657 (2016)
17. Gallus, J.: Fostering public good contributions with symbolic awards: a large-scale natural field experiment at Wikipedia. Manag. Sci. **63**, 3999–4015 (2017)

18. Gorwa, R.: What is platform governance? Inf. Commun. Soc. **22**(6), 854–871 (2019)
19. Green, T., Spezzano, F.: Spam users identification in Wikipedia via editing behavior. In: Proceedings of the International AAAI Conference on Web and Social Media, vol. 11 (2017)
20. Halfaker, A., Geiger, R.S., Morgan, J.T., Riedl, J.: The rise and decline of an open collaboration system: how Wikipedia's reaction to popularity is causing its decline. Am. Behav. Sci. **57**(5), 664–688 (2013)
21. Halfaker, A., Kittur, A., Riedl, J.: Don't bite the newbies: how reverts affect the quantity and quality of Wikipedia work. In: Proceedings of the 7th International Symposium on Wikis and Open Collaboration, pp. 163–172 (2011)
22. Kairam, S.R., Wang, D.J., Leskovec, J.: The life and death of online groups: predicting group growth and longevity. In: Proceedings of the Fifth ACM International Conference on Web Search and Data Mining, pp. 673–682 (2012)
23. Kiesel, J., Potthast, M., Hagen, M., Stein, B.: Spatio-temporal analysis of reverted Wikipedia edits. In: Proceedings of the International AAAI Conference on Web and Social Media, vol. 11 (2017)
24. Kittur, A., Kraut, R.E.: Harnessing the wisdom of crowds in Wikipedia: quality through coordination. In: Proceedings of the 2008 ACM Conference on Computer Supported Cooperative Work, pp. 37–46 (2008)
25. Konieczny, P.: Volunteer retention, burnout and dropout in online voluntary organizations: stress, conflict and retirement of Wikipedians. In: Research in Social Movements, Conflicts and Change. Emerald Publishing Limited (2018)
26. Lemmerich, F., Sáez-Trumper, D., West, R., Zia, L.: Why the world reads Wikipedia: beyond English speakers. In: Proceedings of the Twelfth ACM International Conference on Web Search and Data Mining, pp. 618–626 (2019)
27. Li, A., Yao, Z., Yang, D., Kulkarni, C., Farzan, R., Kraut, R.E.: Successful online socialization: lessons from the Wikipedia education program. Proc. ACM Hum. Comput. Interact. **4**(CSCW1), 1–24 (2020)
28. Loper, E., Bird, S.: NLTK: the natural language toolkit. In: ACL 2006 (2006)
29. Maaten, L.V.D., Hinton, G.E.: Visualizing data using t-SNE. J. Mach. Learn. Res. **9**, 2579–2605 (2008)
30. Maki, K., Yoder, M., Jo, Y., Rosé, C.: Roles and success in Wikipedia talk pages: identifying latent patterns of behavior. In: Proceedings of the Eighth International Joint Conference on Natural Language Processing (Volume 1: Long Papers), pp. 1026–1035 (2017)
31. Morgan, J.T., Halfaker, A.: Evaluating the impact of the Wikipedia teahouse on newcomer socialization and retention. In: Proceedings of the 14th International Symposium on Open Collaboration, pp. 1–7 (2018)
32. Mowery, D., Bryan, C., Conway, M.: Feature studies to inform the classification of depressive symptoms from twitter data for population health. arXiv preprint arXiv:1701.08229 (2017)
33. Murić, G., Abeliuk, A., Lerman, K., Ferrara, E.: Collaboration drives individual productivity. Proc. ACM Hum. Comput. Interact. **3**(CSCW), 1–24 (2019)
34. Nembrini, S., König, I., Wright, M.N.: The revival of the Gini importance? Bioinformatics **34**, 3711–3718 (2018)
35. Panciera, K.A., Halfaker, A., Terveen, L.: Wikipedians are born, not made: a study of power editors on Wikipedia. In: GROUP 2009 (2009)
36. Pinto, J.M.G., Kiehne, N., Balke, W.-T.: Towards semantic quality enhancement of user generated content. In: Dobreva, M., Hinze, A., Žumer, M. (eds.) ICADL 2018. LNCS, vol. 11279, pp. 28–40. Springer, Cham (2018). https://doi.org/10.1007/978-3-030-04257-8_3

37. Proffitt, M.: Leveraging Wikipedia: Connecting Communities of Knowledge. American Library Association (2018)
38. Ribeiro, M.T., Singh, S., Guestrin, C.: "Why should I trust you?": explaining the predictions of any classifier. In: Proceedings of the 22nd ACM SIGKDD International Conference on Knowledge Discovery and Data Mining (2016)
39. Samoilenko, A., Lemmerich, F., Zens, M., Jadidi, M., Génois, M., Strohmaier, M.: (Don't) Mention the war: a comparison of Wikipedia and Britannica articles on national histories. In: Proceedings of the 2018 World Wide Web Conference, pp. 843–852 (2018)
40. Sarasua, C., Checco, A., Demartini, G., Difallah, D., Feldman, M., Pintscher, L.: The evolution of power and standard Wikidata editors: comparing editing behavior over time to predict lifespan and volume of edits. Comput. Support. Cooper. Work (CSCW) **28**(5), 843–882 (2019)
41. Suzuki, Y., Yoshikawa, M.: Mutual evaluation of editors and texts for assessing quality of Wikipedia articles. In: Proceedings of the Eighth Annual International Symposium on Wikis and Open Collaboration, pp. 1–10 (2012)
42. Yang, D., Halfaker, A., Kraut, R., Hovy, E.: Who did what: editor role identification in Wikipedia. In: Proceedings of the International AAAI Conference on Web and Social Media, vol. 10 (2016)
43. Yazdanian, R., Zia, L., Morgan, J., Mansurov, B., West, R.: Eliciting new Wikipedia users' interests via automatically mined questionnaires: for a warm welcome, not a cold start. In: Proceedings of the International AAAI Conference on Web and Social Media, vol. 13, pp. 537–547 (2019)
44. Zhang, D., Prior, K., Levene, M., Mao, R., van Liere, D.: Leave or stay: the departure dynamics of Wikipedia editors. In: Zhou, S., Zhang, S., Karypis, G. (eds.) ADMA 2012. LNCS (LNAI), vol. 7713, pp. 1–14. Springer, Heidelberg (2012). https://doi.org/10.1007/978-3-642-35527-1_1
45. Zhang, Y., Sun, A., Datta, A., Chang, K., Lim, E.P.: Do Wikipedians follow domain experts? A domain-specific study on Wikipedia knowledge building. In: JCDL 2010 (2010)

Federating Scholarly Infrastructures with GraphQL

Muhammad Haris[1(✉)], Kheir Eddine Farfar[2], Markus Stocker[2], and Sören Auer[1,2]

[1] L3S Research Center, Leibniz University Hannover, 30167 Hannover, Germany
haris@l3s.de

[2] TIB—Leibniz Information Centre for Science and Technology, Hannover, Germany
{kheir.farfar,markus.stocker,auer}@tib.eu

Abstract. A plethora of scholarly knowledge is being published on distributed scholarly infrastructures. Querying a single infrastructure is no longer sufficient for researchers to satisfy information needs. We present a GraphQL-based federated query service for executing distributed queries on numerous, heterogeneous scholarly infrastructures (currently, ORKG, DataCite and GeoNames), thus enabling the integrated retrieval of scholarly content from these infrastructures. Furthermore, we present the methods that enable cross-walks between artefact metadata and artefact content across scholarly infrastructures, specifically DOI-based persistent identification of ORKG artefacts (e.g., ORKG comparisons) and linking ORKG content to third-party semantic resources (e.g., taxonomies, thesauri, ontologies). This type of linking increases interoperability, facilitates the reuse of scholarly knowledge, and enables finding machine actionable scholarly knowledge published by ORKG in global scholarly infrastructures. In summary, we suggest applying the established linked data principles to scholarly knowledge to improve its findability, interoperability, and ultimately reusability, i.e., improve scholarly knowledge FAIR-ness.

Keywords: Open Research Knowledge Graph · Federated query · GraphQL · Metadata exchange · Scholarly communication · Machine actionability · Federated scholarly infrastructures

1 Introduction

Scholarly articles are static unstructured text documents [23] and datasets are published in a plethora of formats [18] with heterogeneous metadata on diverse repositories. It is thus difficult to interlink the heterogeneous collection of scholarly knowledge artefacts. Yet, complex information needs rely on retrieval from multiple infrastructures. However, querying multiple scholarly infrastructures and integrating retrieved information manually is a laborious and time-consuming task [38,43]. This problem motivates the need for unified access to

H.-R. Ke et al. (Eds.): ICADL 2021, LNCS 13133, pp. 308–324, 2021.
https://doi.org/10.1007/978-3-030-91669-5_24

and integrated retrieval from numerous, heterogeneous scholarly infrastructures. In other words, we need approaches that support formulating complex information needs via a single endpoint and retrieving integrated answers in a distributed manner [38]. As an example, consider the following scenario: A researcher wants to discover all research outputs (articles) published under a particular grant that have impact (high number of citations) as well as significant results (in a statistical sense of $p < .001$). It is currently impossible to formulate such an information need as a single query.

Towards this goal, we leverage the Open Research Knowledge Graph[1] (ORKG) [20] and present a GraphQL-based endpoint[2] to access its data i.e., machine actionable descriptions of scholarly knowledge and knowledge comparisons. Furthermore, we build on this endpoint and propose a GraphQL-based federated endpoint[3] that allows unified access to data from other scholarly infrastructures. Thus, we propose that the machine actionable scholarly knowledge published in ORKG can be leveraged along with scholarly content of other scholarly infrastructures to answer complex user queries on bibliographic metadata, article content, or both. Specifically, we propose the following generic approaches:

1. *Federated access* to scholarly infrastructures to retrieve and integrate the fragmented scholarly content via a single endpoint. Here, we leverage the DataCite PID Graph[4] [13] and the GeoNames[5] REST API, and link them virtually with ORKG using shared metadata.
2. *Methods for enabling federated access*
   (a) *Persistently identifying and publishing* artefacts by leveraging DOI services. This enables cross-walking between artefact metadata and artefact content and the discovery of content. Here, we leverage DataCite[6] DOI-based identification of ORKG comparisons.
   (b) Linking content to existing, third-party semantic resources (e.g., taxonomies, thesauri, ontologies) to improve the interoperability and reusability of scholarly knowledge. Here, we leverage GeoNames to enrich the description of locations in ORKG.

We thus address the following research question: How can existing, heterogeneous scholarly infrastructures be federated to support complex (meta)data-driven analysis?

2 Related Work

Persistent Identifiers. Persistent identifiers (PID) are used to uniquely and persistently identify research articles, software, datasets and other digital artefacts,

[1] https://orkg.org.
[2] https://www.orkg.org/orkg/graphql.
[3] https://www.orkg.org/orkg/graphql-federated.
[4] https://api.datacite.org/graphql.
[5] https://www.geonames.org/.
[6] https://datacite.org.

as well as people and physical objects such as samples [29]. They were introduced to decouple the identifier from the location on the Web of a digital resource and is thus an approach to address the volatility of locations, which is an issue for the stable reference (e.g., citation) required in the scholarly record, including in print material.

Several organizations provide services to persistently identify research artefacts. Most prominently, *Crossref* and *DataCite* identify research articles and datasets, respectively, by means of DOI while ORCID [14] enables the persistent identification of researchers. There exist a number of emerging identification schemes including for organizations by the *Research Organization Registry* (ROR), samples by the *International Geo Sample Number*[7] (IGSN), and instruments [39]. Persistent identifiers thus enable the unambiguous and stable reference of research artefacts and contextual entities. A structured ORKG comparison of these persistent identifier systems for scholarly content can be found in Auer et al. [6].

Since persistent identifiers have associated metadata, information about artefacts exists independently of the identified artefact. This metadata layer is typically standardized and supports the findability and accessibility of artefacts as well as enables opportunities for metadata linking and sharing among scholarly infrastructures [25]. The importance of persistent identification of artefacts used in research is stressed by the FAIR data principles [41]. Persistent identification is applied and adapted to numerous entity types beyond articles and datasets and their implementation is considered essential for research infrastructures. Richards et al. [33] discussed the persistent identification of datasets; Stocker et al. [39] of instruments; Farjana et al. [12] of geometric and topological entities. Bellini et al. [7] presented an interoperability framework for PID systems. Their approach is to ontologically refine the metadata sets contained in the PID records, which allows to set general information for different PID systems.

Semantic Resources for Scholarly Knowledge. Ontologies are a solution for formal description of content with domain knowledge and for combining data from multiple sources [11] in data integration and classification [34,36]. For the scholarly domain, several ontologies have been proposed for numerous disciplines, including computer science [16,19,35], immunology [4], maritime research [37], construction management [42] and agronomy [21]. Moreover, to describe biomedical content semantically, different bio-ontologies such as the gene ontology [3], protein ontology [27], among others, have been proposed. A suite of ontologies known as Semantic Publishing and Referencing (SPAR) [30,31] have been proposed for the creation of comprehensive metadata for all aspects of semantic publishing and referencing (e.g., document description and bibliographic references).

Scholarly Communication. Several frameworks have been developed to improve scholarly communication. Martin et al. [24] suggested that semantic linking is

[7] https://www.igsn.org.

an important aspect to achieve interoperability among research infrastructures (RIs). They also discussed various ways to enhance the interoperability of RIs, including semantic contextualization, enrichment, mapping and bridging. Hajra et al. [15] presented a way to enhance scholarly communication by linking data from different repositories. They considered bibliographic Linked Open Data (LOD) repositories to compute the semantic similarity between two resources. With the Scholix framework [9], the Research Data Alliance (RDA) Publishing Data Services Working Group (PDS-WG) [8] developed an approach for data-literature interlinking. This framework enables interoperability of metadata about the links between articles and datasets created and exchanged among publishers and data repositories as well as scholarly infrastructures, specifically DataCite, OpenAIRE and Crossref. Another approach to interlink OpenAIRE research metadata and datasets and making metadata accessible for end users was proposed by Ameri et al. [1]. Assante et al. [5] introduced Science 2.0 repositories to enhance the scholarly communication workflow by overcoming the gap between publishing research articles and research lifecycle. The structured comparison of these scholarly infrastructures can be found in [17].

Federated Scholarly Infrastructures. Schwarte et al. [38] proposed a framework, named FedX, which enables efficient processing of SPARQL queries on heterogeneous data sources and also demonstrated the practicability and efficiency of the proposed framework on a set of real-world queries. Similarly, Mosharraf and Taghiyareh [26] proposed a SPARQL-based federated search engine to retrieve Open Educational Resources (OERs) published on the web of data. Arya et al. [2] proposed a personalized federated search framework to retrieve information such as user profiles, jobs, or professional groups from diverse sources.

Several frameworks are proposed to access scholarly data in a federated manner, but they do not consider the content of artefacts while processing user queries. To the best of our knowledge, this is the first attempt towards powering federated queries by incorporating machine-readable form of scholarly artefacts so that the results can be filtered not only at the metadata level but also at the content level.

3 Approach

In this section, we present the approach for federating scholarly infrastructures. We cover three key aspects. First, federated access to ORKG, DataCite and GeoNames infrastructures to retrieve and integrate the fragmented scholarly content to enable complex data-driven analysis. Second, linking ORKG content with third-party semantic resources to ensure ORKG content is interoperable and reusable. Third, DOI-based persistent identification of ORKG artefacts by using DataCite services, specifically state-of-the-art ORKG comparisons, to ensure artefact findability in global scholarly infrastructures as well as the linking of these ORKG artefacts with other artefacts, specifically articles.

3.1 Federated Access to Scholarly Infrastructures

This section describes the proposed approach to federate scholarly infrastructures, in particular ORKG, DataCite and GeoNames. The approach leverages GraphQL as the common interface. DataCite provides a GraphQL endpoint for the PID Graph, which connects persistently identified resources from DataCite, ORCID, ROR, etc., and serves standardized metadata for these resources. Similarly, we implemented a GraphQL endpoint to enable access to ORKG content. Additionally, we also integrated the GeoNames REST API[8] to enable fetching information regarding continents, countries and cities using the same interface. The GeoNames API allows access to geographical data such as fetching all countries of a particular continent by specifying its continent code. For example, retrieving the list of the countries which belong to Asian (AS) continent. Thus, our proposed federation enables crosswalks between metadata about artefacts (e.g., articles and datasets) with their context (e.g., people and organizations) and the content of articles. We thus propose a GraphQL-based federation that virtually integrates ORKG, the DataCite PID Graph, and GeoNames to retrieve integrated information from these scholarly infrastructures through a single search query. Arguably, these data sources can be easily extended with additional sources.

Figure 1 illustrates the proposed federated architecture. The gateway layer virtually integrates the distinct graphs of ORKG, DataCite, and GeoNames to create a unified GraphQL endpoint, enabling the execution of a single query across these infrastructures. As shown in the figure, a single query is posed on the federated gateway in a declarative manner while query parts seamlessly execute on the respective infrastructures. The figure also depicts the underlying methods powering federated access between scholarly infrastructures, i.e., DOI-based persistent identification of machine actionable artefacts and linking the content with third-party semantic resources.

Our approach is based on a virtual integration that exhibits federation as if all scholarly infrastructures are integrated on a single endpoint. In fact, the distributed (meta)data still resides on individual infrastructures, and only the parts relevant to the query are retrieved. Hence, the proposed approach meets the goal to make global scholarly infrastructures interoperable thus serving more complex information needs. As shown in Fig. 1, through our federated architecture, it is now possible to formulate requests with constraints on metadata about entities as well as on data, i.e. published scholarly knowledge.

A federated graph as proposed here plays an important role in enabling complex scholarly (meta)data-driven analysis. For instance, in addition to metadata analysis (e.g., citation networks) research information systems can leverage article-content (data) to power entirely new kinds of data analysis, e.g., citation networks that only include work with highly significant results, $p < .001$.

User Scenario. A researcher reads the COVID-19 article with DOI name `10.1101/2020.03.08.20030643` and discovers that information about the virus'

[8] https://www.geonames.org/export/ws-overview.html.

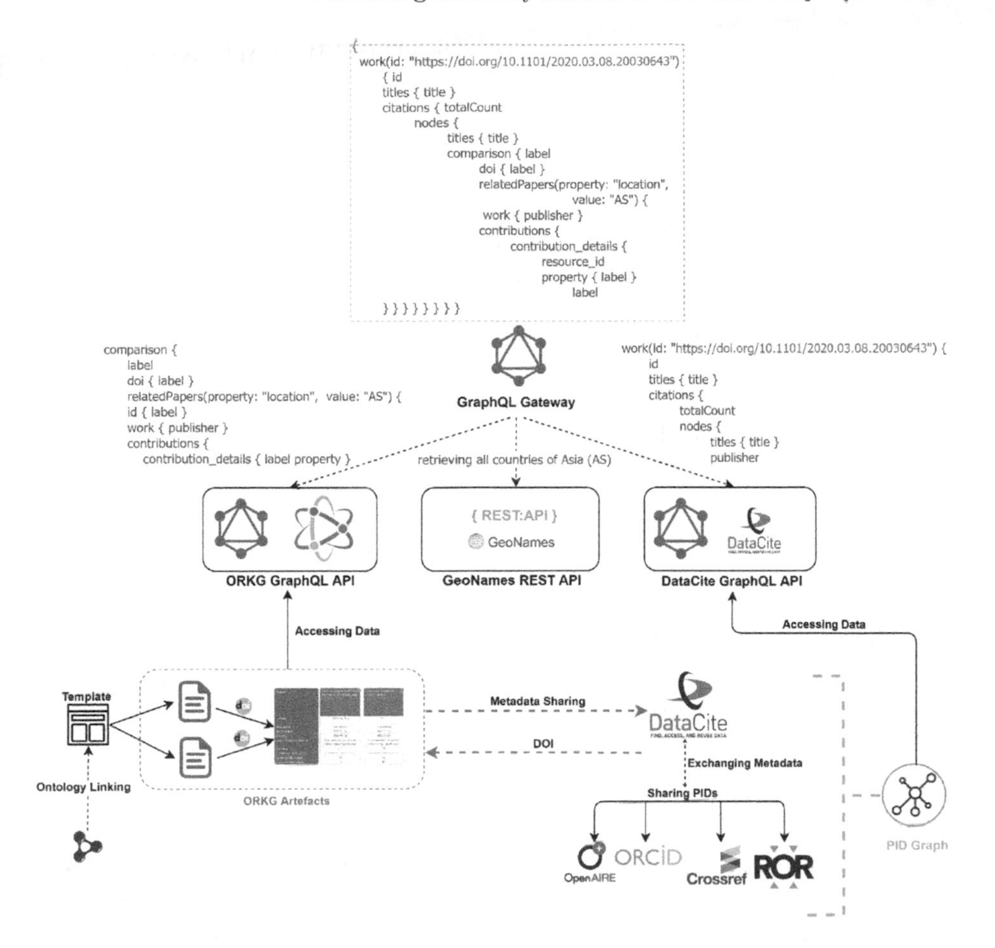

Fig. 1. Overview of the virtually integrated APIs of multiple scholarly infrastructures (ORKG, DataCite, and GeoNames). The figure also shows the execution of a federated query designed to retrieve from DataCite the number of citations of the paper with DOI name `10.1101/2020.03.08.20030643`, from ORKG the data of an ORKG comparison that cites the given paper, and filtering studies conducted on populations in Asia (AS). The example query thus leverages GeoNames and the proposed methods of DOI-based persistent identification of ORKG artefacts and linking content with ontologies to enable cross-walking between metadata and data served by distributed, heterogeneous scholarly infrastructures.

basic reproductive number (R0) published in the article was used in an ORKG comparison. The researcher is interested in calculating the average R0 of studies in a particular region in order to conduct regional analysis. The proposed federated endpoint enables answering such a complex query by executing the relevant parts of the query on the respective endpoints, thus enabling the (meta)data-based analysis required for the research at hand. Figure 1 includes the federated query that implements this user query. First, the paper is retrieved by DOI

(10.1101/2020.03.08.20030643) on the DataCite PID graph. As this paper is cited in an ORKG comparison, the PID Graph provides us the link between the paper DOI and the DOI of the comparison. Second, using the DOI of the comparison, the query retrieves the machine actionable comparison data from ORKG. Third, by leveraging GeoNames we filter for studies in Asia.

With the results obtained from such a complex query, we can perform (bibliographic) metadata-based analysis. For instance, we can compute the distribution of studies across all publishers (Fig. 2(a)). Such bibliographic metadata processing and analysis is a well-known and often performed activity.

The proposed approach also enables article-content (data) analysis. In our user scenario, we can for instance plot the R0 estimates reported in the literature compared in ORKG and compute the average (Fig. 2(b)). The computed average is 3.52. For interested readers, the (meta)data analysis discussed here is available online as a Jupyter notebook[9].

3.2 Methods for Linking Scholarly Knowledge

We present two methods for linking scholarly knowledge, underlying the proposed query federation among scholarly infrastructures.

Linking Semantic Resources. Structuring and semantically representing scholarly knowledge is generally non-trivial. The difficulty strongly depends on the granularity of the description as well as the existence of relevant and reusable schemes, among other reasons.

To facilitate structuring scholarly knowledge, we have proposed ORKG templates[10]. Their purpose is to ease creating comparable content in ORKG. For recurrent information types, e.g., time intervals or quantity values, their unit and confidence interval, templates specify the required properties and apply validation rules on value types to ensure quality and comparable data. This is similar to how SHACL [22] or ShEx [32] set constraints and validate RDF data. Once having been specified, templates can be used to create ORKG content, in particular to describe research contributions. Templates not only ease content creation but, even more importantly, also standardize the description of scholarly knowledge in the ORKG.

Of most relevance here, templates not only support structuring scholarly knowledge but also allow for linking the terms used (classes, properties and individuals) with third-party semantic resources, thereby ensuring that ORKG content created using templates is interoperable and reusable. Hence, to make structured content in ORKG semantic, i.e., machine actionable, we developed a mechanism that supports the efficient linking of ORKG terms with third-party semantic resources. These resources can be accessed and linked using two generic data exchange approaches: REST API and SPARQL. ORKG provides run time

[9] https://gitlab.com/TIBHannover/orkg/orkg-notebooks/-/blob/master/graphql/COVID-19_R0_estimate/COVID-19_R0_meta-data_analysis.ipynb.

[10] https://www.orkg.org/orkg/templates.

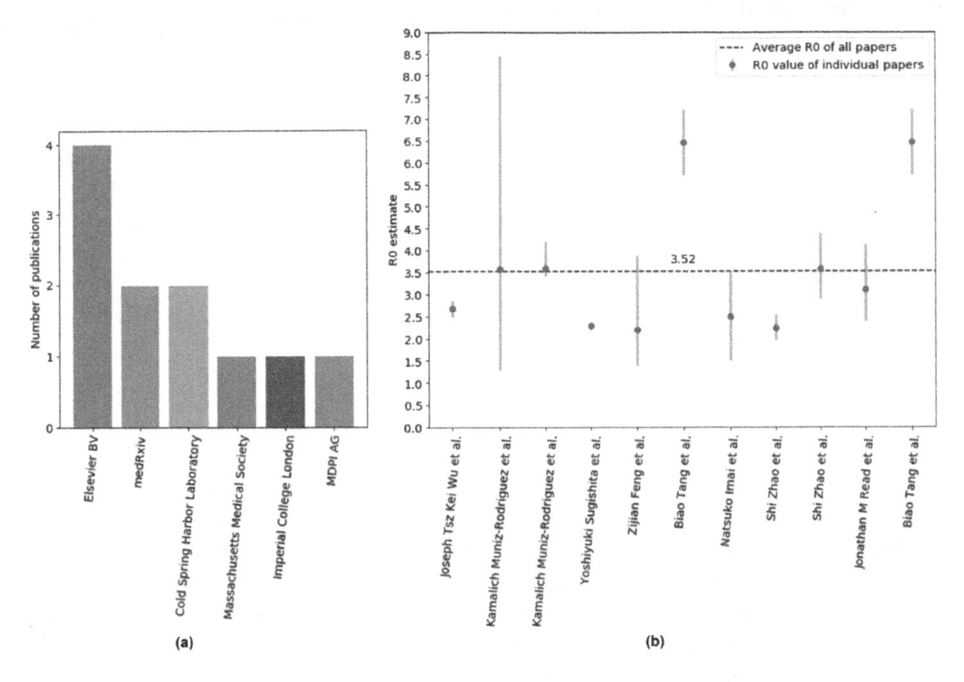

Fig. 2. Results of (meta)data-analysis with data retrieved using our complex user query. (a) Metadata-based analysis: Distribution of studies across publishers. (b) Data-driven analysis: Plot of R0 estimates reported in the literature compared in ORKG and their average.

support to find and access semantic resources while adding and curating content in ORKG.

While specifying a template, it is thus possible to link the template to a particular class, either user-defined or defined by an existing semantic resource. Whenever a particular template is used for describing research contributions in ORKG, the semantics specified by the template are applied to the created content. We present the advantages of linking semantic resources with the following examples.

Example 1. In this example, we leverage the EMBL-EBI Ontology Lookup Service [10] (OLS) and its REST API. While specifying templates, it is possible to lookup classes of semantic resources served by the OLS and use these to specify value ranges. For instance, if different studies mention the confidence interval of measurements in experiments, a respective template can refer to the Statistical Methods Ontology (STATO). Whenever a paper contribution description uses that template to specify properties and values, these will be automatically associated with the ontology. The creation of a confidence interval class and definition of constraints for template properties is shown in Fig. 3 and Fig. 4.

Example 2. In this example, we leverage the GeoNames gazetteer [40] and its REST API to enable the lookup of GeoNames resources when specifying

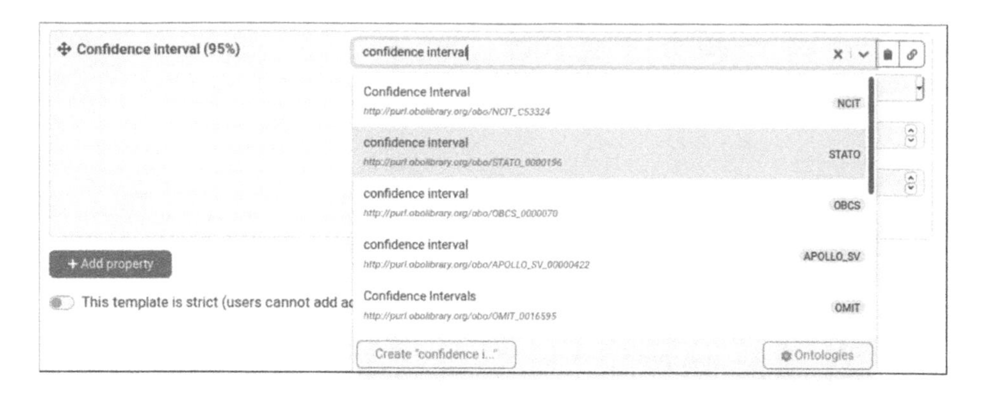

Fig. 3. Lookup and linking externally defined classes using the EBI-OLS semantic resource.

locations in ORKG content. We define a template using the Dublin Core (DC) Location ontology to represent locations. When using this template in ORKG research contribution descriptions, the class `DCLocation`[11] is automatically associated with locations entered as values. While typing location names, a list is fetched from the GeoNames Rest API and the user can select the country from the list, thus creating a dynamic location resource in ORKG which refers to the corresponding GeoNames resource using the `Same as` property (e.g., Fig. 5 for the country of Iran).

Persistent Identification and Linking of ORKG Comparisons. ORKG comparisons are machine actionable tabular overviews of essential information published in the literature w.r.t. a specific research problem. They provide a condensed overview of the state-of-the-art for the respective research problem. Comparisons can be persistently stored with added metadata, including title, description, research field and creators. Figure 6 shows the comparison of research contributions that estimate the COVID-19 basic reproductive number whereas Fig. 7 shows the form displayed to users to publish a comparison (including assignment of a DOI). A detailed description of ORKG comparisons can be found in Oelen et al. [28].

ORKG supports the DOI-based persistent identification of its comparisons to make these artefacts citeable and findable in scholarly infrastructures. A DOI is assigned to a comparison by leveraging DataCite services and publishing metadata following the DataCite metadata schema[12] through its REST API[13].

ORKG ensures that the published metadata associated with the comparison DOI includes links between the DOI of the comparison and DOIs of the compared literature. Similarly, other PIDs (e.g., contributor ORCID IDs and organization

[11] https://www.orkg.org/orkg/class/DCLocation.
[12] https://schema.datacite.org.
[13] https://support.datacite.org/docs.

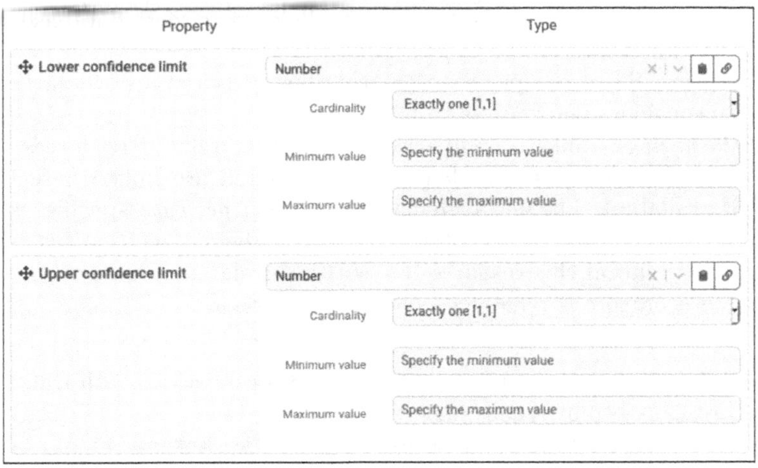

Fig. 4. Confidence interval template specification with constraints on property values.

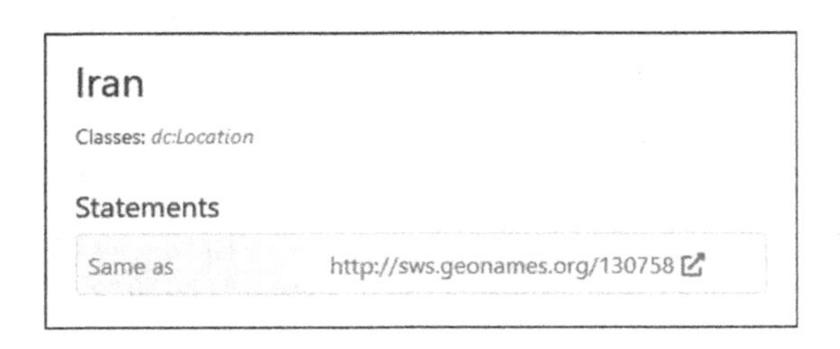

Fig. 5. ORKG location with automatically created `Same as` relation to the corresponding GeoNames resource.

IDs) are also included in the metadata. This metadata is shared with DataCite, which itself shares relevant elements with other infrastructures. By leveraging this existing mechanism, we can ensure that ORKG content and thus machine actionable scholarly knowledge is findable in global scholarly infrastructures. This sharing of metadata enables the cross-walking between ORKG and other scholarly infrastructures based on shared persistent identifiers. Listing 1.1 shows the metadata shared when registering a comparison DOI with DataCite and illustrates the linking between the DOI of the comparison and the DOIs of compared literature. The most important metadata elements are:

- *identifier*: Represents the DOI of the comparison.
- *creators*: Has two sub elements: *creatorName* and *nameIdentifier*, which represent the name and the ORCID of the creator respectively. If the ORCID is included, DataCite exchanges metadata about the link between the comparison DOI and the creator ORCID with ORCID, thus acknowledging the creator's contribution to ORKG in the contributor's ORCID record.

- *subject*: The research field the comparison belongs to, e.g., database systems or information systems.
- *resourceType*: Since comparisons are tabular overviews of scholarly knowledge, we consider comparisons to be datasets.
- *relatedIdentifiers*: Links a comparison with other related resources, in particular articles. Articles included in a comparison are linked here by their DOI (if available). The sub element *relatedIdentifierType* specifies the type of the related identifier (e.g., DOI, URL, etc.) whereas *relationType* specifies the kind of relation this resource has with the related resource. In our case, compared literature is referenced by the comparison.

Listing 1.1. Metadata used to publish an ORKG comparison with DataCite.

```
1 <?xml version="1.0" encoding="UTF-8"?>
2 <resource xmlns="http://datacite.org/schema/kernel-4"
3   xmlns:xsi="http://www.w3.org/2001/XMLSchema-instance"
4   xsi:schemaLocation="http://datacite.org/schema/kernel-4"
5   "http://schema.datacite.org/meta/kernel-4.3/metadata.xsd">
6
7 <identifier identifierType="DOI">10.48366/r44930</identifier>
8
9 <creators>
10   <creator>
11     <creatorName nameType="Personal">Haris, Muhammad
12     </creatorName>
13     <nameIdentifier schemeURI="http://orcid.org/"
14                     nameIdentifierScheme="ORCID">
15       0000-0002-5071-1658</nameIdentifier>
16   </creator>
17 </creators>
18
19 <titles>
20   <title xml:lang="en">COVID-19 Reproductive Number Estimates
21   </title>
22 </titles>
23 <publisher xml:lang="en">Open Research Knowledge Graph
24 </publisher>
25 <publicationYear>2020</publicationYear>
26
27 <subjects>
28   <subject xml:lang="en">Virology</subject>
29 </subjects>
30
31 <language>en</language>
32 <resourceType resourceTypeGeneral="Dataset">Comparison
33 </resourceType>
34
35 <relatedIdentifiers>
36   <relatedIdentifier relationType="References"
37         relatedIdentifierType="DOI">
38         10.1101/2020.03.08.20030643</relatedIdentifier>
39 </relatedIdentifiers>
40 <descriptions>
```

```
41    <description descriptionType="Abstract">
42    Comparison of published reproductive number estimates
43    for the COVID-19 infectious disease</description>
44 </descriptions>
45 </resource>
```

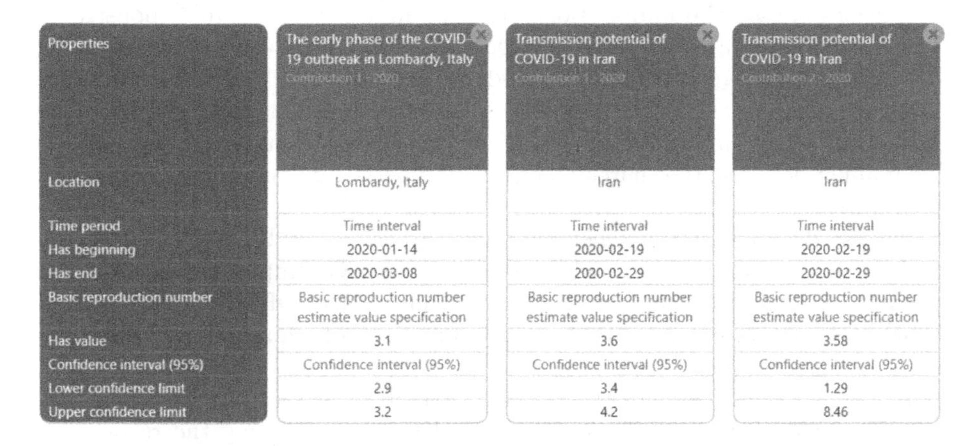

Properties	The early phase of the COVID-19 outbreak in Lombardy, Italy (Contribution 1 - 2020)	Transmission potential of COVID-19 in Iran (Contribution 1 - 2020)	Transmission potential of COVID-19 in Iran (Contribution 2 - 2020)
Location	Lombardy, Italy	Iran	Iran
Time period	Time interval	Time interval	Time interval
Has beginning	2020-01-14	2020-02-19	2020-02-19
Has end	2020-03-08	2020-02-29	2020-02-29
Basic reproduction number	Basic reproduction number estimate value specification	Basic reproduction number estimate value specification	Basic reproduction number estimate value specification
Has value	3.1	3.6	3.58
Confidence interval (95%)	Confidence interval (95%)	Confidence interval (95%)	Confidence interval (95%)
Lower confidence limit	2.9	3.4	1.29
Upper confidence limit	3.2	4.2	8.46

Fig. 6. ORKG comparison of studies estimating the COVID-19 basic reproductive number.

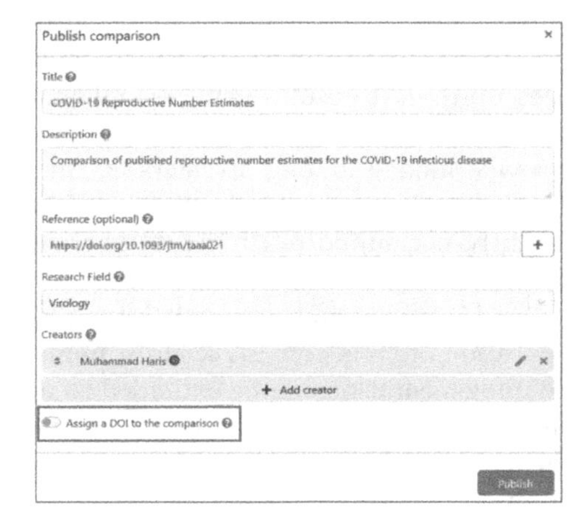

Fig. 7. ORKG form to persistently publish an ORKG comparison with metadata and DOI.

4 Discussion

We have presented a GraphQL-based federated data retrieval system architecture that supports cross-walks between (bibliographic) metadata and article's content (data) across multiple scholarly infrastructures. We suggested that such seamless cross-walking enables formulation of complex user queries. To enable cross-walks, we also presented two crucial means of linking scholarly knowledge, namely (i) DOI-based persistent identification of ORKG artefacts and (ii) linking terms to third-party semantic resources. Our work suggests that Persistent Identifiers (PIDs) are not only interesting to persistently identify scholarly documents but can also be leveraged to persistently identify machine actionable representations of essential information contained in documents. We demonstrated this by leveraging DataCite services to persistently identify ORKG comparisons, thus making ORKG content broadly findable in global scholarly infrastructures. Moreover, software agents can also discover ORKG content using the DataCite PID Graph, which further supports fetching machine-actionable data in a federated manner based on the shared metadata and performing complex (meta)data-driven analysis.

We also demonstrated a generic approach to efficiently link ORKG terms to third-party semantic resources while specifying ORKG templates used to create linked ORKG content. Such kind of linking makes the ORKG content unambiguous, interoperable and reusable, and supports both humans and machines in knowledge discovery.

To address our research question, we virtually integrate the DataCite PID Graph and GeoNames REST API with ORKG, thus ensuring the retrieval of required information in a federated manner. In addition to classical bibliographic metadata analysis, we emphasize that using artefact bibliographic metadata in combination with artefact machine actionable content greatly empowers the federation to enable complex (meta)data-driven analysis needed in research. The integrated access to multiple scholarly infrastructures realizes two main advantages. First, it provides up-to-date results and, second, it provides interesting insights into the connections in research (meta)data. While state-of-the-art scholarly infrastructures provide linking among artefacts at the metadata level, the linked access to artefact content proposed here enables novel (meta)data-driven analysis, which is currently not supported by state-of-the-art scholarly infrastructures.

While the proposed architecture can be extended, the presented implementation is limited to a federation of three infrastructures, namely DataCite, GeoNames and ORKG. The implementation can be extended to additional infrastructures, but doing so relies on additional software development. A further challenge is writing GraphQL queries, which can be difficult for untrained users.

To address these limitations, we aim to enhance the scope of federated queries by virtually integrating other (scholarly) infrastructures (OpenAIRE[14], Wikidata[15], Zenodo[16], etc.). We also plan to develop a user interface to allow users to pose their queries in the form of facets, thus automatically generating GraphQL queries to retrieve the required results in a federated manner.

[14] https://www.openaire.eu/.
[15] https://www.wikidata.org/wiki/Wikidata:Main_Page.
[16] https://zenodo.org/.

5 Conclusion

Federated search is an established practice in retrieving data from disparate sources. However, to the best of our knowledge, to date, no federated architecture allows scholarly metadata-to-data cross-walks, including structured scholarly knowledge in the form presented here for complex (meta)data-driven analysis of scholarly knowledge.

As the main contribution of this work, we presented a federated architecture that supports cross-walks from metadata to data of scholarly artefacts. Our implementation leverages DataCite, ORKG and GeoNames infrastructures.

We also presented two important approaches for linking scholarly knowledge to empower the federation among scholarly infrastructures, i.e., DOI-based persistent identification of ORKG artefacts and linking content with third-party semantic resources. We demonstrated this kind of linking with an implementation in ORKG.

Persistent identification and linking of machine actionable scholarly knowledge with related artefacts, agents, organizations, etc. is also an essential component towards linked scholarly knowledge. As global scholarly infrastructures already extensively share persistent identifier metadata, we have exploited such kind of linking to improve ORKG content findability.

As the amount of scholarly knowledge published by global scholarly infrastructures relentlessly increases, yet scholarly knowledge remains poorly machine processable, we argue that federated and retrieval with constraints on metadata and data should receive more attention in the community. We suggest that the proposed federated system can support researchers in conducting science by effectively enabling them in the formulation of complex information needs that are typical for modern science.

Acknowledgment. This work was co-funded by the European Research Council for the project ScienceGRAPH (Grant agreement ID: 819536) and TIB–Leibniz Information Centre for Science and Technology. The authors thank Mohamad Yaser Jaradeh for his valuable input and comments.

References

1. Ameri, S., Vahdati, S., Lange, C.: Exploiting interlinked research metadata, 3–14, September 2017. https://doi.org/10.1007/978-3-319-67008-9_1
2. Arya, D., Ha-Thuc, V., Sinha, S.: Personalized federated search at linkedin. In: Proceedings of the 24th ACM International on Conference on Information and Knowledge Management, CIKM 2015, New York, NY, USA, pp. 1699–1702. Association for Computing Machinery (2015). https://doi.org/10.1145/2806416.2806615
3. Ashburner, M., et al.: Gene ontology: tool for the unification of biology. The gene ontology consortium. Nat. Genet. **25**, 25–29 (2000)
4. Asiaee, A.H., Minning, T., Doshi, P., Tarleton, R.L.: A framework for ontology-based question answering with application to parasite immunology. J. Biomed. Semant. **6**(1), 31 (2015)

5. Assante, M., Candela, L., Castelli, D., Manghi, P., Pagano, P.: Science 2.0 repositories: time for a change in scholarly communication. D-Lib Mag. **21**, 1–14 (2015). https://doi.org/10.1045/january2015-assante
6. Auer, S., Stocker, M.: Comparison of scholarly identifier systems (2021). https://doi.org/10.48366/R73210. https://www.orkg.org/orkg/comparison/R73210
7. Bellini, E., et al.: Interoperability knowledge base for persistent identifiers interoperability framework. In: 2012 Eighth International Conference on Signal Image Technology and Internet Based Systems, pp. 868–875. IEEE (2012)
8. Burton, A., et al.: The data-literature interlinking service: towards a common infrastructure for sharing data-article links. Program **51**, 75–100 (2017). https://doi.org/10.1108/PROG-06-2016-0048
9. Burton, A., et al.: The Scholix framework for interoperability in data-literature information exchange. D-Lib Mag. **23**, January 2017. https://doi.org/10.1045/january2017-burton
10. Côté, R., Reisinger, F., Martens, L., Barsnes, H., Vizcaino, J., Hermjakob, H.: The ontology lookup service: bigger and better. Nucleic Acids Res. **38**(Suppl_2), W155–W160 (2010)
11. Ding, L., Kolari, P., Ding, Z., Avancha, S.: Using ontologies in the semantic web: a survey. In: Sharman, R., Kishore, R., Ramesh, R. (eds.) Ontologies. Integrated Series in Information Systems, vol. 14, pp. 79–113. Springer, Boston (2007). https://doi.org/10.1007/978-0-387-37022-4_4
12. Farjana, S.H., Han, S., Mun, D.: Implementation of persistent identification of topological entities based on macro-parametrics approach. J. Comput. Des. Eng. **3**(2), 161–177 (2016). https://doi.org/10.1016/j.jcde.2016.01.001
13. Fenner, M., Aryani, A.: Introducing the PID Graph (2019). https://doi.org/10.5438/JWVF-8A66. https://blog.datacite.org/introducing-the-pid-graph/
14. Haak, L., Fenner, M., Paglione, L., Pentz, E., Ratner, H.: ORCID: a system to uniquely identify researchers. Learn. Publ. **25**, 259–264 (2012). https://doi.org/10.1087/20120404
15. Hajra, A., Tochtermann, K.: Linking science: approaches for linking scientific publications across different LOD repositories. Int. J. Metadata Semant. Ontol. **12**(2–3), 124–141 (2017)
16. Happel, H.J., Seedorf, S.: Applications of ontologies in software engineering, January 2006
17. Haris, M.: Comparison of scholarly infrastructures (2021). https://doi.org/10.48366/R73195. https://www.orkg.org/orkg/comparison/R73195
18. Hendler, J.: Data integration for heterogenous datasets. Big Data **2**, 205–215 (2014). https://doi.org/10.1089/big.2014.0068
19. Iannacone, M., et al.: Developing an ontology for cyber security knowledge graphs, 1–4, April 2015. https://doi.org/10.1145/2746266.2746278
20. Jaradeh, M.Y., et al.: Open research knowledge graph: next generation infrastructure for semantic scholarly knowledge. In: Proceedings of the 10th International Conference on Knowledge Capture, K-CAP 2019, New York, NY, USA, pp. 243–246. Association for Computing Machinery (2019). https://doi.org/10.1145/3360901.3364435
21. Jonquet, C., Dzalé-Yeumo, E., Arnaud, E., Larmande, P.: Agroportal: a proposition for ontology-based services in the agronomic domain, June 2015
22. Knublauch, H., Kontokostas, D.: Shapes constraint language (SHACL). W3C Candidate Recomm. **11**(8) (2017)
23. Kuhn, T., et al.: Decentralized provenance-aware publishing with nanopublications. PeerJ Comput. Sci. **2**, e78 (2016)

24. Martin, P., Magagna, B., Liao, X., Zhao, Z.: Semantic linking of research infrastructure metadata. In: Zhao, Z., Hellström, M. (eds.) Towards Interoperable Research Infrastructures for Environmental and Earth Sciences. LNCS, vol. 12003, pp. 226–246. Springer, Cham (2020). https://doi.org/10.1007/978-3-030-52829-4_13
25. Meadows, A., Haak, L., Brown, J.: Persistent identifiers: the building blocks of the research information infrastructure. Insights UKSG J. **32**, March 2019. https://doi.org/10.1629/uksg.457
26. Mosharraf, M., Taghiyareh, F.: Federated search engine for open educational linked data. Bull. IEEE Tech. Comm. Learn. Technol. **18**(6), 6–10 (2016)
27. Natale, D., et al.: The protein ontology: a structured representation of protein forms and complexes. Nucleic Acids Res. **39**, D539–D545 (2010). https://doi.org/10.1093/nar/gkq907
28. Oelen, A., Jaradeh, M.Y., Stocker, M., Auer, S.: Generate fair literature surveys with scholarly knowledge graphs. In: Proceedings of the ACM/IEEE Joint Conference on Digital Libraries in 2020, JCDL 2020, New York, NY, USA, pp. 97–106. Association for Computing Machinery (2020). https://doi.org/10.1145/3383583.3398520
29. Paskin, N.: Digital object identifier (DOI) system. Encyclopedia of Library and Information Sciences, Technical report (2010)
30. Peroni, S., Shotton, D.: FaBiO and CiTO: ontologies for describing bibliographic resources and citations. J. Web Semant. **17**, 33–43 (2012). https://doi.org/10.1016/j.websem.2012.08.001
31. Peroni, S., Shotton, D., et al.: The SPAR ontologies. In: Vrandečić, D. (ed.) ISWC 2018. LNCS, vol. 11137, pp. 119–136. Springer, Cham (2018). https://doi.org/10.1007/978-3-030-00668-6_8
32. Prud'hommeaux, E., Labra Gayo, J.E., Solbrig, H.: Shape expressions: an RDF validation and transformation language. In: Proceedings of the 10th International Conference on Semantic Systems, pp. 32–40 (2014)
33. Richards, K., White, R., Nicolson, N., Pyle, R.: A beginner's guide to persistent identifiers. GBIF (2011)
34. Salatino, A., Thanapalasingam, T., Mannocci, A., Osborne, F., Motta, E.: Classifying research papers with the computer science ontology. In: International Semantic Web Conference (2018)
35. Salatino, A.A., Thanapalasingam, T., Mannocci, A., Osborne, F., Motta, E., et al.: The computer science ontology: a large-scale taxonomy of research areas. In: Vrandečić, D. (ed.) ISWC 2018. LNCS, vol. 11137, pp. 187–205. Springer, Cham (2018). https://doi.org/10.1007/978-3-030-00668-6_12
36. Sanchez-Pi, N., Martí, L., Bicharra Garcia, A.C.: Improving ontology-based text classification: an occupational health and security application. J. Appl. Logic **17**, 48–58 (2016). https://doi.org/10.1016/j.jal.2015.09.008. sOCO13
37. Santipantakis, G., Kotis, K., Vouros, G.: Ontology-based data integration for event recognition in the maritime domain, July 2015. https://doi.org/10.1145/2797115.2797133
38. Schwarte, A., Haase, P., Hose, K., Schenkel, R., Schmidt, M., et al.: FedX: optimization techniques for federated query processing on linked data. In: Aroyo, L. (ed.) ISWC 2011. LNCS, vol. 7031, pp. 601–616. Springer, Heidelberg (2011). https://doi.org/10.1007/978-3-642-25073-6_38
39. Stocker, M., et al.: Persistent identification of instruments. Data Sci. J. **19**, 1–12 (2020). https://doi.org/10.5334/dsj-2020-018
40. Vatant, B., Wick, M.: Geonames ontology. Dostupné, January 2012. http://www.geonames.org/ontology/ontology_v3

41. Wilkinson, M.D., et al.: The fair guiding principles for scientific data management and stewardship. Sci. Data **3**(1), 1–9 (2016)
42. Zhang, S., Boukamp, F., Teizer, J.: Ontology-based semantic modeling of construction safety knowledge: towards automated safety planning for job hazard analysis (JHA). Autom. Constr. **52**, 29–41 (2015). https://doi.org/10.1016/j.autcon.2015.02.005
43. Zhou, Y., De, S., Moessner, K.: Implementation of federated query processing on linked data. In: 2013 IEEE 24th Annual International Symposium on Personal, Indoor, and Mobile Radio Communications (PIMRC), pp. 3553–3557 (2013). https://doi.org/10.1109/PIMRC.2013.6666765

Simple DL: A Toolkit to Create Simple Digital Libraries

Hussein Suleman(✉)

University of Cape Town, Cape Town, South Africa
hussein@cs.uct.ac.za
http://dl.cs.uct.ac.za/, https://github.com/slumou/simpledl

Abstract. Digital library systems are not always successfully implemented and sustainable in low resource environments, such as in poor countries and in organisations without resources. As a result, some archives with important collections are short-lived while others never materialise. This paper presents a new toolkit for the creation of simple digital libraries, based on a long trajectory of research into architectural styles. It is hoped that this system and approach will lower the barrier for the creation of digital libraries and provide an alternative architecture for experiments and the exploration of new design ideas.

Keywords: Simple · Low-resource · Digital library architecture

1 Introduction

The earliest examples of digital libraries (e.g., Project Gutenberg [8], arXiv.org) were based on custom software systems developed to meet what was at the time a very specific goal. Over time, however, it was increasingly recognised that the needs of a specific project could be generalised to a larger community. An example of this was when the open-source EPrints software was created to support increasing interest in self-archiving in the research community [7]. This shift from custom solutions to general toolkits resulted in a proliferation of digital libraries around the world to meet the needs of various communities. Variations of the same underlying architecture have been used to create repositories based on tools such as DSpace [1], AtoM [2] and Omeka [9]. This does not, however, meet the needs of all archivists.

In 2006, the Digital Bleek and Lloyd Collection was created, based on a custom-developed software system [13]. Given that the project was based in a country with relatively poor Internet connectivity, the collection was packaged onto a DVD-ROM so that it could be distributed as part of a related book and in keeping with the LOCKSS [12] principle that many copies keeps information safe. The system was designed on the basis of an atypical set of principles: that the network could not be assumed; that mediation via a software system should be avoided; and that the pre-processing of data to create static representations was always preferable.

H.-R. Ke et al. (Eds.): ICADL 2021, LNCS 13133, pp. 325–333, 2021.
https://doi.org/10.1007/978-3-030-91669-5_25

It can be hypothesized that pre-processed collections in digital libraries that are largely offline are better suited to low-resource environments. They require less ongoing technical maintenance as there are fewer things to break. They also require fewer computational resources for access as there is no software middleware layer. Finally, it can be argued that they are more rescuable, as the digital objects are already in a familiar hierarchical file organisation and do not require APIs to extract data.

This paper presents a software toolkit designed according to these principles - the Simple DL toolkit. The toolkit is designed to be as simple as possible, to enable long-term access to digital libraries even when there is no active preservation and when there is computer system or network failure. It is designed for disaster or, if there is no disaster, to enable easy migration to the next generation of solutions.

Some would argue that digital library software systems are a solved problem. The lack of sustainable digital libraries in poor countries, and failures with current systems, suggest that there is still scope for experimental systems that test alternative design ideas. Simple DL is exactly that - an experimental system with a radically different design meant for interrogation by researchers and practitioners.

2 Related Work

Digital library architecture has evolved to encompass both the granular level of individual systems, as well as collections and systems and how these systems are interconnected and made interoperable at national and international levels [14].

DSpace [18] and EPrints [6] remain among the most popular toolkits for creating repositories. Both are Web-based systems, with Web applications and databases as back-end services. In contrast, Greenstone [19] was designed to function both online and offline. Greenstone collections could be distributed on CDROM, but required installation of software in order to access collections.

Some attempts have been made to avoid software installation altogether. OpenDlib [4] and OpenDL [5] were early efforts to define digital library systems as collections of components, thus reducing the problem to component assembly rather than monolithic software installation. Diligent [3] was a generalisation of the component model of digital library systems to arbitrary instantiations on a high performance grid system. In contrast, Lumpa [10] demonstrated that entire instances of DSpace could be managed within a private cloud on-demand. These grid and cloud solutions attempt to make it easier for end-users by the use of sophisticated high performance computing frameworks.

As an alternative, the Digital Bleek and Lloyd [13] was designed to be easy for end users to use by changing the fundamental architecture and removing the need for network access and computation at the time of access. This simplification was still a custom solution, though variations of the idea showed promise for institutional repositories [11] and systems with non-static collections [17].

While many aspects of offline collections have been investigated, no previous attempts have been made to create a reusable toolkit for simple offline digital libraries to support experimentation and explore different models and principles for digital libraries [16].

3 System Design

This paper reports on the initial design of Simple DL, a toolkit for creating simple pre-generated digital libraries.

3.1 Features

The major features of Simple DL (in its 2021 release) are as follows:

- Metadata is stored in spreadsheets and in XML files. The system can support any metadata format but the default configuration is based on either Dublin Core or ICA-AtoM [2].
- There is no database management system and no database. All unstructured data is stored as flat files, and all structured data is stored as XML.
- There is minimal use of Web applications.
- Sites can be generated and then served locally or via a Web server or shared drive. This allows access from a mobile device (phone or tablet) with no Internet connectivity.
- The site's appearance can be customised using standard XSLT and CSS.
- User profiles are stored for users who make contributions online. Entities are also extracted.
- Submitted items, comments and new user registrations can be moderated.
- Search and browse is implemented as a faceted search that is in-browser.

3.2 Storage Layout

Given that Simple DL relies on file-based stores, the storage of data and software is a key part of the design. This default arrangement can also be customised, as the configuration can specify different locations for the different components. The default locations, and their purposes, are as follows:

- **simpledl** is the core software toolkit. This is meant to be stable across systems and configurations. It contains bin and template. **bin** is the location for the applications/scripts that are core to Simple DL, while **template** contains a template for a new collection's website.
- **public_html** is the self-contained offline website that can be served to users through a Web server or opened directly in a browser. Its contents look like a typical website and the figure does not indicate standard directories for styles, thumbnails, etc. Some directories are, however, specific to Simple DL. **metadata** contains the metadata, in both XML and HTML format, for all items in the collection. **collection** contains all the digital objects. **indices** contains the XML indices for the faceted search. Finally, **cgi-bin** is the default space for Web applications.

– **data** stores all data needed to configure the system and generate the website. **config** stores the core configuration information as well as the XSLT templates to transform XML to HTML pages. **website** is a supplementary template for a collection's website, in order to override and add onto the default template. **comments** and **uploads** store all contributions from users, whose profiles are stored in **users**. Finally, **spreadsheets** stores all spreadsheets containing metadata for the collection. In principle, this **data** directory completely defines the configuration and all data for the system such that its website can be regenerated from scratch, if given only the digital objects in the **collection** directory.
– **db** is a temporary store for working and cached versions of files. **entities**, **comments** and **fulltext** are all caches to speed up the processing. **counter** keep track of identifiers. **moderation** is a set of directories where submissions are stored temporarily before/after being moderated; when a comment or item is accepted, it is added to the relevant **data** directory.

3.3 Import, Index and Generate

The main operations of Simple DL centre around ingesting a metadata collection and creating a website representation of the collection. This is done using 3 steps that corresponds to applications/scripts in the system. This process is similar to that used in Greenstone [19] but is different in that the target is: in the first instance, a file-based store; and in the second instance, a static website.

Step 1 - Import. Metadata is read in from source spreadsheets and source XML files and individual target XML files are created for metadata. A hierarchy of directories is constructed to correspond to the original structure of the source directories, and according to nesting rules defined in the AtoM metadata entries. Entities are extracted if necessary and used to generate user profiles automatically, also as XML files.

Step 2 - Index. All XML metadata and user files are indexed for an information retrieval engine that supports faceted search. Fulltext is extracted from PDF files as needed (and cached).

Step 3 - Generate. All XML files (metadata, users, website pages, etc.) are converted to HTML by applying the XSLT stylesheet. In addition, the template website is copied over and thumbnails are created for granular objects and subcollections.

Each time a new metadata sub-collection is added to the system, these 3 steps need to be invoked. The applications have parameters to control which processing occurs, for greater efficiency. They will also do automatic dependency management, so if a spreadsheet has not changed it will not be imported again and if an HTML file is up-to-date, it will not be regenerated from its XML source.

3.4 Web Management Interface

While all the core applications are meant to be used off-line in the first instance, some archivists may prefer not to use a command-line to interact with the system. As such, there is a rudimentary Web interface that serves mostly as a front-end to the command-line applications.

An administrator is able to log into the system using Google credentials for authentication and authorisation handled by a specification of authorised administrators in the configuration.

Administrators are also able to manage the files through a Web interface and authorise new items, comments and requests for user accounts. Figure 1 shows the default administrator interface.

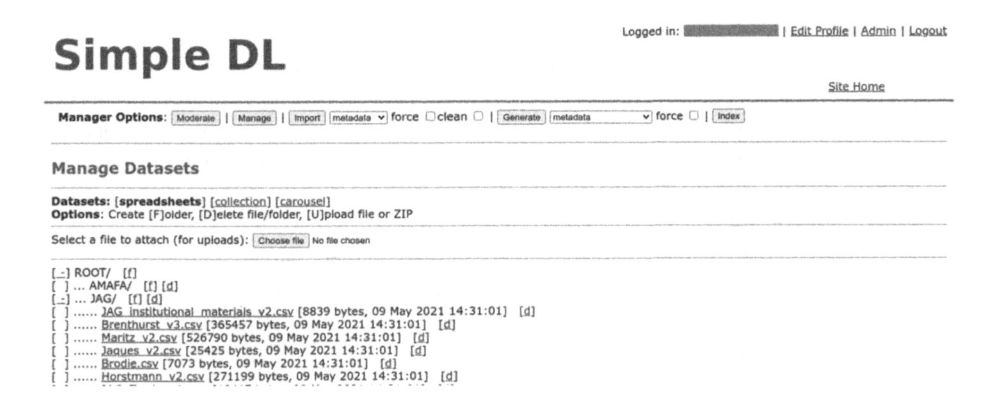

Fig. 1. Simple DL web administrator interface

3.5 Users and Entities

The system has 2 types of user profiles.

Automatically-generated profiles are extracted from the metadata for named entities in defined fields (creator in Dublin Core, eventActor in ICA-AtoM). These are then linked to all items where the entity has been mentioned.

Users can also request permission to make contributions to the digital library (if it is online). Once this is approved by an administrator, a user profile is created and handled similarly to the automatically-generated profiles. The key difference is that a contributor can log into the system while this is not possible for an extracted entity. One open philosophical question is how to link these, or if linking of these should even be allowed.

3.6 Comments and Submissions

Contributors are able to add comments to any metadata item (which are posted after approval by an administrator). As part of a comment, it is possible to attach a new digital object. If approved, this then becomes a part of the collection in its own right, and it can be commented on as well, etc.

New digital objects can also be uploaded as standalone items, without a link to an existing item. These too, must first be approved by an administrator.

The configuration of the system defines: where new items will be placed; and what metadata will be required of new submissions and comments.

3.7 Information Retrieval System

The Information Retrieval subsystem comprises 3 key components: an offline indexer; a Javascript query engine; and a Javascript user interface. Together these implement a classical tf.idf search system, with faceted searching and multiple indices for different data subsets.

The performance of this faceted search system was tested extensively in prior work [15], where it was demonstrated that sub-seconds responses were possible in typical cases for up to 100000 items, which is sufficient for many smaller archives. Given that the goal of this project is not scalability but support for smaller low-resource archives, this was deemed more than sufficient.

4 Case Studies

Three case studies are presented in the next section to illustrate how Simple DL is being applied to different scenarios, where all the projects have a common need for solutions that do not require large amounts of resources for sustainability.

4.1 Emandulo

Emandulo is a project from the Archives and Public Culture (APC) Initiative to gather digital material from related to pre-colonial Southern African history and organise and assemble this as a tool for researchers. Thus, there are subcollections from different institutions, each with its own system of hierarchical organisation.

APC staff assembled all the metadata in spreadsheets, and painstakingly edited this to recontextualise items within the various collections. Given the focus on archives, and archival culture, metadata was considered most important and was therefore highlighted throughout the site.

User contributions are expected and there is a strong requirement for entities and entity management through authority files.

Many additional features were desired within the website. A professional design team was hired to design the look and feel. A carousel was used on the front page and on item pages to show multiple items with different views. Some tables of items on the website (such as contributions made by one contributor) are sortable by columns - this was implemented in Javascript.

Simple DL was effectively used as a replacement for AtoM, which was the previous system used by the project. Figure 2 shows a typical item listing with metadata and composite item thumbnails.

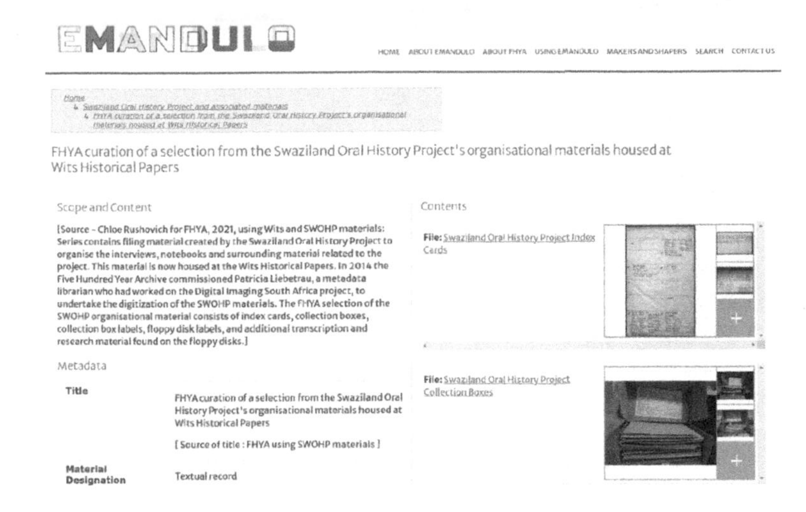

Fig. 2. Emandulo item listing

4.2 Digital Bleek and Lloyd

The Digital Bleek and Lloyd is a project of the Centre for Curating the Archive, to digitise and make available the Bleek and Lloyd Collection of books, drawings, and other historical documents on the language, culture and history of the |Xam and !kun speakers and other early South Africans.

The emphasis on this project was to a larger degree on the visual rather than the metadata, as the visual elements (such as pages of books and annotated drawings) contained most of the information desired by researchers. Also, the older form of text in the books is arguably not representable in Unicode and not understood by any living person so the original form is needed for ongoing study.

Figure 3 shows the faceted search interface used to drill down to images on a particular topic created by a particular contributor.

4.3 NDLTD Document Archive

The Networked Digital Library of Theses and Dissertations (NDLTD) hosts a series of annual symposia around the world. The papers and presentations serve to document the evolution of the community over a long period of time.

Simple DL was therefore configured to serve the content in a minimal manner, giving access to metadata and digital objects for symposia as sub-collections. The metadata was originally in spreadsheets (for ingest into other systems) so could easily be re-purposed for Simple DL. The advantage is that this can be copied and archived offline and will never fail as long as the current variation of HTML is supported.

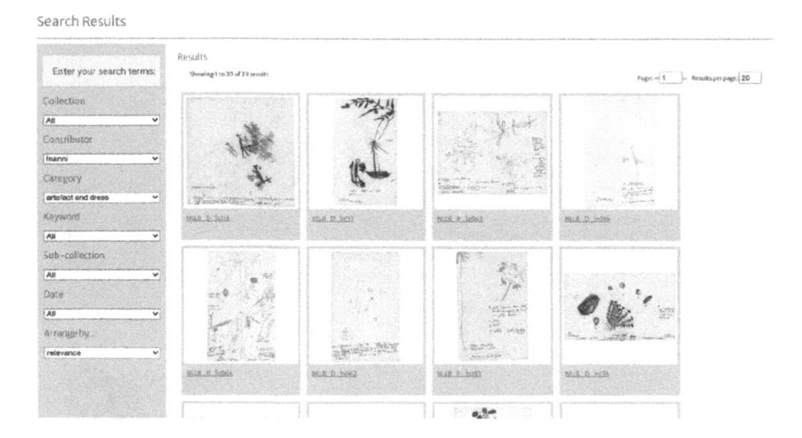

Fig. 3. Bleek and Lloyd faceted search

5 Reflections

Not all archivists need the functionality of popular digital library toolkits. Arguably some archivists need sustainability and the ability to recover from disaster as a paramount requirement. Archivists in low resources environments need systems that will work without much computational power and without much maintenance.

Simple DL has been proposed in this paper as an exemplar of an alternative model of digital library system to meet these objectives. One size does not fit all. This is a solution for those who do not need to store millions of digital objects, but who need a system that is as simple as possible. This is a solution for when the network fails or the operating system upgrades fail, that allows entire digital libraries to be copied as easily as individual items.

Ongoing developments with the toolkit include: making it easier to install and test; improving the performance and stability of various scripts; authoring of complex objects; and archiving of entire digital libraries at a higher level. Ultimately, by keeping the core technology simple, it may even be possible to do a lot more.

Acknowledgements. This research was partially funded by the National Research Foundation of South Africa (Grant numbers: 105862, 119121 and 129253) and University of Cape Town. The authors acknowledge that opinions, findings and conclusions or recommendations expressed in this publication are that of the authors, and that the NRF accepts no liability whatsoever in this regard.

References

1. Bollini, A., Cortese, C., Groppo, E., Mornati, S.: Extending DSpace to fulfil the requirements of digital libraries for cultural heritage management. In: IRCDL 2017 Conference. IRCDL, Modena (2017)

2. Bushey, J.: International council on archives (ICA) "access to memory" (AtoM): open-source software for archival description. Archivi Comput. **1** (2012)
3. Candela, L., Castelli, D., Pagano, P., Simi, M.: The evolution of digital library systems: from OpenDLib to diligent. In: Post-proceedings of the First Italian Research Conference on Digital Library Management Systems (IRCDL 2005), p. 23 (2005)
4. Castelli, D., Pagano, P.: OpenDLib: a digital library service system. In: Agosti, M., Thanos, C. (eds.) ECDL 2002. LNCS, vol. 2458, pp. 292–308. Springer, Heidelberg (2002). https://doi.org/10.1007/3-540-45747-X_22
5. Fox, E.A., Suleman, H., Luo, M., et al.: Building digital libraries made easy: toward open digital libraries. In: Lim, E.P. (ed.) ICADL 2002. LNCS, vol. 2555, pp. 14–24. Springer, Heidelberg (2002). https://doi.org/10.1007/3-540-36227-4_2
6. Gutteridge, C.: GNU EPrints 2 overview (2002)
7. Harnad, S.: The self-archiving initiative. Nature **410**(6832), 1024–1025 (2001)
8. Hart, M.: The history and philosophy of project Gutenberg. Project Gutenberg **3**, 1–11 (1992)
9. Kucsma, J., Reiss, K., Sidman, A.: Using Omeka to build digital collections: the metro case study. D-Lib Mag. **16**(3/4), 1–11 (2010)
10. Lumpa, M., Suleman, H.: Investigating the feasibility of digital repositories in private clouds. In: Jatowt, A., Maeda, A., Syn, S.Y. (eds.) ICADL 2019. LNCS, vol. 11853, pp. 151–164. Springer, Cham (2019). https://doi.org/10.1007/978-3-030-34058-2_15
11. Phiri, L., Williams, K., Robinson, M., Hammar, S., Suleman, H.: Bonolo: a general digital library system for file-based collections. In: Chen, H.-H., Chowdhury, G. (eds.) ICADL 2012. LNCS, vol. 7634, pp. 49–58. Springer, Heidelberg (2012). https://doi.org/10.1007/978-3-642-34752-8_6
12. Reich, V., Rosenthal, D.S.: LOCKSS: a permanent web publishing and access system. D-Lib Mag. **7**(6), 1082–9873 (2001)
13. Suleman, H.: Digital libraries without databases: the Bleek and Lloyd collection. In: Kovács, L., Fuhr, N., Meghini, C. (eds.) ECDL 2007. LNCS, vol. 4675, pp. 392–403. Springer, Heidelberg (2007). https://doi.org/10.1007/978-3-540-74851-9_33
14. Suleman, H.: Design and architecture of digital libraries. In: Digital Libraries and Information Access: Research Perspectives. Facet Publishing (2012)
15. Suleman, H.: Investigating the effectiveness of client-side search/browse without a network connection. In: Jatowt, A., Maeda, A., Syn, S.Y. (eds.) ICADL 2019. LNCS, vol. 11853, pp. 227–238. Springer, Cham (2019). https://doi.org/10.1007/978-3-030-34058-2_21
16. Suleman, H.: Reflections on design principles for a digital repository in a low resource environment. In: Proceedings of HistoInformatics Workshop 2019. CEUR (2019). http://pubs.cs.uct.ac.za/id/eprint/1331/
17. Suleman, H., Bowes, M., Hirst, M., Subrun, S.: Hybrid online-offline digital collections. In: Proceedings of the 2010 Annual Research Conference of the South African Institute of Computer Scientists and Information Technologists, SAICSIT 2010, New York, NY, USA, pp. 421–425. ACM (2010). https://doi.org/10.1145/1899503.1899558. http://doi.acm.org/10.1145/1899503.1899558
18. Tansley, R., et al.: The DSpace institutional digital repository system: current functionality. In: Proceedings of 2003 Joint Conference on Digital Libraries, pp. 87–97. IEEE (2003)
19. Witten, I.H., Boddie, S.J., Bainbridge, D., McNab, R.J.: Greenstone: a comprehensive open-source digital library software system. In: Proceedings of the Fifth ACM Conference on Digital Libraries, pp. 113–121 (2000)

Towards Approximating Population-Level Mental Health in Thailand Using Large-Scale Social Media Data

Krittin Chatrinan[1], Anon Kangpanich[1], Tanawin Wichit[1], Thanapon Noraset[1], Suppawong Tuarob[1](✉), and Tanisa Tawichsri[2]

[1] Faculty of Information and Communication Technology, Mahidol University, Nakhon Pathom, Thailand
{krittin.cha,anon.kan,tanawin.wic}@student.mahidol.edu, {thanapon.nor,suppawong.tua}@mahidol.edu

[2] Puey Ungphakorn Institute for Economic Research, Bangkok, Thailand
TanisaT@bot.or.th

Abstract. Mental health is one of the pressing issues during the COVID-19 pandemic. Psychological distress can be caused directly by the pandemic itself, such as fear of contracting the disease, or by stress from losing jobs due to the disruption of economic activities. In addition, many government measures such as lockdown, unemployment aids, subsidies, or vaccination policy also affect population mood, sentiments, and mental health. This paper utilizes deep-learning-based techniques to extract sentiment, mood, and psychological signals from social media messages and use such aggregate signals to trace population-level mental health. To validate the accuracy of our proposed methods, we cross-check our results with the actual mental illness cases reported by Thailand's Department of Mental Health and found a high correlation between the predicted mental health signals and the actual mental illness cases. Finally, we discuss potential applications that could be implemented using our proposed methods as building blocks.

Keywords: Mental health · Natural language processing · Deep learning · Social networks

1 Introduction

The interruption of the worldwide supply chain caused by the spread of COVID-19 subsequently led to several cities' lockdowns in Thailand since late March 2020. Many businesses were forced to shut down, while some advocated for working from home temporarily. Although the efforts were effective in controlling the epidemics, they negatively impact civilians on a country-wide scale. For example, the bankruptcy of several businesses varied in size increased unemployment and poverty, or mental health problems that originated from the previous examples such as stress, anxiety, or depression [18]. In extreme cases, such mental

H.-R. Ke et al. (Eds.): ICADL 2021, LNCS 13133, pp. 334–343, 2021.
https://doi.org/10.1007/978-3-030-91669-5_26

health problems subsequently drove a widespread of suicidal attempts in the Thai population [4]. Therefore, the ability to automatically monitor population-level mental health so that interventions could be implemented in a timely manner during such a crisis should prove crucial [9].

This research proposes to use big organic data, such as tweets and Facebook posts, to create a mental health and sentiments index which can be compiled quickly and will offer more timely and cost-effectively indicators than surveyed or administrative data. Social media has been used to monitor and trace sentiment anomalies and mental illness in the psychological health domain both at the collective and individual levels. Most aggregate-level monitoring techniques aim to monitor public moods or sentiments to analyze the impacts or predict outcomes of target events in a macro fashion. Tariq et al. proposed a co-training-based technique to discover social messages that indicate at-risk mental illnesses such as anxiety, depression, bipolar, and ADHD [16].

This paper also investigates the relationship between the official statistical data of mental illness cases and the statistical data inferred from online social media. We applied deep learning models to classify individual messages relevant to the mental health issue to infer the overall statistics. Instead of creating a new dataset for mental illness, we experiment with different aspects of mental health, including depression, emotion, sentiment, and suicidal tendency. Our models were trained using available datasets in the English language. To enable the trained models to work with the Thai language, the Thai social media messages are first translated into a corresponding English messages to obtain their predictions. The statistics of the aggregation of individual predictions have a high correlation with the ground-truth mental illness cases. The source code is made available for research purposes at: https://github.com/krittintey/psimilan_exp.

2 Methodology

We aimed to automatically extract the statistical trends of mental health issues from social media messages. To achieve this, we formulate the problem into a two-step approach. First, we predict individual messages for their relevance to mental stress. Then, we aggregate the numbers of relevant messages to estimate the population of mental illness cases. The overview of our approach is illustrated in Fig. 1.

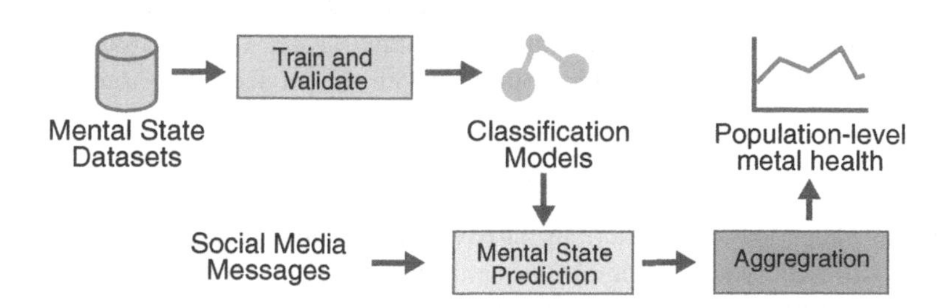

Fig. 1. The overall approach that we use to estimate the trends of mental illness cases.

This section explains the detail of our approach. We first describe the mental state classification and then the aggregation methods.

2.1 Mental State Classification

Since there is a vast amount of social media messages that are not related to mental health, our first step is to classify whether a social media message is related to mental states, especially the negative states. This paper experiments with several text classification models ranging from classical machine learning techniques to pre-trained deep learning models. Instead of collecting and annotating relevant messages, we focus on using relevant existing datasets to create a collection of classifiers. For each dataset, we experiment with the same feature extraction and classification.

Feature Extraction. The representation of a message in this work depends largely on the type of model. Primarily, for the classical machine learning methods, we use a bag-of-word representation with the standard TF-IDF weighting [26]. For the LSTM-based models, we initialize the word representation with the FastText embedding [1]. Lastly, for BERT, we use a pre-trained model, "BERT-Base, Multilingual Cased", from the official release [8].

Classification Models. There are two main types of models for classifying messages based on properties related to mental states: classical machine learning models and deep learning-based models. The classical machine learning models will be used with feature extraction to create baseline models for this experiment. The classification models include Random Forest [3] and Support Vector Machine (SVM) [12]. For deep learning models, Bi-LSTM [15] will be used as another baseline model. CNN-LSTM, a combination of CNN layers and LSTM layers, is another architecture for the experiment [25]. Finally, a fine-tuned BERT is another model that is used [8]. LSTM, Bi-LSTM, CNN-LSTM, and BERT will be connected to classifier layers for the classification approach.

Bi-LSTM is a model that will not be combined with other layers except for a dense layer. As for CNN-LSTM, most of its architecture consists of convolutional layers, pooling layers, dropout layers, LSTM, and dense layers. As for the dense layer, the activation functions used are sigmoid and softmax, which will output accordingly into the predefined classes of mental states. Lastly, for the fine-tuned BERT model, its classifier consists of dense layers for its input and output layers. In this circumstance, the ReLU activation function and Adam Optimizer with weight decay are used.

2.2 Aggregation

To aggregate the individual predictions into overall statistics, we have selected messages relevant to different types of mental states and grouped them on a monthly basis based on the date that the messages were posted. Then, we have taken the number of those monthly messages and divided it by the total number of messages in those months for normalization.

3 Datasets

There are three groups of datasets: mental state datasets for building classifiers, large-scale social media data to infer the statistical trends, and ground-truth-reported mental illness cases.

3.1 Datasets for Training Classifiers

In this study, we experiment with four mental state classifiers. The classifiers were trained and validated with four existing datasets, each of which focuses on different aspects of human mental states, namely depression, emotion, sentiment, and suicidal tendency. The sources and statistics of these datasets are shown in Table 1.

Table 1. Datasets used for creating classifiers for mental illness related messages.

Classes	# of Messages
Depression (D) [14]	
depressed	16,000 (51.61%)
non-depressed	15,000 (48.39%)
Emotion (E) [17]	
joy	5,362 (31.51%)
sadness	4,666 (29.16%)
anger	2,159 (13.49%)
fear	1,937 (12.11%)
love	1,304 (8.15%)
surprise	572 (3.58%)
Sentiment (S) [10]	
positive	800,000 (50%)
negative	800,000 (50%)
Suicide Tendency (ST) [23]	
suicidal	175 (57.76%)
normal	128 (42.24%)

3.2 Social Media Data

Social media data was collected from Twitter (via Twitter API) and Facebook (manually scraped from various Facebook news pages), where we treat each tweet and a Facebook comment/post a *message* to be classified. In this study, we have 3,365,136 messages from the Facebook pages and 26,047 messages from Twitter. For the Facebook messages, we manually scraped them from Thai Facebook news pages, such as Work-PointTODAY[1] and Voice TV[2]. For the Twitter messages, we used Twitter API with Thai keywords that represented the government policies to collect Thai messages that expressed the opinion on those government policies such as Let's Go Halves, Paotang, and so on.

3.3 Ground-Truth Mental Illness Cases

We cross-validated the predicted aggregate mental signals from social networks with the actual reported mental illness cases. Such ground-truth data was extracted from the reports on access to services of depression patients from the Department of Mental Health of Thailand. Such data is recorded on a monthly basis and classified by the fiscal year in different years. A Thailand's fiscal year spans from October to September of the following year [5].

4 Experiments and Results

All the experiments are conducted on a Linux machine with 10 CPU cores (20 threads), two RTX 2080 GPUs, and 128 GB of RAM. This section discusses the experiment results in detail.

4.1 Mental State Classification

We train a set of classifiers on the train data and report performance on the test data for each dataset. All of the datasets have a pre-defined train-test split. Additionally, since some of the datasets have different classes, we used the data balancing technique to combat this problem. We used SMOTE to synthesize more examples in minority classes [6]. The detail of the training process is as follow. For TF-IDF, the maximum feature is set to 200 most frequent words. As for word embedding, we have used pre-trained word vectors from FastText, an open-source model for text representation and text classifier [1]. We use default parameter sets of Scikit-learn for all classical machine learning models [13]. As for the deep learning models, all parameters are set differently according to the model architectures. The batch size is 32. Early stopping, which monitors the trend of classification accuracy, is also used to prevent overfitting.

Table 2 shows precision, recall, and F1 score measured from all baseline models, both classical machine learning and deep learning. As illustrated in Table 2, we can see that the performance for most datasets is quite saturated in which BERT performs best. The F1 scores of the classical machine learning models are approximately close to the F1 score from deep neural network models. We speculated that such variations might be caused by the keywords that are different in each classification task as well as the varied number of training examples.

---

[1] https://www.facebook.com/workpointTODAY/.

[2] https://www.facebook.com/VoiceOnlineTH.

4.2 Population-Level Mental Health

According to the result from Table 2, we select BERT to classify the social media messages. Most of the messages are in Thai, but there is no dataset in the Thai language suitable for this experiment. We employed the Thai-to-English machine translation model from VISTEC-depa Thailand Artificial Intelligence Research Institute is to translate messages to English messages [24]. After the translation, the translated messages will be classified based on the four tasks.

The analysis of the correlation between the ground-truth mental illness cases and our aggregated messages consists of two major intervals.

1. The fiscal year 2019 (December 2018–September 2019)
2. The fiscal year 2020 (December 2019– March 2020) - However, only statistics from December 2019 to March 2020 are considered due to the incompleteness of the statistics. Please note that January 2020 is the period where the COVID-19 started to spread in Thailand.

Table 2. Precisions (P), Recalls (R), and F1 scores the classifiers for the four mental state datasets.

	Models	P	R	F1
D	TF-IDF RF	0.994	0.994	0.994
	TF-IDF SVM	0.99	0.99	0.99
	Bi-LSTM	0.997	0.998	0.998
	CNN-LSTM	0.997	0.997	0.997
	BERT	0.999	0.998	**0.999**
E	TF-IDF RF	0.88	0.881	0.88
	TF-IDF SVM	0.889	0.887	0.887
	Bi-LSTM	0.931	0.89	0.889
	CNN-LSTM	0.94	0.87	0.89
	BERT	0.929	0.928	**0.929**
S	TF-IDF RF	0.749	0.748	0.748
	TF-IDF SVM	0.761	0.761	0.761
	Bi-LSTM	0.808	0.809	0.809
	CNN-LSTM	0.8	0.8	0.8
	BERT	0.819	0.819	**0.819**
ST	TF-IDF RF	0.938	0.934	0.935
	TF-IDF SVM	0.952	0.951	0.951
	Bi-LSTM	0.934	0.933	0.934
	CNN-LSTM	0.967	0.966	0.966
	BERT	0.967	0.966	**0.967**

In the correlation evaluation, we use the statistics of the patients diagnosed with depression from Thailands' the Department of mental health under the Ministry of Public Health to compare and calculate the correlation with the normalized number of depression-detected messages.

Table 3. Correlation results with actual depression patient cases.

	Classes	Correlation	
		2019	2020 (Mar)
D	depressed	0.156	0.88
E	joy	0.58	−0.617
	sadness	0.043	0.455
	anger	−0.339	0.678
	fear	−0.511	0.781
	love	0.625	−0.717
	surprise	−0.275	0.707
S	negative	−0.514	0.713
ST	suicidal	−0.563	0.491

From Table 3, we can see that the normalized number of depression messages has a high correlation with the ground-truth with the value of 0.156 and 0.88 for the period before the COVID-19 started and after, respectively. While the other prediction tasks do not have a strong correlation. We analyze the trends of the normalized number of depression messages further in Figs. 2 and 3. From Fig. 2, we can see that there was no abrupt change in the actual number of mental illness cases, and our normalized numbers of messages follow this trend with slightly increasing toward the end of the period. During the pandemic, we can see from Fig. 3 that the numbers of cases were

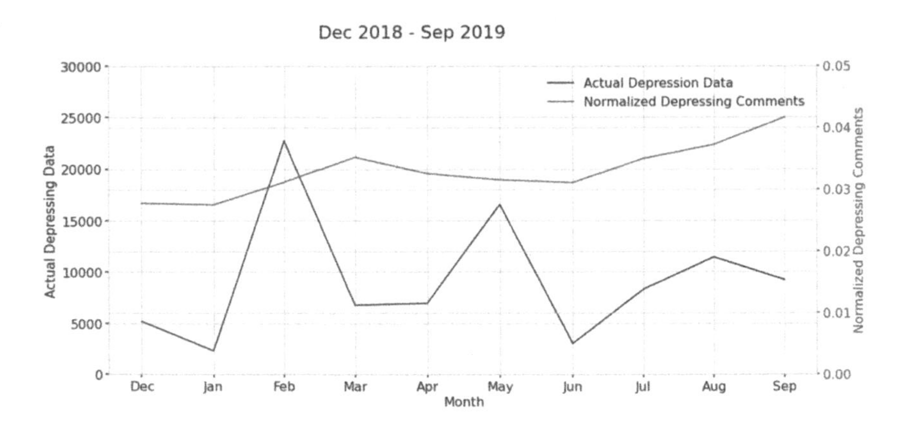

Fig. 2. Graph comparison for the fiscal year 2019

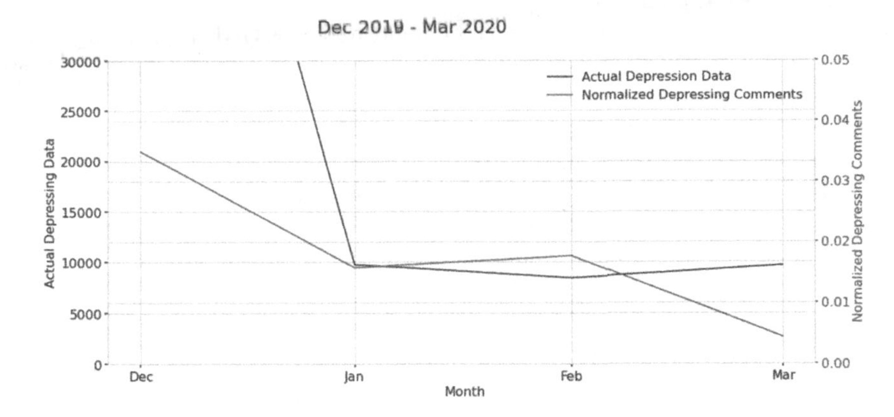

Fig. 3. Graph comparison for the fiscal year 2020

sharply dropped, and the numbers of depressing messages were also decreased. These results suggest that we may be able to use the number of depressing messages from social media to estimate the mental illness trend prior to the traditional statistics.

5 Potential Applications

Large-scale social media data has been established as alternative sources of knowledge that could potentially reflect real-world events [19–22]. This research proposes to use big organic data, such as tweets and Facebook posts, to create a mental health and sentiments index which can be compiled quickly and will offer more timely and cost-effectively indicators than surveyed or administrative data. In response to the worrisome mental health issues, the Department of Mental Health of Thailand has recently issued an online mental health survey with a sample size of 1,500. However, such a small sample size could raise skepticism about the representativeness of the whole population and the lack of comparison data from the "normal" state. To address such issues, the proposed method of constructing mental health index could have a much broader coverage both in terms of population and temporal span, with real-time compilation ability.

Due to the under-utilization of mental health services in Thailand, proactive strategies in providing targeted mental health supports and other ancillary programs that alleviate mental distress are needed. The proposed rapid index can help related agencies to coordinate with social workers, primary health care providers, or other government agencies working with the high-risk group such as debt reliefs, cash transfer, or unemployment-benefit programs to target high-risk groups and provide mental support more proactively. Accurate characterization of the high-risk groups can also help with precise targeting that potentially reduces dead-weight loss incurred from under-utilized healthcare practitioners and resources. Furthermore, successful identification of early signs of mental health from tweets or Facebook posts could give rise to possible targeting technology using online platforms.

Other applications include using the index as an essential input used for Nowcasting [7]. Given the real-time frequency, it would be a good candidate for predicting stock indexes [2], job losses, or macroeconomics indicators like GDP [11], though these would need a relatively long retrospective sequence of the index to validate its prediction efficacy. This index could also supplement official indexes, such as consumer sentiment or consumer confidence indexes. Specifically, the social media signals may have co-movements with these survey-based indexes but provide incremental information as the underlying population, timing, and data-generating processes are different. The ability to filter data by sentiment-oriented keywords used in constructing the index will also allow for customization of indexes later on for specific issues.

6 Conclusions

This paper shows that the detection of population-level mental states from social networks can be used to explain the aggregated level of mental health in Thailand. We trained BERT models to detect mental health signals from Thailand's social media messages. We found that depression has a higher correlation with the actual mental illness case. We believe that it can also be used to provide more immediate feedback to government policies. For future work, we would like to extract and analyze the key events from social media messages that indicate the abrupt change in the mental illness cases.

Acknowledgments. This research project is supported by Puey Ungphakorn Institute for Economic Research (PIER).

References

1. Bojanowski, P., Grave, E., Joulin, A., Mikolov, T.: Enriching word vectors with subword information. Trans. Assoc. Comput. Linguist. **5**, 135–146 (2017). https://transacl.org/ojs/index.php/tacl/article/view/999
2. Bollen, J., Mao, H., Zeng, X.: Twitter mood predicts the stock market. J. Comput. Sci. **2**(1), 1–8 (2011)
3. Breiman, L.: Random forests. Mach. Learn. **45**(1), 5–32 (2001). https://doi.org/10.1023/A:1010933404324
4. Brooks, S., et al.: The psychological impact of quarantine and how to reduce it: rapid review of the evidence. Lancet **395** (2020). https://doi.org/10.1016/S0140-6736(20)30460-8
5. Thailand's Department of Mental Health: Report on access to services of patients with depression (2020). https://thaidepression.com/www/report/main_report/. Accessed 15 Jan 2021
6. Chawla, N.V., Bowyer, K.W., Hall, L.O., Kegelmeyer, W.P.: Smote: synthetic minority oversampling technique. J. Artif. Intell. Res. **16**, 321–357 (2002). https://doi.org/10.1613/jair.953
7. Choi, H., Varian, H.: Predicting the present with Google trends. Econ. Rec. **88**, 2–9 (2012)
8. Devlin, J., Chang, M.W., Lee, K., Toutanova, K.: Bert: pre-training of deep bidirectional transformers for language understanding (2019)
9. Frank, R., McGuire, T.G.: Mental health treatment and criminal justice outcomes. Working Paper 15858, National Bureau of Economic Research, April 2010. https://doi.org/10.3386/w15858, http://www.nber.org/papers/w15858
10. Go, A.: Sentiment classification using distant supervision. cS224N Project Report (2009)
11. Higgins, P.C.: GDPNow: a model for GDP 'nowcasting' (2014)

12. Noble, W.S.: What is a support vector machine? Nat. Biotechnol. **24**(12), 1565–1567 (2006)
13. Pedregosa, F., et al.: Scikit-learn: machine learning in Python. J. Mach. Learn. Res. **12**, 2825–2830 (2011)
14. Samrat, S.: DepressionTweets (2020). https://www.kaggle.com/samrats/depressiontweets/. Accessed 10 Dec 2020
15. Schuster, M., Paliwal, K.K.: Bidirectional recurrent neural networks. IEEE Trans. Signal Process. **45**(11), 2673–2681 (1997). https://doi.org/10.1109/78.650093
16. Tariq, S., et al.: A novel co-training-based approach for the classification of mental illnesses using social media posts. IEEE Access **7**, 166165–166172 (2019). https://doi.org/10.1109/ACCESS.2019.2953087
17. Elvis & Hugging Face Team: Emotions dataset for NLP (2020). https://www.kaggle.com/praveengovi/emotions-dataset-for-nlp. Accessed 8 Dec 2020
18. Torales, J., O'Higgins, M., Castaldelli-Maia, J.M., Ventriglio, A.: The outbreak of covid-19 coronavirus and its impact on global mental health. Int. J. Soc. Psychiatry **66**(4), 317–320 (2020). https://doi.org/10.1177/0020764020915212. pMID: 32233719
19. Tuarob, S., Lim, S., Tucker, C.S.: Automated discovery of product feature inferences within large-scale implicit social media data. J. Comput. Inf. Sci. Eng. **18**(2), 021017 (2018)
20. Tuarob, S., Mitrpanont, J.L.: Automatic discovery of abusive Thai language usages in social networks. In: Choemprayong, S., Crestani, F., Cunningham, S.J. (eds.) ICADL 2017. LNCS, vol. 10647, pp. 267–278. Springer, Cham (2017). https://doi.org/10.1007/978-3-319-70232-2_23
21. Tuarob, S., Tucker, C.S., Salathe, M., Ram, N.: Modeling individual-level infection dynamics using social network information. In: Proceedings of the 24th ACM International on Conference on Information and Knowledge Management, pp. 1501–1510 (2015)
22. Tuarob, S., et al.: Davis: a unified solution for data collection, analyzation, and visualization in real-time stock market prediction. Financ. Innov. **7**(1), 1–32 (2021). https://doi.org/10.1186/s40854-021-00269-7
23. Patel, V., Shah, H., Farooqui, Y.: Hybrid feature based prediction of suicide related activity on Twitter. In: 2020 4th International Conference on Intelligent Computing and Control Systems (ICICCS), pp. 590–595 (2020)
24. VISTEC-depa: English-Thai Machine Translation Models (2020). https://airesearch.in.th/releases/machine-translation-models/. Accessed 25 Dec 2020
25. Wang, J., Yu, L.C., Lai, K.R., Zhang, X.: Dimensional sentiment analysis using a regional CNN-LSTM model. In: Proceedings of the 54th Annual Meeting of the Association for Computational Linguistics (Volume 2: Short Papers), pp. 225–230. Association for Computational Linguistics, Berlin, August 2016. https://doi.org/10.18653/v1/P16-2037, https://www.aclweb.org/anthology/P16-2037
26. Waykole, R.N., Thakare, A.D.: A review of feature extraction methods for text classification. Int. J. Adv. Eng. Res. Dev. **5**(04) (2018). e-ISSN (O): 2348-4470, p-ISSN (P): 2348-6406

Analysis of the Usefulness of Critique Documents on Musical Performance: Toward a Better Instructional Document Format

Masaki Matsubara[1(✉)], Rina Kagawa[1], Takeshi Hirano[2], and Isao Tsuji[3,4]

[1] University of Tsukuba, Tsukuba, Japan
masaki@slis.tsukuba.ac.jp
[2] University of Electro-Communications, Tokyo, Japan
[3] Senzoku Gakuen College of Music, Kawasaki, Japan
[4] Kunitachi College of Music, Tokyo, Japan

Abstract. Today, with the COVID-19 pandemic, the demand for remote and asynchronous lessons for musical performance is rapidly increasing. In these lessons, teachers listen to recordings of musical performances and then return textual critique documents to the performers. However, the common document formats that exist in other fields are not widely known in the field of performance instruction. To address this issue, we launched a project in 2020 to collect and publish a dataset of critique documents. This study describes a statistical analysis of the dataset to investigate which types of elements are useful for performers. The multilevel modeling results revealed that the content of the critiques differed more depending on the teacher than on the musical piece or the student. Particularly, the number of sentences about giving practice advice is a key factor for useful critique documents. These findings would lead to improved forms of critique documents and, eventually, to the development of educational programs for teachers.

Keywords: Textual document format · Digital archive · Knowledge management

1 Introduction

Instructions for playing musical instruments have traditionally been given face-to-face, and teaching in a virtual space has been considered unsuitable. However, today, with the coronavirus disease 2019 (COVID-19) pandemic, the demand for remote and asynchronous lessons is growing rapidly [1,13]. For such lessons, teachers listen to recordings of musical performances write on textual critique documents, and then return them to performers. Today, some music colleges adopt this approach.

M. Matsubara and R. Kagawa—Two authors equally contributed to this research.

H.-R. Ke et al. (Eds.): ICADL 2021, LNCS 13133, pp. 344–353, 2021.
https://doi.org/10.1007/978-3-030-91669-5_27

Remote and asynchronous lessons have an advantage in many aspects, such as facilitation of instruction to remote areas, flexible scheduling, and reduced travel. In particular, accumulated textual critique documents from the lessons have the potential for knowledge transfer and reuse; for example, students can use them for their practice, other students can also use them as references, and teachers can use them to improve their teaching methods. Although previous studies have focused on the transcription of speech in interactive instruction [2, 5–7, 9, 28, 29, 34], only a few have investigated textual documents for musical performance instruction. Hence, the format (i.e., description and arrangement) of textual documents for performance instruction [23] has not been clarified. In the digital era, the collection and utilization of textual critique documents for performance instruction are expected, but their reusability is low.

Therefore, as the first step in investigating useful document formats for performance education, this study aims to clarify what elements are effective for performers. Toward this goal, we launched a project for accumulating instructional textual data with people from a college of music in 2020. We have already collected and published a dataset that consists of 239 textual critique documents for 90 performances of 10 pieces by nine players [19, 20]. The analysis procedure was as follows. First, the usefulness of the critique documents was evaluated by the performers via crowdsourcing. Second, each sentence was categorized into six types by annotators. Third, multilevel modeling was conducted to clarify which types are useful.

The result with the best fitting model revealed the following two points. (1) Number of sentences that are describing practice strategy, objective information, feedback, and advice is effective for improving the usefulness of critique documents. (2) The usefulness of the critiques differed more depending on the teacher than the piece or the student. These findings can be applied to developing a improved document format by explicitly providing recommendations to input effective types of critique in the form.

2 Related Work

2.1 Music Database for Research

Several public datasets or digital archives have been constructed as knowledge resources for music with various perspectives[24], (e.g., performance recordings data [10], metadata [genre, composer, lyrics, etc. [11, 27, 30]], musical scores [MIDI [16], piano notation [8, 31]], information associated with [fingering [22], music analysis [12]], other multimodal information [18, 33], emotions [3, 35], listening history [26], and performers' interpretations [14, 15, 21, 25]). Most of the datasets are the data of the sound source itself. To the best of our knowledge, no datasets has shared human cognition, such as how they played or how they sounded.

2.2 Document Format

In some areas other than music, document formats have been designed. Regarding existing frameworks for document formats, IMRaD (introduction, method,

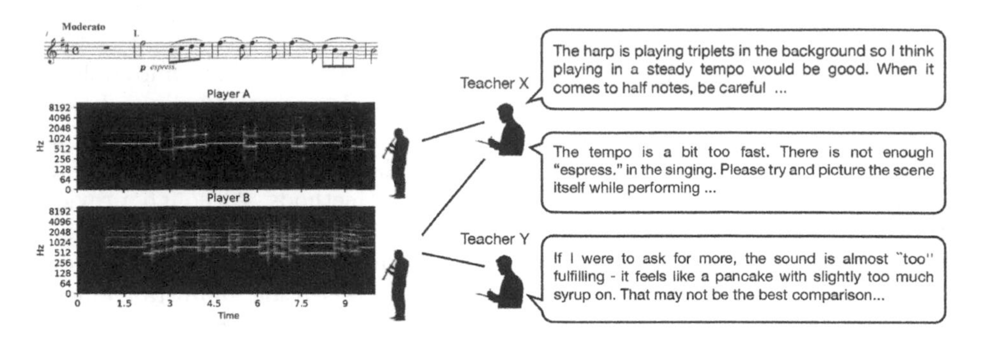

Fig. 1. Examples of a CROCUS dataset [19,20]. (Left) spectrogram of performance recordings of Tchaikovsky's piece by players A and B. (Right) Critique documents from teachers X and Y. From the observation, the tempo and the expression vary among players. Different teachers have different points of view.

result, and discussion) [4] has been a widely employed format for scientific writing. SOAP (subjective data, objective data, assessment, and plan) [32] is widely accepted worldwide in hospitals for clinical notes. In these research fields, document formats have been established, and document datasets are publicly available and actually utilized for research and education [17]. However, all of these are summaries of the opinions of experts in each field, and cannot be applied to performance education.

3 Materials and Methods

3.1 Materials

CROCUS Dataset. In our project, we have already published a *CROCUS* (CRitique dOCUmentS of musical performance) dataset consisting 239 textual critique documents for performance instruction (in Japanese) for 90 performances of 10 pieces by nine players [19,20]. Figure 1 shows examples of the dataset.

Questionnaire Survey of the Usefulness of the Documents. To evaluate the usefulness of the critique documents, we conducted an online questionnaire survey via a crowdsourcing platform. We recruited 200 people who had musical experience outside of school and asked them to provide their demographics and answer the question "Do you think that this document is useful for future performances?" with an 11-point Likert scale (10: useful – 0: useless). Each participant responded to 25 randomly selected textual documents for performance instruction.

Annotation. To clarify what types of element are effective for the usefulness of critique documents, annotators categorized every sentences into six types

Table 1. Types of annotation for each sentence

Types	Example of sentence
Giving Subjective Information (GSI)	It is a very light and springy performance
Giving Objective Information (GOI)	Tempo is late in the second bar
Asking Question (AQ)	Is there a problem with the tuning of the instrument?
Giving Feedback (GF)	The pitch unconsciously moves during a vibrato
Giving Practice (GP)	Please practice this phrase using the metronome
Giving Advice (GA)	The first bar should have no crescendo

(Table 1). According to the Simones' definition [29], this study focused on the following six types: giving subjective information (GSI), giving objective information (GOI), asking question (AQ), giving feedback (GF), and giving advice (GA). Note that, giving information in Simones' definition is divided into GSI and GOI in this study. Other types such, as demonstrating, modeling, and listening/observing, were omitted because these cannot be implemented in remote and asynchronous teaching.

One of these six types was annotated for each sentence. Sentence breaks were periods or exclamation marks. When it was judged that one sentence consisted of descriptions of multiple types, it was separated by a comma. Each of the two annotators annotated for all 239 documents. If the annotations did not match, the final annotation was decided through discussion. The Cohen's Kappa coefficient was 0.96.

3.2 Method: Statistical Models for Analysis

Procedure. Multilevel modeling was conducted for the analysis. Multilevel modeling enables analysis assuming that the behavior of individual data changes depending on the hierarchy of data, that is, the group to which each data belongs. In other words, in this study, not only the change in usefulness among documents but also the influence of the teachers could be analyzed. We first tested the hierarchy of the characteristics of documents and then devised four models for analysis. R 4.1.0, brms 2.15.0, lme4 1.1–27, and lattice 0.20–38 were used.

Hierarchy of the Usefulness. For teachers, the intraclass correlation coefficient (ICC) was 0.45, and the design effect (DE)[1] was 9.43. For players or pieces, ICCs were 0.0, and DEs were 1.0. Therefore, the usefulness scores showed hierarchy among teachers.

[1] DE is a criterion that takes into account both the average number of data in the group and ICC. A DE of over two suggests that the data are hierarchical. $DE = 1 + (k^* - 1)ICC$. k^* means the average number of data of group.

Models. Based on hierarchy, we devised the following four models:

Model I: The usefulness of the documents is affected by the presence or absence of each type.
Model II: The usefulness of the documents is affected by the number of descriptions of each type.
Model III: The usefulness of the documents is affected by the presence or absence of each type and varies depending on teachers.
Model IV: The usefulness of the documents is affected by the number of descriptions of each type and varies depending on teachers.

Let α be intercept, k be a content category, $\beta_k (k = 1, \ldots, 6)$ be coefficient of n_{ki}, n_{ki} be the number of descriptions for each type. In i-th document of the j-th participants, the usefulness of the k-th content is designated as follows:

$$usefulness_{ij} = \alpha + \sum_{k=1}^{6} \beta_k n_{jk} + \sum_{k=1}^{6} \eta_k^{(z_{ijk})} + \sum_{k=1}^{6} \gamma_k^{(z_{ijk})} n_{ik} + e_{ij}$$

Here, z_{ijk} indicates each teacher who wrote the i-th document. $\beta_k^{(z_{ijk})}$ is the random effect of the presence of unknown words on the intercept for the k-th content category of the i-th document. $\gamma_k^{(z_{ijk})}$ is the random effect of the presence of unknown words on the coefficient for n_{ik}.

The model parameters were fitted with four Markov chain Monte Carlo chains with 2,000 iterations and 1,000 burn-in samples with a thinning parameter of one. Non-informative priors were used for all estimations. Specifically, we used $\beta_k \sim N(0, 100)$, $\alpha \sim StudentT(3, 0, 2.5)$, and $\sigma_e \sim StudentT(3, 0, 2.5)$ as the prior distributions of the fixed effects, $StudentT(3, 0, 2.5)$ as the prior distribution of SD of random effects, and $LKJCholesky(1)$ as the prior distribution of the correlation matrix between $\gamma_k^{(g)}$ and $\eta_k^{(g)}$ for $k \in \{1, \ldots, 6\}$ and $g \in \{1, \ldots, 12\}$. The models were compared based on the widely applicable information criterion (WAIC). A smaller WAIC corresponds to a better model.

4 Results

4.1 Usefulness Evaluation and Types Annotation

Figure 2 shows the average scores of the usefulness for each teacher, student, and piece. This result implies that the usefulness of the critiques differed more depending on the teacher than on the piece or the student.

For the annotation results, the percentages of documents containing GSI, GOI, AQ, GF, GP, and GA were 47.28%, 54.81%, 3.34%, 39.33%, 22.18%, and 93.72%, and the average (and standard deviation) of the number of each category per document was, 0.70 (0.90), 0.85 (1.00), 0.03 (0.18), 0.61 (0.88), 0.33 (0.70), and 3.33 (2.50), respectively.

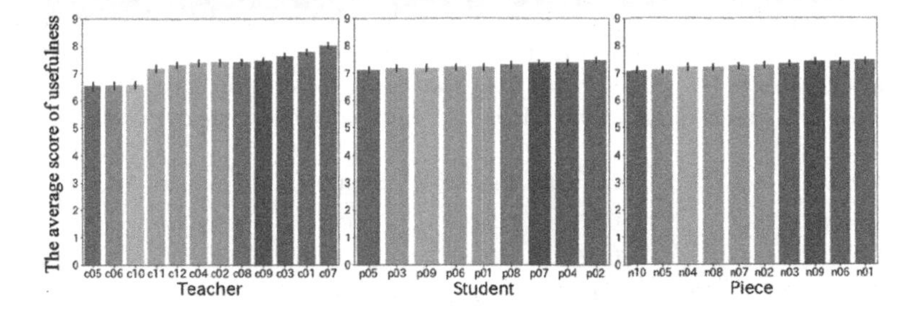

Fig. 2. Average scores of usefulness based on teacher, student, and piece.

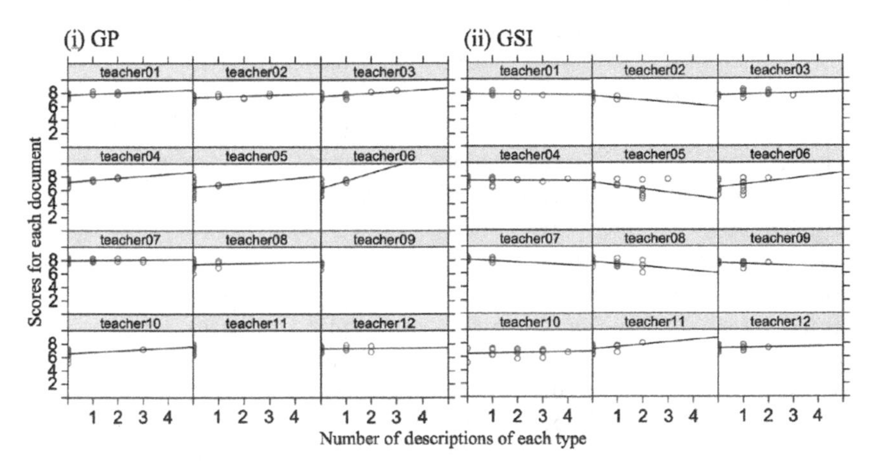

Fig. 3. Relationship between the number of description of GP and GSI in the document written by each of the 12 teachers and the usefulness score for each document. The relationship differed depending on the teachers. For example, the number of GP particularly increases the usefulness of the documents for teachers 06, 05, and 04 but not in teachers 07 and 12. The number of GSI decreased the usefulness in all teachers, except teachers 06 and 11, where the usefulness was increased.

4.2 Multilevel Modeling

As fitting indices, the WAIC values for models I–IV were 449.6, 381.7, 320.0, and 292.1, respectively. All $\hat{R}$ were 1.01 or less. These results indicated that the model IV was the best model; that is, the number of descriptions of all types affects the usefulness, and these influences are affected by the teachers in all types. Table 2 shows the effect score of each type and teacher on usefulness. The results suggest that the more descriptions of GOI ($\beta_2 = 0.13, 95\% CI[0.05 - 0.20]$), GF ($\beta_4 = 0.13, 95\% CI[0.04 - 0.23]$), GP ($\beta_5 = 0.27, 95\% CI[0.09 - 0.46]$), and GA ($\beta_6 = 0.15, 95\% CI[0.07 - 0.22]$) significantly increased the usefulness, and GP had the highest usefulness among all the models.

Figure 3 shows the relationship between the amount of each type of document written by each of the 12 teachers and the usefulness score for each document. The results showed that the relationship differs depending on teachers.

Table 2. Effect of the number of descriptions of each type and teacher on document usefulness. GOI, GF, GP, and GA had positive lower-95% CI, among which GP showed the highest Estimate score. This indicates that usefulness will increase if the number of four types is increased, and that GP is the most effective type for instruction.

Population-Level Effects	Estimate	Est.Error	lower-95% CI	upper-95% CI	$\hat{R}$
Intercept	6.55	0.24	6.04	7.01	1.00
GSI	0.08	0.08	−0.08	0.23	1.00
GOI	0.13	0.04	0.05	0.20	1.00
AQ	0.54	0.92	−1.31	2.39	1.00
GF	0.13	0.05	0.04	0.23	1.00
GP	0.27	0.09	0.09	0.46	1.00
GA	0.15	0.04	0.07	0.22	1.00
Group-Level Effects					
sd(Intercept)	0.20	0.17	0.01	0.62	1.00
sd(GSI)	0.22	0.08	0.08	0.40	1.00
cor(Intercept,GSI)	−0.21	0.56	−0.97	0.91	1.01
sd(Intercept)	0.21	0.18	0.01	0.65	1.00
sd(GOI)	0.05	0.05	0.00	0.18	1.00
cor(Intercept,GOI)	−0.15	0.58	−0.97	0.93	1.00
sd(Intercept)	0.19	0.17	0.01	0.63	1.00
sd(AQ)	1.47	0.86	0.50	3.73	1.01
cor(Intercept,AQ)	−0.01	0.57	−0.94	0.95	1.00
sd(Intercept)	0.21	0.18	0.01	0.67	1.00
sd(GF)	0.08	0.07	0.00	0.25	1.01
cor(Intercept,GF)	−0.13	0.57	−0.97	0.92	1.00
sd(Intercept)	0.22	0.19	0.01	0.72	1.00
sd(GP)	0.21	0.11	0.03	0.48	1.00
cor(Intercept,GP)	−0.24	0.56	−0.97	0.89	1.00
sd(Intercept)	0.44	0.25	0.03	1.03	1.00
sd(GA)	0.10	0.04	0.03	0.19	1.00
cor(Intercept,GA)	−0.67	0.42	−1.00	0.62	1.01
Family Specific Parameters					
Sigma	0.40	0.02	0.36	0.45	1.00

For example, the number of GP generally increased the usefulness of the documents for teachers 06, 05, and 04 but not for teachers 07 and 12. The number of GSI decreased for all the teachers, except for teachers 06 and 11, where the usefulness was increased.

5 Discussion

The results of multilevel modeling analysis showed that the number of descriptions of all types tended to improve the usefulness of critique documents. In addition, a high number of descriptions of GA, GF, GOI, and GP significantly improved the usefulness. These findings would lead to better forms of critique documents where teachers are recommended to include GP, and GA, GF, and GOI if possible. We also confirmed that contents of the critiques differed more depending on the teacher than the piece or the student, suggesting that there is a need for supporting writing critique documents for teachers. To address this problem, considering how to express the sentence of the particular type is the future work.

In addition, the interaction between factors was not considered in the current study. For future work, we would like to examine the interaction between types (e.g., When descriptions of AG are numerous, the influence of GSI is small, and vise versa).

In the future, it is also necessary to investigate the effect of the player's knowledge or the relationship between the player and the teacher on usefulness should be investigated (e.g., whether a relationship of trust has been established, whether the relationship is still shallow).

6 Conclusion

This study showed that based on multilevel modeling, the number of descriptions of six types tended to improve the usefulness of the performance instruction document. Furthermore, the larger the number of descriptions of GA, GF, GOI, and GP, the more significant was the increase in the usefulness of the documents. The effect was different depending on the teacher. In the future, we would like to discuss the arrangement of documents, determine their format, and consider the development of educational programs and writing support technologies so that teachers can make these descriptions.

Acknowledgment. This study was partially supported by JST-Mirai Program Grant Number JPMJMI19G8, and JSPS KAKENHI Grant Number JP19K19347 and JP21H03552.

References

1. Bayley, J.G., Waldron, J.: "it's never too late": adult students and music learning in one online and offline convergent community music school. Int. J. Music Educ. **38**(1), 36–51 (2020)
2. Cavitt, M.E.: A descriptive analysis of error correction in instrumental music rehearsals. J. Res. Music Educ. **51**(3), 218–230 (2003)
3. Chen, Y.A., Yang, Y.H., Wang, J.C., Chen, H.: The AMG1608 dataset for music emotion recognition. In: ICASSP, pp. 693–697 (2015)
4. Day, R.A., et al.: The origins of the scientific paper: the IMRaD format. J. Am. Med. Writers Assoc. **4**(2), 16–18 (1989)
5. Dickey, M.R.: A comparison of verbal instruction and nonverbal teacher-student modeling in instrumental ensembles. J. Res. Music Educ. **39**(2), 132–142 (1991)

6. Duke, R.A.: Measures of instructional effectiveness in music research. Bull. Counc. Res. Music. Educ. **143**, 1–48 (1999)
7. Duke, R.A., Simmons, A.L.: The nature of expertise: Narrative descriptions of 19 common elements observed in the lessons of three renowned artist-teachers. Bull. Counc. Res. Music Educ. **170**, 7–19 (2006)
8. Foscarin, F., McLeod, A., Rigaux, P., Jacquemard, F., Sakai, M.: ASAP: a dataset of aligned scores and performances for piano transcription. In: ISMIR, pp. 534–541 (2020)
9. Goolsby, T.W.: Verbal instruction in instrumental rehearsals: a comparison of three career levels and preservice teachers. J. Res. Music Educ. **45**(1), 21–40 (1997)
10. Goto, M., Hashiguchi, H., Nishimura, T., Oka, R.: RWC music database: popular, classical and jazz music databases. In: ISMIR, pp. 287–288 (2002)
11. Goto, M., Hashiguchi, H., Nishimura, T., Oka, R.: RWC music database: music genre database and musical instrument sound database. In: ISMIR, pp. 229–230 (2003)
12. Hamanaka, M., Hirata, K., Tojo, S.: GTTM database and manual time-span tree generation tool. In: SMC, pp. 462–467 (2018)
13. Hash, P.M.: Remote learning in school bands during the COVID-19 shutdown. J. Res. Music Educ. **68**(4), 381–397 (2021)
14. Hashida, M., Matsui, T., Katayose, H.: A new music database describing deviation information of performance expressions. In: ISMIR, pp. 489–494 (2008)
15. Hashida, M., Nakamura, E., Katayose, H.: Constructing PEDB 2nd edition: a music performance database with phrase information. In: SMC, pp. 359–364 (2017)
16. Hawthorne, C., et al.: Enabling factorized piano music modeling and generation with the MAESTRO dataset. In: ICLR (2019)
17. Kagawa, R., Baba, Y., Tsurushima, H.: Publicly available medical text data with authentic quality (2020). https://doi.org/10.5281/zenodo.4064153
18. Li, B., Liu, X., Dinesh, K., Duan, Z., Sharma, G.: Creating a multitrack classical music performance dataset for multimodal music analysis: challenges, insights, and applications. IEEE Tran. Multimedia **21**(2), 522–535 (2018)
19. Matsubara, M.: Crocus: dataset of musical performance critique, June 2021. https://doi.org/10.5281/zenodo.4748243
20. Matsubara, M., Kagawa, R., Hirano, T., Tsuji, I.: Crocus: dataset of musical performance critiques: relationship between critique content and its utility. In: CMMR (2021)
21. Miragaia, R., Reis, G., de Vega, F.F., Chávez, F.: Multi pitch estimation of piano music using cartesian genetic programming with spectral harmonic mask. In: 2020 IEEE Symposium Series on Computational Intelligence (SSCI), pp. 1800–1807. IEEE (2020)
22. Nakamura, E., Saito, Y., Yoshii, K.: Statistical learning and estimation of piano fingering. Inf. Sci. **517**, 68–85 (2020)
23. Reiter, E., Dale, R.: Building Natural Language Generation Systems. Cambridge University Press, Cambridge (2000)
24. Salamon, J.: What's broken in music informatics research? Three uncomfortable statements. In: 36th International Conference on Machine Learning (ICML), Workshop on Machine Learning for Music Discovery, Long Beach, CA, USA (2019)
25. Sapp, C.S.: Comparative analysis of multiple musical performances. In: ISMIR, pp. 497–500 (2007)
26. Schedl, M.: The LFM-1b dataset for music retrieval and recommendation. In: ICMR, pp. 103–110 (2016)

27. Silla Jr., C.N., Koerich, A.L., Kaestner, C.A.: The Latin music database. In: ISMIR, pp. 451–456 (2008)
28. Simones, L., Schroeder, F., Rodger, M.: Categorizations of physical gesture in piano teaching: a preliminary enquiry. Psychol. Music **43**(1), 103–121 (2015)
29. Simones, L.L., Rodger, M., Schroeder, F.: Communicating musical knowledge through gesture: piano teachers' gestural behaviours across different levels of student proficiency. Psychol. Music **43**(5), 723–735 (2015)
30. Sturm, B.L.: An analysis of the GTZAN music genre dataset. In: ACM Workshop MIRUM. MIRUM 2012, pp. 7–12 (2012)
31. Wang, Z., et al.: Pop909: a pop-song dataset for music arrangement generation. In: ISMIR (2020)
32. Weed, L.L.: Medical Records, Medical Education, and Patient Care: The Problem-oriented Record as a Basic Tool. Press of Case Western Reserve University, Cleveland (1969)
33. Weiß, C., et al.: Schubert Winterreise dataset: a multimodal scenario for music analysis. J. Comput. Cult. Herit. **14**(2), 1–18 (2021)
34. Whitaker, J.A.: High school band students' and directors' perceptions of verbal and nonverbal teaching behaviors. J. Res. Music Educ. **59**(3), 290–309 (2011)
35. Zhang, K., Zhang, H., Li, S., Yang, C., Sun, L.: The PMEmo dataset for music emotion recognition. In: ICMR, pp. 135–142 (2018)

Data Modeling

Modelling Archaeological Buildings Using CIDOC-CRM and Its Extensions: The Case of Fuwairit, Qatar

Manolis Gergatsoulis[1](✉), Georgios Papaioannou[1], Eleftherios Kalogeros[1], Ioannis Mpismpikopoulos[1], Katerina Tsiouprou[1], and Robert Carter[2,3]

[1] Department of Archives, Library Science and Museology, Ionian University, Corfu, Greece
{manolis,gpapaioa,kalogero}@ionio.gr
[2] Institute of Archaeology University College London, London, England
[3] Bahrain Authority for Culture and Antiquities, Manama, Bahrain

Abstract. This paper explores the use of CIDOC CRM and its extensions (CRMba, CRMarchaeo) to represent archaeological buildings that have been studied and interpreted by archaeologists during their work in the field. These archaeological observations and reflections usually appear in archaeological reports containing text and visual representations, such as images and photographs, plans and drawings, and maps. For our approach (case study), we used the recent archaeological excavations and other heritage works of the Origins of Doha and Qatar Project in Fuwairit, Qatar. We investigate, explore and review issues related to the application of classes and properties as they appear in the latest versions of the aforementioned models, i.e. CIDOC CRM, and CRMba. We focus on archaeological building construction, on specific building components and materials, on the visual representations of archaeological buildings as well as on issues of archaeological buildings' chronology and buildings' information provenance. The proposed data model contributes towards an automated system for archaeological buildings representations, documentation, and heritage data integration.

Keywords: Archaeology · Archaeology of buildings · CIDOC CRM · CRMarchaeo · CRMba · Ontologies · Digital humanities

1 Introduction

The CIDOC CRM ontology and its extensions offer a tool to model archaeological work including the archaeology of buildings. Archaeology studies human

This work started in 2020 thanks to a research fellowship by UCL Qatar/Qatar Foundation. R. Carter and G. Papaioannou are former academics at UCL Qatar, and M. Gergatsoulis received the research fellowship. The work continues at the authors' current affiliations.

H.-R. Ke et al. (Eds.): ICADL 2021, LNCS 13133, pp. 357–372, 2021.
https://doi.org/10.1007/978-3-030-91669-5_28

societies and cultures of the past through material objects and remains via excavations and surface surveys. Archaeologists discover objects and structures, and with the knowledge offered by other fields of study such as geography, geology, history, and architecture, they approach and explore places, sites, and cultures of the past. Architectural elements and/or whole buildings of any size and function form parts of the archaeological record. CIDOC CRM[1] and its extensions (e.g. CRMba[2], CRMarchaeo[3], CRMgeo[4], CRMsci[5]) contribute to their digital documentation. Can CIDOC CRM and CIDOC CRM based models sufficiently represent archaeological buildings? To what extent are they able to offer a framework to support buildings documentation and archaeological interpretation? We approach these questions by working towards an automated CRM-based system for archaeologists in modeling archaeological buildings in archaeological sites. To this end, we have represented archaeological buildings from recent archaeological excavation works at Fuwairit in Qatar (2016–2018), part of the Origins of Doha and Qatar Project (ODQ), by employing classes and properties of CIDOC CRM and its extensions.

2 Related Work

There is extensive literature in the last 15 years on using the CIDOC CRM data model [13] to document archaeological science, including buildings and architectural remains [18–21,24,42,47]. The ARIADNE project [38] can be considered as a starting post, as it tackled the problem of data heterogeneity in the documentation of historic buildings with CIDOC CRM [44,45] before the development of the CRMba extension [10,46]. The pre-CRMba work describes the mapping of metadata schemas used in different European counties in their efforts to document historic buildings via CIDOC CRM. In specific, it presents the Central Institute for Cataloguing and Documentation (ICCD, by the Italian Ministry of Culture) [4], the MIDAS schema [6] by the English Heritage (U.K.), the Schéma documentaire appliqué au patrimoine et à l'architecture (SDAPA, by the French Ministry of Culture) [5], and the metadata schemas used in the well-known projects CARARE [3] and 3DICONS [1] and their connections to the Europeana Data Model (EDM) [7]. Additionally, the Catalogue Data Model (ACDM) was developed to extend the DCAT vocabulary [17] for best describing the ARIADNE assets. Moreover, the ARIADNE project and the follow-up ARIADNEplus project [2] led to the development of the CIDOC-CRM extensions CRMarchaeo [11], CRMba [10] and CRMgeo [9], to address the need of archaeological data integration. The CRMba and the way it is connected with CIDOC-CRM and its related extensions are presented below (see Sect. 3.2).

[1] http://www.cidoc-crm.org.
[2] http://www.cidoc-crm.org/crmba.
[3] http://new.cidoc-crm.org/crmarchaeo.
[4] http://www.cidoc-crm.org/crmgeo.
[5] http://www.cidoc-crm.org/crmsci.

Another metadata schema for historic buildings description is the Architecture Metadata Object Schema (ARMOS) [16]. The ARMOS acts as an application profile. It was created by extending the Council's Core Data Index to Historic Buildings and Monuments of the Architectural Heritage (CDI) and by taking elements from various mature metadata specifications, especially from the CDWA (Categories of Description of Works of Arts).

The need for better semantic organization of DBpedia's historic buildings instances is studied in [15]. Mappings of cultural heritage metadata expressed through the VRA Core 4.0 schema [8] to CIDOC CRM is studied in [28,29]. Besides, the mapping of the semantics of Dublin Core (DC) metadata to CIDOC CRM is presented in [35]. These mappings consider the CIDOC CRM as the most appropriate conceptual model for interrelations and mappings between different heterogeneous sources [30]. A recent categorization of the Heritage Building Information Modeling (H-BIM) domain is presented in [36]. It includes four categories/sections: Historic and Archaeological data, Geometry, Pathology, and Performance. The historic and archaeological data section is further divided into the sub-categories of Past drawings, Historic records, Historic texts, Archaeological findings, Historic photographs, Oral histories, and Multimedia.

Related work so far has shown that the CIDOC-CRM and its extensions appear to be the most appropriate modeling approach. The alternatives discussed above, such as the ARMOS, the VRA, the Dublin Core and the CDWA, can only partially cover the needs of representing archaeological buildings, while CIDOC-CRM with its CRMarchaeo, CRMgeo and CRMba extension address archaeological building in a more complete way, yet still evolving and under development, as new and advanced versions keep coming up (see Subsect. 3.2 below).

3 Preliminaries

3.1 The Archaeology of Buildings

The archaeology of buildings studies architectural elements and/or whole buildings that form parts of the archaeological record. Buildings can be of any size and function, from palaces, cathedrals, and factories to houses, huts and warehouses. In comparison with the excavation archaeology which concentrates on digging and revealing what has been underground, the archaeology of buildings focuses on analyzing and recording what survived and exists on the ground, i.e., the remains of buildings. Building remains comprise important evidence about people's life and activities; archaeologists gather and record information on building types, materials, and methods of construction, and how people lived and worked in them, and for what purpose(s) people created and/or transformed these buildings in the past. These data can help towards interpretations on societies' cultures, values, and organizations in specific places and periods of times. The archaeology of buildings examines the interior and the exterior of buildings, the fabrics and functions, the history and the phases of construction [40], and it is a developing field of study within archaeology [33,48].

M. Davies [23] proposed a procedure designed to assist archaeologists in collecting a wide range of information regarding buildings. This procedure involves photographs, measured drawings and photogrammetry to record spatial relationships and to merge all the available information, including extra information by architects, curators, and others. This procedure consists of three parts: a) registering structural elements (e.g., walls, doors, windows, etc.), b) registering surface elements (paint, paper, other), and c) registering the structural evolution (development of buildings, construction phases, repairs, etc.). Before starting the procedure, however, it is important to have as much information as possible about the buildings under investigation and research. This involves information about the history, the surroundings, the landscape and the space where buildings are located. Such information is available online and/or can be found in local libraries, archives, research institutions; they relate to maps, documents and other primary and secondary sources on the buildings and the area. Similar procedures have been offered by other organizations and institutions (e.g., [25,34,36]).

Archaeologists can create ground plans and data sheets with structural elements mentioned and numbered, with detailed descriptions about the materials, the features and the techniques that were used, photographs and/or drawings of buildings. Plans and data sheets help towards reports on the buildings and their history, roles, and functions, as well as for interpretations and analyses.

3.2 The CIDOC CRM and Its Extensions Related to the Archaeology of Buildings

The CIDOC Conceptual Reference Model (CIDOC CRM), is a formal ontology for integrating, mediating and interchanging heterogeneous cultural heritage information. It helps documenting the complex and multivariate world of cultural heritage data and information in a strict and explicit language, it is independent of any particular technology, and it is fully compliant with linked data services and solutions. CIDOC CRM has proposed extensions suitable for documenting diverse kinds of cultural information and activities, including CRMba, CRMarchaeo, CRMgeo, and CRMsci.

The CRMba extension is used to document archaeological standing buildings, and the relationships between different building parts. In CRMba [46], a building can be divided into smaller units and the relations among them can also be described. Besides the building's structural parts, CRMba represents building functional spaces, topological relations, and construction phases. Whole buildings and parts can be expressed as instances of the class 'B1 Built Work', which is a subclass of the CIDOC CRM class 'E24 Physical Human-Made Thing'.

The CRMba extension can be connected to the CRMgeo extension for describing topological and geographical elements, the CRMsci extension for representing scientific observations, measurements and processed data from descriptive and empirical sciences, and the CRMarchaeo extension, used for recording archaeological excavation data [32]. Excavation context sheets from the recent archaeological excavations in Fuwairit in Qatar [31] were used to explore the use of CIDOC CRM, CRMarchaeo and CRMsci [12] to represent excavation and stratigraphy work, activities, observations and finds during excavation field work.

Archaeologists use the CRMba extension along with the CRMarchaeo and CRMgeo [9]. This integration [43] relies on the fact that the CRMba introduces the class 'B5 Stratigraphic Building Unit' which represents the minimal construction unit of a built structure as a subclass of the CRMarchaeo class 'A8 Stratigraphic Unit'. Additionally, the CRMba class 'B2 Morphological Building Section' is a geometric feature with volume that occupies a defined space in a time period that can be modelled with CRMgeo. This work applies CIDOC CRM and CRMba to document data and reports on the historic buildings in Fuwairit (Qatar). This research will offer valuable experience concerning the documentation needs of these data and is expected to influence the process of further developing and refining these models. This work is based on CIDOC CRM version 7.1.1 (April 2021), CRMba version 1.4 (December 2016), CRMarchaeo version 1.5.0 (February 2020), and CRMgeo version 1.2 (September 2015).

4 The Origins of Doha and Qatar Project, and the Archaeological Works at Fuwairit

The Origins of Doha and Qatar Project (ODQ) investigates the history, the archaeology, and the cultures of the past in Doha (the country's capital) and other cities and towns in Qatar, including the historic town of Fuwairit and its historic buildings, from their prehistory to their modern development thanks to oil revenues since the mid-20th century [14,26,27]. ODQ was started in 2012 by University College London in Qatar in collaboration with Qatar Museums, thanks to funding by the Qatar Foundation through Qatar National Research Fund, under grants NPRP5-421-6-010 and NPRP8-1655-6-064. The project is multidisciplinary in terms of methodologies employed: archival research, studies of historical documents, recording of historical buildings, excavations, oral histories of local people, GIS analysis for pre-oil and early oil Doha [37,39,41]. In 2016 in Fuwairit, 90 km north of Doha, ODQ undertook an extensive programme of recording the 26 standing but rapidly collapsing historical buildings. Representing and modelling these efforts by using CIDOC CRM and its extensions comprise the aim of this paper.

5 Describing Archaeological Buildings in Fuwairit and Modelling in CIDOC-CRM and CRMba

In this section we present a methodology for describing archaeological buildings using the CRMba along with the CIDOC CRM. We present our model and give several examples of its use. The material used in examples comes from the archaeological works in the historic town of Fuwairit and its historic buildings appearing in [22].

5.1 Archaeological Building Components, Construction Techniques and Materials

In studying an historical/archaeological building, it is of great importance to address its architecture, to describe its structural and/or functional components, to detect the method(s) and determine the time of its construction and subsequent modifications, to identify the materials used, to explore intended and other uses of the building and/or its parts.

To address the above, in this subsection we propose a model (Fig. 1) which is constructed by appropriately combining classes and properties from CIDOC CRM and CRMba. The proposed model can be used to formally encode all available information on historical/archaeological building aspects referenced above. An example of the proposed model is presented in Fig. 2 and Fig. 3.

To apply the proposed model (Fig. 1) and to demonstrate how it can be used to describe archaeological buildings, we used ‘Building 1’ of Fuwairit, Qatar, as it appears in [22] (page 15). We modelled most of the available information on Building 1 as it appears in the extracted and annotated text fragment from [22] in Fig. 2. In this annotated text fragment, one can see the class instances highlighted in yellow and the visual representations of Building 1 (or its parts) highlighted in green. The corresponding class code can be seen above each highlighted class instance.

The use of controlled vocabularies is proposed for several classes, especially ‘E55 Type’ and ‘E62 String’. Values from controlled vocabularies can be used to classify instances of the classes such as ‘B1 Built Work’, ‘B2 Morphological Building Section’, ‘B3 Filled Morphological Building Section’, ‘E57 Material’ and ‘E36 Visual Item’. It is important to note that the use of controlled vocabularies, when possible, is of great importance as it facilitates the automation of the search and the reasoning procedures over the encoded information.

In the example of Fig. 2, the value ‘*brown color*’ was taken from the Art & Architecture Thesaurus (AAT)[6] of the Getty Research Institute, which is also one of the recommended vocabularies of the VRA metadata schema [8] for selecting values for the ‘material’, ‘stylePeriod’, ‘subject’, ‘technique’ and ‘worktype’ elements.

The model proposed in Fig. 1 can be easily extended to incorporate additional properties connecting the classes of the model. For example, the class ‘E80 part removal’ can be associated with the class ‘B1 Built Work’ through the property ‘P113 removed (was removed by)’. In the same way, the class ‘E79 Part Addition’ can be associated with the class ‘B1 Built Work’ through the property ‘P111 added (was added by)’. Besides, the class ‘E64 End of Existence’ can be associated with the class ‘B1 Built Work’ through the property ‘P93 took out of existence (was taken out of existence by)’. Finally, the class ‘E63 Beginning of Existence’ can be associated with the class ‘B1 Built Work’ through the property ‘P92 brought into existence (was brought into existence by)’. To keep the figure as simple as possible we did not include these properties in Fig. 1.

---

[6] https://www.getty.edu/research/tools/vocabularies/aat/.

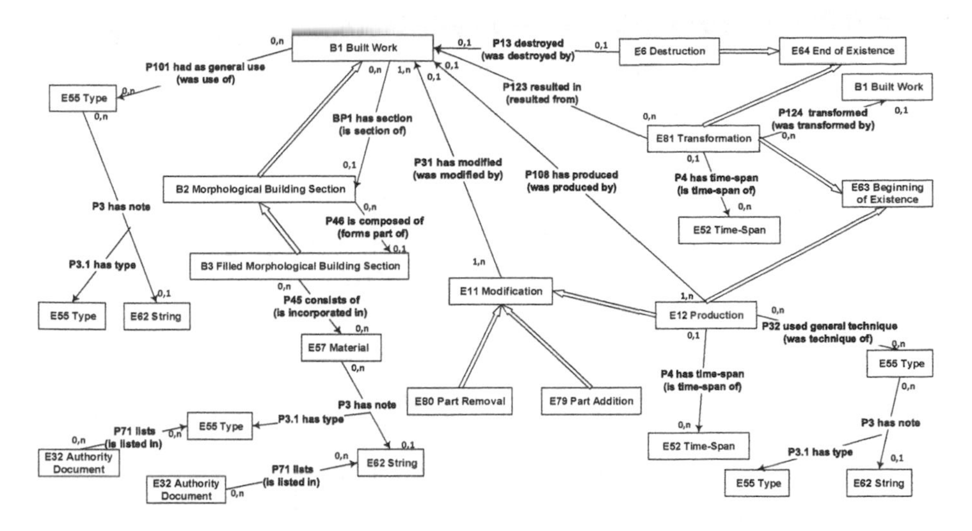

Fig. 1. Modelling the architecture, the components and the construction/modification of a building.

5.2 Describing the Visual Representation of a Building

Several visual representations are used to describe and depict an archaeological building and/or its parts, characteristics, materials, conditions, and details. Traditional visual representations relate to photographs, plans (floor and ceiling plans, sections), drawings, sketches, and artistic paintings. Modern methods include 3D reconstructions, 360° photographs, and video recordings. Archaeological buildings also appear in maps, aerial photographs and drone recordings, while historical documents, oral histories, and past studies and reports confirm (or not) the visual evidence. The above visual representations comprise part of the documentation material of an archaeological building and should also be represented and included in the model.

In this subsection, we present a model to describe visual representations of an archaeological building, parts of a building and/or processes related to a

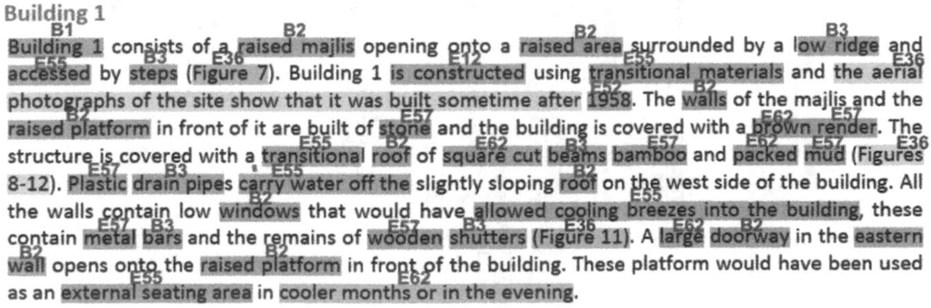

Building 1

Building 1 consists of a raised majlis opening onto a raised area surrounded by a low ridge and accessed by steps (Figure 7). Building 1 is constructed using transitional materials and the aerial photographs of the site show that it was built sometime after 1958. The walls of the majlis and the raised platform in front of it are built of stone and the building is covered with a brown render. The structure is covered with a transitional roof of square cut beams bamboo and packed mud (Figures 8-12). Plastic drain pipes carry water off the slightly sloping roof on the west side of the building. All the walls contain low windows that would have allowed cooling breezes into the building, these contain metal bars and the remains of wooden shutters (Figure 11). A large doorway in the eastern wall opens onto the raised platform in front of the building. These platform would have been used as an external seating area in cooler months or in the evening.

Fig. 2. Description of Building 1 taken from [22].

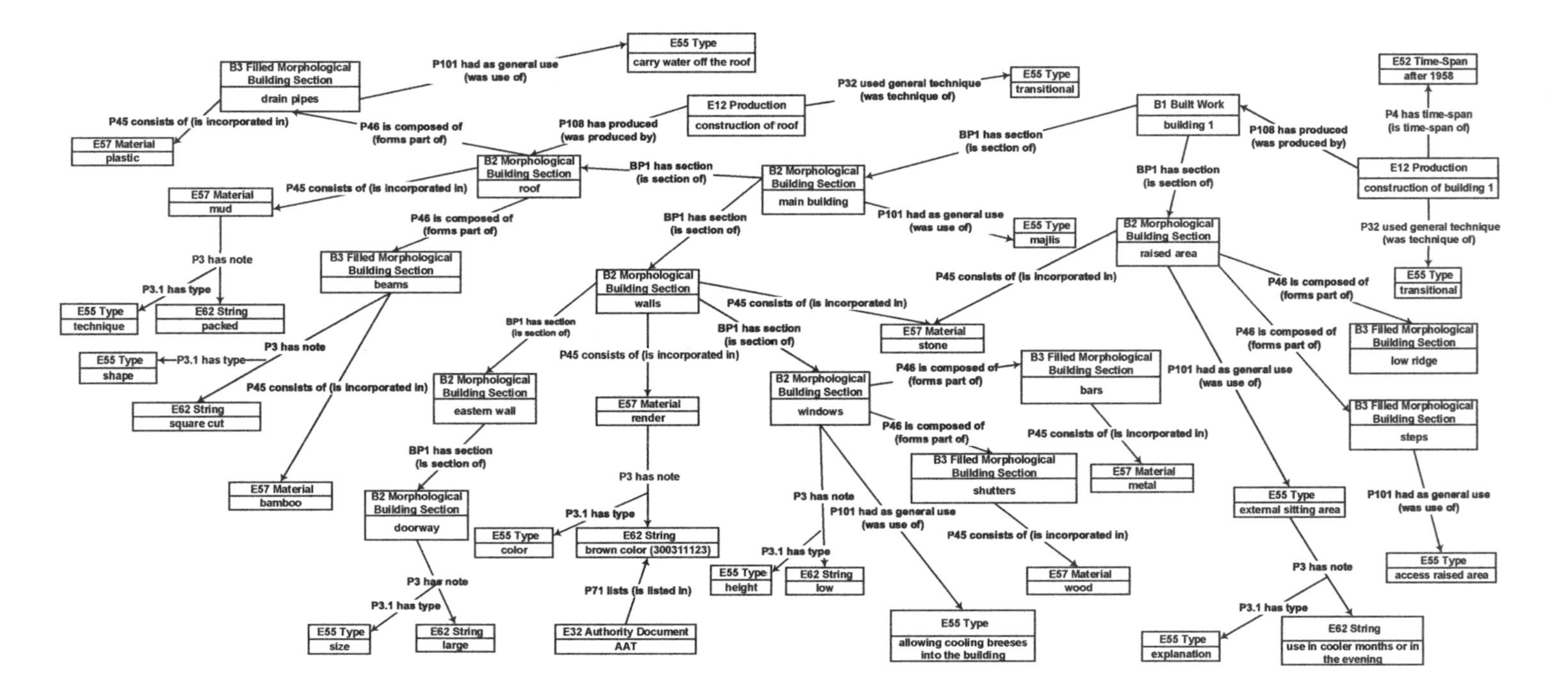

Fig. 3. Description of Building 1 based on the model presented in Fig. 1.

building, such as its construction, reconstruction, demolition, etc. The model is presented in Fig. 4. As an example, the modeling of a photograph depicting a window of 'Building 1' in Fuwairit, Qatar, is in Fig. 5.

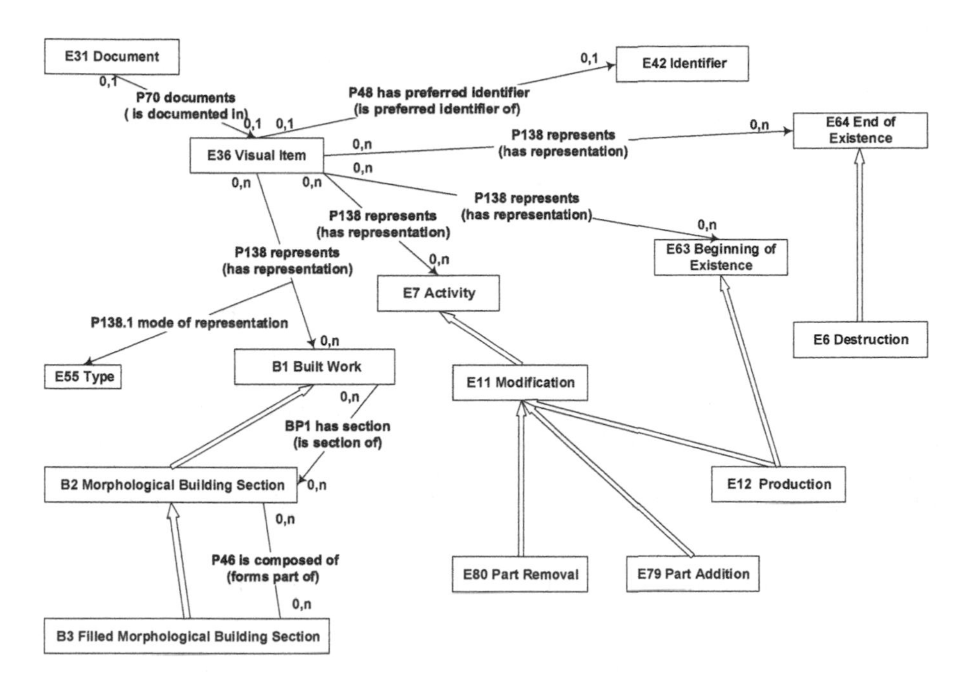

Fig. 4. Modelling the visual representations of buildings and their components.

In Fig. 5, the instance 'Photo of figure 11' of the class 'E36 Visual Item' is directly associated, through the property 'P138 represents (has representation)', with the instance 'windows' of the class 'B2 Morphological Building Section'. The instance 'Photo of figure 11' is not directly associated with the instance 'Building 1' of the class 'B1 Built Work'. Notice, however, that 'Building 1' is connected to 'windows' via a chain of occurrences of the property 'BP1 has section (is section of)'. Following this path we can conclude that 'window' is part of 'Building 1' and therefore that the photo represents (a part of) 'Building 1'. To achieve this connection, we took into account the following:

1. The property 'P138 represents' is inherited through to the classes connected with the property 'BP1i is section of'.
2. The property 'BP1 has section (is section of)' is transitive.

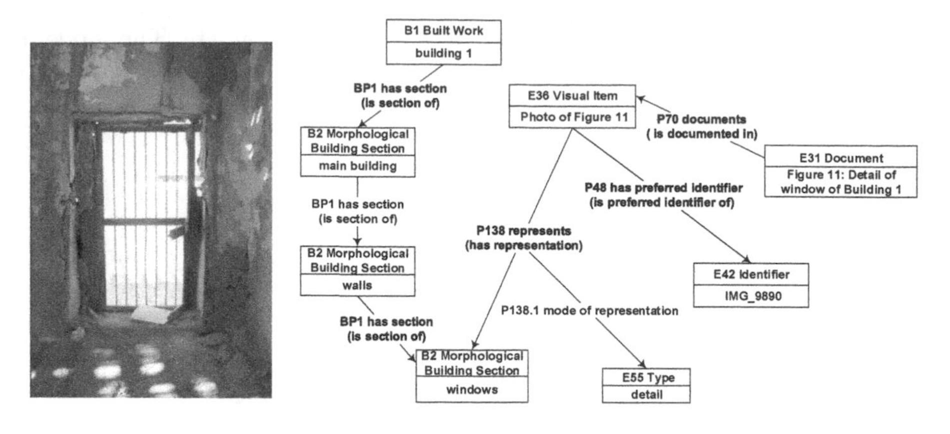

Fig. 5. Example of representing a photo of a window of Building 1.

To accomplish the above reasoning process automatically, the knowledge needed (the propagation of the properties, the transitivity of some properties etc.), should be incorporated in the system. This can be achieved by using a Rule Language such as Semantic Web Rule Language (SWRL)[7] and RuleML[8].

5.3 On the Provenance of Information About Archaeological Buildings

The question of the provenance of information is a key question in archaeology and archaeological finds, including archaeological buildings. Historical accounts and historical documents in general, old and new maps, aerial photographs, oral histories, travelers' accounts and comments, archival records, art works depicting buildings and landscapes, even local myths and traditions can help archaeologists by providing hints to archaeological interpretation. As the provenance of information relates to the validity and the quality of the information content, archaeologists place particular importance in all provenance-related sources and their contents. Information on provenance of information forms a significant part of archaeological documentation and must be represented and included in the model. In this subsection, we present a model for documenting provenance of archaeological information. Our model is depicted in Fig. 6.

A good example is the information concerning the provenance of information about the Building 6 taken from [22](page 27): "*Building 6 was also a school,* ***local residents told us that*** *Building 6 was the girls school and Building 3 was the boys school*". According to this phrase the source of the information that Building 6 was the girls school are local residents. The encoding of this

[7] https://www.w3.org/Submission/SWRL/.
[8] http://wiki.ruleml.org/index.php/RuleML_Home.

information using our model appearing in Fig. 6 is presented in Fig. 7. Notice also that in Fig. 7, the class 'E31 Document' and its instance 'Origins of Doha Project - Season 4 Archive Report: Fuwairit Standing Building Recording', which is associated with the instance 'Building 6' of the class 'B1 Built Work' through the property 'P70 documents (is documented in)', states that the source of information about 'Building 6' is the technical report 'Origins of Doha Project - Season 4 Archive Report: Fuwairit Standing Building Recording'.

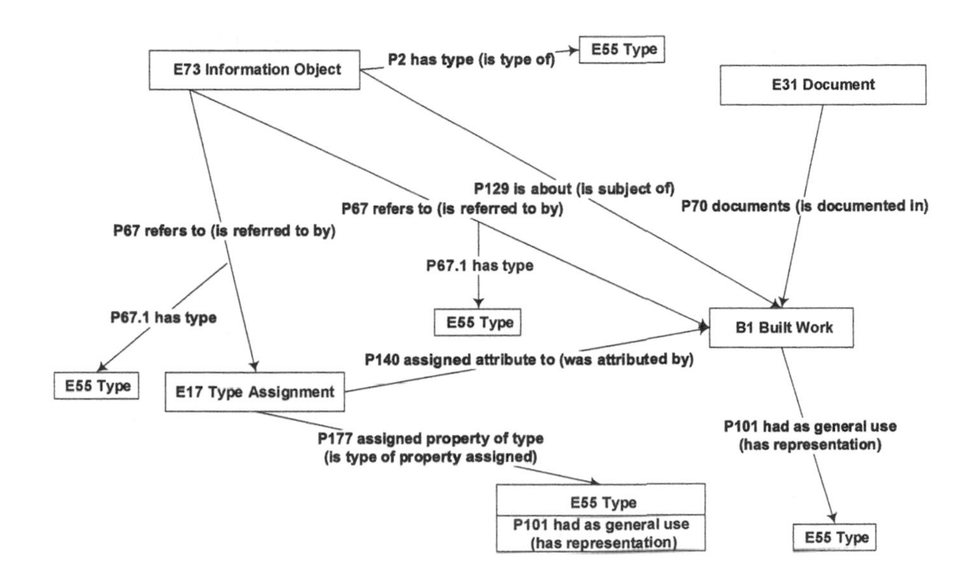

Fig. 6. Modelling the provenance of information about archaeological buildings.

A key question in archaeology is the question 'when'. Archaeological buildings are no exception, as archaeologists' main aim is to secure the chronology of an archaeological building and its phases, in order to relate it to the rest of the archaeological record and proceed to interpretations and conclusions on the building's role(s), use(s) and function(s) within past human societies. Information on chronology forms a significant part of archaeological documentation and must be represented and included in the model. In Fig. 8, we see how we can deduce chronological information concerning a building by taking into account the time an aerial photograph was taken.

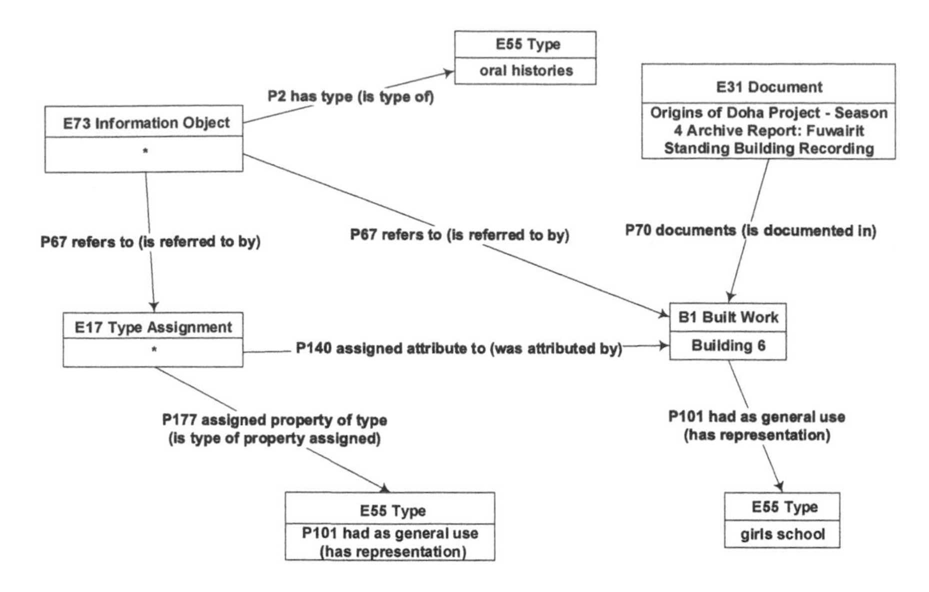

Fig. 7. An example of encoding oral histories about the use of a building.

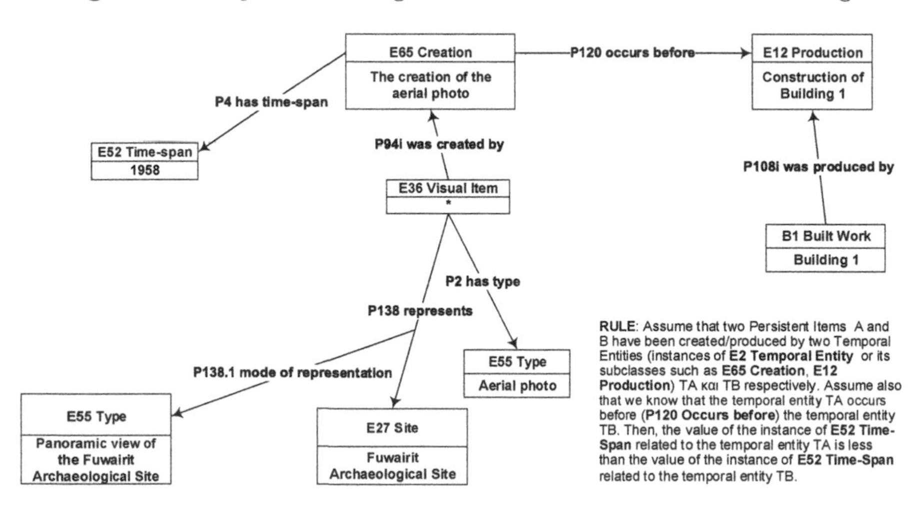

Fig. 8. Inferring the temporal information from the time an aerial photo was taken.

6 Conclusions and Future Work

By using CIDOC CRM and its extension CRMba this work represented aspects of archaeological buildings studied by the archaeologists of the Origins of Doha and Qatar Project in the village of Fuwairit, Qatar. This work supports and contributes to the documentation and management of cultural heritage information, and adds to the discussion on shared interests and endeavors among digital humanities, computing and information studies. We focused on representing specific components of archaeological buildings, as well as on construction techniques and materials. We also addressed the visual representations of buildings and the buildings' chronology and provenance.

As future work, we aim to extend the proposed model with the CRMgeo classes and properties, to describe geographical and topographical information related to the buildings' location and the area's topography. Another next step is to design an automated system for documenting archaeological works (including excavations, post-excavation work and the archaeology of buildings) via CRM ontologies. This will offer archaeologists a tool to document their work in the field in real time, to take advantage of the system's reasoning capabilities, to fully utilize data entry and information retrieval facilities, and to investigate alternative data-driven ways towards heritage management and interpretation.

References

1. 3D-ICONS: 3D Digitisation of Icons of European Architectural and Archaeological Heritage. http://3dicons-project.eu/. Accessed 22 June 2021
2. ARIADNEpus - a data infrastructure serving the archaeological community worldwide. https://ariadne-infrastructure.eu/. Accessed 9 July 2021
3. CARARE. https://www.carare.eu/. Accessed 22 June 2021
4. Home - ICCD - Istituto Centrale per il Catalogo e la Documentazione. http://www.iccd.beniculturali.it/. Accessed 22 June 2021
5. Schéma SDAPA - Schéma documentaire appliqué au patrimoine et á l'architecture. version 1.05. http://www2.culture.gouv.fr/culture/dp/schemaDAPA/. Accessed 22 June 2021
6. MIDAS Heritage - The UK Historic Environment Data Standard, v1.1. Technical report, October 2012
7. Europeana Data Model Primer. Version 1.0. Technical report, July 2013
8. VRA CORE - a data standard for the description of works of visual culture: Official Web Site (Library of Congress) (2014). https://www.loc.gov/standards/vracore. Accessed 22 June 2021
9. CRMgeo: a Spatiotemporal Model: An Extension of CIDOC-CRM to link the CIDOC CRM to GeoSPARQL through a Spatiotemporal Refinement. Version 1.2. Technical report, September 2015
10. Definition of the CRMba: An extension of CIDOC CRM to support buildings archaeology documentation. Version 1.4. Technical report, December 2016
11. Definition of the CRMarchaeo: An Extension of CIDOC CRM to support the archaeological excavation process. Version 1.4.8. Technical report, February 2019
12. Definition of the CRMsci: An Extension of CIDOC-CRM to support scientific observation Version 1.2.7. Technical report, October 2019
13. Definition of the CIDOC Conceptual Reference Model (Vol. A). Version 7.1.1. Technical report, April 2021
14. Adham, K.: Rediscovering the Island: Doha's Urbanity. From pearls to spectacle. In: Elsheshtawy, Y. (eds.) The Evolving Arab City. Tradition, Modernity and Urban Development, pp. 218–257. Routledge (2011)
15. Agathos, M., Kalogeros, E., Kapidakis, S.: A case study of summarizing and normalizing the properties of DBpedia building instances. In: Fuhr, N., Kovács, L., Risse, T., Nejdl, W. (eds.) TPDL 2016. LNCS, vol. 9819, pp. 398–404. Springer, Cham (2016). https://doi.org/10.1007/978-3-319-43997-6_33

16. Agathos, M., Kapidakis, S.: A meta - model agreement for architectural heritage. In: Garoufallou, E., Greenberg, J. (eds.) MTSR 2013. CCIS, vol. 390, pp. 384–395. Springer, Cham (2013). https://doi.org/10.1007/978-3-319-03437-9_37
17. Albertoni, R., Perego, A., Winstanley, P., Cox, S., Beltran, A.G., Browning, D.: Data catalog vocabulary (DCAT) - version 2. W3C recommendation, W3C, February 2020
18. Binding, C., et al.: Implementing archaeological time periods using CIDOC CRM and SKOS. In: Aroyo, L. (ed.) ESWC 2010. LNCS, vol. 6088, pp. 273–287. Springer, Heidelberg (2010). https://doi.org/10.1007/978-3-642-13486-9_19
19. Binding, C., May, K., Souza, R., Tudhope, D., Vlachidis, A.: Semantic technologies for archaeology resources: results from the STAR project. In Contreras, F., Farjas, M., Melero, F.J. (eds.) Proceedings 38th Annual Conference on Computer Applications and Quantitative Methods in Archaeology, BAR International Series 2494, pp. 555–561. BAR Publishing (2013)
20. Binding, C., May, K., Tudhope, D.: Semantic interoperability in archaeological datasets: data mapping and extraction via the CIDOC CRM. In: Christensen-Dalsgaard, B., Castelli, D., Ammitzbøll Jurik, B., Lippincott, J. (eds.) ECDL 2008. LNCS, vol. 5173, pp. 280–290. Springer, Heidelberg (2008). https://doi.org/10.1007/978-3-540-87599-4_30
21. Binding, C., Tudhope, D., Vlachidis, A.: A study of semantic integration across archaeological data and reports in different languages. J. Inf. Sci. **45**(3), 364–386 (2019)
22. Carter, R., Eddisford, D.: Origins of Doha project - season 4 archive report: Fuwairit standing building recording. Technical report, UCL Qatar (2016)
23. Davies, M.: The archaeology of standing structures. Am. J. Hist. Archaeol. **5**, 54–64 (1987)
24. Deicke, A.J.E.: CIDOC CRM-based modeling of archaeological catalogue data. In: De Luca, E.W., Bianchini, P. (eds.) Proceedings of the First Workshop on Digital Humanities and Digital Curation co-located with the 10th Conference on Metadata and Semantics Research (MTSR 2016), volume 1764 of CEUR Workshop Proceedings, Goettingen, Germany. CEUR-WS.org (2016)
25. Fiorini, A.: Archaeology of standing buildings. Teaching and scientific activities. GROMA Documenting archaeology. Open-Access E-J. About Methodol. Appl. Archaeol. **1**, 43–54 (2016)
26. Fletcher, R., Carter, R.A.: Mapping the growth of an Arabian Gulf town: the case of Doha, Qatar. J. Econ. Soc. Hist. Orient **60**, 420–487 (2017)
27. Fuccaro, N.: Pearl towns and oil cities: migration and integration in the Arab coast of the Persian Gulf. In: Freitag, U., Fuhrmann, M., Lafi, N., Riedler, F. (eds.) The City in the Ottoman Empire: Migration and the Making of Urban Modernity, pp. 99–116. Routledge (2010)
28. Gaitanou, P., Gergatsoulis, M.: Defining a semantic mapping of VRA Core 4.0 to the CIDOC conceptual reference model. Int. J. Metadata Semant. Ontol. **7**(2), 140–156 (2012)
29. Gergatsoulis, M., Bountouri, L., Gaitanou, P., Papatheodorou, C.: Mapping cultural metadata schemas to CIDOC conceptual reference model. In: Konstantopoulos, S., Perantonis, S., Karkaletsis, V., Spyropoulos, C.D., Vouros, G. (eds.) SETN 2010. LNCS (LNAI), vol. 6040, pp. 321–326. Springer, Heidelberg (2010). https://doi.org/10.1007/978-3-642-12842-4_37

30. Gergatsoulis, M., Bountouri, L., Gaitanou, P., Papatheodorou, C.: Query transformation in a CIDOC CRM based cultural metadata integration environment. In: Lalmas, M., Jose, J., Rauber, A., Sebastiani, F., Frommholz, I. (eds.) ECDL 2010. LNCS, vol. 6273, pp. 38–45. Springer, Heidelberg (2010). https://doi.org/10.1007/978-3-642-15464-5_6
31. Gergatsoulis, M., Papaioannou, G., Kalogeros, E., Carter, R.: Representing archeological excavations using the CIDOC CRM based conceptual models. In: Garoufallou, E., Ovalle-Perandones, M.-A. (eds.) MTSR 2020. CCIS, vol. 1355, pp. 355–366. Springer, Cham (2021). https://doi.org/10.1007/978-3-030-71903-6_33
32. Giagkoudi, E., Tsiafakis, D., Papatheodorou, C.: Describing and revealing the semantics of excavation notebooks. In: Proceedings of the CIDOC 2018 Annual Conference, Heraklion, Crete, Greece, 19 September–5 October (2018)
33. Giles, K.: Buildings archaeology. In: Smith, C. (ed.) Encyclopedia of Global Archaeology. Lecture Notes in Computer Science, pp. 1033–1041. Springer, Heidelberg (2014). https://doi.org/10.1007/978-1-4419-0465-2_1332 Accessed 27 June 2021
34. Heritage New Zealand Pouhere Taonga (HNZ-PT). Investigation and recording of buildings and standing structures, November 2018. https://www.heritage.org.nz/~/-/media/2793010e741541cfbc24ba23feb22520.ashx. Accessed 27 June 2021
35. Kakali, C., et al.: Integrating Dublin core metadata for cultural heritage collections using ontologies. In Proceedings of the 2007 International Conference on Dublin Core and Metadata Applications, DC 2007, Singapore, August 27–31 2007, pp. 128–139 (2007)
36. Khalil, A., Stravoravdis, S., Backes, D.: Categorisation of building data in the digital documentation of heritage buildings. Appl. Geomat. **13**(1), 29–54 (2021). https://doi.org/10.1007/s12518-020-00322-7
37. Marras, A.: DOHA-Doha online historical atlas. Come le Carte Raccontano un Territorio (2016). https://medium.com/@annamao/doha-doha-online-historical-atlas-come-le-carte-raccontano-un-territorio-bf8c85df5e3d. Accessed 2 Apr 2020
38. Meghini, C., et al.: ARIADNE: a research infrastructure for archaeology. ACM J. Comput. Cult. Herit. **10**(3), 18:1–18:27 (2017)
39. Michalski, M., Carter, R., Eddisford, D., Fletcher, R., Morgan, C.: DOHA-Doha online historical atlas. In: Matsumoto, M., Uleberg, E. (eds.) Oceans of Data. Proceedings of the 44th Annual Conference on Computer Applications and Quantitative Methods in Archaeology CAA 2016, Oslo, 30 March–3 April 2016, pp. 253–260 (2016)
40. Milsted, N.: Buildings Archaeology. Resource booklet and activities. YAC, Council for British Archaeology, in partnership with Historic England (2018)
41. Morgan, C., Carter, R., Michalski, M.: The origins of doha project: online digital heritage remediation and public outreach in a vanishing pearling town in the Arabian Gulf. In Conference on Cultural Heritage and New Technologies (CHMT 2020), November 2–4 2015, Stadt Archaeologie Wien, pp. 1–8 (2016)
42. Niccolucci, F.: Documenting archaeological science with CIDOC CRM. Int. J. Digit. Libr. **18**(3), 223–231 (2017)
43. Ronzino, P.: Harmonizing the CRMba and CRMarchaeo models. Int. J. Digit. Libr. **18**(4), 253–261 (2017)
44. Ronzino, P., Amico, N., Felicetti, A., Niccolucci, F.: European standards for the documentation of historic buildings and their relationship with CIDOC-CRM. In: Practical Experiences with CIDOC CRM and its Extensions (CRMEX 2013) Workshop, 17th International Conference on Theory and Practice of Digital Libraries (TPDL 2013), pp. 70–79 (2013)

45. Ronzino, P., Amico, N., Niccolucci, F.: Assessment and comparison of metadata schemas for architectural heritage. In: Proceedings of CIPA, pp. 71–78 (2011)
46. Ronzino, P., Niccolucci, F., Felicetti, A., Doerr, M.: CRMba a CRM extension for the documentation of standing buildings. Int. J. Digit. Libr. **17**(1), 71–78 (2016)
47. Vlachidis, A., Binding, C., May, K., Tudhope, D.: Automatic metadata generation in an archaeological digital library: semantic annotation of grey literature. In: Przepiórkowski, A., Piasecki, M., Jassem, K., Fuglewicz, P. (eds.) Computational Linguistics. Studies in Computational Intelligence, vol. 458, pp. 187–202. Springer, Heidelberg (2013). https://doi.org/10.1007/978-3-642-34399-5_10
48. Watson, K.: Why buildings archaeology? (2020). https://thecityremains.org/2020/02/21/why-buildings-archaeology/. Article published in the Blog: The city remains - Studies in the archaeology and history of Christchurch. Accessed 27 June 2021

Integrating Heterogeneous Data About Quebec Literature into an IFLA LRM Knowledge Base

Ludovic Font(✉), Dominique Piché, Amal Zouaq, and Michel Gagnon

École Polytechnique de Montréal, Montreal, Canada
{ludovic.font,dominique.piche,amal.zouaq,michel.gagnon}@polymtl.ca

Abstract. In the process of translating cultural heritage data into linked open data, in particular from multiple sources, heterogeneity of data is a crucial issue. In this paper, our aim is to demonstrate the interest of linked open data and more precisely the IFLA LRM model for the digital representation of Quebec literature, based on data from various perspectives: authors, publishers, and books. We propose an IFLA LRM knowledge base to represent the aggregated data from these heterogeneous sources.

We also extend the IFLA LRM model to take into account important aspects including literary awards, a more detailed description of authors, and book series, as well as interoperability with schema.org. Finally, we evaluate the resulting knowledge base using competency questions obtained from our data providers and show the benefit of data integration.

Keywords: Linked open data · Cultural heritage · Literature · Heterogeneous data

1 Introduction

In the last decades, important efforts have been made to turn cultural heritage data into a digital format. The Ministry of Culture and Communications of Quebec (MCCQ) participates in this movement, and is currently working on several projects to exploit the flexibility and strength of linked open data to represent data about various subjects, such as governance systems, museum artifacts or literature [11,20]. When dealing with cultural heritage, one of the biggest challenges is the interoperability issue, as the data is often produced by various actors and perspectives and provided in different formats [2,25].

In this paper, we explore how Semantic Web technologies and the IFLA Library Reference Model (LRM)[1] can help harmonize, distribute and offer rich querying capabilities over a unified knowledge base related to Quebec books and writers. In the context of a collaborative project between MCCQ and Polytechnique Montréal, our task is to demonstrate the applicability and interest of linked

[1] https://www.ifla.org/publications/node/11412.

H.-R. Ke et al. (Eds.): ICADL 2021, LNCS 13133, pp. 373–391, 2021.
https://doi.org/10.1007/978-3-030-91669-5_29

data technologies for the representation of data about literature in Quebec. We obtained a large quantity of bibliographic records from four sources: the Quebec national library and archives (BAnQ), Messageries ADP (a book distributor), the Infocentre Littéraire des écrivains du Québec, a website that provides information about writers in Quebec (designated by ILE in the rest of this paper), and Les Éditions Hurtubise, a publisher based in Montreal, Quebec. These heterogeneous datasets offer different perspectives on the world of literature: the data from ILE is centered on the authors, whereas the data from Hurtubise and ADP are centered on the book as a product. Therefore, to represent this heterogeneous data in a knowledge base, we had to choose a model that enables a multifaceted representation of the concept of "book". For this task, we chose the IFLA LRM, for reasons explained in Sect. 2. Although this model offers an elegant way of representing book-related information, its usage for modeling real-world data requires some extensions. In this paper, we present the issues we encountered in this process and the steps to overcome them, in order to represent data offering different perspectives in a single knowledge base. We extend the IFLA LRM with schema.org to provide additional information on authors and literary awards. To evaluate the validity of the process, we use a set of competency questions from our dataset providers, and we compare our ability to answer these questions on each dataset and using our integrated knowledge base.

The paper is organized as follows. In Sect. 2, we provide an overview of existing works for the digital representation of book-related data. Next, we present our datasets and the model we adopted to integrate the data into a unified knowledge base, and the challenges that complicated this task. In Sect. 3, we present our methodology for creating a knowledge base from varied raw data, including the extensions to the LRM model and our validation process. Then, in Sect. 5, we present our implementation of this process for translation of data into RDF and alignment of LRM entities (authors, works, etc.). Lastly, we show, in Sect. 6, how the obtained knowledge base improves our ability to answer queries about books and authors, and we discuss its limitations, as well as those of the model, in Sect. 7.

2 Related Work

The interest of using linked open data to represent book-related information has repeatedly been proven [1,3,7,14,31]. Many efforts have been made to create bridges between "classical" bibliographic formats, such as MARC [4] and Onix [5,9], and linked open data. This led to the creation of the BIBFRAME model [23,30,33,34], initially as a model to represent MARC data. It also led to the creation of a series of models by the International Federation of Library Associations (IFLA), such as FRBR [35], FRAD [24], FRSAD [10], and, more recently, LRM [27]. Although not explicitly designed for a usage in RDF, these models allow for a flexible and interconnected representation of the various elements describing a book when compared to the more cumbersome formats MARC and Onix. In order to get closer to linked open data, RDA [26] was also developed.

Important efforts must be made to ensure the intercompatibility of those models: between RDA and BIBFRAME [33], between Onix and RDA [9], and between Onix and MARC [13], to provide some examples. This illustrates that converting data from one format to another is a delicate task, especially since we are combining data from four datasets, each having its specific format, into a central and unified knowledge base.

Previous works tackling the integration of heterogeneous data sources in a unified ontology in the world of cultural heritage data include FinnONTO [15] (more specifically CultureSampo [18] and BookSampo [17]), ArCo [2], Europeana [25], and others [12,16,21,22,32]. These works are generally focused on aggregating existing cultural heritage data, typically from governmental open data sources. Indeed, many national libraries are exploring the possibilities of linked open data, in France [29], United Kingdom [6], Spain [36] or Sweden [19]. They generally propose designing unified data models for final exploitation, then manually aligning metadata schemas to the models, either directly or relying on a community-based approach using of data annotation tools. In this work, we use an automatic heuristic-based approach to align entities in our datasets and we propose an enriched IFLA LRM-based model, one of the latest Web standards that turns cataloging formats into an open Semantic Web format. We present our usage of the model in Sect. 3.3.

3 Methodology

3.1 Datasets

As previously mentioned, our datasets offer different perspectives on the concept of “book”. The BAnQ dataset provides us with 400,000 bibliographic records and 13,000 authority records in MARC21, where books are considered as bibliographic entities. The Messageries ADP (shortened as ADP in the remainder of this paper) dataset contains 120,000 records in ONIX format, focused on the publishing aspect, with information such as the publication and printing status of each book. The Hurtubise dataset provides us with 1,400 records, with detailed and varied information on each book, such as its targeted audience or geographical references in textual attributes. Lastly, to build the ILE dataset, we extracted information about 27,000 books and 1,800 authors from their website[2]. In this dataset, each author is related to a list of his or her published books.

Overall, our data is heterogeneous, with differences in terms of quantity of records, types of records (non-bibliographic documents, information about authors), internal consistency, encoding, and available fields. This variability offers opportunities for combining complementary information and different perspectives, but it also complicates the task of creating a model allowing the representation and aggregation of this heterogeneous information.

It should be noted that, since this project is a proof of concept, the data, both in its raw form and translated in RDF, is at the moment confidential. We are however working on making it publicly available at the end of the project. The ontology, however, is available[3].

[2] http://www.litterature.org/.
[3] http://www.labowest.ca/mcc/mcc-litterature.ttl.

3.2 Competency Questions

To assess the capabilities of linked open data for modeling, representing, and interconnecting book-related data, we were given a list of questions (aka competency questions) provided by our partners in this project, ranging from simple ones, such as *What is the place of birth of writers* to more complex ones, such as *Who are the poets from Québec City of less than 40 years old?* In Table 1, we provide a sample of these questions. The complete list is available online[4].

We identified that these questions required 26 available fields in the final knowledge base, such as *Birth date* or *Format.* Other questions could not be answered with the data at our disposal. We then associated each available field with classes and/or properties in the LRM Model, or, if needed, created new ones, as explained in Sect. 3.3. For instance, for the question *Who are the poets from Québec City of less than 40 years old?*, we identified the fields *birth date, birthplace*, and *author specialty.* To represent that last field in LRM, we created the data property *author area of specialty* as a subproperty of *LRM-E6-A2: field of activity.*

3.3 The IFLA LRM Model

The choice of the IFLA LRM[5] model is based on several considerations. First, the book concept is represented at four different levels of abstraction: *work*,

Table 1. Competency questions and associated fields from the UNEQ data source

Competency question	Necessary fields
Which books have been self-published and commercially published during the past year?	Publication type
How many people are currently publishing poetry in Québec? Science-fiction? Culinary books?	Publication place & Theme
Who are the poets from Québec City, less than 40 years old?	Birth place & Birth date & Author specialty
How many books are sold online on foreign platforms (e.g. Amazon)?	Sales by platform
Who are the authors who published digital or audio books during the past five years?	Publication date & Book format
Who are the authors who published books related to digital practices during the past five years?	Publication date & Other content references
Who are the authors who were invited to interviews or participated in articles in Radio Canada in the past 12 months?	Participation to interviews or articles
What are the books dealing with specific themes, such as love between elderly people or fiction about climate change	Theme & Other content references

[4] http://www.labowest.ca/mcc/competency-questions.pdf.
[5] https://www.ifla.org/files/assets/cataloguing/frbr-lrm/ifla-lrm-august-2017_rev201712.pdf.

expression, *manifestation* and *item*. The *work* is the higher conceptual level, representing the work of art as created by the writer (the story of "Alice in Wonderland"); the *expression* represents the embodiment of that work in a written form (the French translation by André Gagnon), the *manifestation* represents the creation of a series of physical entities that correspond to published editions (the 2012 edition published by Hurtubise), and the *item* represents an individual, physical object (the copy of the book owned by someone). This last level is not used in our representation, since we do not have any data about individual items. Second, LRM introduces the concept of *nomen*, which allows for the representation of the appellation or identification of an entity. For instance, the title of a book is a *nomen*, as well as its subtitle and ISBN. The concept of *nomen* can be used for a flexible representation of names, titles, identifiers and categories, including the origin of the information. It also enables the integration of various classification schemas, as each classification can be associated with a reference URL. Third, some work is currently being done to develop a model called LRMoo [28], which makes LRM compatible with CIDOC CRM [8], a widely used model for cultural heritage representation. Choosing LRM as our model ensures a long-term compatibility with other knowledge bases developed for Quebec's cultural heritage. Figure 1 shows an example of a book as represented with the LRM model. In the figures of this paper, *mcc* is the prefix representing our namespace.

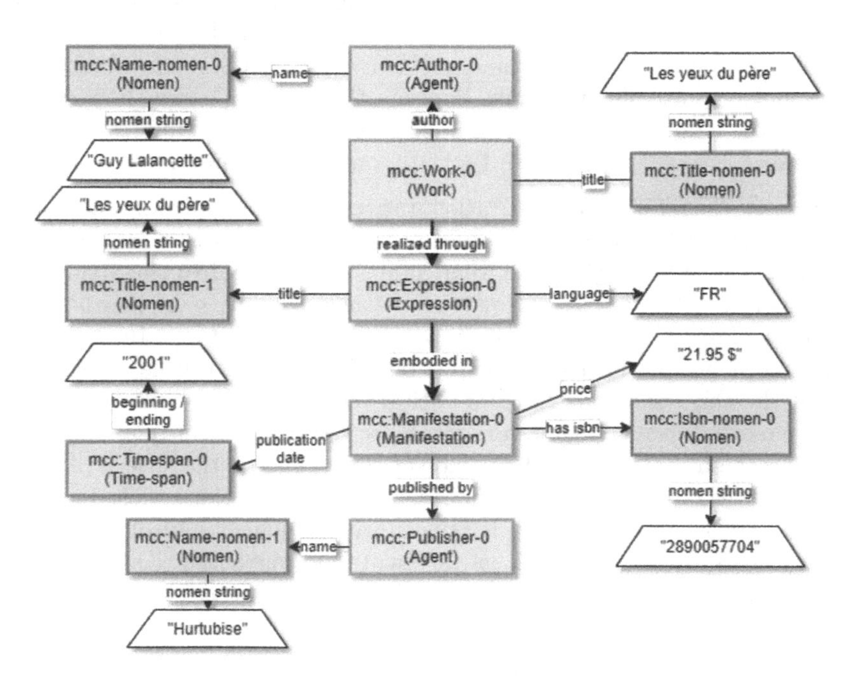

Fig. 1. Example of a book description with LRM

3.4 Data Challenges

Records. Our initial data is structured as a list of bibliographic records, each containing associated information (name of the author, title, date of publication...), either directly in a table or in a more complex representation such as Onix or MARC21. Such records typically represent a manifestation in the LRM model. However, since the model also uses two higher levels (expression and work) for the representation of books, it is necessary to group the records that represent two manifestations of the same expression, and, similarly, to group the expressions of a work. For instance, the Hurtubise dataset uses three separate records for the first volume of the series *À l'ombre du clocher* by Michel David, representing three editions of the book in 2006, 2010 and 2020, named *À l'ombre du clocher - Tome 1*, *À l'ombre du clocher - Tome 1* and *À l'ombre du clocher - Tome 1 (compact)*. These records, since they represent the same expression and work, should be translated into five entities: three manifestations (one for each edition), one expression, and one work. This can only be done if an alignment between these records is performed.

Identifiers. The standard identifier for books is the ISBN, but it is associated to a manifestation, and therefore cannot be used to associate several manifestations to one expression or work. There is no unique identifier at the expression or work level, neither in our data, nor, to our knowledge, in the literature world in general. Regarding authors, the ISNI provides an identifier, but it is only used as a reference when providing information about authors, and never in the records about books. In our data, only one dataset, BAnQ, contains information about authors in a separate file, with links between the author file and the records file using only the name of the author (and no identifier). Lastly, information about the publisher is even sparser, with typically only a name provided.

Alignment. Therefore, the unification process between all our datasets has to be done based on the available information, which consists only in a set of strings (author names, titles, publisher names). This requires the strings to be consistent across the datasets. However, we noticed that it is clearly not the case, with widespread differences, even within the same dataset. In the previous example on the works of Michel David, two of the three records have an identical title, but the third one adds the annotation *(compact)* to the title, making the matching more difficult. These variations are amplified when aligning records from different datasets. In the ILE dataset, for example, the title is *À l'ombre du clocher*, without the volume number, and in ADP, the article is often at the end (*L'ombre du clocher -à*). It is necessary to identify automatically that these variations are actually different facets of the same entity. For this task, we adopted an approach based on heuristics, presented in Sect. 5.3.

4 IFLA LRM Enrichment

Even though LRM is well-suited for representing the various facets of the book concept, it does not cover all the required elements in our datasets. Overall, our extension consists in the addition of the Series and Award classes as well as several data and object properties that are described in the following sections. We also interconnected LRM and schema.org as explained in the next sections.

Integration of Our Model with Schema. An aspect not covered by the LRM model is the representation of additional information about authors (such as birth places and sex). Since our data contains such information, we expanded our model by using schema.org[6]. This results into a representation that allows a full compatibility of our knowledge base with schema.org as well as LRM. This part of our model is however generic, since schema.org does not deal with the various levels of representation of a book (work, expression and manifestation).

To do so, we used subclass relations and property chains to create correspondences between the schema ontology and the LRM models. For instance, we considered that, in our knowledge base, *schema:Book* is a subclass of the *lrm:LRM-E4 (Manifestation)* class from LRM. We also used some property chains, that allow for an equivalence between a succession of LRM properties and a schema.org property. This way, RDF triples regarding books are represented using LRM, but it is also possible to automatically infer the equivalent schema.org triples. However, since it is impossible to specify property chains on data properties such as *schema:isbn*, we also explicitly generated schema.org data properties along with the rest of the knowledge base.

In Figs. 1 and 2, we present some information from a record as represented in the LRM and in schema.org, respectively. The red elements of this second representation correspond to shortcuts of LRM property chains (either automatically inferred or explicitly generated), whereas the black elements have no equivalent in LRM.

The Series Class. The LRM model already enables to represent a series, i.e. a work that is composed of a series of other works, such as *The Lord of the Rings*. In this example, both the entire series and each volume should be represented as *Works* in LRM. However, this representation does not allow for an easy way to distinguish between the two, and obtaining the set of all series in our data, for example, must be done by querying for works containing another work. Since this particular situation happens very often in our data, we expanded the model by creating a subclass of *Work*, *Series*. This allows for an explicit separation between the volumes of a series and the encompassing series.

The Award Attribution Class. The scope of LRM does not directly encompass the concept of literary award, which is an important concept, both in our data and in our competency questions. Schema.org has a schema: award property, but does not allow the multi-sided relationship between, an author, a book or set of books, the name of the award, and the date of the award. Therefore, we created a new class *Award Attribution* to represent this relationship. These are shown in Fig. 3.

Object Properties. We also needed to expand three aspects: the relation between a contributor and a work or expression (*was created by*); the relation

[6] https://schema.org/.

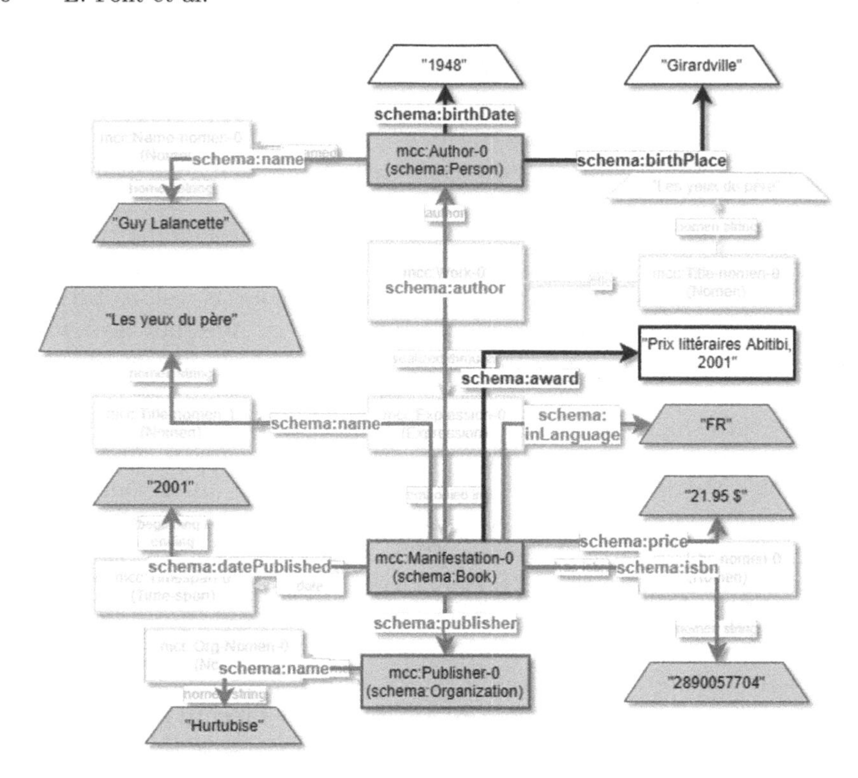

Fig. 2. Example of a book description with Schema

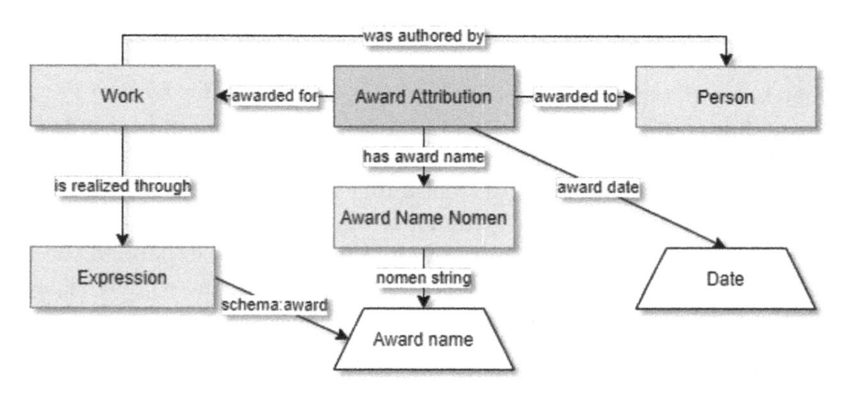

Fig. 3. The award attribution class and associated properties.

between any entity and the various nomens describing it (*has appellation*); and the relation between a nomen and its constituent parts (*has part* and *has derivation*). These additions were considered important to model all our requirements or speed our SPARQL queries.

The first aspect is necessary since LRM encompasses all contributions to a work or expression in the two properties *LRM-R5 was created by* (for works) and *LRM-R6 was created by* (for expressions). This makes it impossible to know if a contributor is an author, a translator or an illustrator.

Regarding the second aspect, even though it is possible to represent most typical use-cases directly in LRM, we decided to create subproperties of *has appellation*, such as *has first name*. This allows for easier SPARQL querying. For instance, to obtain the first name of a person in LRM, it is necessary to query three triples: [?person mcc:LRM-R13 ?nomen], [?nomen mcc:LRM-E9-A3 "First name"] and [?nomen mcc:LRM-E9-A2 ?firstName]. In our extension, only two triple are needed, since there is a specific property *has first name* that links a person to the nomen that represents the first name of this person: [?person mcc:MCC-R13-5-2 ?nomen]. This property makes the triple that indicates the nature of the nomen unnecessary.

Lastly, *has part* and *has derivation* allow for the creation of relationships between nomens, but the nature of these relations only stems from the schemes associated to each nomen. To clarify this aspect, we added several subproperties to *has part*, such as *includes first name* or *includes middle name* to link the nomen representing the full name and the nomens representing the first and middle names. We also added a subproperty of *is derivation*, *is initial of*. These additions clarify relations between nomens without the need to query the nomen scheme. This also simplifies the SPARQL query, for instance, to obtain the ordering of people by last name only.

All our new properties are shown in Table 2. We followed the convention of LRM, where the name of each property begins with *LRM-R* followed by a number, and start our properties with *MCC-R*. The first three columns represent the level of the property in the taxonomy, with elements of the second columns being subproperties of the previous element of the first column (MCC-R5-1 is a subproperty of LRM-R5).

Data Properties. We also refined some data properties to represent specific fields in our data, mainly for attributes of manifestations. First, we created some subproperties of *extent* to represent physical attributes such as *dimension* (*depth*, *height*, and *width*) or *number of pages*. We also added the representation of *price* and *original language*, among others, by creating a subproperty of the existing LRM properties, *Representative expression attribute* and *Language*.

5 Architecture

5.1 Preprocessing

Since our data comes from different sources, each with their specific format and standards, it is necessary, at the beginning of our process, to clean the data and unify its format. This process, although tedious, is quite straightforward. First, we created an intermediate representation containing all the relevant fields, necessary to answer the competency questions, and translated each dataset into

Table 2. Object properties in our extension of LRM

First level	Second level	Third level	Label	Domain	Range
LRM-R5			was created by	Work	Agent
	MCC-R5-1		was authored by	Work	Agent
LRM-R6			was created by	Expression	Agent
	MCC-R6-1		artist	Expression	Agent
	MCC-R6-2		was illustrated by	Expression	Agent
	MCC-R6-3		was written by	Expression	Agent
	MCC-R6-4		photographer	Expression	Agent
	MCC-R6-5		preface writer	Expression	Agent
	MCC-R6-6		postface writer	Expression	Agent
	MCC-R6-7		translator	Expression	Agent
	MCC-R6-8		adapter	Expression	Agent
	MCC-R6-9		editing coordinator	Expression	Agent
LRM-R13			has appellation	Res	Nomen
	MCC-R13-1		has classification	Work	Nomen
		MCC-R13-1-1	has Thema classification	Work	Nomen
		MCC-R13-1-2	has Dewey classification	Work	Nomen
	MCC-R13-2		has title statement	Work	Nomen
		MCC-R13-2-1	has abridged title	Work	Nomen
		MCC-R13-2-2	has uniform title	Work	Nomen
		MCC-R13-2-3	has complete title	Work	Nomen
		MCC-R13-2-4	has subtitle	Work	Nomen
	MCC-R13-3		has ID	Res	Nomen
		MCC-R13-3-1	has ISBN	Manifestation	Nomen
	MCC-R13-4		has URL	Res	Nomen
	MCC-R13-5		has personal name	Person	Nomen
		MCC-R13-5-1	has full name	Person	Nomen
		MCC-R13-5-2	has first name	Person	Nomen
		MCC-R13-5-3	has last name	Person	Nomen
		MCC-R13-5-4	has middle name	Person	Nomen
	MCC-R13-6		has award name	Award attribution	Nomen
LRM-R16			has part	Nomen	Nomen
	MCC-R16-1		includes first name	Nomen	Nomen
	MCC-R16-2		includes last name	Nomen	Nomen
	MCC-R16-3		includes main part of title	Nomen	Nomen
	MCC-R16-4		includes middle name	Nomen	Nomen
	MCC-R16-5		includes subtitle	Nomen	Nomen
LRM-R17			is derivation	Nomen	Nomen
	MCC-R17-1		is abridged form of	Nomen	Nomen

this format. Then, we implemented automated processes to clean the data, for example by removing leftover special characters or separating the field "lifespan" into birth date and death date.

5.2 Translation Based on Competency Questions

The process we followed to create the knowledge base is composed of three steps. The first one was to assess, for each competency question, if the information is available, and in which form, in each dataset. Next, we ensured that the chosen model, IFLA LRM, could represent these elements. Then, we translated the relevant subset of each dataset in RDF, according to the model. This process resulted in four LRM knowledge bases, one for each dataset. Lastly, we interconnected the four resulting knowledge bases by identifying the instances that represent the same real-world objects. For example, if two datasets contain a given book, then at the end of this alignment step, there should be only one entity that represents this book, with triples extracted from both datasets.

5.3 Alignment Strategies

In order to integrate our four IFLA LRM knowledge bases, we had to devise a method for automatically identifying identical entities (authors, books, manifestations, expressions). Given the sparsity of cross-dataset unique identifiers, we chose to rely on string edit distance heuristics and value matching on combinations of entities' attributes.

One unique identifier available across sources is the ISBN attached to manifestations. Exact matches on this identifier allow not only for resolving manifestation entities, but also expressions and works, as two expressions sharing the same manifestation must be the same expression, and two works sharing the same expression must be the same work.

Although LRM allows for linking a single manifestation to several expressions, and a single expression to several works, our data is not complete enough to make use of that possibility. We plan to address this aspect in future work.

For the alignment of the other elements (works, authors, publishers), we rely on heuristic rule-based matching. The rules for matching two entities of the same class are enumerated in Table 3, with Levenshtein ratios (as described in Eq. 1) for matching attributes indicated in parentheses. Pairs of entity records fulfilling at least one matching rule are considered to refer to the same real world entity. Each pair is analysed with all rules associated to its type. Any matched rule can identify a positive match, in no particular order. The alignment step provides us with a list of entity classes, each comprised of a list of URIs corresponding to the same entity.

For Publisher entities, we have at our disposal an authority list of Quebec publisher names, to which we match the publishers extracted from our datasets.

$$Levenshtein(str1,\ str2) = \left(1 - \frac{string\ edit\ distance}{max(len(str1),\ len(str2))}\right) \qquad (1)$$

Table 3. Entity matching rules, with experimentally-defined Levenshtein ratios indicated in parentheses

Entity	Rule	Match	Description
Work	1	Title (0.90) & Subtitle (0.90)	Similar titles and subtitles
	2	Title (0.90) & Author name (0.95)	Similar titles and Author names
	3	Title (0.90) & Publication year (1)	Similar titles and same publication years
Author	1	Name (1)	With sources without Author's Works
	2	Name (0.95) & Birth year (1)	Name and birth year match
	3	ISNI (1)	ISNI is a unique URI
	4	ILE URI (1)	ILE URI is a unique URI
	5	Name (0.95) & Work Title (0.80)	Similar names and one similarly titled Work
Publisher	1	Authority list (0.95)	Name close matched to authority list
	2	Publisher name (0.95)	Publishers with similar names

5.4 The Unified Knowledge Base

Using the list of aligned URIs, the objective is to generate a knowledge base able to answer SPARQL queries, both in general, and specifically on any number of sources ("Who are the authors of less than 40 years old, *according to ADP*?"). Furthermore, since appellations, such as the title of a book, can change depending on the source, even after the cleaning step, it was necessary to create "canonical nomens". To do so, we compared the nomen used in each dataset for this entity, if present, and chose one using a priority order: BAnQ, Hurtubise, Ile, ADP. For instance, if a person has three nomens in BAnQ, Ile and ADP, the canonical nomen is the one from BAnQ since it is higher on the priority list.

Therefore, we created five knowledge bases, one for each source and a canonical knowledge base. Each of these is self-sufficient and can be queried separately. The knowledge graphs specific to a source contain only the RDF translation of the source's data. Unless the user specifically asks for data according to a specific source, all queries are made on the canonical graph. This graph contains all the information available in all the data sources. Nomens in this integrated canonical graph are represented by their canonical values. An example of the separation of information between named graphs is given in Fig. 4. For readability purposes, this figure shows a simplified RDF representation, as the whole set of nomens used to represent a name (complete name, first name, last name, etc.), as well as the actual strings containing these parts, are aggregated in a single element.

In this example, the canonical graph is composed of the left half of the figure, and the Hurtubise graph is composed of the right half. Including the central part in both graphs is necessary to ensure that both are self-sufficient. In most use-cases, only the canonical graph is used, and therefore the only name that appears is "Louise Bail". However, if the user asks specifically for the data according to Hurtubise, then the second graph is used, and only "Bail Louise" appears. In this example, the canonical graph contains the publication place, but not the Hurtubise graph, since this information comes from the ILE dataset.

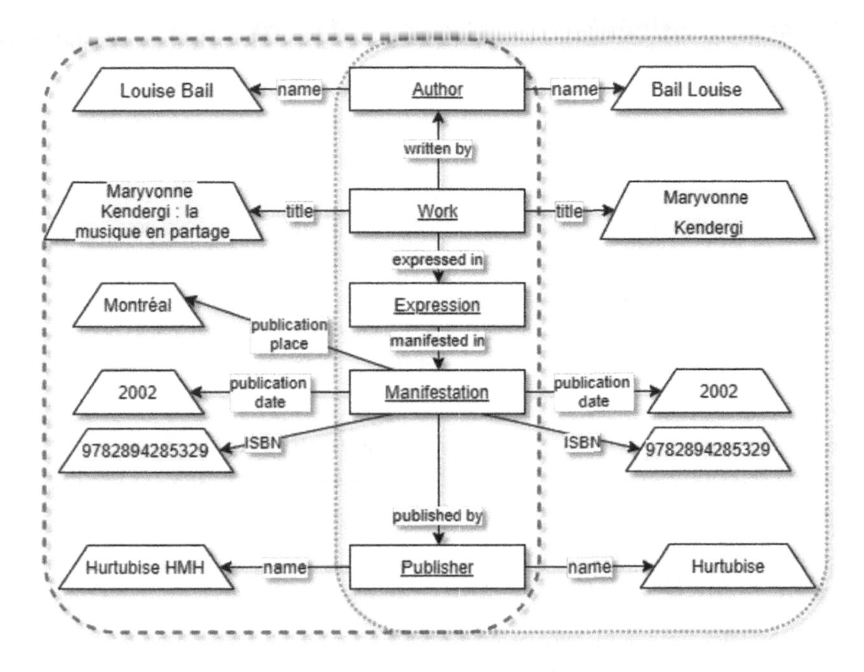

Fig. 4. The separation of information into graphs, after alignment, depending on the source. The section delimited by the red dashed line represents the canonical graph, whereas the section delimited by the green dotted line represents the Hurtubise graph.

6 Results

6.1 Descriptive Statistics of the Datasets

We provide, in Table 4, descriptive statistics of the LRM entities obtained after processing records in various formats, including the number of works, expressions, manifestations, authors and publishers in each source and in the unified knowledge base. We can notice the difference between the number of manifestations and the number of expressions/works. In the model, a single expression can be linked to several manifestations. As we explained in Sect. 3.4, we implemented a system to align records representing a manifestation, and aggregate them in an expression, before representing the resulting expressions in a work. The numbers here illustrate this process.

Table 4. The number of entities in the RDF translation of each dataset and in the integrated knowledge base

Entity type	BAnQ	ADP	Hurtubise	l'Île	Integrated knowledge base	
					Before alignment	After alignment
Work	50,140	13,707	1,012	21,726	86,585	74,764
Expression	58,476	14,772	1,056	26,203	100,507	87,007
Manifestation	79,687	15,399	2,505	27,662	125,283	109,240
Author	16,960	6,235	636	1,766	25,597	22,692
Publisher	3,112	38	1	2,315	5,466	3,921
Total	208,375	50,151	5,210	79,702	343,438	297,624

We can also note the difference between the total number of entities in each source and the number of entities in the unified knowledge base. The difference illustrates the alignment of various entities in the various datasets.

6.2 Alignment Evaluation

Heuristics for Authors and Works were evaluated on test sets made up of pairs of positive and negative matches, with 4 332 labeled pairs of Authors and 5 574 labeled pairs of Works. These sets were constructed from records with matching ISBNs (our highest confidence matches). An F1 score of **0.9955**, with a precision of 0.9945 and recall of 0.9965, was obtained on our Author test set. Although rule-based heuristics can attain high scores on Author matching, this is a significantly more trivial task than Work matching, where our heuristics achieve an F1 score of **0.9119**, with a precision of 0.9962, but a recall of only 0.8407. Our rules sacrifice recall in favour of precision, with a higher rate of false negatives but avoiding more false positives.

Combined with Table 4, since the alignment has a very high accuracy but not many matches are found in the complete knowledge base (as seen in the small difference between the two last columns), it is tempting to conclude that most entities (works, authors, etc.) are present in a few datasets only. To explore this possibility, we created a smaller subset of data. First, we selected 44 authors that are present in all four datasets. Second, we only extracted, from each dataset, information about these 44 authors and related entities (works, expressions, etc.). Third, we used the exact same process on this subset of data and computed the numbers displayed in Table 5.

Table 5. The number of entities in the integrated knowledge base, for a smaller subset, before and after alignment

Entity type	Before alignment	After alignment
Work	3,972	2,053
Expression	4,299	2,546
Manifestation	5,396	3,388

The numbers in this second table indicate that the alignment process is indeed able to find an appropriate number of matches between entities. Since this subset has been explicitly built to contain authors that are present in the four datasets, we would expect an important intersection between the works of these authors among these four. When looking at the Works, there are 2,053 works in the unified database compared to 3,972 before alignment. This confirms the hypothesis emitted earlier, that most entities are only present in one or two datasets.

6.3 Knowledge Base Evaluation

The ability of our knowledge base to answer our competency questions compared to each individual dataset is used to assess the success of our data integration.

In Table 6, we present the competency questions for which several datasets provide an answer, and compare the ability of each source to answer the question with the integrated knowledge base. We do not include in this table the questions for which information is simply not available in the data or not yet translated, and neither the questions for which only one dataset has the answer.

Overall, out of the twenty-nine initial competency questions, twenty can be answered by our data. Among these twenty, thirteen have an answer from a single source. The remaining seven are displayed in this table. The complete list of competency questions, as well as the number of entities for which information is available to answer each question, is provided online. Let n be the total number of entities in the complete knowledge base and $x_{d,q}$ the number of entities for which question q is answerable with the information from dataset d ($d \in \{ADP, BAnQ, Hurtubise, Ile, CompleteKB\}$). Each cell contains the ratio $((x_{d,q}/n) * 100)$ for a given question and dataset.

In this table, we can see that the information available in various datasets is complementary. For instance, ADP is only able to provide themes for 10.1% of works, and BAnQ for 65.6%. However, by combining them in a knowledge base, as well as the 0.7% of Hurtubise, we are able to provide a theme for 75.6% of all works. Even though these seven questions represent only about half of the questions the knowledge base is able to answer, this illustrates the interest of combining data from different sources in order to increase its reach.

Table 6. Comparison of the ability of each dataset and the integrated knowledge base to answer competency questions

Competency question	Proportion of entities with an answer (in %)				
	With only data from one source				With complete KB
	ADP	BAnQ	Hurtubise	Ile	
Questions about works					
What are the main themes of books?	10.1	65.6	0.7	0	75.6
Questions about manifestations					
Is this book published in large font, hardcover, paperback or digital format?	14.1	29.5	1.2	10.9	47.7
How many pages does this book have?	10.8	53.7	1.2	17.2	69.1
How much does this book cost?	14.0	32.4	0	0	44.3
What are the books that have been recently published?	14.1	44.4	0.7	25.2	68.4
How many people are currently publishing poetry in Québec? Science-fiction? Culinary books?	1.2	43.8	0.3	0	44.7
Who are the authors who published digital or audio books during the past five years?	14.1	29.5	1.2	10.9	46.8

7 Discussion and Conclusion

We demonstrated that the LRM is a relevant choice to represent our book-related data. However, it also has some limitations. The properties available to provide information about people, such as authors, and awards, are very limited. We therefore completed it by using the schema.org vocabulary in two ways: first, to provide properties that are out of scope for LRM, such as properties on persons and awards, and, second, as a redundant layer to represent our knowledge about books in boths models.

We also plan on connecting our knowledge base with external sources, mainly Wikidata, both for the completion of our data with Wikidata, and in the other direction, for the completion of Wikidata. Indeed, Wikidata does not contain all the information about Quebec writers. Out of the 1800 writers present in the data from ILE, more than 600 do not have a page at all. However, for those that have a Wikidata page, the information that it contains could be a relevant addition to our knowledge base.

Overall, the choice of IFLA LRM was appropriate for representing heterogeneous book-related data. With the addition of the schema.org vocabulary to deal with some aspects of our data that reach beyond the scope of LRM, we obtain a satisfactory ontology. The most delicate part of the translation process was the alignment step, both internal and external, to identify corresponding entities, because of the lack of both unique identifiers and rigorous standards for the writing of titles and names. The concept of nomen, central in the LRM, allows for a fine representation of the provenance of entities and appellations. We proposed heuristics for the alignement. Our future work will explore the use of transformer models for the alignment task. Finally, with the separation of data into a canonical and source graphs, it becomes possible to easily query the knowledge base on either all available data, or according to a particular source. In our future work, we plan to enrich our knowledge graph with data extracted from readers' social networks.

References

1. Alemu, G., Stevens, B., Ross, P., Chandler, J.: Linked data for libraries: benefits of a conceptual shift from library-specific record structures to RDF-based data models **113**(11), 549–570 (2012). https://doi.org/10.1108/03074801211282920, https://www.emerald.com/insight/content/doi/10.1108/03074801211282920/full/html
2. Carriero, V.A., Gangemi, A., Mancinelli, M.L., Nuzzolese, A.G., Presutti, V., Veninata, C.: Pattern-based design applied to cultural heritage knowledge graphs. Semantic Web (Preprint), pp. 1–45 (2019)
3. Coyle, K.: Designing data for use: from alphabetic order to linked data **24**(2), 154–159 (2011). https://doi.org/10.1629/24154, http://serials.uksg.org/articles/10.1629/24154/, number: 2 Publisher: UKSG in association with Ubiquity Press
4. Coyle, K.: MARC21 as data: A start (14) (2011). https://journal.code4lib.org/articles/5468/comment-page-1
5. Daly, F.: ONIX: The metadata standard for the information and entertainment industries **18**(2), 28–40 (2002). https://doi.org/10.1007/BF02687806
6. Deliot, C.: Publishing the British national bibliography as linked open data. Cat. Index **174**, 13–18 (2014)
7. DeWeese, K.P., Segal, D.: Libraries and the Semantic Web, Synthesis Lectures on Emerging Trends in Librarianship, vol. 1. Morgan & Claypool (2014). http://www.morganclaypool.com/doi/abs/10.2200/S00615ED1V01Y201411ETL003
8. Doerr, M.: The CIDOC conceptual reference module: an ontological approach to semantic interoperability of metadata. AI Mag. **24**(3), 75–75 (2003)
9. Dunsire, G.: Distinguishing content from carrier: the RDA/ONIX framework for resource categorization **13**(1) (2007). https://doi.org/10.1045/january2007-dunsire, http://www.dlib.org/dlib/january07/dunsire/01dunsire.html
10. Dunsire, G.: Interoperability and semantics in RDF representations of FRBR, FRAD and FRSAD. In: Concepts in Context: Proceedings of the Cologne Conference on Interoperability and Semantics in Knowledge Organization, p. 113 (2010)
11. Ferry, F., Zouaq, A., Gagnon, M., et al.: Automatic identification of relations in Quebec heritage data. In: Ioannides, M. (ed.) EuroMed 2018. LNCS, vol. 11196, pp. 188–199. Springer, Cham (2018). https://doi.org/10.1007/978-3-030-01762-0_16
12. Frosterus, M., Hyvönen, E., Laitio, J.: Creating and publishing semantic metadata about linked and open datasets. In: Wood, D. (ed.) Linking Government Data, pp. 95–112. Springer, New York (2011). https://doi.org/10.1007/978-1-4614-1767-5_5
13. Godby, C.: Mapping ONIX to MARC (2021)
14. Hanemann, J., Kett, J.: Linked data for libraries (2010). https://www.ifla.org/past-wlic/2010/149-hannemann-en.pdf
15. Hyvönen, E., Viljanen, K., Tuominen, J., Seppälä, K.: Building a national semantic web ontology and ontology service infrastructure –the FinnONTO approach. In: Bechhofer, S., Hauswirth, M., Hoffmann, J., Koubarakis, M. (eds.) ESWC 2008. LNCS, vol. 5021, pp. 95–109. Springer, Heidelberg (2008). https://doi.org/10.1007/978-3-540-68234-9_10

16. Kim, S., Ahn, J., Suh, J., Kim, H., Kim, J.: Towards a semantic data infrastructure for heterogeneous cultural heritage data-challenges of Korean Cultural Heritage Data Model (KCHDM). In: 2015 Digital Heritage, vol. 2, pp. 275–282. IEEE (2015)
17. Mäkelä, E., Hypén, K., Hyvönen, E., et al.: BookSampo—lessons learned in creating a semantic portal for fiction literature. In: Aroyo, L. (ed.) ISWC 2011. LNCS, vol. 7032, pp. 173–188. Springer, Heidelberg (2011). https://doi.org/10.1007/978-3-642-25093-4_12
18. Mäkelä, E., Hyvönen, E., Ruotsalo, T.: How to deal with massively heterogeneous cultural heritage data-lessons learned in CultureSampo. Semant. Web **3**(1), 85–109 (2012)
19. Malmsten, M.: Making a library catalogue part of the semantic web. Data Anal. Knowl. Discov. **3**(3), 3–7 (2009)
20. Marchand, E., Gagnon, M., Zouaq, A., et al.: Extraction of a knowledge graph from French cultural heritage documents. In: Bellatreche, L. (ed.) TPDL/ADBIS -2020. CCIS, vol. 1260, pp. 23–35. Springer, Cham (2020). https://doi.org/10.1007/978-3-030-55814-7_2
21. Mekhabunchakij, K.: Modeling linked open data for decision support in Thailand tourism. In: 7th International Conference on Restructuring of the Global Economy, pp. 89–94 (2017)
22. Noor, S., et al.: Modeling and representation of built cultural heritage data using semantic web technologies and building information model. Comput. Math. Organ. Theory **25**(3), 247–270 (2019). https://doi.org/10.1007/s10588-018-09285-y
23. Park, J.R., Richards, L.L., Brenza, A.: Benefits and challenges of BIBFRAME: cataloging special format materials, implementation, and continuing educational resources **37**(3), 549–565 (2019). https://doi.org/10.1108/LHT-08-2017-0176, publisher: Emerald Publishing Limited
24. Patton, G.E.: From FRBR to FRAD: extending the model. IFLA: World Library (2009)
25. Peroni, S., Tomasi, F., Vitali, F.: The aggregation of heterogeneous metadata in web-based cultural heritage collections: a case study. Int. J. Web Eng. Technol. **8**(4), 412–432 (2013)
26. Possemato, T.: How RDA is essential in the reconciliation and conversion processes for quality linked data **9**(1), 10 (2018). https://dialnet.unirioja.es/servlet/articulo?codigo=6260088, publisher: Dipartimenti di Scienze dell'Antichità, Medioevo e Rinascimento e Linguistica. Università degli Studi di Firenze Section: JLIS.it: Italian Journal of Library, Archives and Information Science. Rivista italiana di biblioteconomia, archivistica e scienza dell'informazione
27. Riva, P., Le Boeuf, P., Žumer, M., et al.: IFLA library reference model: a conceptual model for bibliographic information (2018)
28. Riva, P., Žumer, M.: FRBRoo, the IFLA library reference model, and now LRMoo: a circle of development (2017)
29. Simon, A., Wenz, R., Michel, V., Di Mascio, A.: Publishing bibliographic records on the web of data: opportunities for the BnF (French National Library). In: Cimiano, P., Corcho, O., Presutti, V., Hollink, L., Rudolph, S. (eds.) ESWC 2013. LNCS, vol. 7882, pp. 563–577. Springer, Heidelberg (2013). https://doi.org/10.1007/978-3-642-38288-8_38
30. Steele, T.D.: What comes next: understanding BIBFRAME **37**(3), 513–524 (2019). https://doi.org/10.1108/LHT-06-2018-0085, publisher: Emerald Publishing Limited
31. Stuart, D.: Facilitating Access to the Web of Data: A Guide for Librarians. Facet Publishing, London (2011)

32. Alma'aitah, W.Z., Talib, A.Z., Osman, M.A.: Opportunities and challenges in enhancing access to metadata of cultural heritage collections: a survey. Artif. Intell. Rev. **53**(5), 3621–3646 (2019). https://doi.org/10.1007/s10462-019-09773-w
33. Taniguchi, S.: Examining BIBFRAME 2.0 from the viewpoint of RDA metadata schema **55**(6), 387–412 (2017). https://doi.org/10.1080/01639374.2017.1322161, publisher: Routledge
34. Tharani, K.: Linked data in libraries: a case study of harvesting and sharing bibliographic metadata with BIBFRAME **34**(1), 5–19 (2015). https://doi.org/10.6017/ital.v34i1.5664, https://ejournals.bc.edu/index.php/ital/article/view/5664, number: 1
35. Tillett, D.B.: What is FRBR? A conceptual model for the bibliographic universe **54**(1), 24–30 (2005). https://doi.org/10.1080/00049670.2005.10721710, publisher: Routledge
36. Vila-Suero, D., Villazón-Terrazas, B., Gómez-Pérez, A.: datos.bne.es: a library linked dataset. Semant. Web **4**(3), 307–313 (2013)

Evaluating the Robustness of Embedding-Based Topic Models to OCR Noise

Elaine Zosa[1(✉)], Stephen Mutuvi[2,3], Mark Granroth-Wilding[1,4], and Antoine Doucet[2]

[1] University of Helsinki, Helsinki, Finland
elaine.zosa@helsinki.fi
[2] L3i Laboratory, University of La Rochelle, La Rochelle, France
[3] Multimedia University of Kenya, Nairobi, Kenya
[4] Silo AI, Helsinki, Finland

Abstract. Unsupervised topic models such as Latent Dirichlet Allocation (LDA) are popular tools to analyse digitised corpora. However, the performance of these tools have been shown to degrade with OCR noise. Topic models that incorporate word embeddings during inference have been proposed to address the limitations of LDA, but these models have not seen much use in historical text analysis. In this paper we explore the impact of OCR noise on two embedding-based models, Gaussian LDA and the Embedded Topic Model (ETM) and compare their performance to LDA. Our results show that these models, especially ETM, are slightly more resilient than LDA in the presence of noise in terms of topic quality and classification accuracy.

Keywords: Topic modelling · Word embeddings · OCR noise

1 Introduction

Large-scale collections of historical documents are becoming more accessible to researchers due to the efforts made to digitize these materials. Digitization pipelines commonly involve passing the material through an optical character recognition (OCR) engine which outputs text that can be used for downstream tasks. Due to various factors such as the printing quality of the original material, font, and layout styles, the output of OCR engines varies in quality. OCR errors stemming from this process can have a significant impact when downstream natural language processing (NLP) tools are used to analyse this data.

Topic modelling is a method to extract latent topics in a collection of documents. It is a popular approach in Digital Humanities and data-driven historical research to analyse large historical collections such as newspaper archives [9,16,18], academic journals [11] and handwritten diaries [3]. Probabilistic topic models such as the Latent Dirichlet Allocation (LDA) [2] model a topic as a distribution over a vocabulary and a document as a mixture of topics. Prior

H.-R. Ke et al. (Eds.): ICADL 2021, LNCS 13133, pp. 392–400, 2021.
https://doi.org/10.1007/978-3-030-91669-5_30

research quantifying the impact of OCR noise on topic modelling shows that the topics and topic mixtures deteriorate in quality as the level of noise increases [12,17].

Word embeddings are distributed representations of words in a dense vector space that encode their usage in a corpus [10,14]. They can capture both syntactic and semantic attributes of words such that words that typically occur in similar contexts are in close proximity to each other in the embedding space. Approaches that combine topic modelling with word embeddings to improve the semantic coherence of topics and address the challenge of scaling topic models to large vocabularies include Gaussian LDA [4], spherical Hierarchical Dirichlet Process (sHDP) [1], and the Embedded Topic Model (ETM) [5]. GLDA and ETM are LDA-like models that use word embeddings and have shown improved topic quality over LDA on clean datasets.

Non-embedded topic models like LDA use word co-occurrence statistics to discover latent topics in a corpus and the negative impact of OCR noise on topic modelling is due to the distortion of the word distributions when words are misspelled [17]. In embedding-based models, word identities are replaced with word *embeddings* that, in principle, can be more resilient to OCR noise, provided misspellings of the same word cluster together in the embedding space. There is, however, no existing work that investigates the robustness of these models on data with OCR noise and whether they show any improvement over LDA.

In this paper we conduct a quantitative assessment of the performance of two embedding-based models, GLDA and ETM, on datasets with OCR noise. Our aim is to test whether embedding-based models can be used to improve the analysis of digitised historical documents.

2 Related Work

Latent Dirichlet Allocation (LDA) [2] is a probabilistic topic modelling method for extracting topics from a document collection. It models a topic as a probability distribution over a fixed vocabulary and a document as a mixture of topics. LDA relies on the co-occurrence of the words in the documents to infer the latent topics and topic mixtures of the documents.

Models that use word embeddings have been proposed to improve topic quality and handle out-of-vocabulary words. Gaussian LDA (GLDA) [4] is the first LDA-based topic model that directly incorporates word embeddings during topic inference. Instead of treating topics as categorical distributions over the vocabulary, GLDA characterizes topics as multivariate Gaussian distributions over the word embedding space whose mean and variance are estimated during topic inference. Words are ranked according to their probability density under the posterior-predictive distribution given the training corpus.

In the Embedded Topic Model (ETM) [5], topics and words share the same embedding space and a topic is a point in the embedding space called a topic embedding. Words are generated from a categorical distribution whose natural parameter is the inner product of the word embeddings associated with a topic and the respective topic embedding. The most probable words in the topic are those with embeddings that are close to the topic embedding.

Various studies have evaluated the impact of OCR errors on unsupervised topic modelling. A comparative study of document clustering and topic modelling on OCRed text indicated that OCR noise had a greater performance impact on topic modelling than on document clustering [17]. Another evaluation revealed that while OCR noise resulted in lower topic coherence, it had little impact on model stability [12]. A more general study on the impact of noisy OCR on historical text analysis using a corpus of eighteenth-century texts found that topics extracted from OCRed texts aligned well with topics from the gold standard texts although the authors hinted that the topic model had trouble with poetry-adjacent topics [7]. These previous evaluations, however, focused on well-established topic models based on word co-occurrence and as far as we are aware embedding-based models have not been tested to analyse OCR-ed data.

3 Methodology

Following [17], we first evaluate the topic models on a corpus of historical documents with real OCR noise that have aligned gold standard (GS) texts. Then we evaluate the models on a larger corpus where synthetic noise has been introduced at increasing levels.

3.1 Datasets

Real Noise. The Overproof dataset consists of 30,301 digitised news articles from the Sydney Morning Herald 1842–1954, from the archives of the National Library of Australia [6][1]. The articles were processed using the ABBYY FineReader OCR tool and additional corrections were done using crowd-sourced annotations. The OCRed articles have a word error rate (WER) of 25% [13]. The OCR and GS articles are aligned on a character level.

Synthetic Noise. To generate data with synthetic noise, we start with a clean dataset and gradually corrupt the data by introducing noise at increasing levels. We use the Reuters RCV1 dataset as the clean dataset. This consists of over 800K English news wire articles with assigned categorical labels [8]. We use a reduced dataset of 50K articles sampled from the largest categories.

We follow a procedure that generates synthetic noise based on a noise model constructed from a dataset with real noise [17]. To build a noise model, we construct a matrix $\mathbf{M}$ where $\mathbf{M}_{x,y}$ is the number of times character x in a GS article is confused with character y in the corresponding OCR article.

To generate parameterised noise, we interpolate the matrix $\mathbf{M}$ such that $\mathbf{M}_\gamma = \gamma\mathbf{M} + (1-\gamma)\mathbf{I}$ where γ is the interpolation parameter. When $\gamma = 0$, no noise is introduced, while at $\gamma = 1.0$, the interpolated matrix is equivalent to $\mathbf{M}$.

[1] http://overproof.projectcomputing.com/datasets/.

We generate corrupted datasets from the Reuters corpus with γ ranging from 0 to 1 in increments of 0.2. This resulted in datasets with character error rates (CER) of 0%, 7%, 14%, 21%, 28% and 35%. Table 1 summarizes the datasets used in our experiments.

Table 1. Datasets used in the experiments.

	#types	#tokens	#art.
Overproof OCR	1.3M	10M	30K
Overproof GS	414K	9.8M	30K
Reuters	414K	12.4M	50K

3.2 Model Training and Word Embeddings

We use LDA as our baseline model. We trained LDA models using the Gensim library[2], leaving the prior parameters to be inferred during training. For ETM, we used the authors' implementation[3] with default hyperparameters. For GLDA, we used the `gaussianlda` package, which implements the algorithm in Python[4]. We ran the sampler for 20 iterations, based on initial experiments with the clean *20-Newsgroups* dataset.

In our experiments with real noise data, we experimented with two different types of word embeddings: (1) pre-trained GloVe embeddings trained on English Wikipedia and Gigaword [14][5]; and (2) word2vec embeddings [10] trained on the Overproof data (we trained separate embeddings for the OCR and GS portions of the data). This is to investigate whether word embeddings trained on a large amount of clean data result in better topic models than embeddings trained on more limited and noisier data. On experiments with synthetic data, we used word2vec embeddings trained on the corrupted Reuters data. We trained separate embeddings for each noise level.

We trained topic models with 50 topics on the OCR and GS portions of the Overproof data and 100 topics for each noise setting of the synthetic Reuters data. To account for the randomness inherent in the models we repeated each experiment ten times and report the averaged results.

3.3 Evaluation Measures

Topic Coherence. Topic coherence quantifies the interpretability of a topic as represented by its most probable terms. We use the C_v coherence measure [15] implemented in the Gensim package[6].

[2] https://radimrehurek.com/gensim.
[3] https://github.com/adjidieng/ETM.
[4] https://pypi.org/project/gaussianlda/.
[5] https://nlp.stanford.edu/projects/glove/.
[6] https://radimrehurek.com/gensim/models/coherencemodel.html.

Topic Diversity. Models that learn more diverse topics are preferable to models with redundant topics. We measure topic diversity as the proportion of unique words out of all the top words representing all the topics in the model [5]. For topic coherence and diversity, we evaluate on the top 20 terms of each topic.

Classification Accuracy. We evaluate the quality of the per-document topic proportions inferred by the models through a supervised document classification task. We train a classifier on a portion of the data using the inferred topic proportions as features and pre-assigned categories as labels, then test the classifier on the unseen portion. As this evaluation requires gold standard labels, we only run this evaluation on the Reuters dataset with synthetic noise. We used a logistic regression classifier with ten-fold cross-validation in our evaluation.

4 Results and Discussion

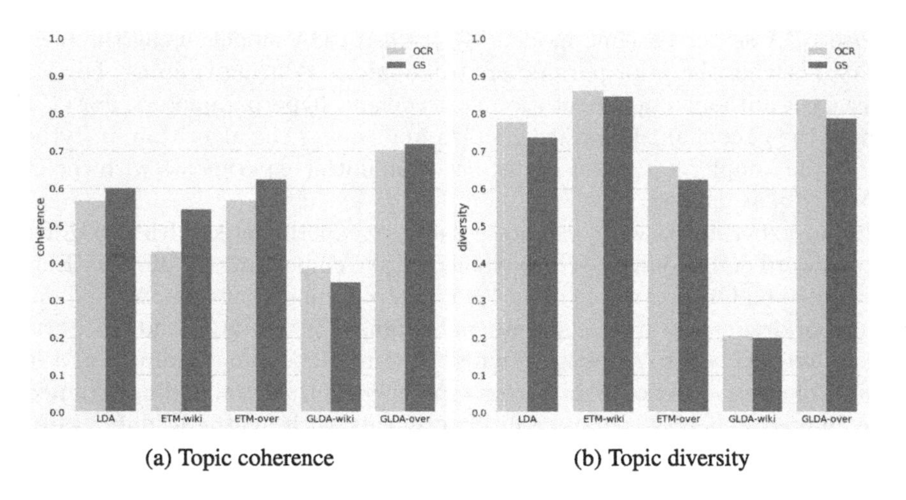

(a) Topic coherence (b) Topic diversity

Fig. 1. Performance on the Overproof dataset averaged over 10 runs. *wiki* models use word embeddings trained on Wikipedia while *over* models use embeddings trained on the Overproof data.

4.1 Performance on Real Noise

Figure 1 shows the results of our experiments on real noise data. In terms of topic coherence, almost all the models perform better on the GS documents than the OCR documents, as would be expected (Fig. 1a). GLDA with Overproof embeddings is the best performing model while GLDA with Wikipedia embeddings is the worst-performing. ETM with Overproof embeddings has similar topic coherences to LDA–both have a coherence of 0.57 for OCR while for GS, ETM is only a little better with a mean coherence of 0.62 and LDA has 0.6. ETM with Wikipedia embeddings performs worse than LDA with a coherence of 0.43 for OCR and 0.54 for GS.

Table 2. Most coherent topics from LDA, ETM, and Gaussian LDA on the Overproof dataset.

Topic No.	*Top words*	*Coh*
LDA-OCR		
33	Petitioner, respondent, nisi, decree, honor, formerly, appeared, ground, marriage, granted	0.95
8	Club, match, team, cricket, played, play, runs, first, association, matches	0.83
LDA-GS		
11	Petitioner, marriage, decree, respondent, formerly, nisi, appeared, married, ground, granted	0.95
9	Accused, prisoner, charged, guilty, charge, court, trial, stealing, months, sessions	0.82
ETM-OCR		
31	Respondent, petitione, nisi, appeared, honor, formerly, decree, ground, issue, foi	0.91
9	Charged, court, fined, john, police, prisoner, two, sentenced, months, guilty	0.81
ETM-GS		
21	Petitioner, marriage, appeared, formerly, respondent, decree, ground, nisi, married, granted	0.95
41	Match, cricket, team, played, wickets, runs, play, second, first, club	0.88
GLDA-OCR		
12	Managers, woiking, administrator, guidance, servlco, goneral, publicity, lenders, bown	0.73
38	Accompanying, pipers, recoived, governors, alio, transmitted, photographs, btato, lag	0.73
GLDA-GS		
47	Parent, outset, sult, cardiff, terror, dawn, tha, alley, biggest, sweepin	0.72
1	Discontinued, livered, forcibly, blacksmith, extracted, interrupted, reopened, sampson, tempted	0.72

These results indicate that for embedding-based topic models, it is preferable to use embeddings trained on the target corpus rather than on a general-knowledge dataset like Wikipedia, despite the latter being larger in size and cleaner, especially when the target corpus is a specialized document collection, such as historical documents. One reason for this could be that Wikipedia is a modern dataset while the Overproof corpus is made up of articles from the mid-nineteenth to the mid-twentieth century.

Now we take a closer look at the characteristics of the topics produced by one run of each of the models (Table 2). We focus on ETM and GLDA with Overproof embeddings. We see that the most coherent ETM and LDA topics are more coherent than the GLDA topics despite GLDA having the best mean topic coherence overall. GLDA is known to produce qualitatively different topics from LDA [4] and we notice that it also produces qualitatively different topics from ETM. Another difference is that topics produced by ETM on the OCR documents show a high degree of correspondence with topics from the GS data, while the same cannot not be said of the GLDA topics. For instance, Topic 31 of ETM-OCR and Topic 21 of ETM-GS are topics on legal matters and show many overlapping terms (they share 17 of their top 20 terms). We found no such correspondences with the GLDA topics.

In terms of topic diversity, OCR topics are more diverse than GS topics for all models (Fig. 1b). We hypothesize that this is primarily due to the higher vocabulary size of the OCR documents resulting from misspellings. While the training data used for word embeddings has a high impact on the coherence of the embedding-based models, it does not seem have a significant influence on topic diversity. ETM with Wikipedia embeddings has the most diverse topics while GLDA with the same Wikipedia embeddings has the most redundant topics.

4.2 Performance on Synthetic Noise

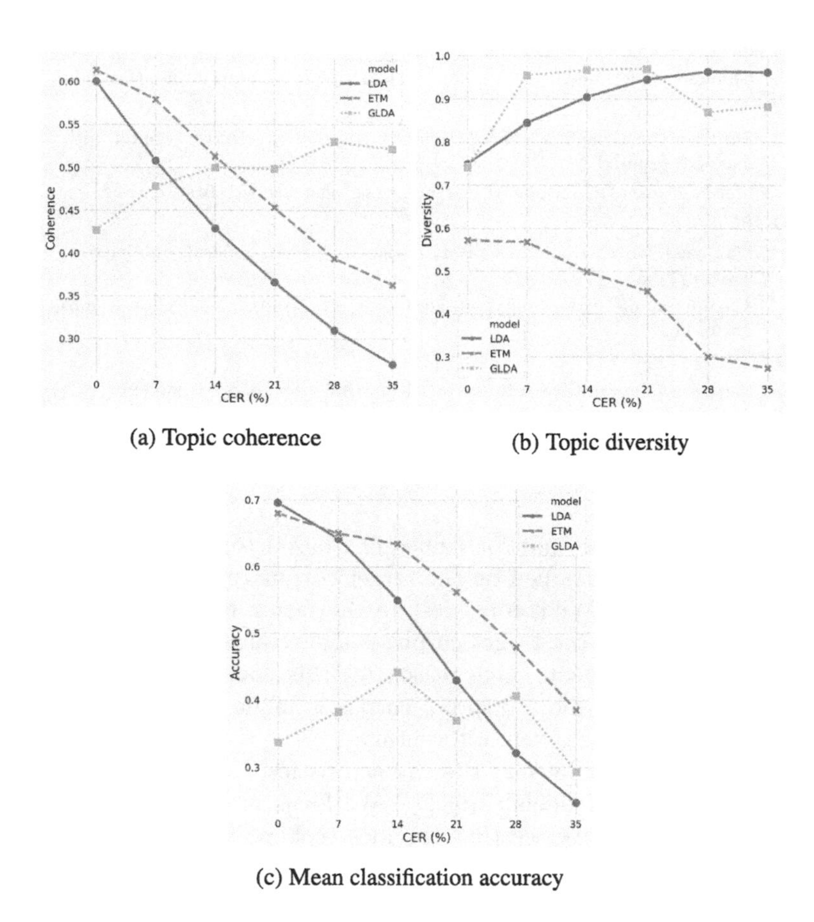

(a) Topic coherence

(b) Topic diversity

(c) Mean classification accuracy

Fig. 2. Performance on the Reuters data with synthetic noise averaged across 10 runs.

Experimental results on the synthetic data are shown in Fig. 2. ETM and LDA degrade linearly in coherence as noise increases, but ETM is more resilient than LDA (Fig. 2a). Interestingly, GLDA improves in coherence as noise increases.

We think one reason for this is that as an effect of the nature of GLDA topics, which are unimodal distributions in the embedding space, GLDA topics have the tendency to cluster misspelled words together. This leads to topics that, while having a high coherence score, are not qualitatively meaningful.

With regards to topic diversity, our results show that ETM produces less diverse topics than LDA or GLDA at all noise levels (Fig. 2b), corroborating our results in the real noise data (Fig. 1b). As noise increases, ETM topics become even less diverse (at 35% CER, diversity is at 0.27, 0.88, and 0.96 for ETM, GLDA and LDA, respectively). It is surprising therefore to find that even though ETM has the lowest topic diversity, it performs better than LDA and GLDA in the document classification task (Fig. 2c). On further investigation we found that for ETM and LDA, the topics that are most relevant in document classification are diverse and, for the most part, are preserved across noise levels while the redundant topics tend to have smaller weights that do not impact the classification performance significantly.

5 Conclusions

In this paper we assessed the impact of real and synthetic OCR noise on two embedding-based topic models, Gaussian LDA and ETM, with LDA as our baseline. We also experimented with different word embeddings for GLDA and ETM.

On real noise, GLDA is the best-performing model in terms of topic coherence while. ETM performs as well as LDA. ETM and GLDA produce more diverse topics than LDA. We note, however, that GLDA produces qualitatively different topics than ETM and LDA. Our experiments on synthetic data revealed that ETM performed better than LDA in terms of topic coherence and classification accuracy across noise levels. On the other hand, GLDA improved in topic coherence with increased noise and produced more varied topics but performed worse in document classification because its topics do not correlate with the gold standard labels in the dataset.

LDA is a popular method for analysing digitised historical collections but it is not without its shortcomings, especially when applied to documents with OCR errors. In our experiments, we have shown that topic models that incorporate information from word embeddings improve slightly over LDA in the presence of OCR noise in terms of coherence, diversity, and document classification.

Acknowledgements. This work has been supported by the European Union Horizon 2020 research and innovation programme under grants 770299 (NewsEye) and 825153 (EMBEDDIA).

References

1. Batmanghelich, K., Saeedi, A., Narasimhan, K., Gershman, S.: Nonparametric spherical topic modeling with word embeddings. In: Proceedings of the Conference. Association for Computational Linguistics. Meeting, vol. 2016, p. 537. NIH Public Access (2016)

2. Blei, D.M., Ng, A.Y., Jordan, M.I.: Latent dirichlet allocation. J. Mach. Learn. Res. **3**(Jan), 993–1022 (2003)
3. Blevins, C.: Topic modeling Martha Ballard's diary. Cameron Blevins (2010)
4. Das, R., Zaheer, M., Dyer, C.: Gaussian LDA for topic models with word embeddings. In: Proceedings of the 53rd Annual Meeting of the Association for Computational Linguistics and the 7th International Joint Conference on Natural Language Processing (Volume 1: Long Papers), pp. 795–804 (2015)
5. Dieng, A.B., Ruiz, F.J., Blei, D.M.: Topic modeling in embedding spaces. Trans. Assoc. Comput. Linguist. **8**, 439–453 (2020)
6. Evershed, J., Fitch, K.: Correcting noisy OCR: context beats confusion. In: Proceedings of the First International Conference on Digital Access to Textual Cultural Heritage, pp. 45–51 (2014)
7. Hill, M.J., Hengchen, S.: Quantifying the impact of dirty OCR on historical text analysis: eighteenth century collections online as a case study. Digit. Scholarsh. Humanit. **34**(4), 825–843 (2019)
8. Lewis, D.D., Yang, Y., Rose, T.G., Li, F.: Rcv1: a new benchmark collection for text categorization research. J. Mach. Learn. Res. **5**(Apr), 361–397 (2004)
9. Marjanen, J., Zosa, E., Hengchen, S., Pivovarova, L., Tolonen, M.: Topic modelling discourse dynamics in historical newspapers. arXiv preprint arXiv:2011.10428 (2020)
10. Mikolov, T., Chen, K., Corrado, G., Dean, J.: Efficient estimation of word representations in vector space. arXiv preprint arXiv:1301.3781 (2013)
11. Mimno, D.: Computational historiography: data mining in a century of classics journals. J. Comput. Cult. Herit. (JOCCH) **5**(1), 1–19 (2012)
12. Mutuvi, S., Doucet, A., Odeo, M., Jatowt, A.: Evaluating the impact of OCR errors on topic modeling. In: Dobreva, M., Hinze, A., Žumer, M. (eds.) ICADL 2018. LNCS, vol. 11279, pp. 3–14. Springer, Cham (2018). https://doi.org/10.1007/978-3-030-04257-8_1
13. Nguyen, T.T.H., Jatowt, A., Coustaty, M., Nguyen, N.V., Doucet, A.: Deep statistical analysis of OCR errors for effective post-OCR processing. In: 2019 ACM/IEEE Joint Conference on Digital Libraries (JCDL), pp. 29–38. IEEE (2019)
14. Pennington, J., Socher, R., Manning, C.D.: Glove: global vectors for word representation. In: Proceedings of the 2014 Conference on Empirical Methods in Natural Language Processing (EMNLP), pp. 1532–1543 (2014)
15. Röder, M., Both, A., Hinneburg, A.: Exploring the space of topic coherence measures. In: Proceedings of the Eighth ACM International Conference on Web Search and Data Mining, pp. 399–408 (2015)
16. Viola, L., Verheul, J.: Mining ethnicity: discourse-driven topic modelling of immigrant discourses in the USA, 1898–1920. Digital Scholarship in the Humanities (2019)
17. Walker, D., Lund, W.B., Ringger, E.: Evaluating models of latent document semantics in the presence of OCR errors. In: Proceedings of the 2010 Conference on Empirical Methods in Natural Language Processing, pp. 240–250 (2010)
18. Yang, T.I., Torget, A., Mihalcea, R.: Topic modeling on historical newspapers. In: Proceedings of the 5th ACL-HLT Workshop on Language Technology for Cultural Heritage, Social Sciences, and Humanities, pp. 96–104 (2011)

Pattern-Based Acquisition of Scientific Entities from Scholarly Article Titles

Jennifer D'Souza[1(✉)] and Sören Auer[1,2]

[1] TIB Leibniz Information Centre for Science and Technology, Hannover, Germany
{jennifer.dsouza,auer}@tib.eu

[2] L3S Research Center at Leibniz University of Hannover, Hannover, Germany

Abstract. We describe a rule-based approach for the automatic acquisition of salient scientific entities from Computational Linguistics (CL) scholarly article titles. Two observations motivated the approach: (i) noting salient aspects of an article's contribution in its title; and (ii) pattern regularities capturing the salient terms that could be expressed in a set of rules. Only those lexico-syntactic patterns were selected that were easily recognizable, occurred frequently, and positionally indicated a scientific entity type. The rules were developed on a collection of 50,237 CL titles covering all articles in the ACL Anthology. In total, 19,799 *research problems*, 18,111 *solutions*, 20,033 *resources*, 1,059 *languages*, 6,878 *tools*, and 21,687 *methods* were extracted at an average precision of 75%.

Keywords: Terminology extraction · Rule-based system · Natural language processing · Scholarly knowledge graphs · Semantic publishing

1 Introduction

Scientists increasingly face the information overload-and-drown problem even in narrow research fields given the ever-increasing flood of scientific publications [19,21]. Recently, solutions are being implemented in the domain of the digital libraries by transforming scholarly articles into "digital-first" applications as machine-interpretable scholarly knowledge graphs (SKGs), thus enabling completely new technological assistance to navigate the massive volumes of data through intelligent search and filter functions, and the integration of diverse analytics tools. There are several directions to this vision focused on representing, managing and linking metadata about articles, people, data and other relevant keyword-centered entities (e.g., Research Graph [3], Scholix [7], Springer-Nature's SciGraph or DataCite's PID Graph [9], SemanticScholar [1]). This trend tells us that we are on the cusp of a great change in the digital technology applied to scholarly knowledge. Notably, next-generation scholarly digital library (DL) infrastructures have arrived: the Open Research Knowledge Graph (ORKG) [18]

Supported by TIB Leibniz Information Centre for Science and Technology, the EU H2020 ERC project ScienceGRaph (GA ID: 819536).

H.-R. Ke et al. (Eds.): ICADL 2021, LNCS 13133, pp. 401–410, 2021.
https://doi.org/10.1007/978-3-030-91669-5_31

digital research and innovation infrastructure by TIB and partner institutions, argues for obtaining a semantically rich, interlinked KG representations of the "content" of the scholarly articles, and, specifically, only *research contributions.*[1] With intelligent analytics enabled over such contributions-focused SKGs, researchers can readily track research progress without the cognitive overhead that reading dozens of articles impose. A typical dilemma then with building such an SKG is deciding the type of information to be represented. In other words, what would be the information constituent candidates for an SKG that reflects the overview? While the scope of this question is vast, in this paper, we describe our approach designed with this question as the objective.

"*Surprisingly useful information can be found with only a very simply understanding of the text.*" [14] The quotation is the premise of the "Hearst" system of patterns which is a popular text mining method in the CL field. It implemented discovering lexical relations from a large-scale corpus simply by looking for the relations expressed in well-known ways. This simple but effective strategy was leveraged in supporting the building up of large lexicons for natural language processing [15], e.g., the WordNet lexical project [24]. Our approach is inspired after the "Hearst" methodology but on scholarly article titles content thereby implementing a pattern-based acquisition of scientific entities. Consider the two paper title examples depicted in Table 1. More fluent readers of English can phrase-chunk the titles based on lexico-syntactic patterns such as the colon punctuation in title 1 and prepositional phrase boundary markers (e.g., 'to' in title 2). Following which, with some domain awareness, the terms can be semantically conceptualized or typed (e.g., as *research problem*, *resource*, *method*, *tool*, etc.). Based on such observations and circling back to the overarching objective of this work, we propose and implement a pattern-based acquisition approach to mine contribution-focused, i.e. salient, scientific entities from article titles. While there is no fixed notion of titles written with the purpose of reflecting an article's contribution, however, this is the generally known practice that it contains salient aspects related to the *contribution* as a single-line summary. To the best of our knowledge, a corpus of only article titles remains as yet comprehensively unexplored as a resource for SKG building. Thus, our work sheds a unique and novel light on SKG construction representing *research overviews.*

In this paper, we discuss CL-Titles-Parser – a tool for extracting salient scientific entities based on a set of lexico-syntactic patterns from titles in Computational Linguistics (CL) articles. Six concept types of entities were identified applicable in CL titles, viz. *research problem*, *solution*, *resource*, *language*, *tool*, and *method.* CL-Titles-Parser when evaluated on almost all titles (50,237 of 60,621 total titles) in the ACL Anthology performs at a cumulative average of 75% IE precision for the six concepts. Thus, its resulting high-precision SKG integrated in the ORKG can become a reliable and essential part of the scientist's workbench in visualizing the overview of a field or even as crowdsourcing signals for authors to describe their papers further. CL-Titles-Parser is released as a standalone program https://github.com/jd-coderepos/cl-titles-parser.

[1] The ORKG platform can be accessed online: https://orkg.org/.

Table 1. Two examples of scholarly article titles with their concept-typed scientific terms which constitutes the IE objective of the CL-TITLES-PARSER

SemEval-2017 Task 5: Fine-Grained Sentiment Analysis on Financial Microblogs and News *research_problem*: ['SemEval-2017 Task 5'] *resource*: ['Financial Microblogs and News'] *method*: ['Fine-Grained Sentiment Analysis']
Adding Pronunciation Information to Wordnets *solution*: ['Adding Pronunciation Information'] *tool*: ['Wordnets']

2 Related Work

Key Summary of Research in Phrasal Granularity. To bolster search technology, the phrasal granularity was used to structure the scholarly record. Thus scientific phrase-based entity annotated datasets in various domains including multidisciplinarily across STEM [4,10,13,22] were released; machine learning systems were also developed for automatic scientific entity extraction [2,5,6,23]. However, none of these resources are clearly indicative of capturing only the salient terms about research contributions which is the aim of our work.

Pattern-Based Scientific Terminology Extraction. Some systems [16] viewed key scholarly information candidates as problem-solution mentions. [8] used the discourse markers "thus, therefore, then, hence" as signals of problem-solution patterns. [12] used semantic extraction patterns learned via bootstrapping to the dependency trees of sentences in Abstracts to mine the research focus, methods used, and domain problems. Houngbo and Mercer [17] extracted the methods and techniques from biomedical papers by leveraging regular expressions for phrase suffixes as "algorithm," "technique," "analysis," "approach," and "method." AppTechMiner [26] used rules to extract application areas and problem solving techniques. The notion of application areas in their model is analogous to research problem in ours, and their techniques are our tool or method. Further, their system extracts research problems from the article titles via rules based on functional keywords, such as, "for," "via," "using" and "with" that act as delimiters for such phrases. CL-TITLES-PARSER also extracts problems from titles but it does so in conjunction with other information types such as tools or methods. AppTechMiner uses citation information to determine term saliency. In contrast, since we parse titles, our data source itself is indicative of the saliency of the scientific terms therein w.r.t. the article's contribution. Finally, [20], like us, use a system of patterns to extract methods from the titles and Abstracts of articles in Library Science research. We differ in that we extract six different types of scientific entities and we focus only on the article titles data source.

Next, in the article, we describe the CL-TITLES-PARSER for its pattern-based acquisition of scientific entities from Computational Linguistics article titles.

3 Preliminary Definitions

We define the six scientific concept types handled in this work. The main aim here is not to provide rigorous definitions, but rather just to outline essential features of the concepts to explain the hypotheses concerning their annotation.

i. Research problem. The theme of the investigation. E.g., "Natural language inference." In other words, the answer to the question "which problem does the paper address?" or "On what topic is the investigation?" ***ii. Resource.*** Names of existing data and other references to utilities like the Web, Encyclopedia, etc., used to address the *research problem* or used in the *solution.* E.g., "Using Encyclopedic Knowledge for Automatic Topic Identification." In this sentence, "Encyclopedic Knowledge" is a *resource* used for *research problem* "Automatic Topic Identification." ***iii. Tool.*** Entities arrived at by asking the question "Using what?" or "By which means?" A *tool* can be seen as a type of a *resource* and specifically software. ***iv. Solution.*** A novel contribution of a work that solves the *research problem.* E.g., from the title "PHINC: A Parallel Hinglish Social Media Code-Mixed Corpus for Machine Translation," the terms 'PHINC' and 'A Parallel Hinglish Social Media Code-Mixed Corpus' are solutions for the *problem* 'Machine Translation.' ***v. Language.*** The natural language focus of a work. E.g., Breton, Erzya, Lakota, etc. *Language* is a pertinent concept w.r.t. an overview SKG about NLP solutions. ***vi. Method.*** They refer to existing protocols used to support the *solution*; found by asking "How?"

4 Tool Description

4.1 Formalism

Every CL title T can be expressed as one or more of the following six elements $te_i = \langle rp_i, res_i, tool_i, lang_i, sol_i, meth_i \rangle$, representing the *research problem, resource, tool, language, solution,* and *method* concepts, respectively. A title can contain terms for zero or more of any of the concepts. The goal of CL-TITLES-PARSER is, for every title t_i, to annotate its title expression te_i, involving scientific term extraction and term concept typing.

4.2 Rule-Based Processing Workflow

CL-TITLES-PARSER operates in a two-step workflow. First, it aggregates titles as eight main template types with a default ninth category for titles that could not be clustered by any of the eight templates. Second, the titles are phrase-chunked and concept-typed based on specific lexico-syntactic patterns that are group-specific. The concept type is selected based on the template type category and some contextual information surrounding the terms such as prepositional and verb phrase boundary markers.

Step 1: Titles clustering based on commonly shared title lexico-syntactic patterns. While our rule-based system implements eight patterns in total, we describe four template patterns as examples.

Template "hasSpecialCaseWord():" applies to titles written in two parts – a one-word solution name, a colon separator, and an elaboration of the solution name. E.g., "SNOPAR: A Grammar Testing System" consisting of the one word 'SNOPAR' solution name and its elaboration 'A Grammar Testing System.' Further, there are other instances of titles belonging to this template type that are complex sentences, i.e. titles with additional prepositional or verb phrases, where mentions of the *research problem*, *tool*, *method*, *language* domain etc. are also included in the latter part of the title. E.g., "GRAFON: A Grapheme-to-Phoneme Conversion System for Dutch" is a complex title with a prepositional phrase triggered by "for" specifying the *language* domain "Dutch."

Template "Using ..." applies to titles that begin with the word "Using" followed by a *resource* or *tool* or *method* and used for the purpose of a *research problem* or a *solution*. E.g., the title "Using WordNet for Building WordNets" with *resource* "WordNet" for *solution* "Building WordNets"; or "Using Multiple Knowledge Sources for Word Sense Discrimination" with *resource* "Multiple Knowledge Sources" for *research problem* "Word Sense Discrimination."

Template "... case study ..." Titles in this category entail splitting the phrase on either side of "case study." The first part is processed by the precedence-ordered rules to determine the concept type. The second part, however, is directly cast as *research problem* or *language* since they were observed as one of the two. The checks for *research problem* or *language* are made by means of regular expressions implemented in helper functions. E.g., the title "Finite-state Description of Semitic Morphology: A Case Study of Ancient Accadian" would be split as "Finite-state Description of Semitic Morphology" and "Ancient Accadian" language domain, where "Ancient Accadian" is typed as *language* based on regex patterns. See Table 2 for examples of some regular expressions employed.

Table 2. Regular expressions in suffix patterns for scientific term concept typing

languages	$reLanguage = (...\|Tigrigna\|Sundanese\|Balinese\|...)$
tool	$reTool = (...\|memory\|controller\|workbench(es)?\|...)$
resource	$reResource = (...\|corp(ora\|us)\|vocabular(ies\|y)\|cloud\|...)$
method	$reMethod = (...\|protocol\|methodolog(ies\|y)\|recipe\|...)$

Template "... : ..." A specialized version of this template is "hasSpecialCaseWord():". Here, titles with two or more words in the phrase preceding the colon are considered. They are split in two parts around the colon. The parts are then further processed to extract the scientific terms. E.g., "Working on the Italian Machine Dictionary: A Semantic Approach" split as "Working on the Italian Machine Dictionary" and "A Semantic Approach" where the second part is a non-scientific information content phrase. By non-scientific information content phrase we mean phrases that cannot be categorized as one of the six concepts.

Step 2: Precedence-ordered scientific term extraction and typing rules. The step is conceptually akin to sieve-based systems that were successfully demonstrated on the coreference resolution [25] and biomedical name normalization [11] tasks. The idea in sieve-based systems is simply that an ordering is imposed on a set of selection functions from the most constrained to the least. Similarly, in this second step of processing titles in our rule-based system, we apply the notion of selection precedence as concept precedence. However, there are various concept precedences in our system that depend on context information seen as the count of the connectors and the type of connectors in any given article title.

In this step, within each template category the titles are generally processed as follows. **Step 1. Counting connectors** – Our connectors are a collection of 11 prepositions and 1 verb defined as: $connectors\_rx$ = $(to|of|on|for|from|with|by|via|through|using|in|as)$. For a given title, its connectors are counted and the titles are phrase chunked as scientific phrase candidates split on the connectors themselves. **Step 2. Concept typing** – This involves selecting the workflow for typing the scientific phrases with concept types among our six, viz. *language*, *tool*, *method*, *resource*, and *research problem*, based on context information. Workflow branches were implemented as a specialized system of rules based on the number of connectors. The next natural question is: after the workflow branch is determined, what are the implementation specifics for typing the scientific terms per our six concepts? We explain this with the following specific case. A phrase with 0 connectors is typed after the following concept precedence order: *language* $\prec$ *tool* $\prec$ *method* $\prec$ *resource* $\prec$ *research problem* where each of the concepts are implemented as regex checks. Some example regexes were shown earlier in Table 2. And it only applies to five of the six concepts, i.e. *solution* is omitted. On the other hand, if a title has one connector, it enters first into the OneConnectorHeu() branch. There, the first step is determining which connector is in the phrase. Then based on the connector, separate sets of concept type precedence rules apply. The concept typing precedence rules are tailored based on the connector context. For instance, if the connector is 'from,' the title subphrases are typed based on the following pattern: *solution* from *resource*.

This concludes a brief description of the working of CL-Titles-Parser.

5 Evaluation

In this section, some results from CL-Titles-Parser are discussed for scientific term extraction and concept typing in Computational Linguistics article titles.

Evaluation Corpus. We downloaded all the article titles in the ACL anthology as the 'Full Anthology as BibTeX' file dated 1-02-2021. See https://aclanthology.org/anthology.bib.gz. From a total of 60,621 titles, the evaluation corpus comprised 50,237 titles after eliminating duplicates and invalid titles.

When applied to the evaluation corpus, the following total scientific concepts were extracted by the tool: 19,799 *research problem*, 18,111 *solution*, 20,033 *resource*, 1,059 *language*, 6,878 *tool*, and 21,687 *method*. These scientific concept lists were then evaluated for extraction precision.

5.1 Quantitative Analysis: Scientific Concept Extraction Precision

First, each of the six scientific concept lists were manually curated by a human annotator to create the gold-standard data. The extracted lists and the gold-standard lists were then evaluated w.r.t. the *precision* metric. Table 3 shows the results. We see that CL-TITLES-PARSER demonstrates a high information extraction precision for all concept types except *research problem*. This can be attributed in part to the long-tailed list phenomenon prevalent in the scientific community as the scholarly knowledge investigations steadily progress. With this in mind, the gold-standard list curation was biased toward already familiar research problems or their derivations. Thus we estimated that at least 20% of the terms were pruned in the gold data because they were relatively new as opposed to being incorrect. Note, recall evaluations were not possible as there is no closed-class gold standard as scientific terms are continuously introduced.

5.2 Qualitative Analysis: Top N Terms

As qualitative analysis, we examine whether the terms extracted by our tool reflect popular research trends. Table 4 shows the top five terms in each of the six concept types. The full scientific concept lists sorted by occurrences are available in our code repository https://github.com/jd-coderepos/cl-titles-parser/tree/master/data-analysis. Considering the *research problem* concept, we see that variants of "machine translation" surfaced to the top accurately reflective of the large NLP subcommunity attempting this problem. As a *tool*, "Word Embeddings" are the most predominantly used. "Machine Translation" itself was the most employed *method*. Note that the concept types are not mutually exclusive. A term that is a *research problem* in one context can be a *method* in a different context. As an expected result, "English" is the predominant *language* researched. "Twitter" showed as the most frequently used *resource*. Finally, predominant *solutions* reflected the nature of the article itself as "overview" or "a study" etc. Then Table 5 shows the *research problem*, *resource*, and *tool* concepts research trends in the 20th vs. 21st century. Contemporarily, we see new predominant neural *research problem* mentions, an increasing use of social media as a *resource*; and various neural networks as *tools*.

Table 3. Precision of CL-TITLES-PARSER for scientific term extraction and concept typing from 50,237 titles in the ACL anthology

Concept type	*Precision*	Concept type	*Precision*
research problem	58.09%	*method*	77.29%
solution	80.77%	*language*	95.12%
tool	83.40%	*resource*	86.96%

Table 4. Top 5 scientific phrases for the six concepts extracted by CL-TITLES-PARSER

research-problem	statistical machine translation (267), machine translation (266), neural machine translation (193), sentiment analysis (99), information extraction (85)
tool	word embeddings (77), neural networks (63), conditional random fields (51), convolutional neural networks (41), spoken dialogue systems (32)
method	machine translation (105), domain adaptation (68), sentiment analysis (68), named entity recognition (67), statistical machine translation (66)
language	English (150), Chinese (87), Japanese (87), German (81), Arabic (74)
resource	Twitter (204), text (173), social media (132), the web (115), Wikipedia (98)
solution	overview (39), a study (23), an empirical study (25), a comparison (21), a toolkit (17)

Table 5. Top 5 *research problem*, *resource*, and *tool* phrases from paper titles reflecting research trends in the 20th (7,468 titles) vs. the 21st (63,863 titles) centuries.

	research problem	*resource*	*tool*
20th	machine translation (56)	text (38)	machine translation system (8)
21st	statistical machine translation (258)	text (251)	word embeddings (87)
20th	information extraction (19)	discourse (17)	natural language interfaces (7)
21st	machine translation (210)	Twitter (204)	neural networks (57)
20th	speech recognition (16)	TAGs (9)	neural networks (6)
21st	neural machine translation (193)	social media (132)	conditional random fields (51)
20th	natural language generation (15)	bilingual corpora (9)	WordNet (3)
21st	sentiment analysis (99)	the web (115)	convolutional neural networks (41)
20th	continuous speech recognition (12)	dialogues (9)	semantic networks (3)
21st	question answering (81)	Wikipedia (98)	spoken dialogue systems (31)

6 Conclusion and Future Directions

We have described a low-cost approach for automatic acquisition of contribution-focused scientific terms from unstructured scholarly text, specifically from Computational Linguistics article titles. Work to extend the tool to parse Computer Science titles at large is currently underway. The absence of inter-annotator agreement scores to determine the reliability with which the concepts can be selected will also be addressed in future work. Evaluations on the ACL anthology titles shows that our rules operate at a high precision for extracting *research problem*, *solution*, *resource*, *language*, *tool*, and *method*. We proposed an incremental step toward the larger goal of generating contributions-focused SKGs.

References

1. Ammar, W., et al.: Construction of the literature graph in semantic scholar. In: Proceedings of the 2018 Conference of the North American Chapter of the Association for Computational Linguistics: Human Language Technologies (Industry Papers), vol. 3, pp. 84–91 (2018)
2. Ammar, W., Peters, M.E., Bhagavatula, C., Power, R.: The AI2 system at SemeEal-2017 task 10 (ScienceIE): semi-supervised end-to-end entity and relation extraction. In: SemEval@ACL (2017)
3. Aryani, A., et al.: A research graph dataset for connecting research data repositories using RD-switchboard. Sci. Data **5**(1), 1–9 (2018)
4. Augenstein, I., Das, M., Riedel, S., Vikraman, L., McCallum, A.: SemEval 2017 task 10: ScienceIE - extracting keyphrases and relations from scientific publications. In: SemEval@ACL (2017)
5. Beltagy, I., Lo, K., Cohan, A.: SciBERT: pretrained language model for scientific text. In: EMNLP (2019)
6. Brack, A., D'Souza, J., Hoppe, A., Auer, S., Ewerth, R.: Domain-independent extraction of scientific concepts from research articles. In: Jose, J.M., et al. (eds.) ECIR 2020. LNCS, vol. 12035, pp. 251–266. Springer, Cham (2020). https://doi.org/10.1007/978-3-030-45439-5_17
7. Burton, A., et al.: The Scholix framework for interoperability in data-literature information exchange. D-Lib Mag. **23**(1/2) (2017)
8. Charles, M.: Adverbials of result: phraseology and functions in the problem-solution pattern. J. Engl. Acad. Purp. **10**(1), 47–60 (2011)
9. Cousijn, H., et al.: Connected research: the potential of the PID graph. Patterns **2**(1), 100180 (2021)
10. D'Souza, J., Hoppe, A., Brack, A., Jaradeh, M.Y., Auer, S., Ewerth, R.: The STEM-ECR dataset: grounding scientific entity references in stem scholarly content to authoritative encyclopedic and lexicographic sources. In: LREC, Marseille, France, pp. 2192–2203, May 2020
11. D'Souza, J., Ng, V.: Sieve-based entity linking for the biomedical domain. In: Proceedings of the 53rd Annual Meeting of the Association for Computational Linguistics and the 7th International Joint Conference on Natural Language Processing (Volume 2: Short Papers), pp. 297–302 (2015)
12. Gupta, S., Manning, C.D.: Analyzing the dynamics of research by extracting key aspects of scientific papers. In: Proceedings of 5th International Joint Conference on Natural Language Processing, pp. 1–9 (2011)
13. Handschuh, S., QasemiZadeh, B.: The ACL RD-TEC: a dataset for benchmarking terminology extraction and classification in computational linguistics. In: COLING 2014: 4th International Workshop on Computational Terminology (2014)
14. Hearst, M.A.: Automatic acquisition of hyponyms from large text corpora. In: Coling 1992 Volume 2: The 15th International Conference on Computational Linguistics (1992)
15. Hearst, M.A.: Automated discovery of wordnet relations. WordNet: An Electronic Lexical Database, vol. 2 (1998)
16. Heffernan, K., Teufel, S.: Identifying problems and solutions in scientific text. Scientometrics **116**(2), 1367–1382 (2018)
17. Houngbo, H., Mercer, R.E.: Method mention extraction from scientific research papers. In: Proceedings of COLING 2012, pp. 1211–1222 (2012)

18. Jaradeh, M.Y., et al.: Open research knowledge graph: next generation infrastructure for semantic scholarly knowledge. In: Proceedings of the 10th International Conference on Knowledge Capture, K-CAP 2019, pp. 243–246. Association for Computing Machinery, New York (2019). https://doi.org/10.1145/3360901.3364435
19. Johnson, R., Watkinson, A., Mabe, M.: The STM Report. An Overview of Scientific and Scholarly Publishing. 5th edn., October 2018. https://www.stm-assoc.org/2018_10_04_STM_Report_2018.pdf
20. Katsurai, M., Joo, S.: Adoption of data mining methods in the discipline of library and information science. J. Libr. Inf. Stud. **19**(1), 1–17 (2021)
21. Landhuis, E.: Scientific literature: information overload. Nature **535**(7612), 457–458 (2016)
22. Luan, Y., He, L., Ostendorf, M., Hajishirzi, H.: Multi-task identification of entities, relations, and coreference for scientific knowledge graph construction. In: EMNLP (2018)
23. Luan, Y., Ostendorf, M., Hajishirzi, H.: Scientific information extraction with semi-supervised neural tagging. arXiv preprint arXiv:1708.06075 (2017)
24. Miller, G.A.: WordNet: An Electronic Lexical Database. MIT Press, Cambridge (1998)
25. Raghunathan, K., et al.: A multi-pass sieve for coreference resolution. In: Proceedings of the 2010 Conference on Empirical Methods in Natural Language Processing, pp. 492–501 (2010)
26. Singh, M., Dan, S., Agarwal, S., Goyal, P., Mukherjee, A.: AppTechMiner: mining applications and techniques from scientific articles. In: Proceedings of the 6th International Workshop on Mining Scientific Publications, pp. 1–8 (2017)

Enriching the Metadata of a Digital Collection to Enhance Accessibility: A Case Study at Practice in Kyushu University Library, Japan

Tetsuya Nakatoh[2](✉), Hironori Kodama[1], Yuko Hori[1], and Emi Ishita[1]

[1] Kyushu University Library, Kyushu University, Fukuoka 819-0395, Japan
[2] Nakamura Gakuen University, Fukuoka 814-0198, Japan
nakatoh@nakamura-u.ac.jp

Abstract. In this practice paper, we report on the enrichment of the metadata of the rare materials digital archive provided by Kyushu University Library and present the results of its effectiveness. We examined the metadata of the rare materials archive. Metadata for items created at an early stage were missing for several fields, although basic fields were filled in, such as the title, language, and rights. We selected the following fields that may improve accessibility to each item: title's pronunciation in Katakana, title's pronunciation using the Roman letters, alternative title, pronunciation of the creator's name in Katakana, creator's identifier (ID), creator's alternative name, bibliographic ID at Kyushu University, material ID of the source material, and publication year. These blank fields were filled in for 7,249 metadata during 2020 and 2021. The number of accesses of these items was examined, and the number of accesses from outside the campus, including Japan and overseas, significantly increased statistically. These results indicate that enriching metadata contributes to enhancing the accessibility of digital archives and collections.

Keywords: Digital archive · Metadata · Accessibility

1 Introduction

With the development of various technologies, the digitization of books has become easier, several software packages for digital libraries have been developed, and high-quality images have become available on the Internet. The digital information world has changed dramatically. At the present time, various digital collections and digital archives are available. Many academic university libraries also provide digital collections or digital archives for their rare book collections. These digital collections could become a feature of university libraries.

Once a digital archive of rare books has been built, an archive focuses on adding new digitization images to enhance it. As a result, digital images and metadata created in the past have not focused on. In many cases, the quality of the digital images created at an early stage is poor or the metadata is incomplete because of technical problems.

H.-R. Ke et al. (Eds.): ICADL 2021, LNCS 13133, pp. 411–418, 2021.
https://doi.org/10.1007/978-3-030-91669-5_32

Addressing these issues is important, in addition to expanding collections for university libraries that have a digital collection or archive.

For this practice paper, we examined the metadata of the rare materials digital archive run by Kyushu University Library in Japan and found that many fields of metadata created at an early stage were blank. Next, we filled in specific metadata fields and analyzed the number of accesses of the updated items to verify the effectiveness of enriching metadata. In this paper, we report on this case study as a practice paper.

The remainder of this paper has the following structure: In Sect. 2, we provide an overview of the rare materials digital archive at Kyushu University Library. In Sect. 3, we describe the metadata enrichment and its effectiveness using the analysis of the number of accesses. Finally, in Sect. 4, we conclude this practice paper.

2 Rare Materials Digital Archive at Kyushu University Library

2.1 Outline of the Digital Collections

Kyushu University Library started to digitize rare materials in 1997 [1] and provides the results as the rare materials digital archive on the library website [2]. As of June 2021, the library also provides digital collections of the ownership seals of books and various types of materials about coal mines. Table 1 shows the names of the collections, their descriptions, and the number of items in each collection.

The JPcoar schema [3] is used for the metadata of the digital collections [4]. The digital images have been available in IIIF (International Image Interoperability Framework) format since 2018. IIIF defines a set of common application programming interfaces that support interoperability between image repositories [5]. Additionally, digital images have been released to the public domain since October 2018, with some exceptions related to copyright issues. Thus, Kyushu University Library provides an environment in which digital images of valuable materials are freely available, and also allows university non-members to use them, including off-campus, in Japan and overseas.

Table 1. Digital collections provided at Kyushu University Library.

Names of collection	Description	#Items
Rare materials	Catalogs and images of rare books and materials held by Kyushu University. The number in parentheses is the number of items contained digital images	113,788 (7,249)
Ownership seals	Digital images of ownership seals of books provided in the Kyushu University collection	830
Coal images	A collection of digital images about coal mines. This includes private paper money in coal miners, postcards, maps, and the bulletins of each miner's company and miners' union	7,307

2.2 Rare Materials Digital Archive

We focused on the metadata of the rare materials digital archive. The archive contains over 70 collections [6]. However, it contains digitized images of only some of the materials in the collections. Various types of materials are included, such as stories, essays, poems, and maps. For example, the archive contains the Gazoku Bunko (Bunko means a collection in Japanese), Hama Bunko, and Kasuga Bunko. The archive also contains collections of foreign books, such as the old medical books and materials collection, and the Thomas collection. All the materials are highly valuable as collections of rare books, as well as research materials. Therefore, providing digital images of these rare collections should make a great contribution to the academic world.

2.3 Total Number of Accesses of the Rare Materials Digital Archive

To understand the overall access counts for the rare materials digital archive, we examined the number of accesses of the archive from January 2020 to March 2021, as shown in Fig. 1. The access location, such as on-campuses, in Japan, overseas, or unknown, was identified using the IP address. The number of accesses from Japan was the highest, followed by overseas and on-campus. This indicates that the archive is in high demand from outside Kyushu University.

Compared with the number of accesses in the same month last year, the number of accesses was higher in April, June, and July 2020. The number of accesses in all other months was higher last year. The highest increase ratio was 208.8% in March 2021 compared with March 2020. The number of accesses is easily affected by several factors, such as the semester schedule and examination terms. In 2020 and 2021 particularly, library services were limited because of the COVID-19 pandemic. We supposed that is the reason that the number of accesses increased.

3 Metadata Enrichment

3.1 Enriching Metadata

We reviewed 7,249 metadata of items with digital images in the rare materials digital archive. The metadata of each item comprised many types of fields. We found that many fields of metadata that were registered recently contained rich information. However, metadata created at an early stage had many blank fields, although fundamental fields contained information, such as the title, language, and rights. Therefore, for this practice paper, we selected several fields that should be filled in and updated them to improve the accessibility of items in the archive.

We filled in the following nine fields; title's pronunciation in Katakana (Japanese phonogram), title's pronunciation in Roman letters (phonogram in alphabets), alternative titles, creator's pronunciation in Katakana, creator's identifier (ID), alternative creator's name, publication year (based on the Japanese calendar), bibliographic ID at Kyushu University Library, and original material ID.

The pronunciation of the title indicates how the written Kanji (Chinese characters) words are pronounced. In the case of the Japanese language, the same Kanji notation can

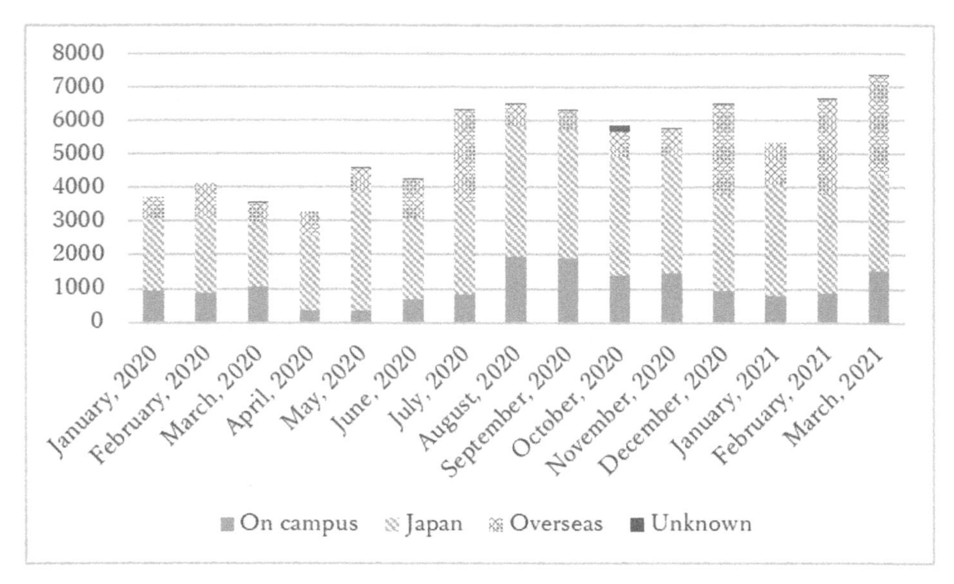

Fig. 1. Number of accesses of the Rare Materials Digital Archive.

be pronounced in various ways. Phonetic symbols, for example, Katakana or an alphabet (Roman letters) are sometimes used to express the pronunciation of words in Kanji. The titles and creators of many rare books published a very long time ago are written in Kanji characters. It is often difficult to know the correct pronunciation without specialized knowledge. When users search the archive of rare materials, the pronunciation might assist them. It is useful to add a pronunciation in Roman letters for foreigners that search the archive. This makes searching easier not only for users who understand Japanese but also foreign researchers who are not able to input Japanese. We also added the Japanese calendar for the publication year field. Japan uses the Japanese calendar. For example, 2021 is "Reiwa 3rd (令和3年)." The publication years of many rare books are unknown. We input years using the Japanese calendar and the Christian era for books for which we identified the publication years. We also added the Kyushu University Library bibliographic ID and the original material ID for internal management.

Table 2 shows an example of content input into each field of metadata for『神路手引草/ [増穂残口]』.

We started to review the metadata in 2019 and updated the metadata in 2019 and 2020. In total, we updated 1,182 items by May 2021. Table 3 shows the number of updated metadata items and the dates.

Table 2. Example of content in each metadata field.

Fields	Metadata format	Example
Title's pronunciation in Katakana	/dc:title#2	カミジノテビキグサ
Title's pronunciation in Roman letters	/dc:title#3	Kamiji no tebikigusa
Alternative title	/dcterms:alternative#1	手引艸
Creator's pronunciation in Katakana	/jpcoar:creator#1/jpcoar:creatorName#2	マスホ, ザンコウ
Creator's identifier	/jpcoar:creator#1/jpcoar:nameIdentifier#1	DA05025971
Alternative creator's name	/jpcoar:creator#1/jpcoar:creatorAlternative#1	増穂, 最仲
Publication year (based on a Japanese calendar)	/dcterms:temporal#1	享保4年

Table 3. Number of updated items and dates.

Date	Number of items	Date	Number of items
May, 2020	3	November, 2020	68
August, 2020	3	December, 2020	894
September, 2020	3	January, 2021	3
October, 2020	201	February, 2021	7

3.2 Number of Accesses

As shown in Fig. 1, the number of accesses to the rare materials digital archive increased. To examine the effectiveness of metadata enrichment, we examined the number of accesses of updated items whose blank fields were filled in for each item of metadata and compared them with the number of accesses of the non-updated item group.

As shown in Table 3, the number of updated items varied each month. In this section, we analyze the items updated in October and December 2020. We updated many items in both months. We created three groups for the analysis. Group A contained 201 items that were updated in October 2020. Group B contained 894 items that were updated in December 2020. Group C contained 2,468 items that were never updated for comparison. Additionally, because the update timing was different, we examined the number of accesses for the same period before and after the updated month.

We examined the number of accesses of Groups A and C six months before and after the updates, excluding the updated month (October 2020): from April to September 2020, and from November 2020 to April 2021, respectively. Table 4 shows the average number of accesses of items on-campuses; outside, including Japan and overseas; bots, and the total. The bots were identified by the domain names from which they were accessed. For both groups, the number of accesses on-campuses decreased with statistical significance using the Mann–Whitney U test ($p < .001$). Conversely, the number of accesses from outside and bots, and the total increased with statistical significance ($p < .001$ for outside and bots, and $p = 0.01$ for the total). Bots increased considerably.

Table 4. Average access counts of Groups A (updated in October 2020) and C (non-updated).

	Item	Campuses		Outside		Bots		Total	
		Before	After	Before	After	Before	After	Before	After
Group A (updated)	201	1.19	0.35	2.77	2.86	0.25	17.95	3.71	21.13
Group C (non-updated)	2,486	0.15	0.07	0.58	0.82	0.06	4.90	0.80	5.79

We also examined the updated metadata in Group B for five months before and after the updates: from July to November 2020, and from January to May 2020, respectively. Table 5 shows the average number of accesses of items. The results are similar to those for Group A. The number of accesses from outside and bots, and the total increased with statistical significance ($p < .001$ for outside, bots, and the total).

Table 5. Average access counts for Groups B (updated in December 2020) and C (non-updated).

	Item	Campuses		Outside		Bots		Total	
		Before	After	Before	After	Before	After	Before	After
Group B (updated)	894	0.52	0.13	1.47	1.72	0.05	20.27	2.03	22.12
Group C (non-updated)	2,486	0.17	0.05	0.48	0.67	0.06	5.50	0.71	6.22

3.3 Discussion

The number of accesses of updated items from outside increased for Groups A and B. These results indicate that enriching metadata contributed to the increase in the number of accesses. Additionally, the number of accesses from bots also significantly increased, which can be considered to be an effect of the enrichment.

We enriched the metadata of digital image items to improve the accessibility of digital collections. In particular, we focused on adding pronunciations in Katakana and Roman letters. This would allow foreign researchers in environments where it is difficult to input Japanese or students without sufficient specialized knowledge to search the digital collections. It is significant to add pronunciations in Roman letters. It has been shown to be a problem that Japanese digital archives should solve in terms of multilingual support for sharing data internationally [7]. Providing and sharing digital data, such as open data, should be promoted, but it is also important to provide digital images in IIIF format. Thus, it is a very important result that metadata enrichment led to increasing the number of accesses.

The results indicate that increasing the number of accesses from outside is highly necessary for these collections. As future work, we aim to further improve accessibility by adding the title in English, in addition to pronunciations in Katakana and Roman letters. However, highly specialized knowledge is required to determine these titles. Additionally, experts would need to check the English titles after librarians add them. This would be very time-consuming and expensive. The library has created Cute.Guides for the archive [8]. It would be useful to create an English version of this navigation as a first step.

4 Conclusion

In this practice paper, we focused on the metadata of the rare materials digital archive and filled in several blank metadata fields in the archive. We updated the metadata of 1,180 items. As a result, the number of accesses to these items from outside increased. This indicates that metadata enrichment contributes to enhancing the accessibility of rare book collections.

When we build or run digital collections, once metadata and digital images are created and registered in the collections, further review or examination is often not performed. However, as shown here, it is possible that these collections can be enhanced if we review metadata and digital images constantly.

This practice paper represents a joint research project between researchers and librarians. On the basis of this paper, we will continue to review and maintain metadata to improve their quality. Additionally, we will discuss an environment for providing materials and digital content that meet users' needs.

Acknowledgements. This work was supported in part by JSPS KAKENHI Grant Number JP18K18508.

References

1. Yoshimatsu, N., et al.: Self-digitalizing of rare book collections: FY2010 project report of creating rare-book's images and their metadata database. Kyushu Univ. Libr. Res. Dev. Division Annual Report 2010/2011, 42–47 (2011). https://doi.org/10.15017/20109. (in Japanese)
2. Kyushu University Library: Rare Materials Digital Archive (2021). https://catalog.lib.kyushu-u.ac.jp/opac_browse/rare/?lang=1. Accessed 13 July 2021
3. Japan Consortium for Open Access Repository. JPCOAR Schema Guidelines (2021). https://schema.irdb.nii.ac.jp/en. Accessed 13 July 2021
4. Kyushu University Library: Metadata no teikyo (2021). https://www.lib.kyushu-u.ac.jp/ja/metadata. Accessed 13 July 2021

5. International Image Interoperability Framework. About IIIF (2021). https://iiif.io/about/. Accessed 19 Sept 2021
6. Kyushu University Library: Kityo Shiryo (Kyudai collection) (2021). https://guides.lib.kyushu-u.ac.jp/rare/top. Accessed 13 July 2021
7. Digital Archive no renkei ni kansuru kankei shocho to renrakukai & Jitumusha kyogikai. Digital Archive no koutiku & kyoyu & katsuyo guidelines (2017). http://www.kantei.go.jp/jp/singi/titeki2/digitalarchive_kyougikai/guideline.pdf. Accessed 13 July 2021
8. Kyushu University Library. Cute.Guides (2021). https://guides.lib.kyushu-u.ac.jp/cuteguides/home. Accessed 13 July 2021

Neural-Based Learning

PEERAssist: Leveraging on Paper-Review Interactions to Predict Peer Review Decisions

Prabhat Kumar Bharti[1(✉)], Shashi Ranjan[1], Tirthankar Ghosal[2], Mayank Agrawal[1], and Asif Ekbal[1]

[1] Department of Computer Science and Engineering, Indian Institute of Technology Patna, Patna, India
{prabhat_1921cs32,shashi.cs17,mayank265,asif}@iitp.ac.in
[2] Institute of Formal and Applied Linguistics, Faculty of Mathematics and Physics, Charles University, Prague, Czech Republic
ghosal@ufal.mff.cuni.cz

Abstract. Peer review is the widely accepted method of research validation. However, with the deluge of research paper submissions accompanied with the rising number venues, the paper vetting system has come under a lot of stress. Problems like dearth of adequate reviewers, finding appropriate expert reviewers, maintaining the quality of the reviews are steadily and strongly surfacing up. To ease the peer review workload to some extent, here we investigate how an Artificial Intelligence (AI)-powered review system would look like. We leverage on the paper-review interaction to predict the decision in the reviewing process. We do not envisage an AI reviewing papers in the near-future, but seek to explore a human-AI collaboration in the decision-making process where the AI would leverage on the human-written reviews and paper full-text to predict the fate of the paper. The idea is to have an assistive decision-making tool for the chairs/editors to help them with an additional layer of confidence, especially with borderline and contrastive reviews. We use cross-attention between the review text and paper full-text to learn the interactions and henceforth generate the decision. We also make use of sentiment information encoded within peer-review texts to guide the outcome. Our initial results show encouraging performance on a dataset of paper+peer reviews curated from the ICLR openreviews. We make our codes and dataset (https://github.com/PrabhatkrBharti/PEERAssist) public for further explorations. We re-iterate that we are in an early stage of investigation and showcase our initial exciting results to justify our proposition.

Keywords: Peer reviews · Decision prediction · Deep neural network · Cross attention

H.-R. Ke et al. (Eds.): ICADL 2021, LNCS 13133, pp. 421–435, 2021.
https://doi.org/10.1007/978-3-030-91669-5_33

1 Introduction

The exponential growth of the article submissions to top-tier and popular conferences is posing a significant challenge to the chairs to recruit sufficient expert reviewers [7]. Sometimes they are forced to hire non-expert, novice reviewers to cope up with the reviewing load which is contributing towards less-informed and poor reviews resulting in inconsistent judgements. The chairs are usually overwhelmed with the above editorial burden which additionally includes justified paper-reviewer allocation [14], chasing the missing or non-responding reviewers [9], participating in the review discussions [11], etc. However, the climax lies in deciding whether the manuscripts will be accepted or reject based on the peer reviews [24]. It is becoming increasingly difficult to ensure the integrity of review process, especially when reviewers fail to respond on time. Here in this work, we take up this important problem and investigate if an Artificial Intelligence (AI) could help the editors/program chairs in these situations, especially when non-responding/missing reviewers are common. *Our curiosity is driven by the question, what if the AI-based system could help editors get rid of this burden to some extent?*

Another problem is the *inconsistency or arbitrariness* in peer review decisions. Bornmann et al. [2] discussed this critical problem in the peer-review process. Again, to study the arbitrariness in peer-review decision making, the NeurIPS 2014 organizers conducted an experiment[1] in which 10% of submissions were delegated to two independent program committees, instead of one. The outcomes indicated that the two committees had different opinions and about 57% of the papers accepted by the first committee were rejected by the second one and vice versa. Something similar was pursued by a team of researchers at ICML 2020 [20] where they identified multiple avenues to address scarcity of qualified reviewers in large conferences. However, despite all the inherent flaws [19] and criticisms [21], the peer review process is still probably the widely-accepted gatekeeper of scientific knowledge and wisdom we have at this moment. One hindrance for research in peer reviews is the non-availability of the confidential and proprietary peer-review texts and metadata. However, recently certain scholarly venues (e.g., ICLR) have started making their peer review data public to foster transparency and restore trust on the reviewing process. Mention could be made of the PeerRead [13] dataset. Here the authors release a collection of peer reviews and associated meta-data from premier NLP/ML conferences. They also present baseline models for the tasks of review-rating prediction and acceptance-decision prediction. For predicting the acceptance decision they train a binary classifier with hand-crafted features. Another significant effort in this direction is DeepSentiPeer [8]. Here the authors propose a deep neural model leveraging on paper, review, and reviewer-sentiment information to predict the peer review outcome in terms of decision and recommendation scores. In this paper, we propose a novel cross-attention-based deep neural architecture that effectively incorporates a sectional summary of the paper, associated

[1] http://blog.mrtz.org/2014/12/15/the-nips-experiment.html.

review text and the sentiment of reviewers encoded in peer review texts as an input. The proposed architecture combines four segments into an end-to-end relationship: paper-section encoder, review-sentence encoder, review-attended paper-encoder, paper-attended review-encoder. Each encoder firstly derives contextual level representation then generates a higher-level review-attended paper and paper-attended review representation. The output of review-attended paper-encoder and paper-attended review-encoder are leveraged as features to build the model for the decision prediction. We perform experiments with ICLR data from 2017, 2018, 2019, and 2020 in a year-wise fashion. We significantly outperform *PeerRead* and *DeepSentiPeer*. We attribute this improvement to our attention-based modeling of interactions between the paper and the review to generate the judgement. The proposed system is never intended to replace expert reviewers from the peer-review process. Instead, to assist the editors/program chairs and provide additional perspective to make more informed decisions.

2 Related Work

Artificial intelligence in peer review is an essential but understudied topic. However, nowadays topic is getting community attention due to recent advancements in artificial intelligence research. This section discusses some of the specific studies conducted involving AI in peer review systems.

Charlin et al. [6] proposed Toronto Paper Matching System to automatically assign reviewers to papers. Price et al. [17] conducted a thorough study of the various computational support tools of the peer review system. To assess the quality, tone, and quantity of comments on the review. Wang et al. [24] proposed a multi-instance learning-based approach to predict the sentiment of sentences in peer reviews. Hua et al. [3] assess the efficiency and efficacy of the reviewing process by automatically detecting argumentative propositions put forward by reviewers and their types. They collected 14.2K reviews from major machine learning (ML) and natural language processing (NLP) conference venues and annotated 400 of them based on different propositions. Li et al. [15] developed a neural model to predict citations in accepted papers. Qiao et al. [18] used a recurrent convolutional network model incorporating both modularity and attention mechanisms to predict the aspect scores of an academic paper. Ghosal et al. [10] investigated the impact of various characteristics on the pre-screening of research papers. Superchi et al. [22] provide a comprehensive overview of criteria tools used to assess the quality of peer review reports in the biomedical field. The CiteTracked dataset [16] is a collection of peer reviews and citation information spanning six years and includes scientific articles from the machine learning community. Yuan et al. [26] proposed ReviewAdvisor, a dataset with an ambitious goal of automating scientific peer review. Wicherts et al. [25] designed a tool to allow various stakeholders to assess the transparency of the peer-review process and have proposed that transparency of the peer-review process can be considered an indicator of the quality of the peer-review process. Thelwall et al. [23] leveraged sentiment encoded in peer review texts to automatically assessing peer review praise and criticism. Chakraborthy et al. [5] demonstrated that the distribution of aspect-based sentiment acquired from a review differed considerably for accepted and rejected papers. Kang et al. [13] proposed the PeerRead dataset where they present models for the tasks of review rating prediction and acceptance decision prediction. Ghosal et al. [8] proposed DeepSentiPeer, a deep neural architecture to classify a research paper (accept/reject) and predict recommendation scores from the peer reviews. This paper takes forward the problem defined in Kang et al. [13] by splitting the channels for review and paper text and augmenting sentiment features(via a pre-trained features extractor). In contrast to this work, we use cross-attention between review and paper to learn their interactions. Our motivation is to leverage on these interactions just as the editors/area-chairs while arriving to their decision in peer review.

3 Data Description

For this work, we curate a dataset of 4.5k papers submitted to ICLR conference from 2017 to 2020 along with their associated accept/reject decision with ratings from OpenReview Platform, an open peer-reviewing portal. The data for ICLR 2017 was taken from PeerRead [13] dataset while we crawl the data for ICLR 2018, 2019 and 2020 using on openreview[2] website. We report the detailed year-wise dataset statistics in Table 1. ICLR 2017 reviews that we have taken from PeerRead [13] suffer from various data quality issues. It contains many blanks and duplicates review and marked the final decision of the area chair as a meta-review (marked the boolean field as false). We removed above mentioned issue and add a new column named final decision (0,1) where 0 signifies reject, 1 signifies accept, and explicitly treat them as meta-reviews. Notably, later studies that use PeerRead [13], such as DeepSentiPeer, do not mention whether they have addressed the issue.

Ethical Statement: The reviews from ICLR are publicly available and we scraped using the official OpenReview API.

Table 1. Data statistics and analysis

Conference edition	# Papers	# Reviews	Avg length of reviews (sentences)	Avg length of review sentences (words)	Acc. Rate
ICLR-2017	354	1062	17.40	302.56	39.5%
ICLR-2018	806	2418	23.24	398.09	40.8%
ICLR-2019	1337	4011	25.30	430.81	35.8%
ICLR-2020	1970	5910	25.29	433.04	31.0%
Total	4467	13401			

[2] https://openreview.net/.

3.1 Analysis and Potential Use Cases

We perform qualitative analysis through a comprehensive study of papers submitted to ICLR between 2017 and 2020 to comprehensively understand the relationship between peer review texts, rating, and review decisions to provide significant insights to the authors. We hypothesize that the answers to the stated questions indicate they bring out the reviewer's stand on the work. To this end, we found the following insights.

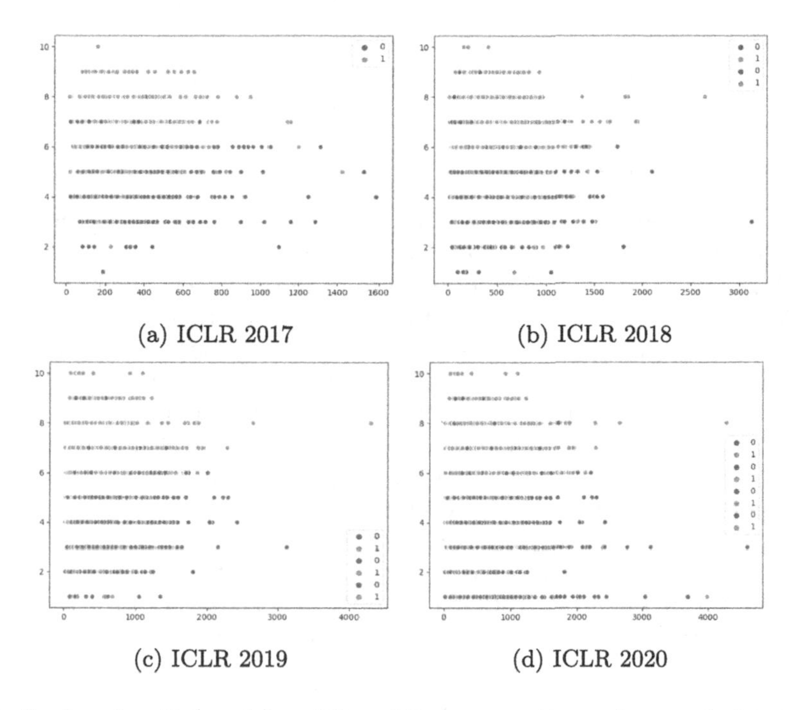

(a) ICLR 2017 (b) ICLR 2018

(c) ICLR 2019 (d) ICLR 2020

Fig. 1. Reviews length (words) partitioned By paper rating and paper decision, where x (axis) represent→ Reviews length (words), y (axis) represent→ Rating of paper, and 0/1 → Reject/Accept

1. Strongly opinionated reviews texts with ratings (1 and 8) having longer reviews to justify their decision. Figure 1 shows that this is only true for reviews that reject a paper.
2. Accepted paper with rating 8 that has short review, in which case, perhaps reviewer feels that the paper's acceptance does not need to justify as much as its rejection.

4 Approach Description

In this section, we present our proposed approach for the decision prediction problem.

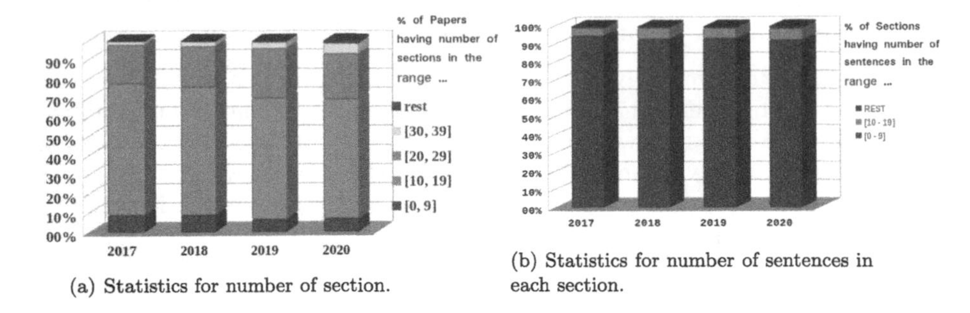

(a) Statistics for number of section.

(b) Statistics for number of sentences in each section.

Fig. 2. Statistics of a number of sections/sentences in paper across the dataset.

4.1 Data Pre-processing

Initially, we convert the paper PDFs to JSON encoded files using the science-parse[3] library. After that, we divide the paper into number of sections ($\max\{N_{sec}\} = 30$), Nsec = number of sections in the paper (empirically driven from dataset statistics shown in the Fig. 2a). The intuition is to make a number of section similar across the input data. Then we generate the summary for each section of the paper using summa[4] python library, and now the summarized section of the paper is referred to as the section of the paper. Similarly, by looking into data, we fix number of sentences in each section as ($\max\{N_{sent}\}$ $= 19$), where N_{sent} = number of sentences in each summarized section of the paper (shown in the Fig. 2b). Now we get papers made up of $\max\{N_{sec}\}$ sections, where each section has $\max\{N_{sent}\}$ sentences.

DeepSentiPeer uses 1000 tokens from paper, and 98 tokens from review each. As expected, the use of the sectional summary of the paper, associated review text significantly enhance the performance of our classifier.

4.2 Embedding Layer

This layer is the representation layer and here we describe how we convert the text into vectors to be fed to the neural network.

Document Encoding. For document encoding, we are using the following state-of-the-art sentence embeddings.

1. Transformer variant of SciBert [1]: We extract $\max\{N_{sent}\}$ from each section of the paper and represent each sentence $s_i \in \mathbb{R}^d$ using SciBert [1], where $d = 768$ is the sentence embedding dimension. Now the encoded paper is represented as $P = sec_1 \oplus sec_2 \oplus sec_3 \oplus \ldots \oplus sec_{N_{sec}}$, and encoded i^{th} section is represented as $sec_i = s_1 \oplus s_2 \oplus s_3 \oplus \ldots \oplus s_{N_{sent}}$, where $\oplus$ is the concatenation operator (we did padding wherever necessary).

[3] https://github.com/allenai/science-parse/.
[4] https://pypi.org/project/summa/.

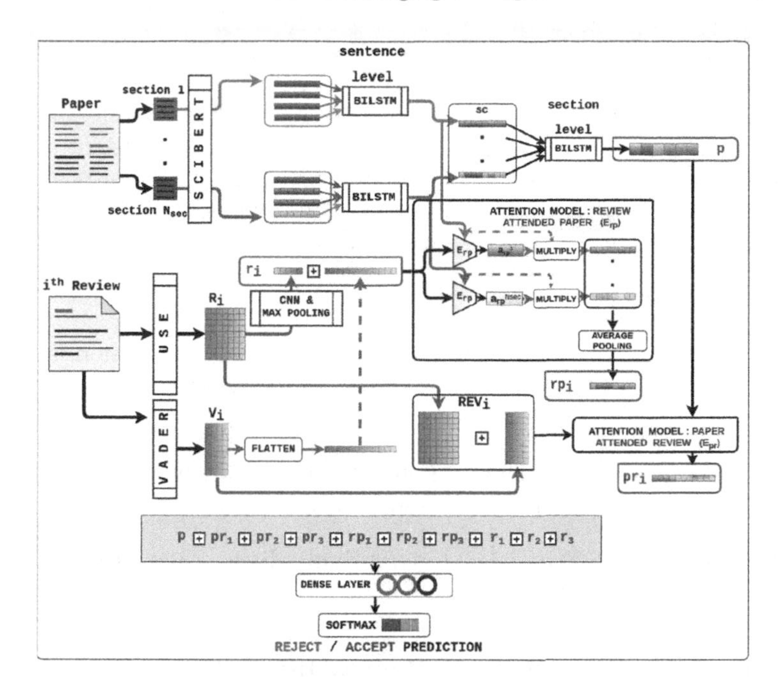

Fig. 3. PEERAssist: cross-attention based deep neural architecture for decision prediction.

2. Universal Sentence Encoder [4]: For review encoding, we use the Universal Sentence Encoder(USE). The input is review's sentences, and the output is a sentence embedding of dimension d = 512. It creates a review representation as $R = S_1 \oplus S_2 \oplus S_3 \oplus \ldots \oplus S_m, R \in \mathbb{R}^{m \times d}$, where m is the maximum number of sentences in the review.

4.3 VADER Sentiment Analyzer

For sentiment features in review text, we use VADER (Valence Aware Dictionary for Sentiment Reasoning) [12], which gives a vector of dimension 4 for each sentence s_i, $v_i \in \mathbb{R}^4$. Then sentiment of review sentences is represented as $V = v_1 \oplus v_2 \oplus v_3 \oplus \ldots \oplus v_m, V \in \mathbb{R}^{m \times 4}$.

4.4 Deep Learning-Based Features Extraction

Paper-Feature Extraction: We use the hierarchical Bidirectional Long Short-Term Memory (BiLSTM) to extract features from the paper's texts. The first layer of BiLSTM shown in Fig. 3 is used to extract the features from the sentence embedding generated by SciBert. The second layer BiLSTM is used to obtain the final paper representation p from sc as shown in Fig. 3.

Review-Feature Extraction: We use Convolutional Neural Network (CNN) to extract features from the review's texts. We use F different sets of CNN filters $[f_1, f_2, \ldots, f_F]$, where each filter set f_k contains filters of different sizes $[1 \times R_{embdim}, 2 \times R_{embdim} \ldots Z \times R_{embdim}]$, where R_{embdim} represents the embedding dimension of review sentences generated by USE. symbol f_k^t represent a kth set filter of size $(t \times R_{embdim})$ and the output of hth window of f_k^t applied on R_1 (first review embedding matrix) is given as.

$$y_h^{f_k^t} = g\left(W_{f_k^t}.R1_{h:h+t-1} + b_{f_k^t}\right)$$

$R1_{h:h+t-1}$ means h^{th} window of size $(t \times R_{embdim})$, g() is the non-linear function, $b_{f_k^t}$ is the bias for the filter f_k^t. Afterwards, we are applying max pooling on y (output) calculated in a particular filter application on R_1. $\hat{f_k^t} = \max\left(y_1^{f_k^t}, y_2^{f_k^t}, y_3^{f_k^t}, \ldots\right)$ And then, we are taking all the max-pooled value, and then we finally get first review CNN representation i.e. R_1^{cnn}

$R_1^{cnn} = (\hat{f_k^t}, 1 \le t \le Z, and\ 1 \le k \le F)$ Similarly, we calculate R_2^{cnn}, R_3^{cnn} $R^{cnn} = R_1^{cnn} \oplus R_2^{cnn} \oplus R_3^{cnn}$.

4.5 Cross-Attention for Paper-Review Interaction

The core contribution of our work lies in the fact that we leverage on paper-review interaction via a cross-attention network to predict the paper acceptance. This layer acts as a bridge which allows information to flow between the paper and reviews. It consists of two processes:

1. *Reviews attending paper* process allows us to see the paper content with review information in hand.
2. *Paper attending reviews* process allows getting the reviewer's thrust from the paper's point of view.

The main idea of using cross-attention between paper and review is to generate a collection of paper-aware features of review sentences and vice-versa to simulate the editorial workflow just as human reviewer does.

1. Reviews attending paper process E_{rp}
   We are finding importance of different sections $(sc[i], 1 \le i \le N_{sec})$ of the paper according to the each review information $(r[i], 1 \le i \le 3)$ separately. E_{rp} is defined as the softmax working upon tanh activation, and it is the core function of this attention. It is formulated as follows

$$\beta_{rp}^{<i,j>} = Tanh(W_{rp}.(r[i] \oplus sc[j]) + b_{rp}) \tag{1}$$

$$a_{rp}^{<i,j>} = \frac{\exp(\beta_{rp}^{<i,j>})}{\sum_{k=1}^{N_{sec}} \exp \beta_{rp}^{<i,k>}} \tag{2}$$

   where E_{rp} is the Eqs. (1) & (2) altogether, and from this we get attention values i.e. a_{rp}.

Now, a_{rp} values act as a factor to pay due importance (alignment of paper-review achieved via attention) to the section of the paper.

$$rp[i] = \frac{1}{N_{sec}} \sum_{j=1}^{N_{sec}} (a_{rp}^{<i,j>}).(sc[j]) \tag{3}$$

$$rp = (rp[1] \oplus rp[2] \oplus rp[3]) \tag{4}$$

Here, rp is referred to as reviews attended paper representation.

2. Paper attending reviews process E_{pr}
   E_{pr} is the core function to produce attention values for sentences of the reviews $(rev[i][j], 1 \leq i \leq 3, \&1 \leq j \leq m)$, where rev[i][j] represents j^{th} sentence of the ith review. E_{pr} is basically a sigmoid activation. We represent as:

$$\beta_{pr}^{<i,j>} = (W_{pr}.(p \oplus rev[i][j]) + b_{pr}) \tag{5}$$

$$a_{pr}^{<i,j>} = \frac{\exp(\beta_{pr}^{<i,j>})}{1 + \exp \beta_{pr}^{<i,j>}} \tag{6}$$

The Eqs. (5) & (6) altogether becomes the notation E_{pr}, which produces attention value i.e. a_{pr}. These attention value a_{pr} becomes the factor to scale or diminish the sentence representation of the review.

$$pr[i] = \frac{1}{m}.\sum_{j=1}^{m} (a_{pr}^{<i,j>}).(rev[i][j]) \tag{7}$$

$$pr = (pr[1] \oplus pr[2] \oplus pr[3]) \tag{8}$$

pr is referred to as paper attended review representation.

4.6 Fusion and Accept/Reject Decisions

The obtained feature vectors from paper-review interaction i.e. pr and rp along with review and paper final representation vector i.e. r and p are concatenated for getting all the basic feature information i.e. $base_{info} = (p \oplus pr \oplus rp \oplus r)$. Then, $base_{info}$ is passed through a dense layer (we use different activations to get different insights in $base_{info}$), and output of dense layer ($dense_{out}$) is passed through a softmax layer to predict the probabilities associated with accept/reject class.

$$\rho_{(acc/rej)} = softmax(W_\rho \cdot dense_{out} + b_\rho) \tag{9}$$

5 Experimental Setup

To compare with DeepSentiPeer [8], and PeerRead [13] for the mentioned task, we keep the training, testing split identical across the task and run their official implementations provided by the authors to generate the comparing figures. We select the model with the highest accuracy on the validation set for evaluation. We train the model for decision prediction using the Binary Cross-Entropy loss function.

5.1 Implementation Details

We use Keras on top of tensorflow-2.4.1 to build the model and use sentence-level BiLSTM's output of 460 units, section-level BiLSTM's output of 358 units. To decide the parameter's value of CNN, as m (maximum sentences in review) is changing across the datasets, we keep the F variable, proportional to m, and set z's value equal to 7. As $base_{info}$ units keep changing across the dataset, we keep neurons in a dense layer, proportional to the number of $base_{info}$ units. Moreover, we train the model with Adam optimizer, with batch size 32 and dropout set to 0.7. We keep dropout high intentionally to prevent overfitting and prevent the model from learning the main class only. We also do an oversampling to mitigate the imbalance problem and keep each batch balanced while training. Table 2 provides the hyper-parameters details of PEERAssist.

Table 2. Hyper-parameters details of PEERAssist

Layer	Parameter name	Value
Embedding	Embedding dimension	Paper = 768, Review = 512
CNN	Kernal size Number of filters	Different kernel sizes [1, 2, 3, 4, 5, 6, 7] Each kernel size has number of filters = num_filters num_filters = int((3*rev_max_sents)//7 *hyper_factor)
Sentence level BILSTM	Hidden units	int(lstm_base*hyper_factor) + int(lstm_base*0.2)
Section level BiLSTM	Hidden units	int(lstm_base * hyper_factor)
Dense	Hidden units	int((total_out//15)* hyper_factor)
Dropout	Dropout rate	0.7
Training params	Batch size Optimizer Loss function	32 Adam Binary_cross entropy
Others	Hyper_factor Lstm_base Total_out	Dropout rate 0.7 256 Size of vector (concatenated [p + pr + rp + r]

5.2 Comparing Systems

1. **DeepSentiPeer** [8]: The authors proposed a deep neural architecture that considers the Review and Paper text and augmented it with sentiment features (via a pre-trained CNN-based feature extractor) to predict the acceptance decision.
2. **PeerRead(Baseline)** [13]: To predict the acceptance decision, the authors trained a binary classifier with some hand-crafted features, meta features (e.g., number of authors, number of references, etc.) and textual features (e.g., number of unique words, number of sections, average sentence length, etc.)

Table 3. Results on the decision prediction task

Accuracy (Accept/Reject)					
Baseline	Model types	Dataset			
		ICLR 2017	ICLR 2018	ICLR 2019	ICLR 2020
	PeerRead Baseline (only Paper)	55.26*	57.92*	56.71*	57.82*
Comparing system	DeepSentiPeer (Paper + Review+ Sentiment)	71.05	73.31*	72.02*	72.97*
Proposed architecture	Paper + Review	**69.66**	**71.64**	**76.65**	**75.82**
	Paper + Review + Sentiment	**76.04**	**76.12**	**77.84**	**76.23**

5.3 Results and Discussion

We experimented with the (2017,18,19, and 2020) ICLR datasets, and computed Accuracy for the evaluation. The comparison results for input variants (paper+review), (paper+review+sentiment) are reported in Table 3. Our proposed approach outperforms with around 22.78% higher accuracy than the compared approaches on ICLR 2017 dataset, only using paper + review variant still our model can outperform 14.4% higher accuracy than compared approaches. DeepSentiPeer [8] and PeerRead [13] performed their experiments on ICLR 2017, ACL 2017, and CoNLL 2016 dataset only. However, we report our approach's performance on the recent dataset, namely ICLR 18, 19, and 2020. Thus, we ran PeerRead publicly available code[5] on this dataset, and found that our approach outperforms by a margin of 18.02%(ICLR-18), 21.13%(ICLR-19), and 18.4%(ICLR-2020), respectively. The reason behind this is that the work done by PeerRead relies only on the hand-designed features extracted from the paper, whereas we consider paper features, associated review text and the sentiment of the reviewer encoded in peer review in our proposed architecture. It strongly suggests that interaction between paper and review, and the associated sentiment helps the model for accurate prediction.

Similarly, we also ran the DeepSentiPeer[6] on our recent data and found our approach outperforming their performance. We show the results in Table 3. The reason behind this is that DeepSentiPeer [8] uses review text and paper, sentiment within the review with sentence embedding (using Universal Sentence Encoder) to predict acceptance decisions. Here, we use a sectional summary of the paper with a section embedding (SciBert, which gives a good weightage to scientific words while encoding). Afterwards, we used cross attention between a sectional summary of the paper and review text. Expectedly, according to our experiments, the use of cross attention enhances our model performance.

5.4 Ablation Study

To validates the effectiveness of our cross attention-based proposed framework. We analyze the outputs of our framework by the accuracy shown in Table 4. We

[5] https://github.com/allenai/PeerRead.git.
[6] https://github.com/aritzzz/DeepSentiPeer.git.

observe the drop in the accuracy when we use the paper+ review + sentiment without cross attention as an input variant. When we use the paper + review + sentiment with cross attention, we obtain the significant improvement in accuracy 4.64%(ICLR-17), 2.51%(ICLR-18), 4.86%(ICLR-19) and 2.72%(ICLR-20). It strongly suggests that cross attention would efficiently guide the framework to make a good prediction between (accepted or rejected) classes.

Table 4. Comparison between our proposed architecture with its variants

Input varients	Dataset			
	ICLR 2017	ICLR 2018	ICLR 2019	ICLR 2020
	Accuracy	Accuracy	Accuracy	Accuracy
Proposed Architecture (with cross attention)	76.04	76.12	77.84	76.23
Paper + Review + Sentiment (without cross attention)	71.40	73.61	72.98	73.51

5.5 Observation

- For decision prediction (classification) problem, we analyze the confusion matrix as shown in Fig. 5. The true reject is above 75%, and the true accept is above 60%. This signifies the better performance of our proposed model. But for 2020, our proposed model's performance degrades for identifying the false samples. The possible reason could be the decline in paper acceptance rate over this year, see Table 1.
- The majority of review texts are objective and represent the reviewer's sentiment about the work (as indicated in the Fig. 4, the score for neutral polarity is high compared to positive and negative).
- Strongly opinionated comments are those texts which reflect the sentiment of the reviewer and clearly bringing out the reviewer's stand on the work. By knowing which comments are significant would also provide signals to estimate the reviewer's knowledge, confidence, and strength of the review. Also, identifying crucial comments could help by provide significant insights to the editors to draft a meta-review for the paper. With these objectives we use VADER to calculate the sentiment polarity of a review texts which generates a compound sentiment score for a review texts between +1(extreme positive) and −1(extreme negative).

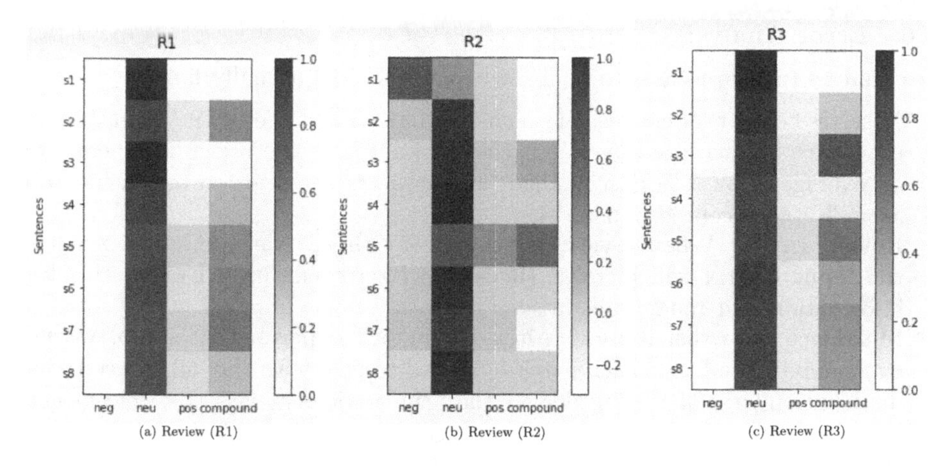

Fig. 4. Depicting Sentiment Polarity in Review texts. neu→Neutral Sentiment Score, neg→Negative Sentiment Score, pos→Positive Sentiment Score, and compound→Compound Sentiment Score (we take the average of the sentence-wise sentiment scores). s1...s8→are the sentences in the review texts.

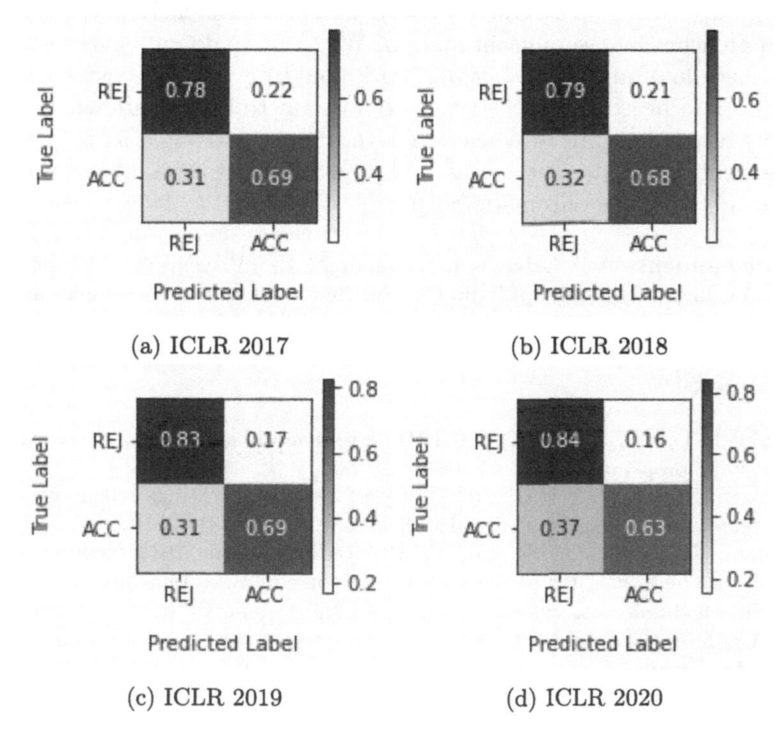

Fig. 5. Normalized confusion matrix on ICLR 2017, 18, 19 and 2020 test data for decision prediction with our proposed architecture (Paper+ Review + Sentiment).

5.6 Error Analysis

We analyse the error cases to realize where our model actually fails.

- Misclassified actual classes: Fig. 5 shows that an average of 19% misclassified paper are the accepted ones, but they are actually the rejected papers, and an average of 32.75% misclassified instances are the rejected ones which are actually the accepted papers.
- Review length: When review length is very short, the number of zeros in the input matrix will increase (because of zero padding). Due to this less information, our model fails in such cases.
- Imbalance data distribution: Although in our experimental setup, we use oversampling and keep each batch balanced to deal with the imbalanced class distribution, it can not be denied that the majority class (i.e. reject) gets benefited in borderline cases, which cause our model to misclassify.

6 Conclusion and Future Work

In this work, we propose a deep neural architecture that leverages on interaction between paper and reviews via cross-attention; sentiment of reviewers, to predict the acceptance of a paper. Our proposed model consistently outperforms the earlier approaches by a significant margin. Whereas we do not envisage an AI to take up the role of an editor, but our work could be a step towards human-AI collaboration in peer reviews. Next we would want to investigate which sections or aspects of a paper are responsible for the final decision so as to provide an iterative feedback to the authors as well as make the editors more informed of the reasons behind the automatic output.

Acknowledgement. Asif Ekbal is a recipient of the Visvesvaraya Young Faculty Award and acknowledges Digital India Corporation, Ministry of Electronics and Information Technology, Government of India for supporting this research.

References

1. Beltagy, I., Lo, K., Cohan, A.: SciBERT: a pretrained language model for scientific text. arXiv preprint arXiv:1903.10676 (2019)
2. Bornmann, L., Daniel, H.D.: Reliability of reviewers' ratings when using public peer review: a case study. Learn. Publish. **23**(2), 124–131 (2010)
3. Burstein, J., Doran, C., Solorio, T.: Proceedings of the 2019 conference of the North American chapter of the association for computational linguistics: human language technologies, volume 1 (long and short papers). In: Proceedings of the 2019 Conference of the North American Chapter of the Association for Computational Linguistics: Human Language Technologies, (Long and Short Papers), vol. 1 (2019)
4. Cer, D., et al.: Universal sentence encoder arXiv preprint arXiv:1803.11175 (2018)
5. Chakraborty, S., Goyal, P., Mukherjee, A.: Aspect-based sentiment analysis of scientific reviews. In: Proceedings of the ACM/IEEE Joint Conference on Digital Libraries in 2020, pp. 207–216 (2020)
6. Charlin, L., Zemel, R.: The Toronto paper matching system: an automated paper-reviewer assignment system (2013)

7. Ghosal, T., Sonam, R., Ekbal, A., Saha, S., Bhattacharyya, P.: Is the paper within scope? Are you fishing in the right pond? In: 2019 ACM/IEEE Joint Conference on Digital Libraries (JCDL), pp. 237–240. IEEE (2019)
8. Ghosal, T., Verma, R., Ekbal, A., Bhattacharyya, P.: DeepSentiPeer: harnessing sentiment in review texts to recommend peer review decisions. In: Proceedings of the 57th Annual Meeting of the Association for Computational Linguistics, pp. 1120–1130 (2019)
9. Ghosal, T., Verma, R., Ekbal, A., Bhattacharyya, P.: A sentiment augmented deep architecture to predict peer review outcomes. In: 2019 ACM/IEEE Joint Conference on Digital Libraries (JCDL), pp. 414–415. IEEE (2019)
10. Ghosal, T., Verma, R., Ekbal, A., Saha, S., Bhattacharyya, P.: Investigating impact features in editorial pre-screening of research papers. In: Proceedings of the 18th ACM/IEEE on Joint Conference on Digital Libraries, pp. 333–334 (2018)
11. Huisman, J., Smits, J.: Duration and quality of the peer review process: the author's perspective. Scientometrics **113**(1), 633–650 (2017). https://doi.org/10.1007/s11192-017-2310-5
12. Hutto, C., Gilbert, E.: VADER: a parsimonious rule-based model for sentiment analysis of social media text. In: Proceedings of the International AAAI Conference on Web and Social Media, vol. 8 (2014)
13. Kang, D., et al.: A dataset of peer reviews (PeerRead): collection, insights and NLP applications. arXiv preprint arXiv:1804.09635 (2018)
14. Kelly, J., Sadeghieh, T., Adeli, K.: Peer review in scientific publications: benefits, critiques, & a survival guide. Ejifcc **25**(3), 227 (2014)
15. Li, S., Zhao, W.X., Yin, E.J., Wen, J.R.: A neural citation count prediction model based on peer review text. In: Proceedings of the 2019 Conference on Empirical Methods in Natural Language Processing and the 9th International Joint Conference on Natural Language Processing (EMNLP-IJCNLP), pp. 4914–4924 (2019)
16. Plank, B., van Dalen, R.: CiteTracked: a longitudinal dataset of peer reviews and citations. In: BIRNDL@ SIGIR, pp. 116–122 (2019)
17. Price, S., Flach, P.A.: Computational support for academic peer review: a perspective from artificial intelligence. Commun. ACM **60**(3), 70–79 (2017)
18. Qiao, F., Xu, L., Han, X.: Modularized and attention-based recurrent convolutional neural network for automatic academic paper aspect scoring. In: Meng, X., Li, R., Wang, K., Niu, B., Wang, X., Zhao, G. (eds.) WISA 2018. LNCS, vol. 11242, pp. 68–76. Springer, Cham (2018). https://doi.org/10.1007/978-3-030-02934-0_7
19. Smith, R.: Peer review: a flawed process at the heart of science and journals. J. R. Soc. Med. **99**(4), 178–182 (2006)
20. Stelmakh, I., Shah, N.B., Singh, A., Daumé III, H.: A novice-reviewer experiment to address scarcity of qualified reviewers in large conferences. CoRR abs/2011.15050 (2020). https://arxiv.org/abs/2011.15050
21. Sun, M.: Peer review comes under peer review. Science **244**(4907), 910–913 (1989)
22. Superchi, C., González, J.A., Solà, I., Cobo, E., Hren, D., Boutron, I.: Tools used to assess the quality of peer review reports: a methodological systematic review. BMC Med. Res. Methodol. **19**(1), 48 (2019)
23. Thelwall, M., Papas, E.R., Nyakoojo, Z., Allen, L., Weigert, V.: Automatically detecting open academic review praise and criticism. Online Information Review (2020)
24. Wang, K., Wan, X.: Sentiment analysis of peer review texts for scholarly papers. In: The 41st International ACM SIGIR Conference on Research & Development in Information Retrieval, pp. 175–184 (2018)
25. Wicherts, J.M.: Peer review quality and transparency of the peer-review process in open access and subscription journals. PLoS ONE **11**(1), e0147913 (2016)
26. Yuan, W., Liu, P., Neubig, G.: Can we automate scientific reviewing? arXiv preprint arXiv:2102.00176 (2021)

ContriSci: A BERT-Based Multitasking Deep Neural Architecture to Identify Contribution Statements from Research Papers

Komal Gupta[1(✉)], Ammaar Ahmad[1], Tirthankar Ghosal[2], and Asif Ekbal[1]

[1] Department of Computer Science and Engineering, Indian Institute of Technology Patna, Patna, India
{komal_2021cs16,1801cs08,asif}@iitp.ac.in

[2] Institute of Formal and Applied Linguistics, Faculty of Mathematics and Physics, Charles University, Prague, Czech Republic
ghosal@ufal.mff.cuni.cz

Abstract. With the rapid growth of scientific literature, it is becoming increasingly difficult to identify scientific contribution from the deluge of research papers. Automatically identifying the specific contribution made in a research paper would help quicker comprehension of the work, faster literature survey, comparison with the related works, etc. Here in this work, we investigate methods to automatically extract the *contribution statements* from research articles. We design a multitask deep neural network leveraging *section identification* and *citance classification* of scientific statements to predict whether a given scientific statement specifies a contribution or not. In the long-run, we envisage to create a knowledge graph of scientific contributions for machine comprehension and more straightforward navigation of research contributions in a particular domain. Our approach achieves the best performance over earlier methods (a relative improvement of 8.08% in terms of F_1 score) for contributing sentence identification over a dataset of Natural Language Processing (NLP) papers. We make our code available at here (https://github.com/ammaarahmad1999/Sem-Eval-2021-Task-A).

Keywords: Contribution identification · Multitask learning · Information extraction

1 Introduction

The exponential growth in the scientific literature in recent years [16] has lead to a serious scholarly information overload on researchers. It is increasingly becoming challenging for researchers to navigate the gigantic scholarly knowledge to find the relevant and desired answers to their questions. *Arxiv* estimates that it

K. Gupta and A. Ahmad—Equal Contribution.

H.-R. Ke et al. (Eds.): ICADL 2021, LNCS 13133, pp. 436–452, 2021.
https://doi.org/10.1007/978-3-030-91669-5_34

received around ~400 submissions daily in 2018 [1]. It is now almost impossible for a researcher to go through the entire volume of scientific literature, even in their domain. For some other stakeholders, such as doctors who are super busy with their day-to-day job but want to keep abreast with the recent developments in medicine and biology, they have very little time to devote for search, retrieval, and ingest relevant scholarly developments. However, it is imminent that researchers are updated with the latest progress in science to take the science forward. Hence the need for automated support to sift through scholarly contributions is increasingly becoming evident. Using Natural Language Processing (NLP) and Machine Learning (ML), there are now attempts to comprehend the scholarly document discourse [10] by machines. A growing scholarly document processing community [6] seeks to ease the scholarly burden on researchers and develop automated mechanisms to support them in various stages of the research life cycle. However, one big disadvantage of mining scholarly documents and subsequent computation is that most of the scholarly knowledge is encoded in PDFs, making information extraction difficult. One of the significant attempts to make research knowledge computable is the *Open Research Knowledge Graph* [13]. The core idea in building this artifact is to extract contributions from research articles and integrate those contributions to a knowledge graph to automatically make the scholarly knowledge navigable, discoverable, and computable by algorithms. A knowledge graph of scientific contributions would allow machines to navigate through the prior knowledge in literature, make meaningful comparisons, comprehend the *novelty* of a new research article, etc. However, the first step towards building this knowledge graph of scientific contributions is to develop NLP techniques to automatically extract contribution statements (i.e., sentences where the authors specify their exact contribution) from research articles. Extraction of contribution sentences is not straightforward as researchers write in different ways to express their contributions. Underlying semantics and context play a crucial role; scientific texts hold background information that makes semantic processing challenging.

Here in this work, we develop an NLP-based approach to identify the contribution sentences from research articles automatically. We develop a multitasking deep neural architecture that inputs a research statement and labels if the statement signifies a research contribution. We make use of the dataset [13] released as part of the NLP Contribution Graph challenge [2] to investigate this specific objective. On analyzing the data (Sect. 6.1) we reveal that contribution statements in a research paper are concentrated on certain sections and appear less frequently on other sections. Hence, identifying the section of the research statement is an important signal to the problem. Also, one does not expect a scientific statement which states an original contribution to cite other works. Hence a *contribution sentence* might not be a *citance* (sentence containing a citation). Building on these two signals, we propose ContriSci, a SciBERT [7] based multitasking architecture that leverages on *section identification* and *citance classification* to conclude whether a given research statement is a contributing one or not. We develop two scaffold tasks for ContriSci that help the main task. **Subtask 1: Section Identification** - *to predict the section title of the research* *statement* and **Subtask 2: Citance Classification** - *to classify the research statement as citance or non-citance.* Our proposed model outperforms all the submissions in the SemEval 2021 NLP Contribution Graph (NCG) challenge [2] on the NCG dataset [13] by a significant margin (4.8% F_1 score). We use two additional datasets to train and boost the scaffold tasks: the ACL Anthology Sentence Corpus [4] for Subtask 1 and the SciCite Corpus [11] for Subtask 2. Our proposed model is novel, and to the best of our knowledge, multitasking is not applied so far to address this specific task.

2 Related Work

Earlier literature presents very less work on this problem. Gupta et al. [14] presented a method for characterizing a research work in terms of its *focus*, *domain* of application, and *techniques* used. They classify sentences from the abstract into the above categories via matching semantic patterns. Brack et al. [8] present a novel method that extracts domain independent scientific concepts from the abstract of research articles. Although a novel problem, some interesting submissions came through in the NCG challenge. Shailabh et al. [20] propose a classifier for the sentence using SciBERT with Bidirectional Long Short-Term Memory. Liu et al. [16] present a BERT-based classifier for the classification of contribution sentences. The authors also include position features in the classifier. D'Souza et al. [13] compiled the SemEval-2021 Shared Task *NLP Contribution Graph* task participants' systems.

Now-a-days multitask learning is popular in NLP to leverage on the interactions of the related tasks. Carauna et al. [9] showed that multitask learning is a technique that improves generalization by using the domain information contained in the training signals of the related scaffold tasks. Liu et al. [17] use Recurrent Neural Network and propose three different multitasking model structures and shared information to the model with both task-specific and shared layers. All the architecture of multitasking is trained jointly on all these scaffolds tasks. Liu et al. [18] propose an adversarial multitask learning framework, which focuses on shared layers to extract the common and task-specific features. In processing scholarly documents, Cohan et al. [11] propose a multitask model to identify the intent of a citation in research articles with two structural scaffolds tasks: section identification and citation worthiness. They use structural information of research articles for classification of citation intents.

We leverage on the idea of multitask learning and use it to learn to identify contributing sentences in research papers automatically.

3 Task Definition

Given a scientific paper P, the objective is to extract contribution sentences $S_1, S_2, S_3, S_4 ... S_n$ (where n is the number of contribution sentences in the research article). The following is an example of a contribution sentence from the NCG dataset [13]:

On Web Questions, not specifically designed as a simple QA dataset, 86% of the questions can now be answered with a single supporting fact, and performance increases significantly (from 36.2% to 41.0% F1-score).

It is evident that the above statement is not a citance, nor does it deserve a citation. It is from the *results* section of a certain paper.

4 Dataset Description

As discussed earlier, we use the *NLPContributionGraph* (NCG) [12] dataset. The dataset contains approximately 442 NLP papers spread across 24 different NLP topics. It has three layers of annotations : 1) Contribution sentences, 2) Scientific words and predicate phrases from the contribution sentences, 3) Subject-predicate-object triple statements from the phrases motivated towards building a knowledge graph. We present the raw statistics of the training, testing and trial folds in Table 1.

Table 1. Data statistics of NCG corpus

Analysis	Train set	Trial set	Test set
Document	237	50	155
Total No of Sentences	54964	11433	33645
Contribution Sentences	5064	1029	2720
Non-Contribution Sentences	49900	10404	30925
Average Length of Sentences	20.61	21.07	20.43
Maximum Length of Sentence	389	188	377
Contribution Sentences having Citation	97	14	45
Non Contribution Sentences having Citation	526	107	484

5 Data Preprocessing

The NCG dataset contains three types of files for each paper: the actual paper in PDF, plain text of the PDF parsed using *Grobid* PDF parser [3], and another text file of the paper generated using *Stanza* [19]. We perform the following pre-processing.

5.1 Combining Incomplete Sentences in the Stanza File

We found that the dataset contains many incomplete sentences in the stanza files, which do not provide proper context and the baseline model [20] fails to classify those incomplete sentences. For example under the topic *Paraphrase Generation*, paper number *0*, the following two lines are incorrectly truncated due to special characters (replaced by? in the stanza file).

1. *164: The Critical Difference (CD) for Nemenyi test depends upon the given?*
2. *165: (confidence level, which is 0.05 in our case) for average ranks and N (number of tested datasets).*

Therefore, we combine sentences that end with any of the following symbols "?" ";" "?:" "," . Also, there are sentences that break on the citation. For example in *Natural Language Inference* paper number *58* stanza file,

1. *63. Chen et al.*
2. *64. propose using a bilinear term similarity function to calculate attention scores with pre trained word embeddings .*

We combined these kinds of sentences as well.

5.2 Extraction of Main Section and Sub Section Titles

For the *Section Identification* task, we annotate our NCG dataset by extracting the section titles to which a sentence belongs. Moreover, to add more context to the sentence, we extract the corresponding subtitles of the sentences as well. We use grobid and stanza files to identify them by following rule-based heuristics.

1. If the sentence length is ≤ 4 and it contains a substring like *Abstract, Introduction, Related Work, Background, Experiment, Implementation* e.t.c then it is a main heading.
2. We identify the sentences following blank lines in the Grobid files as potential section titles and subtitles. These sentences are check for the following conditions:
   (a) If the sentence length is <10 and there is a substring (length ≥ 2) of the title of the paper present in a sentence which does not end with any of the English stopwords, then it is a main heading and named as *method.*
   (b) If the above criteria *2a)* fails then it is not a main heading, and all such kinds of sentences are considered as subheading provided they do not end with a stopword like [*by, as, in, and, that*] nor it contains only numbers like for example "2." "4.1".

5.3 Extracting Previous and Next Sentence

In addition to the original sentence and subsection, we concatenate the previous and the following sentence of the current sentence to the input representation to provide more context to the model. If the sentence is the first sentence of the subsection, then the previous sentence is blank. Similarly, if the sentence is the last one of the subsection, then the following sentence is blank.

6 Data Analysis After Pre-processing

After pre-processing, we found that number of contribution sentences in sections (such as *Related Work, Background, Previous Work, Future Work, Conclusion,* or *Discussion*) is negligible. Therefore, we remove the sentences belonging to these sections from the dataset. Then, we combine the train and trial dataset for training as well as for the scaffold task analysis.

6.1 Analysis of Contribution Sentences in Sections

Figure 1a shows the distribution of all the sentences and the Fig. 1b shows the distribution of contribution sentences across sections. Most of the contribution sentences are present in *Experiment* followed by *Result* and *Introduction* section. Around 20% of sentences in *Experiment* are contribution sentences, whereas <5% of sentences in the *Method* section are contribution sentences. This uneven distribution helps us in the identification of contribution sentences.

6.2 Analysis of Contribution Sentences Having Citation

We analyze the cited sentences in the dataset. The analysis of the cited sentences are shown in Fig. 1c and 1d. The total number of non cited sentences in the dataset is 46980, and the number of cited sentences is 538. Out of these 538 sentences, only 109 sentences are contribution sentences which are <2% of total contribution sentences. Thus, the cited contribution sentences are negligible across the dataset.

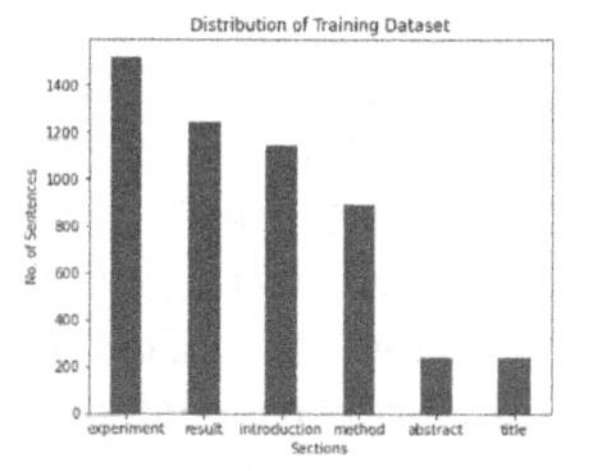

(a) Shows the distribution of all training sentences into the sections.

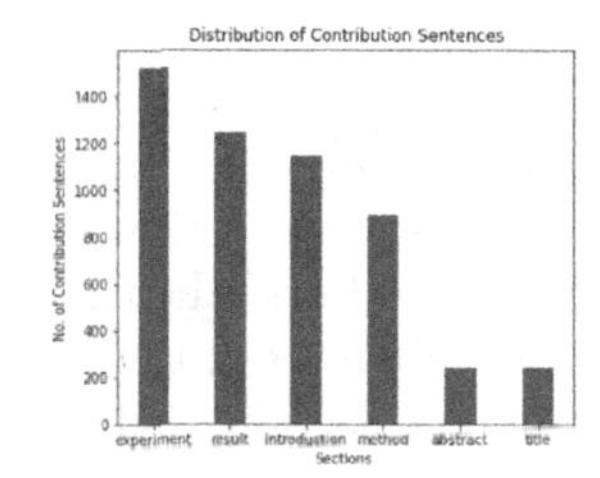

(b) Shows the distribution of training contribution sentences into the sections.

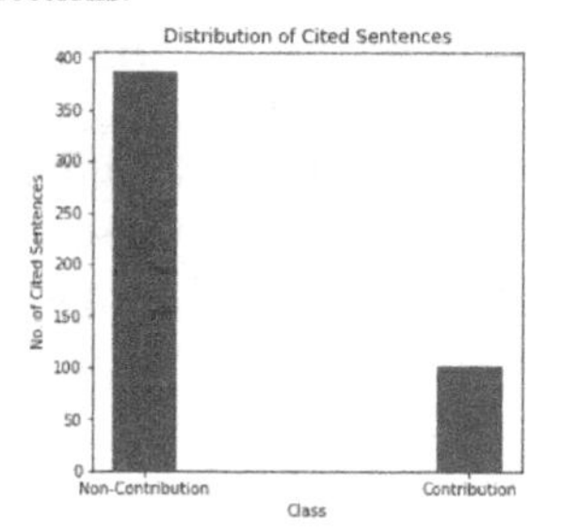

(c) Shows total non contributed cited sentences and contributed cited sentences in the training set.

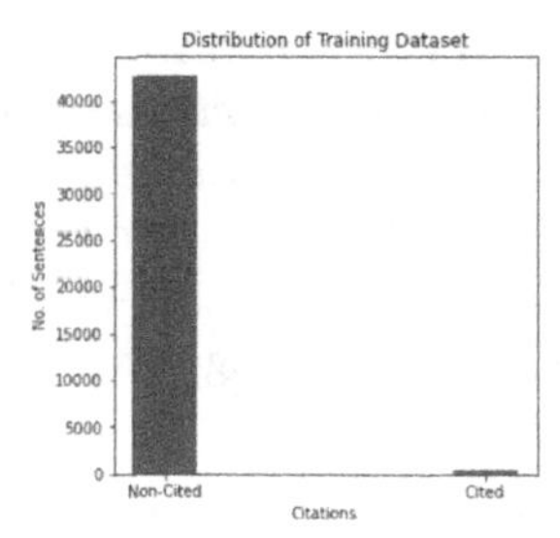

(d) Shows total cited sentences and total contributed cited sentences in training set.

Fig. 1. Shows the analysis graph of NCG data (training + trial) set. Figures *a* and *b* showing the analysis of the distribution of all the sentences and contribution sentences in the section. Figures *c* and *d* shows the occurrence of the cited and non-cited contributing sentences across the dataset.

7 Methodology

We propose a BERT-based multitask learning (ContriSci) model for extracting contribution sentences from the research articles. This multitask model has two scaffold tasks that are related to the structure of research articles. These scaffold tasks help the main task in identifying contribution sentences. We train both the scaffold tasks with NCG data as well as the extra data.

7.1 ContriSci Model

Figure 2 shows the architecture of the ContriSci model. All the tasks share the SCIBERT [7] layer. We are giving input to the model in two ways. If the sentence belongs to NCG dataset then it is in the form of *Current Sentence + # + Subheading + # + Previous Sentence + # + Next Sentence*. Whereas if the sentence belongs to the scaffold dataset, it is in the form of a single sentence. We have tokenized the input and set the maximum length to 256. If the length is greater than the maximum length, we truncate those inputs from the right-hand side to maximum length.

Section Identification. The first scaffold task is to predict the section heading of the sentences. Semantic structure, as well as distribution of contribution sentences, varies from section to section. Contribution sentences in the *Introduction* section introduced their approach, whereas sentences in the *Result* section describe the outcomes of the experiments. Learning these linguistic patterns of sentences across sections helps the model in the identification of contribution sentences.

Citance Classification. The second scaffold task of the ContriSci model is to classify whether a sentence contains a citation or not. The main goal of this scaffold task is to separate the cited sentences from the non cited sentences. The number of cited sentences in the research article is $<5\%$. Out of these sentences, the number of contributed cited sentences is negligible. This information is of great help in separating contribution and non-contribution sentences.

7.2 ContriSci Architecture Description

As defined by [9], multitask learning is an approach to improve the learning of one task by transferring knowledge from related tasks. It requires some shareable parameters among all the tasks. In multitask learning, T_0 is the main task and (n−1) auxiliary tasks T_i. For our ContriSci model, n = 3. We use an independent Multi-Layer Perceptron (MLP) for each task and *Softmax* layer on the top of it for classification. In particular, given the SCIBERT [CLS] output vector x, we pass it to n MLPs followed by softmax to obtain prediction probabilities for each task:

$$y^{(i)} = softmax(MLP^{(i)}(x)) \tag{1}$$

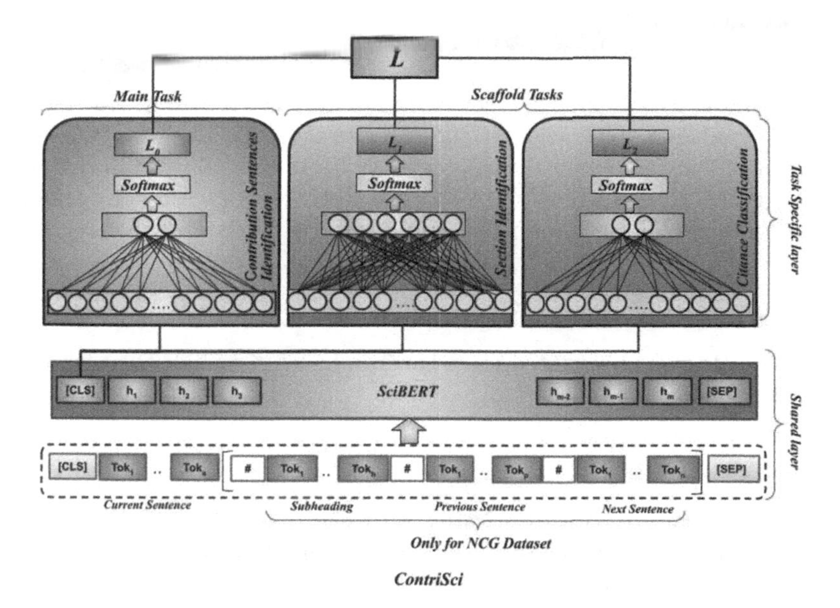

Fig. 2. Architecture of the ContriSci. Here, ContriSci model aims for the identification of contribution sentences. The main task is predicting contribution sentences, and two scaffolds predict the section title (section identification) and predict citation in the sentences (citance classification).

Only the main task output $y^{(0)}$ is of our interest. The rest of the output $(y^{(1)}, y^{(2)})$ is only used while training to improve the performance of the main task.

7.3 Data for Scaffold Tasks

We make use of the additional data for enhancing the weight of the scaffold tasks.

1. **Citance Classification:** For *citance classification* scaffold task, we use the scicite [11] dataset. In the dataset, there are sample sentences from papers; sentences with citance are labeled as positive and others as negative. The size of this dataset is 73K examples with approximately 1:6 positive to negative examples.
2. **Section Identification:** We use the *ACL Anthology Sentence Corpus (AASC)* [4] dataset for *section identification.* AASC dataset is a large corpus of sentences with 10 labeled sections containing more than 2 million examples. Our dataset is a subset of this corpus containing 75K samples from the 5 sections: *abstract*, *introduction*, *result*, *background*, and *method.* Since there is no *experiment* section in AASC dataset and distribution of contribution sentences in *result* and *experiment* section are similar. So, we combine the section label for these sentences into *result*, as we are only interested in distribution of sentences for the scafffold task.

7.4 Training Procedure

We train the multitask model according to Algorithm 1. Each dataset (NCG, Scicite [11], and ACL anthology [4]) has it's own data loader. A batch is selected from either of the three data loaders sequentially in a 3:5:5 ratio in each epoch. The ratio is according to the size of each dataset. After each batch training, we update the weights according to the task-specific objectives of that batch.

7.5 Loss

We use the categorical weighted cross-entropy loss for each of the tasks. Cross-entropy is defined as:

$$L = -\sum_{i=0}^{n-1} w_i t_i \log(P_i) \tag{2}$$

where n is a number of batches, w_i are class weights, t_i is truth label and P_i is a softmax probability of ith. Thus, the overall loss function is a linear combination of loss for each task defined by:

$$L = L_0 + \lambda_1 * L_1 + \lambda_2 * L_2 \tag{3}$$

Here L_0, L_1, L_2 is a loss for the main task, section identification task, and citance classification task, respectively. Here, λ_1 and λ_2 are the hyperparameters. Each class is assigned equal weightage for L_1 loss since each label in the ACL Anthology dataset has an equal number of examples. The distribution of cited and non cited sentences in the Scicite dataset is 1:6. So for L_2 loss, we set the class weight as the inverse ratio of the number of examples in each class in the citance classification scaffold dataset. Finally, for the main loss L_0, we set the class weight as a hyperparameter.

Algorithm 1: Multi-task learning for sentence identification

Create 3 separate data loaders (D_0, D_1, D_2) for 3 dataset respectively each of mini batch of 16.
D_0 - NCG Dataset, D_1 - AASC Dataset D_2 - Scicite Dataset
for *epoch in 1, 2, ..., epoch*$_{max}$ **do**
  $N = length(D_0)/3$
  for *i in range*(N) **do**
    Model train with 3 batches of D_0
    Compute $L_0(\theta), L_1(\theta), L_2(\theta)$
    Compute Gradient ***grad***
    Update Model ($\theta = \theta - \alpha * \textbf{\textit{grad}}$)
    Model train with 5 batches of D_1
    Compute $L_1(\theta)$
    Compute Gradient ***grad***
    Update Model ($\theta = \theta - \alpha * \textbf{\textit{grad}}$)
    Model train with 5 batches of D_2
    Compute $L_2(\theta)$
    Compute Gradient ***grad***
    Update Model ($\theta = \theta - \alpha * \textbf{\textit{grad}}$)
  end
end

8 Evaluation

8.1 Experimental Setup

We implement our proposed model with *allenai/scibert_scivocab_uncased* (SCIBERT has word piece vocabulary (scivocab)) [5]. Each task has an independent MLP layer comprising of a fully connected layer of dimension 768 followed by a classification layer. Batch size is set to 16 and the model is trained with *AdamW* optimizer. We use the activation function *Tanh* and *Softmax* on the fully connected layer and classification layer, respectively. We train the model for two epochs on the GPU using the pytorch framework.

8.2 Hyperparameters

Learning rate(α), λ_1, λ_2, dropout, and class weights are hyperparameters tuned for best performance. We varied α between 1e-6 to 2e−5 and experimented with a dropout of 0.1 and 0.2. λ_1: weightage for section identification loss varied between 0 to 0.3. λ_2: weightage for citance classification loss varied between 0 to 0.3. Class weights: loss weights for each class in the main task varied between 0.5 to 0.88. We use the random search technique for hyperparameter tuning. ContriSci model achieves the best performance on the following parameters: α - 1e−5, λ_1 - 0.18, λ_2 - 0.09, *dropout* - 0.2, and *class weights* - 0.75.

8.3 Baseline

We compare our model with the baseline as proposed in [20] and [16]. Shailabh et al. [20] used the pre-trained SCIBERT with (Bidirectional Long Short-Term Memory) (BiLSTM) [15] as a sentence-level binary classifier. Sentences of the stanza file are input into the SCIBERT model. The last layer of SCIBERT is the input into the stacked BiLSTM layers. Liu et al. [16] presents a SCIBERT-based binary sentence classifier with features to handle the sentence characteristics. They also process topmost and innermost section header and position as well in the articles.

Table 2. Results on NCG test set. Table also shows the comparison of the proposed model with the top-performing models' results reported in the SemEval 2021 competition.

Models	F1	P	R
SCIBERT + BiLSTM [20]	0.4680	0.3669	0.5701
SCIBERT + Positional Feature [16]	0.5727	0.5361	0.6146
Previous SOTA Model [2]	0.5941	0.5519	0.6433
Proposed Model(ContriSci)	**0.6421**	0.5943	0.6943

8.4 Results and Analysis

We compare our proposed model's performance with the results reported in the SemEval 2021. In Table 2, the first three results are taken from the leaderboard of SemEval 2021. The first comparison is with [20] SCIBERT+BiLSTM model. They have applied the BiLSTM layer over the SCIBERT. The best score of this model is 0.4680. The second comparison is with the SCIBERT + Positional Feature [16]. The score of this model is 0.5727. The third model [2] achieves a score of 0.5941. Our multitask model surpasses all these previous state-of-the-art score by 4.8%, achieving the F1 score of 0.6421. Figure 3a shows the Precision, Recall, and F1 score of the main task, i.e. identifying contributing sentences across the different sections. In the NCG training dataset, most of the titles of the paper correspond to the contribution sentences, and hence the title section has recall 1.0. In the training dataset, sentences in the *Method* section are highly skewed (1:19) towards the non-contribution class. This could be one of the reasons why our model fails to differentiate between the contribution and non-contribution sentences belonging to this section. Figure 3b shows the results of cited sentences indicating that such kinds of sentences have a higher F1 score of around 0.7. Therefore *citance classification* scaffold task plays an important role.

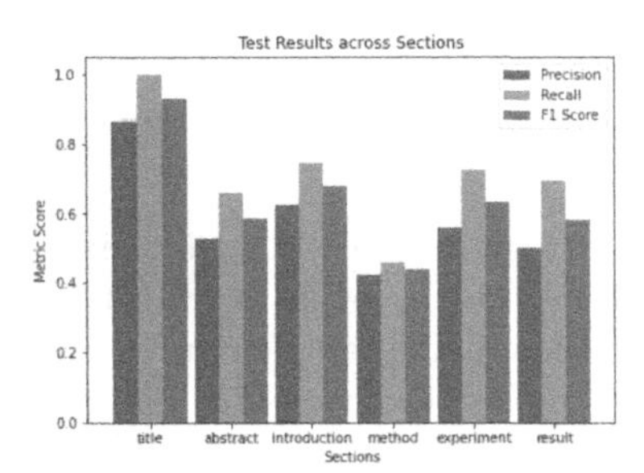

(a) **Evaluation results w.r.t F1, Precision, Recall for section-wise contribution sentence identification in our proposed model.**

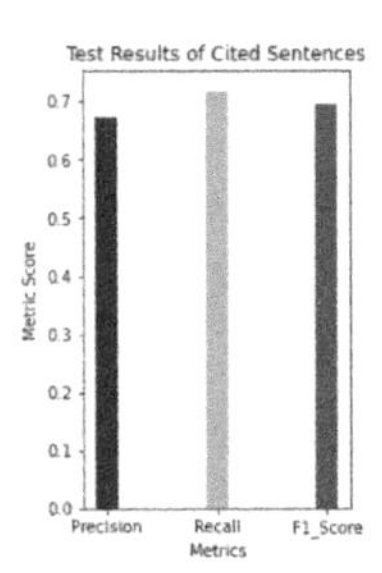

(b) **Evaluation results w.r.t. F1, Precision, Recall for the identification of contributed cited sentences in our proposed model.**

Fig. 3. Shows the graph of testing results analysis on the NCG testing set. (Where Green, Orange, and Blue colors show F1, Recall, Precision respectively). Figures *a* and *b* showing the graph analysis for both the scaffold (Section Identification and Citance Classification) separately. (Color figure online)

8.5 Error Analysis

Table 3 shows the misclassified sentences of the proposed model. Out of which examples 1, 2 are false negative and 3, 4 are false positive.

Table 3. Error Analysis in our proposed model (ContriSci)

S.No	Misclassified sentences	Explanation
1	*Sentence:* The key insights in our approach are 1. to jointly predict short and long answers in a single model rather than using a pipeline approach, 2. to split each document into multiple training instances by using overlapping windows of tokens, like in the original BERT model for the SQuAD task, 3. to aggressively downsample null instances (i.e. instances without an answer) at training time to create a balanced training set, 4. to use the "[CLS]" token at training time to predict null instances and rank spans at inference time by the difference between the span score and the "[CLS]" score	This is a contribution sentence, and this sentence is classified as a non- contribution sentence. Our proposed model is not able to learn long sentences well. The given sentence has about 100 tokens and is also divided into points. In which scientific words are less, and these words are not relevant to the proposed model of the respective paper
2	*Sentence:* 1) Part of Speech Tagging 2) Evaluation on FDDB Database	Smaller sentences which are usually subheading, our model is not able to correctly classify due to lack of contributing contextual information in these subtitles
3	*Sentence:* Therefore, we propose two different methods for building this subset and we call them sense vocabulary compression methods	The sentence given in this example is *False Positive* because the word *we propose* has come in this sentence, so it has been declared as a contribution sentence. However, we are not getting any necessary information about the paper from this sentence, so it is a non-contribution sentence
4	*Sentence:* The model was implemented using Python and Theano	This is also a non-contribution sentence. The reason for its misclassification is that it gives information about the model, but the contribution of this sentence in the article is significantly less. The sentence, whether the contribution in the article is less or more, the Proposed model is not known well

8.6 Annotation Anomalies

When we analyze the NCG training dataset, we find many anomalies in the dataset. We will describe some anomalies of the NCG dataset in the Table 6.

8.7 Ablation Analysis

Tables 4 show ablation studies of our proposed model on the NCG testing set. Table 4 shows the results on SCIBERT's input sequence 128 and 256. When we train the main model with surrounding sentences, we obtained the score of 0.5816. Both scaffold tasks: *section identification*, and *citance classification*, contribute towards the model performance. The length of 93.91% input sentences from the training dataset is less than 128, and the length of 99.8% input sentences is less than 256. Hence, there is a slight improvement in the result when the maximum input length is 256.

Table 4. Performance of our proposed model along with individual scaffold tasks when the model train with 128 and 256 tokens.

Model (128 Tokens)	F1	P	R
Proposed Model - Both Scaffold Tasks	0.5816	0.5629	0.6014
Proposed Model - Section Identification	0.5989	0.5917	0.6063
Proposed Model - Citance Classification	0.5996	0.5862	0.6136
Proposed Model	0.6327	0.5977	0.6720
Model (256 Tokens)	F1	P	R
Proposed Model - Both Scaffold Tasks	0.5694	0.5199	0.6294
Proposed Model - Section Identification	0.5998	0.5534	0.6548
Proposed Model - Citance Classification	0.6052	0.5799	0.6327
Proposed Model	**0.6421**	0.5943	0.6943

9 Error Analysis in Individual Scaffold Task

In Table 5, we analyze the scaffold task separately. The first example is incorrectly predicted as non-contribution by a simple scibert and multitask model with one scaffold. In contrast, our proposed model makes the correct prediction. Only a simple scibert model fails to classify the second example correctly. In the table, for the third and fourth examples, one of the models with a single scaffold makes an incorrect prediction. In general, a model with *section identification* scaffold performs better on sentences with more numerical information, as evident by the second and third examples. Our proposed model ContriSci correctly classified all the examples shown in Table 5 with the help of both the scaffold tasks.

Table 5. The Scaffold task is analyzed separately and compared with the proposed model. Where PM - Proposed Model, CC - Citance Classification, SI - Section Identification

S. No	Misclassified sentences	PM-Both Scaffold	PM-CC	PM-SI	PM
1	Language model pretraining has recently been shown to provide significant performance gains for a range of challenging language understanding problems	Yes	Yes	Yes	No
2	2) Larger layer size, hidden state dimension, and beam size have little impact on the performance; our setting, L = 2, H = 200, and B = 5 looks adequate in terms of speed/performance trade - off	Yes	No	No	No
3	An ensemble of 5 LSTM+ A models further improves this score to 92.8	Yes	No	Yes	No
4	The computational complexity of this network is bounded to be no more than twice that of one convolution block	Yes	Yes	No	No

Table 6. Annotation anomalies in NCG training set

S. No	Annotation anomaly	Example
1	Citation Removed in Stanza and Grobid File **Explanation:** When we compared the sentences in PDF file with Grobid and stanza file, we noticed that some sentences have been removed. In the example 1 sentence one is taken from Grobid, and sentence two is taken from the original sentence PDF. All the citations (such as (graves 2013) and (mhil 2014)) are not in sentence one	**Annotated Sentence:** Attentive neural networks have recently demonstrated success in a wide range of tasks ranging from handwriting synthesis, digit classification, machine translation, image captioning, speech recognition and sentence summarization, to geometric reasoning **Original sentence:** Attentive neural networks have recently demonstrated success in a wide range of tasks ranging from hand writing synthesis (Graves, 2013), digit classification (Mnih et al. 2014), machine translation (Bahdanau et al. 2015), image captioning (Xu et al. 2015), speech recognition (Chorowski et al. 2015) and sentence summarization (Rush et al. 2015), to geometric reasoning (Vinyals et al. 2015)
2	Citation Break the Sentence into Two Parts **Explanation:** The sentence in example 2 is taken from the stanza file. This sentence is divided into two parts by the dot of citation in the stanza file. Similarly, there are many sentences in which citation is there. In the stanza, the file is divided into two parts	**Annotated Sentence:** 1) For example, Yu et al. 2) used CNN representations as feature inputs to a logistic regression model **Original sentence:** For example, Yu et al. [36] used CNN representations as feature inputs to a logistic regression model
3	Sentences are Break with Question Mark **Explanation:** In some sentences in grobid, special symbols have changed into question marks, and in the stanza file, the sentences in which the question mark has come in the middle, those sentences are divided into two parts from there. example 3 shows two parts of the same sentence and the original sentence below	**Annotated Sentence:** 1) Furthermore, let e L ? 2)R L be a vector of 1s and h N be the last output vector after the premise and hypothesis were pricessed by the two LSTMs respectively **Original sentence:** Furthermore, let e L∈R L be a vector of 1s and h N be the last output vector after the premise and hypothesis were processed by the two LSTMs respectively

(*continued*)

Table 6. (*continued*)

S. No	Annotation anomaly	Example
4	Some Sentences are Wrongly Annotated **Explanation:** In example 4, the sentence given in example 4 tells about the result of the Single Model and Ensemble Model and shows how much improvement is in Ensemble Model, so this sentence has a high probability of being a contribution. However, in the NCG dataset, it is the non-contribution is annotated	On SQuAD, our single model obtains an exact match (EM) score of 79.5% and F1 score of 86.6%, while our ensemble model further boosts the result to 82.3% and 88.5% respectively
5	Issues in Length of <4 Sentences **Explanation:** In the NCG dataset, The length of less than 4 sentences in length is wrongly annotated. In example 5, the first two contribution sentences have only word, then how will the model recognize from a single word that this sentence contributes to the article and the last three sentences are abbreviations. If the model does not even know the full form of these short-form keys, how will the model recognize whether it is a contribution sentence or a non-contribution sentence?	1) Sudoku 2) Subtask A 3) NQG *(abbreviation)* 4) ATSA *(abbreviation)* 5) s 2s+ att *(abbreviation)*
6	Inconsistency in Labeling same Subtitles in different Article: **Explanation:** The subheading *Natural Language Inference* exists in a total of six papers. Four of them have been annotated as one, and two of them have been annotated as zero	In *Natural Language Inference* Paper number 10, Line 139 Label is one, but in Natural Language Inference Paper 60, Line 93 Label is zero
7	Section Title Removed and some Sentences are Jumbled **Explanation:** Section titles are missing in the Grobid file of some articles. That is why our section identification scaffold task is mispredicting the sections of some sentences. Apart from this, in some sections in some Grobid files, there are jumbled sentences in the section; if compared with PDF, then the order of sentences in the Grobid file is wrong	The order of sentences is incorrect in the *Experiment, Document Modeling* section of the Grobid file of *Natural language Inference* papers number 18 in the NCG dataset and Introduction heading in Natural language Inference paper number 21

10 Conclusion and Future Work

In this task, we propose a deep multitask architecture for classifying the contribution sentences with the help of two supporting tasks, *viz.* section identification and citance classification. Evaluation on the SemEval 2021 shared task dataset show that our model achieves the state-of-the-art performance with an F1 score of 0.6421. An exciting line of future work is to find subtasks that have similarities to the main task and help the main task classify sentences. For example, another subtask we could add in the proposed model that helps the main task by calculating the relative distance of each sentence from the main heading and subheading. In the future, we want to add SCIBERT on the task-specific layer as well.

Acknowledgement. Asif Ekbal is a recipient of the Visvesvaraya Young Faculty Award and acknowledges Digital India Corporation, Ministry of Electronics and Information Technology, Government of India for supporting this research.

References

1. Arxiv submission rate statistics arxiv e-print repository. https://arxiv.org/help/stats/2018_by_area. Accessed 15 July 2021
2. Codalab - competition. https://competitions.codalab.org/competitions/25680#results. Accessed 15 July 2021
3. Github - kermitt2/grobid: a machine learning software for extracting information from scholarly documents. https://github.com/kermitt2/grobid. Accessed 15 July 2021
4. Overview—aasc. https://kmcs.nii.ac.jp/resource/AASC/AASC.html. Accessed 15 July 2021
5. Scibert-allenai. https://huggingface.co/allenai/scibert_scivocab_uncased. Accessed 15 July 2021
6. Beltagy, I., et al.: Proceedings of the second workshop on scholarly document processing. In: Proceedings of the Second Workshop on Scholarly Document Processing (2021)
7. Beltagy, I., Lo, K., Cohan, A.: SciBERT: a pretrained language model for scientific text. arXiv preprint arXiv:1903.10676 (2019)
8. Brack, A., D'Souza, J., Hoppe, A., Auer, S., Ewerth, R.: Domain-independent extraction of scientific concepts from research articles. Adv. Inf. Retrieval **12035**, 251 (2020)
9. Caruana, R.: Multitask learning. Mach. Learn. **28**(1), 41–75 (1997)
10. Chandrasekaran, M.K., et al.: Overview of the first workshop on scholarly document processing (SDP). In: Proceedings of the First Workshop on Scholarly Document Processing, pp. 1–6 (2020)
11. Cohan, A., Ammar, W., Van Zuylen, M., Cady, F.: Structural scaffolds for citation intent classification in scientific publications. arXiv preprint arXiv:1904.01608 (2019)
12. D'Souza, J., Auer, S.: NLPContributions: an annotation scheme for machine reading of scholarly contributions in natural language processing literature. arXiv preprint arXiv:2006.12870 (2020)

13. D'Souza, J., Auer, S., Pedersen, T.: SemEval-2021 task 11: NLPContributionGraph-structuring scholarly NLP contributions for a research knowledge graph. arXiv preprint arXiv:2106.07385 (2021)
14. Gupta, S., Manning, C.D.: Analyzing the dynamics of research by extracting key aspects of scientific papers. In: Proceedings of 5th International Joint Conference on Natural Language Processing, pp. 1–9 (2011)
15. Hochreiter, S., Schmidhuber, J.: Long short-term memory. Neural Comput. **9**, 1735–1780 (1997)
16. Liu, H., Sarol, M.J., Kilicoglu, H.: Uiuc_bionlp at semeval-2021 task 11: a cascade of neural models for structuring scholarly NLP contributions. arXiv preprint arXiv:2105.05435 (2021)
17. Liu, P., Qiu, X., Huang, X.: Recurrent neural network for text classification with multi-task learning. arXiv preprint arXiv:1605.05101 (2016)
18. Liu, P., Qiu, X., Huang, X.: Adversarial multi-task learning for text classification. arXiv preprint arXiv:1704.05742 (2017)
19. Qi, P., Zhang, Y., Zhang, Y., Bolton, J., Manning, C.D.: Stanza: a python natural language processing toolkit for many human languages. arXiv preprint arXiv:2003.07082 (2020)
20. Shailabh, S., Chaurasia, S., Modi, A.: Knowgraph@ iitk at semeval-2021 task 11: building knowledge graph for NLP research. arXiv preprint arXiv:2104.01619 (2021)

Automated Mining of Leaderboards for Empirical AI Research

Salomon Kabongo[1(✉)], Jennifer D'Souza[2(✉)], and Sören Auer[1,2(✉)]

[1] L3S Research Center, Leibniz University of Hannover, Hannover, Germany
kabenamualu@l3s.de

[2] TIB Leibniz Information Centre for Science and Technology, Hannover, Germany
{jennifer.dsouza,soeren.auer}@tib.eu

Abstract. With the rapid growth of research publications, empowering scientists to keep an oversight over scientific progress is of paramount importance. In this regard, the *leaderboards* facet of information organization provides an overview on the state-of-the-art by aggregating empirical results from various studies addressing the same research challenge. Crowdsourcing efforts like PapersWithCode among others are devoted to the construction of *leaderboards* predominantly for various subdomains in Artificial Intelligence. Leaderboards provide machine-readable scholarly knowledge that has proven to be directly useful for scientists to keep track of research progress – their construction could be greatly expedited with automated text mining.

This study presents a comprehensive approach for generating *leaderboards* for knowledge-graph-based scholarly information organization. Specifically, we investigate the problem of automated *leaderboard* construction using state-of-the-art transformer models, viz. Bert, SciBert, and XLNet. Our analysis reveals an optimal approach that significantly outperforms existing baselines for the task with evaluation scores above 90% in F1. This, in turn, offers new state-of-the-art results for *leaderboard* extraction. As a result, a vast share of empirical AI research can be organized in the next-generation digital libraries as knowledge graphs.

Keywords: Table mining · Information extraction · Scholarly text mining · Knowledge graphs · Neural machine learning

1 Introduction

Our present rapidly amassing wealth of scholarly publications [29] poses a crucial dilemma for the research community. A trend that is only further bolstered in a number of academic disciplines with the sharing of PDF preprints ahead (or even instead) of peer-reviewed publications [13]. The problem is: *How to stay on-track with the past and the current rapid-evolving research progress?* In this era of the publications deluge [45], such a task is becoming increasingly infeasible even within one's own narrow discipline. Thus, the need for novel technological infrastructures to support intelligent scholarly knowledge access

H.-R. Ke et al. (Eds.): ICADL 2021, LNCS 13133, pp. 453–470, 2021.
https://doi.org/10.1007/978-3-030-91669-5_35

models is only made more imminent. A viable solution to the dilemma is if *research results* were made skimmable for the scientific community with the aid of advanced knowledge-based information access methods. This helps curtail the time-intensive and seemingly unnecessary cognitive labor that currently constitute the researcher's task of searching just for the results information in full-text articles to track scholarly progress [9]. Thus, *strategic reading* of scholarly knowledge focused on core aspects of research powered by machine learning may soon become essential for all users [43]. Since the current discourse-based form of the *results* in a static PDF format do not support advanced computational processing, it would need to transition to truly digital formats (at least for some aspects of the research).

In this regard, an area gaining traction is in empirical Artificial Intelligence (AI) research. There, *leaderboards* are being crowdsourced from scholarly articles as alternative machine-readable versions of performances of the AI systems. *Leaderboards* typically show quantitative evaluation results (reported in scholarly articles) for defined machine learning tasks evaluated on standardized datasets using comparable metrics. As such, they are a key element for describing the state-of-the-art in certain fields and tracking its progress. Thus, empirical results can be benchmarked in online digital libraries. Some well-known initiatives that exist to this end are: PapersWithCode (PwC) [4],[1] NLP-Progress [3], AI-metrics [1], SQUaD explorer [6], Reddit SOTA [5], and the Open Research Knowledge Graph.[2]

Expecting scientists to alter their documentation habits to machine-readable versions rather than human-readable natural language is unrealistic, especially given that the benefits do not start to accrue until a critical mass of content is represented in this way. For this, the retrospective structuring from pre-existing PDF format of results is essential to build a credible knowledge base. Prospectively, machine learning can assist scientists to record their results in the *leaderboards* of next-generation digital libraries such as the Open Research Knowledge Graph (ORKG) [27]. In our age of the "deep learning tsunami", [38] there are many studies that have used neural network models to improve the construction of automated scholarly knowledge mining systems [7,12,28,36]. With the recent introduction of language modeling techniques such as transformers [44], the opportunity to obtain boosted machine learning systems is further accentuated.

In this work, we empirically tackle the *leaderboard* knowledge mining machine learning (ML) task via a detailed set of evaluations involving a large dataset and several ML models. The *leaderboard* concept varies wrt. the domains or the captured data. Inspired by prior work [24,26,40], we define a *leaderboard* as comprising the following three scientific concepts: 1. *Task*, 2. *Dataset*, and 3. *Metric* (TDM). However, this base *leaderboard* structure can be extended to include additional concepts such as method name, code links, etc. In this work, we restrict our evaluations to the core TDM triple. Thus, constructing a *leader-*

[1] https://paperswithcode.com.
[2] https://www.orkg.org/orkg/benchmarks.

board in our evaluations entails the extraction of all related TDM statements from an article. E.g., (Language Modeling, Penn Treebank, Test perplexity) is a *leaderboard* triple of an article about the 'Language Modeling' *Task* on the 'Penn Treebank' *Dataset* in terms of the 'Test perplexity' *Metric*. Consequently, the construction of comparisons and visualizations over such machine-interpretable data can enable summarizing the performance of empirical findings across systems.

While prior work [24,40] has already initiated the automated learning of *leaderboards*, these studies were mainly conducted under a single scenario, i.e. only one learning model was tested, and over a small dataset. For stakeholders in the Digital Library (DL) community interested in leveraging this model practically, natural questions may arise: *Has the optimal learning scenario been tested? Would it work in the real-world setting of large amounts of data?* Thus, we note that it should be made possible to recommend a technique for knowledge organization services from observations based on our prior comprehensive empirical evaluations [28]. Our ultimate goal with this study is to help the DL stakeholders to select the optimal tool to implement knowledge-based scientific information flows w.r.t. *leaderboards*. To this end, we evaluated three state-of-art transformer models, viz. Bert, SciBert, and XLNet, each with their own respective unique strengths. The automatic extraction of *leaderboards* presents a challenging task because of the variability of its location within written research, lending credence to the creation of a human-in-the-loop model. Thus, our *leaderboard* mining system will be prospectively effective as intelligent helpers in the crowdsourcing scenario of structuring research contributions [42] within knowledge-graph-based DL infrastructures. Our approach called ORKG-TDM is developed and integrated into the scholarly knowledge organization platform Open Research Knowledge Graph (ORKG) [27].

In summary, the contributions of our work are:

1. we construct a large empirical corpus containing over 4,500 scholarly articles and *leaderboard* TDM triples for the development of text mining systems;
2. we empirically evaluate three different transformer models and leverage the best model, i.e. ORKG-TDM, for the ORKG benchmarks curation platform;
3. in a comprehensive empirical evaluation of ORKG-TDM we obtain 93.0% micro and 92.8% macro F1 scores which outperform existing systems by over 20 points.

To the best of our knowledge, ORKG-TDM obtains state-of-the-art results for *leaderboard* extraction defined as (*Task*, *Dataset*, *Metric*) triples extraction from empirical AI research articles. Thus ORKG-TDM can be readily leveraged within KG-based DLs and be used to comprehensively construct *leaderboards* with more concepts beyond the TDM triples. To facilitate further research, our data and code is made publicly available.[3]

[3] https://github.com/Kabongosalomon/task-dataset-metric-nli-extraction, http://doi.org/10.5281/zenodo.5105798.

2 Related Work

Organizing scholarly knowledge extracted from scientific articles in a Knowledge Graph has been viewed from various Information Extraction (IE) perspectives.

Digitalization Based on Textual Content Mining. Building a scholarly knowledge graph with text mining involves two main tasks: 1. scientific term extraction and 2. extraction of scientific or semantic relations between the terms.

Addressing the first task, several dataset resources have been created with scientific term annotations and the term concept typing to foster the training of supervized machine learners. For instance, the ACL RD-TEC dataset [22] annotates computational terminology in Computational Linguistics (CL) scholarly articles and categorizes them simply as *technology* and *non-technology* terms. The ScienceIE SemEval 2017 shared task [10] annotates the full text in articles from Computer Science, Material Sciences, and Physics domains for *Process*, *Task* and *Material* types of keyphrases. SciERC [37] annotates articles from the machine learning domain with six concepts *Task*, *Method*, *Metric*, *Material*, *Other-ScientificTerm* and *Generic*. The STEM-ECR [18] corpus annotates *Process*, *Method*, *Material*, and *Data* concepts in article abstracts interdisciplinarily across ten STEM disciplines.

For the identification of relations between scientific terms in the natural language processing (NLP) community, within the context of human annotations on the abstracts of scholarly articles [10,20], seven relation types between scientific terms have been studied. They are HYPONYM-OF, PART-OF, USAGE, COMPARE, CONJUNCTION, FEATURE-OF, and RESULT. The annotations are in the form of generalized relation triples: ⟨experiment⟩ COMPARE ⟨another experiment⟩; ⟨method⟩ USAGE ⟨data⟩; ⟨method⟩ USAGE ⟨research task⟩. Since human language exhibits the paraphrasing phenomenon, identifying each specific relation between scientific concepts is impractical. In the framework of an automated pipeline for generating knowledge graphs over massive volumes of scholarly records, the task of classifying scientific relations (i.e., identify the appropriate relation type for each related concept pair from a set of predefined relations) is therefore indispensable.

In other text mining initiatives, comprehensive knowledge mining themes are being defined on scholarly investigations. The recent NLPContributionGraph Shared Task [16,17,19] released KG annotations of contributions including the facets of research problem, approach, experimental settings, and results, in an evaluation series that showed it a challenging task. Similarly, in the Life Sciences, comprehensive KGs from reports of biological assays, wet lab protocols and inorganic materials synthesis reactions and procedures [7,8,31–33,41] are released as ontologized machine-interpretable formats for training machine readers.

Digitalization Based on Table Mining. In the earlier subsection, we discussed information extraction models defined for retrieving the relevant structured information from the textual body of articles. Recent efforts are geared to mining information from the semi-structured format of information in articles as tables. Unlike the high performances seen in information mining systems applied to textual data, text mining performances over tables are relatively much lower.

Milosevic et al. [39] tested methods for extracting numerical (number of patients, age, gender distribution) and textual (adverse reactions) information from tables identified by the *<table>* tag in the clinical literature as XML articles. Further, another line of work examined the classification of tables from HTML pages as entity, relational, matrix, list, and nondata leveraging specialized table embeddings called TabVec [21]. Wei et al. [46] defined a question answering task with data in Table cells as the answers over two different datasets, i.e. web data tables and news articles in text format tables. Another model called TAPAS [23] also addressed question answering over tabular data by extending Bert's architecture to encode tables as input and training it end-to-end over tables crawled from Wikipedia. TableSeer [34] is a comprehensive tables mining search engine that crawls digital libraries, detects tables from documents, extracts their metadata, and indexes and ranks tables in a user-friendly search interface.

Digitalization Based on Textual and Tabulated Content Mining. IBM's science result extractor [24] first defined the $(Task, Dataset, Metric)$ extraction task from articles.[4] They trained a Bert classification model leveraging context data from the abstract, tables, and from table headers and captions. Their dataset comprised pdf-to-text converted articles from PapersWithCode (PwC). Following which, AxCell [30] presented an automated machine learning pipeline for extracting results from papers. It used several novel components, including table segmentation, to learn relevant structural knowledge to aid extraction. Unlike the first system, AxCell was trained and tested over LaTeX source code of machine learning papers from https://arxiv.org. Furthermore, the SciRex corpus creation endeavor [26] defines mostly similar information targets as science result extractor. However, SciRex is evaluated on clean LaTeX sources unlike the IBM extractor and our objective of trying to identify robust machine learning pipelines over articles in PDF format. Another recent system, SciNLP-KG [40], reformulated the *leaderboard* extraction task as one with relation *evaluatedOn* between tasks and datasets, *evaluatedBy* between tasks and metrics, as well as coreferent and related relations between the same type of entities. Like us, they comprehensively investigated several transformer model variants. Nevertheless, owing to their task reformulation, relation-based evaluation, and different dataset our results cannot be compared. Finally, Hou et al., the developers of IBM's science result extractor, recently released the TDMSci corpus [25] as a sequence labeling task for extracting *Task*, *Dataset*, and *Metric* at the sentence-level. We maintain the original document-level inference task definition.

[4] They also evaluated the extracting the best score as an automated task which proved very challenging owing to inconsistency with which the best scores are reported and thereby the inability of pdf-to-text extractors to mine the data effectively.

In this paper, we investigate the science result extractor system [24], however, with a detailed empirical perspective. We comprehensively evaluate the potential of transformer models for the task by testing Bert, SciBert, and XLNet. We also test the models over a much larger empirical dataset emulating their application in real-world settings within scholarly digital libraries as https://www.orkg.org.

3 Our Leaderboards Labeled Corpus

To facilitate supervised system development for the extraction of *leaderboards* from scholarly articles, we built an empirical corpus that encapsulates the task. *Leaderboard* extraction is essentially an inference task over the document. To alleviate the otherwise time-consuming and expensive corpus annotation task involving expert annotators, we leverage distant supervision from the available crowdsourced metadata in the PwC KB. In the remainder of this section, we explain our corpus creation and annotation process.

Scholarly Papers and Metadata from the PwC Knowledge Base. We created a new corpus as a collection of scholarly papers with their TDM triple annotations for evaluating the *leaderboards* extraction task inspired by the original IBM science result extractor [24] corpus. The collection of scholarly articles for defining our *leaderboard* extraction objective is obtained from the publicly available crowdsourced leaderboards PwC.[5] It predominantly represents articles in the Natural Language Processing and Computer Vision domains, among other AI domains such as Robotics, Graphs, Reasoning, etc. Thus, the corpus is representative for empirical AI research. The original downloaded collection (timestamp 2021-05-10 at 12:30:21)[6] was pre-processed to be ready for analysis. While we use the same method here as the science result extractor, our corpus is different in terms of both labels and size, i.e. number of papers, as many more *leaderboards* have been crowdsourced and added to PwC since the original work.

PDF Pre-processing. While the respective articles' metadata in machine-readable form was directly obtained from the PwC data release, the document itself being in PDF format needed to undergo pre-processing for pdf-to-text conversion so that its contents could be mined. For this, the GROBID parser [35] was applied to extract the title, abstract, and for each section, the section title and its corresponding content from the respective PDF article files. Each article's parsed text was then annotated with TDM triples via distant labeling to create the final corpus.

[5] https://paperswithcode.com/.

[6] Our corpus was downloaded from the PwC Github repository https://github.com/paperswithcode/paperswithcode-data and was constructed by combining the information in the files *All papers with abstracts* and *Evaluation tables* which included article urls and TDM crowdsourced annotation metadata.

Table 1. Ours vs. the original science result extractor [24] corpora statistics. The "unknown" labels were assigned to papers with no TDM-triples after the label filtering stage.

	Ours		Original	
	Train	Test	Train	Test
Papers	3,753	1,608	170	167
"unknown" annotations	922	380	46	45
Total TDM-triples	11,724	5,060	327	294
Avg. number of TDM-triples per paper	4.1	4.1	2.64	2.41
Distinct TDM-triples	1,806	1,548	78	78
Distinct *Tasks*	288	252	18	18
Distinct *Datasets*	908	798	44	44
Distinct *Metrics*	550	469	31	31

Paper Annotation via Distant Labeling. Each paper was associated with its *leaderboard* TDM triple annotations. These were available as the crowdsourced metadata of each article in the PwC knowledge base (KB). The number of triples per article varied between 1 (minimum) and 54 (maximum) at an average of 4.1 labels per paper. The corpus was thus annotated as a distant labeling task since the labels for each paper were directly imported from the PwC KB without additional human curation of the varying forms of label names. Additionally, *leaderboards* that appeared in less than five papers were ignored. Consequently to the TDM labels filtering stage, some articles were without TDM triples and these articles were annotated with the label "unknown".

Our overall corpus statistics are shown in Table 1. We adopted the 70/30 split for the Train/Test folds for the empirical system development (described in detail in Sect. 6). In all, our corpus contained 5,361 articles split as 3,753 training data and 1,608 test data instances. There were unique 1,850 TDM-triples overall. Note that since the test labels were a subset of the training labels, the unique labels overall can be considered as those in the training data. Table 1 also shows the distinct *Tasks*, *Datasets*, *Metrics* in the last three rows. Our corpus contains 288 *Tasks* defined on 908 *Datasets* and evaluated by 550 *Metrics*. This is significantly larger than the original corpus which had 18 *Tasks* defined on 44 *Datasets* and evaluated by 31 *Metrics*.

4 Leaderboard Extraction Task Definition

The task is defined on the dataset described in previous section. The dataset can be formalized as follows. Let p be a paper in the collection P. Each p is annotated with at least one triple (t_i, d_j, m_k) where t_i is the ith task defined, d_j the jth dataset and m_k the kth system evaluation metric. The number of triples per paper vary.

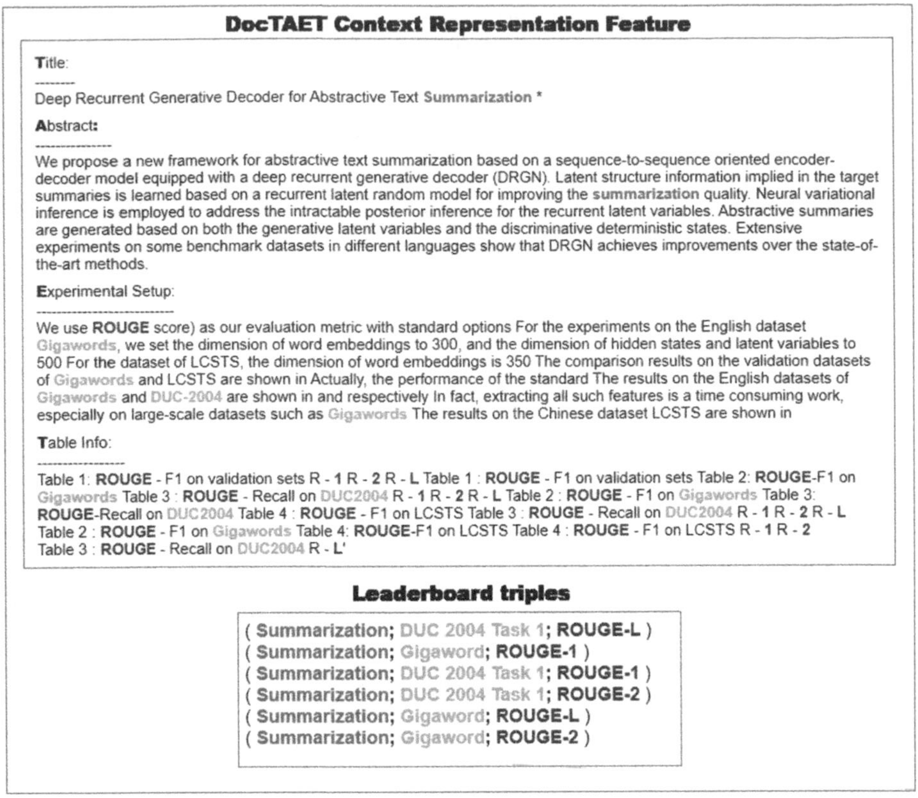

DocTAET Context Representation Feature

Title:

Deep Recurrent Generative Decoder for Abstractive Text **Summarization** *

Abstract:

We propose a new framework for abstractive text summarization based on a sequence-to-sequence oriented encoder-decoder model equipped with a deep recurrent generative decoder (DRGN). Latent structure information implied in the target summaries is learned based on a recurrent latent random model for improving the **summarization** quality. Neural variational inference is employed to address the intractable posterior inference for the recurrent latent variables. Abstractive summaries are generated based on both the generative latent variables and the discriminative deterministic states. Extensive experiments on some benchmark datasets in different languages show that DRGN achieves improvements over the state-of-the-art methods.

Experimental Setup:

We use **ROUGE** score) as our evaluation metric with standard options For the experiments on the English dataset Gigawords, we set the dimension of word embeddings to 300, and the dimension of hidden states and latent variables to 500 For the dataset of LCSTS, the dimension of word embeddings is 350 The comparison results on the validation datasets of Gigawords and LCSTS are shown in Actually, the performance of the standard The results on the English datasets of Gigawords and DUC-2004 are shown in and respectively In fact, extracting all such features is a time consuming work, especially on large-scale datasets such as Gigawords The results on the Chinese dataset LCSTS are shown in

Table Info:

Table 1: **ROUGE** - F1 on validation sets R - **1** R - **2** R - **L** Table 1 : **ROUGE** - F1 on validation sets Table 2: **ROUGE**-F1 on Gigawords Table 3 : **ROUGE** - Recall on DUC2004 R - **1** R - **2** R - **L** Table 2 : **ROUGE** - F1 on Gigawords Table 3: **ROUGE**-Recall on DUC2004 Table 4 : **ROUGE** - F1 on LCSTS Table 3 : **ROUGE** - Recall on DUC2004 R - **1** R - **2** R - **L** Table 2 : **ROUGE** - F1 on Gigawords Table 4: **ROUGE**-F1 on LCSTS Table 4 : **ROUGE** - F1 on LCSTS R - **1** R - **2** Table 3 : **ROUGE** - Recall on DUC2004 R - **L**'

Leaderboard triples

(**Summarization;** DUC 2004 Task 1; **ROUGE-L**)
(**Summarization;** Gigaword; **ROUGE-1**)
(**Summarization;** DUC 2004 Task 1; **ROUGE-1**)
(**Summarization;** DUC 2004 Task 1; **ROUGE-2**)
(**Summarization;** Gigaword; **ROUGE-L**)
(**Summarization;** Gigaword; **ROUGE-2**)

Fig. 1. The DocTAET model with context features as a concatenation of the Scholarly **Doc**ument's **T**itle, **A**bstract, **E**xperimental-Setup, and **T**able content/captions for training NLI transformer models on a set of *leaderboard* triples. The figure illustrates specifically six (*task*, *dataset*, *metric*) triples and their context in the original article text extracted as the feature for transformer models.

In the supervised inference task, the input data instance corresponds to the pair: a paper p represented as the DocTAET context feature $p_{DocTAET}$ and its TDM-triple (t, d, m). The inference data instance, then is $(c; [(t, d, m), p_{DocTAET}])$ where $c \in \{true, false\}$ is the inference label. Thus, specifically, our *leaderboard* extraction problem is formulated as a natural language inference task between the DocTAET context feature $p_{DocTAET}$ and the (t, d, m) triple annotation. (t, d, m) is $true$ if it is among the paper's TDM-triples, otherwise $false$. The $false$ instances are artificially created by random selection of (t, d, m) annotations from another paper. Cumulatively, *leaderboard* construction is a multi-label, multi-class inference problem.

4.1 DocTAET Context Feature

In Fig. 1, we depict the DocTAET context feature [24]. Essentially, the *leaderboard* extraction task is defined on the full document content. However, the respective (*task*, *dataset*, *metric*) label annotations are mentioned only in specific places in the full paper such as in the Title, Abstract, Introduction, Tables. The DocTAET feature was thus defined to capture the targeted context information to facilitate the (*task*, *dataset*, *metric*) triple inference. It focused on capturing the context from four specific places in the text-parsed article, i.e. from the title, abstract, first few lines of the experimental setup section as well as table content and captions.

5 Transformer-Based Leaderboard Extraction Models

For *leaderboard* extraction [24], we employ deep transfer learning modeling architectures that rely on a recently popularized neural architecture – the transformer [44]. Transformers are arguably the most important architecture for natural language processing (NLP) today since they have shown and continue to show impressive results in several NLP tasks [15]. Owing to the self-attention mechanism in these models, they can be fine-tuned on many downstream tasks. These models have thus crucially popularized the transfer learning paradigm in NLP. We investigate three transformer-based model variants for *leaderboard* extraction in a Natural Language Inference configuration.

Natural language inference (NLI), generally, is the task of determining whether a "hypothesis" is true (entailment), false (contradiction), or undetermined (neutral) given a "premise" [2]. For *leaderboard* extraction, the slightly adapted NLI task is to determine that the (*task*, *dataset*, *metric*) "hypothesis" is true (entailed) or false (not entailed) for a paper given the "premise" as the DocTAET context feature representation of the paper.

Currently, there exist several transformer-based models. In our experiments, we investigated three core models: two variants of Bert, i.e. the vanilla Bert [15] and the scientific Bert (SciBert) [11]. We also tried a different type of transformer model than Bert called XLNet [47] which employs Transformer-XL as the backbone model. Next, we briefly describe the three variants we use.

5.1 Bert Models

Bert (i.e., Bidirectional Encoder Representations from Transformers), is a bidirectional autoencoder (AE) language model. As a pre-trained language representation built on the deep neural technology of transformers, it provides NLP practitioners with high-quality language features from text data simply out-of-the-box and thus improves performance on many NLP tasks. These models return contextualized word embeddings that can be directly employed as features for downstream tasks [28].

The first Bert model we employ is $\text{Bert}_{\text{base}}$ (12 layers, 12 attention heads, and 110 million parameters) which was pre-trained on billions of words from the BooksCorpus (800M words) and the English Wikipedia (2,500M words).

The second Bert model we employ is the pre-trained scientific Bert called SciBert [11]. SciBert was pretrained on a large corpus of scientific text. In particular, the pre-training corpus is a random sample of 1.14M papers from Semantic Scholar[7] consisting of full texts of 18% of the papers from the computer science domain and 82% from the broad biomedical domain. For both $\text{Bert}_{\text{base}}$ and SciBert, we used their uncased variants.

5.2 XLNet

XLNet is an autoregressive (AR) language model [47] that enables learning bidirectional contexts using Permutation Language Modeling. This is unlike Bert's Masked Language Modeling strategy. Thus in PLM all tokens are predicted but in random order, whereas in MLM only the masked (15%) tokens are predicted. This is also in contrast to the traditional language models, where all tokens are predicted in sequential order instead of randomly. Random order prediction helps the model to learn bidirectional relationships and therefore better handle dependencies and relations between words. In addition, it uses Transformer XL [14] as the base architecture, which models long contexts unlike the Bert models with contexts limited to 512 tokens. Since only cased models are available for XLNet, we used the cased $\text{XLNet}_{\text{base}}$ (12 layers, 12 attention heads, and 110 million parameters).

6 Automated Leaderboard Mining

6.1 Experimental Setup

Parameter Tuning. We used the Hugging Transfomer libraries (https://github.com/huggingface/transformers) with their Bert variants and XLNet implementations. In addition to the standard fine-tuned setup for NLI, the transformer models were trained with a learning rate of $1e^{-5}$ for 14 epochs; and used the *AdamW* optimizer with a weight decay of 0 for *bias*, *gamma*, *beta* and 0.01 for the others. Our models' hyperparameters details can be found in our code repository online at https://github.com/Kabongosalomon/task-dataset-metric-nli-extraction/blob/main/train_tdm.py.

In addition, we introduced a task-specific parameter that was crucial in obtaining optimal task performance from the models. It was the number of *false* triples per paper. This parameter controls the discriminatory ability of the model. The original science result extractor system [24] considered $|n| - |t|$ *false* instances for each paper, where $|n|$ was the distinct set of triples overall and $|t|$ was the number of *true leaderboard* triples per paper. This approach would not generalize to our larger corpus with over 2,500 distinct triples. In other words,

[7] https://semanticscholar.org.

considering that each paper had on average 4 *true* triples, it would have 2,495 *false* triples which would strongly bias the classifier learning toward only *false* inferences. Thus, we tuned this parameter in a range of values in the set {10, 50, 100} which at each experiment run was fixed for all papers.

Finally, we imposed an artificial trimming of the DocTAET feature to account for Bert's maximum token length of 512. For this, the token lengths of the experimental setup and table info were initially truncated to roughly 150 tokens, after which the DocTAET feature is trimmed at the right to 512 tokens.

Two-Fold Cross Validation. To evaluate robust models, we performed two-fold cross validation experiments. In each fold experiment, we train a model on 70% of the overall dataset, and test on the remaining 30% ensuring that the test data splits are not identical between the folds. Thus, all cumulative results reported are averaged over the two folds. Also, Table 1 corpus statistics are averaged total counts over the two experimental folds.

Evaluation Metrics. Within the two-fold experimental settings, we report macro- and micro-averaged precision, recall, and F1 scores for our *leaderboard* extraction task on the test dataset. The macro scores capture the averaged class-level task evaluations, whereas the micro scores represent fine-grained instance-level task evaluations.

Further, the macro and micro evaluation metrics for the overall task have two evaluation settings: 1) considers papers with *leaderboards* and papers with "unknown" in the metric computations; 2) only papers with *leaderboards* are considered while the papers with "unknown" are excluded.

Table 2. *leaderboard* triple extraction task, comparison of our ORKG-TDM models versus the original science result extractor model (first row in parts (a) and (b)) on the original corpus (see last two columns in Table 1 for the original corpus statistics).

	Macro P	Macro R	Macro F1	Micro P	Micro R	Micro F1
(a) Task + Dataset + Metric Extraction						
$\text{TDM-IE}_{\text{Bert}}$	62.5	75.2	65.3	60.8	76.8	67.8
$\text{ORKG-TDM}_{\text{Bert}}$	68.1	67.5	65.5	79.6	63.3	70.5
$\text{ORKG-TDM}_{\text{SciBert}}$	65.7	77.2	68.3	65.7	76.8	70.8
$\text{ORKG-TDM}_{\text{XLNet}}$	71.7	73.9	**70.6**	77.1	70.9	**73.9**
(b) Task + Dataset + Metric Extraction (without "Unknown" annotation)						
$\text{TDM-IE}_{\text{Bert}}$	54.1	65.9	56.6	60.2	73.1	66.0
$\text{ORKG-TDM}_{\text{Bert}}$	59.0	55.4	54.7	79.5	57.6	66.8
$\text{ORKG-TDM}_{\text{SciBert}}$	57.6	68.7	60.1	65.3	73.1	69.0
$\text{ORKG-TDM}_{\text{XLNet}}$	63.5	64.1	**61.4**	76.4	66.4	**71.1**

6.2 Experimental Results

In this section, we discuss our experimental results shown in Tables 2, 3, and 4 with respect to four research questions elicited as **RQ1**, **RQ2**, **RQ3**, and **RQ4**.

RQ1: How Well do Our Transformer Models Perform for *leaderboard* Extraction Compared to the Original Science Result Extractor When Trained in the Identical Original Experimental Setting? In the last two columns in Table 1, we showed statistics of the comparatively smaller original corpus that defined and evaluated the *leaderboard* extraction task [24]. As a quick recap, the original corpus had 78 distinct TDM-triples including "unknown", and a distribution of 170 papers in the train dataset and 167 papers in the test dataset as fixed partitions; there were 46 and 45 "unknown" papers in the train and test sets, respectively. We evaluate all the three transformer models on this original corpus to compare our model performances. These results are shown in Table 2. As we can see in the table, all three of our models outperform the original with XLNet reporting the best score. We obtain a 70.6 macro F1 versus 65.3 in the baseline. The Bert model is only a few fractional points better at 65.5. Further, we obtain a 73.9 micro F1 versus 67.8 in the baseline, with the Bert model at 70.5. With these results our models outperform the original system in the original settings reported for this task.

Table 3. Top results for Bert (10neg; 1,302unk), SciBert (10neg; 1,302unk), and XLNet (10neg; 1,302unk)

	Macro P	Macro R	Macro F1	Micro P	Micro R	Micro F1
	Average evaluation accross 2-fold					
ORKG-TDM$_{\text{Bert}}$	**92.8**	93.9	92.4	**95.5**	89.1	92.1
ORKG-TDM$_{\text{SciBert}}$	90.9	93.4	91.1	94.1	88.5	91.2
ORKG-TDM$_{\text{XLNet}}$	**92.8**	**94.8**	**92.8**	94.9	**91.2**	**93.0**
	Average evaluation accross 2-fold (without "Unknown" annotation)					
ORKG-TDM$_{\text{Bert}}$	**91.7**	92.1	90.8	**95.7**	88.3	91.8
ORKG-TDM$_{\text{SciBert}}$	89.7	91.4	89.4	94.4	87.6	90.9
ORKG-TDM$_{\text{XLNet}}$	91.6	**93.1**	**91.2**	95.0	**90.5**	**92.7**

RQ2: How Do the Transformer Models Perform on a Large Corpus for *leaderboard* Construction? We examine this question in light of the results reported in Table 3. We find that again, consistent with the observations made in Table 2, XLNet outperforms Bert and SciBert. Note that we have leveraged XLNet with the limited context length of 512 tokens (by truncating parts of the context described in Sect. 6.1) and thus the potential of leveraging the full context in XLNet models remains untapped. Nevertheless, XLNet still distinguishes itself from the Bert models with its PLM language modeling objective versus Bert's MLM. This in practice has also shown to perform better [47]. Thus XLNet with limited context encoding can still outperform the Bert models. Among all

Table 4. Performance of our best model, i.e. ORKG-TDM$_{\text{XLNet}}$, for *Task*, *Dataset*, and *Metric* concept extraction of the *leaderboard*

Entity	Macro			Micro		
	P	R	F_1	P	R	F_1
TDM	91.6	93.1	91.2	95.0	90.5	92.7
Task	**93.7**	**94.8**	**93.6**	**97.4**	**93.6**	**95.5**
Dataset	**92.9**	93.6	92.4	96.6	91.5	94.0
Metric	92.5	94.2	92.5	96.0	92.5	**94.2**

Table 5. Ablation results of our best model, i.e. ORKG-TDM$_{\text{XLNet}}$, for *leaderboard* extraction as $(task, dataset, metric)$ triples

Document representation	Macro P	Macro R	Macro F1	Micro P	Micro R	Micro F1
Title + Abstract	88.6	92.9	89.4	92.6	90	91.3
Title + Abstract + ExpSetup	89.2	91.5	89.2	94.2	89	91.5
Title + Abstract + TableInfo	90.5	**94.4**	91.2	93.5	93.2	93.3
Title + Abstract + ExpSetup + TableInfo	**92.3**	93.5	**91.7**	**95.1**	92	**93.5**

the models, we find SciBert shows a slightly lower performance compared to Bert. This is a slight variation on performance observations obtained in the original smaller dataset (results in Table 2) where Bert was slightly lower than SciBert. These results indicate that in smaller datasets, SciBert should be preferred since its underlying science-specific pretrained corpus would compensate for signal absences in the task training corpus. However, with larger training datasets for finetuning the Bert models, the underlying pretraining corpus domains do not show to be critical to the overall model performance.

Further, we observe the macro and micro evaluations for all three systems (ORKG-TDM$_{\text{Bert}}$, ORKG-TDM$_{\text{SciBert}}$, ORKG-TDM$_{\text{XLNet}}$) have evenly balanced, similar scores. This tells us that the models handle the majority and minority TDM classes evenly. Thus given our large training corpus, the transformers remain unaffected by the underlying dataset varying class distributions.

RQ3: Which of the Three *leaderboard* Concepts are Easy or Challenging to Extract? As a fine-grained examination of our best model, i.e. ORKG-TDM$_{\text{XLNet}}$, we examined its performance for extracting each of three concepts $(Task, Dataset, Metric)$ separately. These results are shown in Table 4. From the results, we observe that *Task* is the easiest concept to extract, followed by *Metric*, and then *Dataset*. We ascribe the low performance for extracting the *Dataset* concept due to the variability in its naming seen across papers even when referring to the same real-world entity. For example, the real-world dataset entity 'CIFAR-10' is labeled as 'CIFAR-10, 4000 Labels' in some papers and 'CIFAR-10, 250 Labels' in others. This phenomenon is less prevalent for *Task* and the

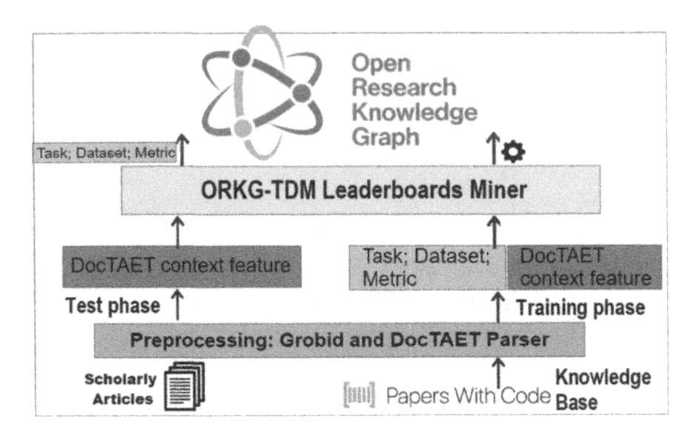

Fig. 2. Overview of the *leaderboard* extraction machine learning process flow for ORKG-TDM in the open research knowledge graph digital library

Metric concepts. For example, the *Task* 'Question Answering' is rarely referenced differently across papers addressing the task. Similarly, for *Metric*, 'accuracy' as an example, has very few variations.

RQ4: Which Aspect of the DocTAET Context Feature Representation had the Highest Impact for *leaderboard* Extraction? Further, in Table 5, we breakdown performance of our best model, i.e. ORKG-TDM$_{\text{XLNet}}$, examining the impact of the features as the shortened context from the articles for TDM inference. We observe, that adding additional contextual information in addition to title and abstract increases the performance significantly, while the actual type of additional information (i.e. experimental setup, table information or both) impacts the performance to a lower extend.

6.3 Integrating ORKG-TDM in Scholarly Digital Libraries

Ultimately, with *leaderboard* construction we aim to give researchers oversight over the state-of-the-art wrt. certain research questions (i.e. *Tasks*). Figure 2 shows how our ORKG-TDM *leaderboard* mining is integrated into the Open Research Knowledge Graph scholarly knowledge platform. The fact that the automated *leaderboard* mining results are incomplete – note, we do not extract the best score owing to noisy signals in the pdf-to-text parser output – and partially imprecise can be alleviated by the ORKG crowdsourcing and curation features. In ORKG, *leaderboards* can be dynamically visualized, versioned etc.

Although, the experiments of our study targeted empirical AI research, we are confident, that the approach is transferable to similar scholarly knowledge extraction tasks in other domains. For example in chemistry or material sciences, experimentally observed properties of substances or materials under certain conditions could be obtained from various papers.

7 Conclusion and Future Work

In this paper, we investigated the *leaderboard* extraction task w.r.t. three different transformer-based models. Our overarching aim with this work is to build a system for comparable scientific concept extractors from scholarly articles. Therefore as a next step, we will extend the current triples (*Task*, *Dataset*, *Metric*) model with additional concepts that are suitable candidates for a *leaderboard* such as *score* or *code urls*, etc. In this respect, we will adopt a hybrid system wherein some elements will be extracted by the machine learning system as discussed in this work while other elements will be extracted by a system of rules and regular expressions. Also, we plan to combine the automated techniques presented herein with a crowdsourcing approach for further validating the extracted results and providing additional training data. Our work in this regard is embedded in a larger research and service development agenda, where we build a comprehensive knowledge graph for representing and tracking scholarly advancements [27]. We also envision the TDM extraction approach to be transferable to other domains (such as materials science, engineering simulations etc.). Our ultimate target is to create a comprehensive structured knowledge graph tracking scientific progress in various scientific domains, which can be leveraged for novel machine-assistance measures in scholarly communication, such as question answering, faceted exploration and contribution correlation tracing.

Acknowledgements. This work was co-funded by the Federal Ministry of Education and Research (BMBF) of Germany for the project LeibnizKILabor (grant no. 01DD20003) and by the European Research Council for the project ScienceGRAPH (Grant agreement ID: 819536).

References

1. AI metrics. https://www.eff.org/ai/metrics. Accessed 26 Apr 2021
2. Natural Language Inference. https://paperswithcode.com/task/natural-language-inference. Accessed 22 Apr 2021
3. Nlp-progress. http://nlpprogress.com/. Accessed 26 Apr 2021
4. paperswithcode.com. https://paperswithcode.com/. Accessed 26 Apr 2021
5. Reddit sota. https://github.com/RedditSota/state-of-the-art-result-for-machine-learning-problems. Accessed 26 Apr 2021
6. Squad explorer. https://rajpurkar.github.io/SQuAD-explorer/. Accessed 26 Apr 2021
7. Anteghini, M., D'Souza, J., Dos Santos, V.A.M., Auer, S.: SciBERT-based semantification of bioassays in the open research knowledge graph. In: EKAW-PD 2020, pp. 22–30 (2020)
8. Anteghini, M., D'Souza, J., Martins dos Santos, V.A.P., Auer, S.: Representing semantified biological assays in the open research knowledge graph. In: Ishita, E., Pang, N.L.S., Zhou, L. (eds.) ICADL 2020. LNCS, vol. 12504, pp. 89–98. Springer, Cham (2020). https://doi.org/10.1007/978-3-030-64452-9_8
9. Auer, S.: Towards an open research knowledge graph, January 2018. https://doi.org/10.5281/zenodo.1157185

10. Augenstein, I., Das, M., Riedel, S., Vikraman, L., McCallum, A.: SemEval 2017 task 10: ScienceIE - extracting keyphrases and relations from scientific publications. In: SemEval@ACL (2017)
11. Beltagy, I., Lo, K., Cohan, A.: SciBERT: a pretrained language model for scientific text. arXiv preprint arXiv:1903.10676 (2019)
12. Brack, A., D'Souza, J., Hoppe, A., Auer, S., Ewerth, R., et al.: Domain-independent extraction of scientific concepts from research articles. In: Jose, J.M. (ed.) ECIR 2020. LNCS, vol. 12035, pp. 251–266. Springer, Cham (2020). https://doi.org/10.1007/978-3-030-45439-5_17
13. Chiarelli, A., Johnson, R., Richens, E., Pinfield, S.: Accelerating scholarly communication: the transformative role of preprints (2019)
14. Dai, Z., Yang, Z., Yang, Y., Carbonell, J.G., Le, Q., Salakhutdinov, R.: Transformer-XL: attentive language models beyond a fixed-length context. In: Proceedings of the 57th Annual Meeting of the Association for Computational Linguistics, pp. 2978–2988 (2019)
15. Devlin, J., Chang, M.W., Lee, K., Toutanova, K.: BERT: pre-training of deep bidirectional transformers for language understanding. arXiv preprint arXiv:1810.04805 (2018)
16. D'Souza, J., Auer, S., Pedersen, T.: SemEval-2021 task 11: NLPcontributiongraph - structuring scholarly NLP contributions for a research knowledge graph. In: Proceedings of the Fifteenth Workshop on Semantic Evaluation. Association for Computational Linguistics, Bangkok, August 2021
17. D'Souza, J., Auer, S., Pederson, T.: SemEval-2021 task 11: NLPContributionGraph - structuring scholarly NLP contributions for a research knowledge graph, May 2021. https://zenodo.org/record/4737071
18. D'Souza, J., Hoppe, A., Brack, A., Jaradeh, M.Y., Auer, S., Ewerth, R.: The STEM-ECR dataset: grounding scientific entity references in stem scholarly content to authoritative encyclopedic and lexicographic sources. In: LREC, Marseille, France, pp. 2192–2203, May 2020
19. D'Souza, J., Auer, S.: Sentence, phrase, and triple annotations to build a knowledge graph of natural language processing contributions–a trial dataset. J. Data Inf. Sci. 20210429 (2021)
20. Gábor, K., Buscaldi, D., Schumann, A.K., QasemiZadeh, B., Zargayouna, H., Charnois, T.: SemEval-2018 task 7: semantic relation extraction and classification in scientific papers. In: Proceedings of The 12th International Workshop on Semantic Evaluation, pp. 679–688 (2018)
21. Ghasemi-Gol, M., Szekely, P.: TabVec: table vectors for classification of web tables. arXiv preprint arXiv:1802.06290 (2018)
22. Handschuh, S., QasemiZadeh, B.: The ACL RD-TEC: a dataset for benchmarking terminology extraction and classification in computational linguistics. In: COLING 2014: 4th International Workshop on Computational Terminology (2014)
23. Herzig, J., Nowak, P.K., Mueller, T., Piccinno, F., Eisenschlos, J.: TaPas: weakly supervised table parsing via pre-training. In: Proceedings of the 58th Annual Meeting of the Association for Computational Linguistics, pp. 4320–4333 (2020)
24. Hou, Y., Jochim, C., Gleize, M., Bonin, F., Ganguly, D.: Identification of tasks, datasets, evaluation metrics, and numeric scores for scientific leaderboards construction. arXiv preprint arXiv:1906.09317 (2019)
25. Hou, Y., Jochim, C., Gleize, M., Bonin, F., Ganguly, D.: TDMSci: a specialized corpus for scientific literature entity tagging of tasks datasets and metrics. In: Proceedings of the 16th Conference of the European Chapter of the Association for Computational Linguistics: Main Volume, pp. 707–714 (2021)

26. Jain, S., van Zuylen, M., Hajishirzi, H., Beltagy, I.: SciREX: a challenge dataset for document-level information extraction. In: Proceedings of the 58th Annual Meeting of the Association for Computational Linguistics, pp. 7506–7516 (2020)
27. Jaradeh, M.Y., et al.: Open research knowledge graph: next generation infrastructure for semantic scholarly knowledge. In: Proceedings of the 10th International Conference on Knowledge Capture, pp. 243–246 (2019)
28. Jiang, M., D'Souza, J., Auer, S., Downie, J.S.: Improving scholarly knowledge representation: evaluating BERT-based models for scientific relation classification. In: Ishita, E., Pang, N.L.S., Zhou, L. (eds.) ICADL 2020. LNCS, vol. 12504, pp. 3–19. Springer, Cham (2020). https://doi.org/10.1007/978-3-030-64452-9_1
29. Jinha, A.E.: Article 50 million: an estimate of the number of scholarly articles in existence. Learn. Publ. **23**(3), 258–263 (2010)
30. Kardas, M., et al.: AxCell: automatic extraction of results from machine learning papers. In: Proceedings of the 2020 Conference on Empirical Methods in Natural Language Processing (EMNLP), pp. 8580–8594 (2020)
31. Kononova, O., et al.: Text-mined dataset of inorganic materials synthesis recipes. Sci. Data **6**(1), 1–11 (2019)
32. Kulkarni, C., Xu, W., Ritter, A., Machiraju, R.: An annotated corpus for machine reading of instructions in wet lab protocols. In: NAACL: HLT, Volume 2 (Short Papers), New Orleans, Louisiana, pp. 97–106, June 2018. https://doi.org/10.18653/v1/N18-2016
33. Kuniyoshi, F., Makino, K., Ozawa, J., Miwa, M.: Annotating and extracting synthesis process of all-solid-state batteries from scientific literature. In: LREC, pp. 1941–1950 (2020)
34. Liu, Y., Bai, K., Mitra, P., Giles, C.L.: TableSeer: automatic table metadata extraction and searching in digital libraries. In: Proceedings of the 7th ACM/IEEE-CS Joint Conference on Digital Libraries, pp. 91–100 (2007)
35. Lopez, P.: GROBID: combining automatic bibliographic data recognition and term extraction for scholarship publications. In: Agosti, M., Borbinha, J., Kapidakis, S., Papatheodorou, C., Tsakonas, G. (eds.) ECDL 2009. LNCS, vol. 5714, pp. 473–474. Springer, Heidelberg (2009). https://doi.org/10.1007/978-3-642-04346-8_62
36. Luan, Y., He, L., Ostendorf, M., Hajishirzi, H.: Multi-task identification of entities, relations, and coreference for scientific knowledge graph construction. arXiv preprint arXiv:1808.09602 (2018)
37. Luan, Y., He, L., Ostendorf, M., Hajishirzi, H.: Multi-task identification of entities, relations, and coreference for scientific knowledge graph construction. In: EMNLP (2018)
38. Manning, C.D.: Computational linguistics and deep learning. Comput. Linguist. **41**(4), 701–707 (2015)
39. Milosevic, N., Gregson, C., Hernandez, R., Nenadic, G.: A framework for information extraction from tables in biomedical literature. Int. J. Doc. Anal. Recognit. **22**(1), 55–78 (2019). https://doi.org/10.1007/s10032-019-00317-0
40. Mondal, I., Hou, Y., Jochim, C.: End-to-end NLP knowledge graph construction. arXiv preprint arXiv:2106.01167 (2021)
41. Mysore, S., et al.: The materials science procedural text corpus: annotating materials synthesis procedures with shallow semantic structures. In: Proceedings of the 13th Linguistic Annotation Workshop, pp. 56–64 (2019)
42. Oelen, A., Stocker, M., Auer, S.: Crowdsourcing scholarly discourse annotations. In: 26th International Conference on Intelligent User Interfaces, pp. 464–474 (2021)
43. Renear, A.H., Palmer, C.L.: Strategic reading, ontologies, and the future of scientific publishing. Science **325**(5942), 828–832 (2009)

44. Vaswani, A., et al.: Attention is all you need. In: Advances in Neural Information Processing Systems, pp. 5998–6008 (2017)
45. Ware, M., Mabe, M.: The STM report: an overview of scientific and scholarly journal publishing, March 2015
46. Wei, X., Croft, B., Mccallum, A.: Table extraction for answer retrieval. Inf. Retr. **9**(5), 589–611 (2006). https://doi.org/10.1007/s10791-006-9005-5
47. Yang, Z., Dai, Z., Yang, Y., Carbonell, J., Salakhutdinov, R., Le, Q.V.: XLNet: generalized autoregressive pretraining for language understanding. arXiv preprint arXiv:1906.08237 (2019)

A Prototype System for Monitoring Emotion and Sentiment Trends Towards Nuclear Energy on Twitter Using Deep Learning

Snehameena Arumugam[1], Likai Peng[1], Jin-Cheon Na[1](✉), Guangze Lin[1], Roopika Ganesh[1], Xiaoyin Li[1], Qing Chen[1], Shirley S. Ho[1], and Erik Cambria[2]

[1] Wee Kim Wee School of Communication and Information, Nanyang Technological University, Singapore 637718, Singapore
{Snehamee001,Peng0106,Glin006,Roopika001,W200031, Qchen015}@e.ntu.edu.sg, {tjcna,TSYHo}@ntu.edu.sg

[2] School of Computer Science and Engineering, Nanyang Technological University, Singapore 637718, Singapore
cambria@ntu.edu.sg

Abstract. Nuclear energy is one of controversial topics that affects people's lives, and it is important for policy makers to analyze what people feel towards the subject. But manual analysis of related user-generated contents on social media platforms is a daunting task, and automatic data analysis and visualization come to help. So, in this research, we firstly developed a model for classifying the emotion of nuclear energy related tweets and another model for the aspect-based sentiment analysis of nuclear energy tweets using the BERT (Bidirectional Encoder Representations from Transformers). After that, we developed a prototype system for visualization of the analyzed results stored in the database. The user interface dashboards of the system allow users to monitor emotion and sentiment trends towards nuclear energy by analyzing recent nuclear energy tweets crawled weekly.

Keywords: Emotion analysis · Aspect based sentiment classification · Deep learning · BERT · Nuclear energy

1 Introduction

Social media have served as the digital mediator to share and exchange ideas and information to a larger audience [6]. And Twitter is one of trending social media platforms in the current era when it comes to opinionating people's choices. Therefore, automatic analysis of user-generated content on platforms like Twitter help relevant organizations not only to understand people's emotions and opinions on the diverse aspects of target subjects but also to make effective decisions.

Energy has been harnessed by humans for millennia. Especially, nuclear energy is a seriously disputed topic that has been on the ground for years. Though there are various types of power plants arising across the globe, some governments are building nuclear power plants for better power supply, yet diverse groups of people have different opinions

H.-R. Ke et al. (Eds.): ICADL 2021, LNCS 13133, pp. 471–479, 2021.
https://doi.org/10.1007/978-3-030-91669-5_36

on it. Some people worry that it might cause adverse effects, and some others are happy to have such inventions in their country leading the country to develop. With platforms like Twitter, more opinions arise across the world, making emotion and sentiment analyses play a vital role in understanding deeper insights about users' views towards the disputed topic.

So, emotion classification and aspect-based sentiment analysis (ABSA) of nuclear energy related tweets, and visualization of the classified results are major tasks of our research work. The emotion classifier is a text-based classification model that predicts the emotion of users among the predefined emotion categories. In this study, ABSA is a combination of two subtasks, i.e., aspect category detection (ACD) which aims to identify the pre-defined aspect categories in a tweet and aspect category sentiment analysis (ACSA) which aims to classify the sentiment polarity of a specified aspect category in the tweet. For instance, with the input tweet *"I'm afraid that to construct a power plant with all safety precautions makes our country's budget to increase to millions"*, our emotion classification model would predict it as "fear" and the ABSA model would identify "safety" and "cost" aspects and predict their sentiments as negative. In addition, the user interface dashboards of the prototype system provide visualization of the analysis results. In later sections, first, we present a review of related works, followed by the descriptions of the emotion classification and ABSA models. Next, we describe the system architecture and visualization dashboards of the protype system. The concluding section summarizes our research work.

2 Related Work

Emotion and sentiment analyses have been applied to various real time applications such as brand monitoring, reputation management, media monitoring, and market research [1, 5]. This research as well contributes the same ideology of analyzing what people feel about a target subject, i.e., nuclear energy, and what aspects fuel them to take the emotion and sentiment. In the past, emotion and sentiment analyses were conducted with traditional machine learning approaches such as Naive Bayes (NB) classifiers and support vector machine (SVM), but effective results were not obtained. Nowadays, deep learning has come as a helping hand to produce more accurate results for emotion and sentiment analyses. Therefore, various types of deep learning algorithms have been developed over time to perform emotion and aspect-based sentiment analyses of textual data. One of relevant studies is an ABSA model using recurrent neural network (RNN) with an attention mechanism [4] that analyzed nuclear energy tweets. For emotion classification, Kant and his colleagues [3] investigated various deep learning approaches including the transformer model. They concluded that the transformer model with a byte-pair-encoding (BPE) and fine tuning with an additional single linear layer with an activation function made a good model to capture relevant contexts and achieved around 60% accuracy with 8 emotion categories.

In this research, we applied a popular transformer model, called the BERT (Bidirectional Encoder Representations from Transformers) [2], for the development of deep learning models for the emotion and aspect-based sentiment analyses of nuclear energy tweets. BERT is pre-trained with unlabeled text by combining the left and right contexts to obtain a deep two-direction representation. Therefore, with only one additional output layer, the pre-trained BERT model can be easily fine-tuned to produce state-of-the-art results for various natural language processing tasks.

3 Emotion Classification

3.1 Dataset Preparation

To collect nuclear energy related tweets, Twitter-API was used with the following Boolean search query: "Nuclear Energy" AND ("Nuclear Cost | Nuclear Emission | Nuclear radiation | Nuclear Waste | Nuclear Efficiency | Nuclear Sustainability | Nuclear Safety"). 1,373 tweets were collected with the query, and then each tweet was labelled as one of five emotion categories, i.e., Happy, Fear, Sad, Anger, or Neutral. For example, when a tweet contains nuclear, safe, and clean words, it was labelled as "Happy". Since the collected tweets were not enough to train a model effectively, external labelled emotion dataset (11,327 tweets) [8] was added to have a larger and more balanced emotion dataset. The final dataset has a combination of nuclear energy related tweets and generic domain tweets, which is a total of 12,700 tweets. Table 1 shows the emotion category distribution of the tweets in train and test datasets. Before the tweets were applied to the model, they were preprocessed and cleaned by removing punctuations, user handles, and website links in them.

Table 1. Train and test data split across emotion categories.

	Happy	Sad	Fear	Anger	Neutral	Total Count
Training dataset	1952 (333)*	1878 (237)	1577 (85)	1645 (79)	2055 (439)	**9107** (1173)
Test dataset	783 (76)	689 (13)	691 (12)	707 (14)	723 (85)	**3593** (200)

*: Number of our collected tweets excluding the external labeled tweets

3.2 Model Building and Results

The BERT base model was fine-tuned with an additional single linear layer to build the emotion classifier. The input to the model is a cleaned tweet and the output is one of five emotion categories. The final model was trained with the following hyper parameters: learning rate = 2e–5, epochs = 3, and batch size = 6. Test accuracy was 81%, and detailed test results are shown in Table 2.

4 Aspect Based Sentiment Analysis

4.1 Dataset Preparation

Instead of collecting a dataset from scratch, we started with the nuclear energy dataset used in [4], which includes 5,208 tweets created between 2013 and 2017. These tweets

Table 2. Test results of the emotion classifier.

	Precision	Recall	F1-Score
Happy	0.85	0.81	0.83
Sad	0.77	0.82	0.80
Fear	0.86	0.83	0.84
Anger	0.80	0.77	0.78
Neutral	0.76	0.82	0.79
Average	0.81	0.81	0.81
Accuracy			0.81

are labelled with 6 aspect categories ("cost": development and maintenance price and its subsidy; "efficiency": energy productive efficiency; "emission": CO2, carbon emission and green energy; "radiation": nuclear radiation and radio-active pollution; "safety": nuclear safety, reliability, risk, and disaster; "waste": nuclear waste and its dump) and 2 sentiment polarities (positive and negative). In the dataset, a few labelling errors and irrelevant tweets mentioning nuclear weapons were encountered and corrected. To cover the topics of the nuclear energy tweets more comprehensively, we added the "sustainability" (nuclear power development sustainability) aspect and expanded the aspects into 7 categories. At the same time, the sentiment polarity is expanded to 3 categories: positive, neutral, and negative. Each tweet belongs to one or more aspects, and each aspect in the tweet is labelled with a sentiment polarity value.

After relabeling, the dataset had the problem of unbalanced distribution of aspects and sentiment polarities. So, to balance the number of each aspect's tweets as much as possible, a Boolean search was used to crawl additional 237 tweets related to the "sustainability" aspect on the Twitter official website in the last two years. After labeling these tweets, they were added into the previously relabeled dataset, and finally, a total of 7,178 labelled tweets were obtained, of which 5,258 tweets are unique (see Table 3).

Table 3. Breakdown of Positive, Negative and Neutral tweets across all aspects.

	Cost	Efficiency	Emission	Radiation	Safety	Sustainability	Waste	Total
Positive	534	662	879	136	744	132	329	**3416**
Negative	727	301	278	325	786	63	726	**3206**
Neutral	140	173	50	5	68	55	65	**556**
Total	1401	1136	1207	466	1598	250	1120	**7178**

Before the tweets were applied to the model, they were preprocessed as follows: (1) Convert the currency symbols ($, £, and €) into corresponding English words (dollars, pounds, and euros); (2) Decode and convert emoji expressions into corresponding text representations, such as "✌" would be converted to "victory_hand"; (3) Filter out unrecognizable non-English words, symbols, and hashtags; (4) Replace contractions in the string of text, such as "can't" is converted to "cannot"; (5) Remove non-ASCII characters and punctuation; (6) Convert Arabic numerals into corresponding textual representation, such as "22" would be converted into "twenty-two".

Lastly, the processed dataset was divided into 2,950, 1,519 and 789 datasets according to the ratio of 56%, 29% and 15% as the training set, test set and validation set. To balance the distribution of aspects further in the dataset, tweets with "radiation" and "sustainability" aspects were data augmented using synonym substitution and back translation approaches. We finally added 714 records through synonym substitution and 654 records through back-translation.

4.2 Model Building and Results

The BERT-based ABSA model proposed by Sun and his colleagues [7] was applied to build our ABSA model. Sentence pair classification approach, called BERT-pair, was used to input a tweet and a corresponding aspect to predict its sentiment. The final output of the model is the probability of each sentiment category (positive, negative, neutral, and none). For instance, with the example tweet *"As Americans balance a clean environment and reliable energy, one solution advances toward both adding in safety enhancements and efficient waste. Recycling nuclear energy offers a sustainable way forward #nuclearenergy."*, when the target aspect is "waste", the aspect part is represented as {nuclear - waste} with the "positive" label. Similarly, when the aspect is "safety/sustainability", the aspect is represented as {nuclear – safety/sustainability} with positive. However, for the "cost/efficiency/radiation/emission" aspect, the aspect is represented as {nuclear - cost/efficiency/radiation/emission} with the "none" label since the aspect is not mentioned in the tweet. For each tweet, we generated 7 (tweet, aspect) input pairs for getting an output prediction for each aspect, which allows to detect the relevant aspect categories and their sentiment polarities of the tweet at the same time.

Table 4 shows the performance of the BERT model that was trained with the following hyper parameters: epoch = 3, batch size = 24, and learning rate = 2e−5. For aspect category detection accuracy, only if the model predicted all the aspects correctly in a tweet, then it was considered a correct prediction. Any of the wrong predictions in the tweet was not regarded as a right prediction. In sentiment classification, the "none" category appeared in most of the prediction results because most of the tweets contain only one or two aspects. So, for sentiment accuracy calculation, we only considered the predictions on positive, negative, and neutral categories.

Table 4. Model performance on the test set.

Model	Aspect category detection accuracy	Aspect-based sentiment classification accuracy
BERT	0.69	0.82

Since automatic aspect category detection is not perfect, the BERT model can miss relevant aspects in a tweet. To overcome this issue, we built another BERT model with only positive, negative, and neutral categories (0.82 accuracy). Given a tweet, if the first BERT model returns the "none" category for all 7 aspects, it is put into the second model, which uses predefined keywords to detect the relevant aspect categories of the tweet and predicts the sentiment polarities of the detected aspects. The "other" aspect category was added to the second model for handling relevant tweets but not belonging to the existing 7 aspect categories. Also, irrelevant tweets were detected and ignored.

5 Prototype System for Monitoring Emotion and Sentiment Trends

5.1 System Architecture and Description

Figure 1 shows the system architecture of the prototype system. At first, new tweets are scrapped weekly from Twitter by a scheduler. These scrapped tweets are then cleaned and preprocessed and stored to the database. These stored tweets are then retrieved from the database as inputs for both the emotion analysis model and ABSA models. The prediction results from these models are then stored in the database. These results are then visualized as dashboards which are generated with the Jinja Framework and JavaScript. Currently the dashboards are only accessible within a local machine as a protype system.

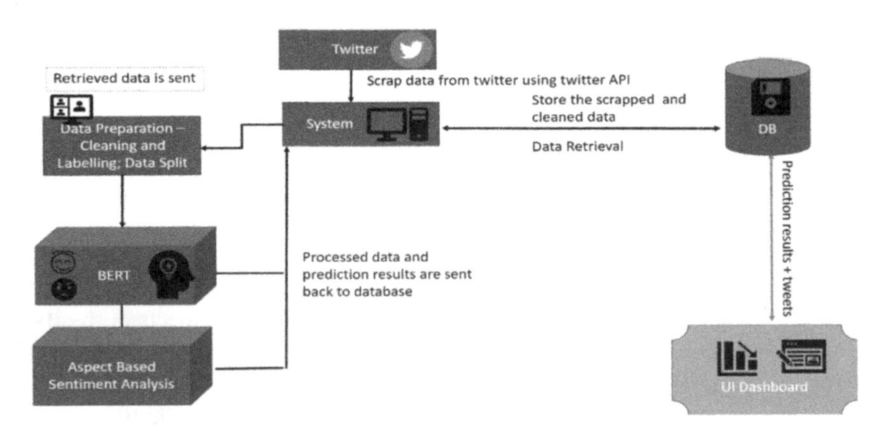

Fig. 1. The system architecture

5.2 Visualization

UI dashboards are used to visualize the prediction results of the emotion and ABSA models. The first default panel depicts the prediction results of the emotion classifier (Fig. 2), and when the "Sentiment" button is clicked (#1 in Fig. 2), the second panel shows the results of the ABSA classifiers (Fig. 3). Especially, the sentiment panel has a separate sub-panel for each aspect, and the "Overall" button is used to show the results

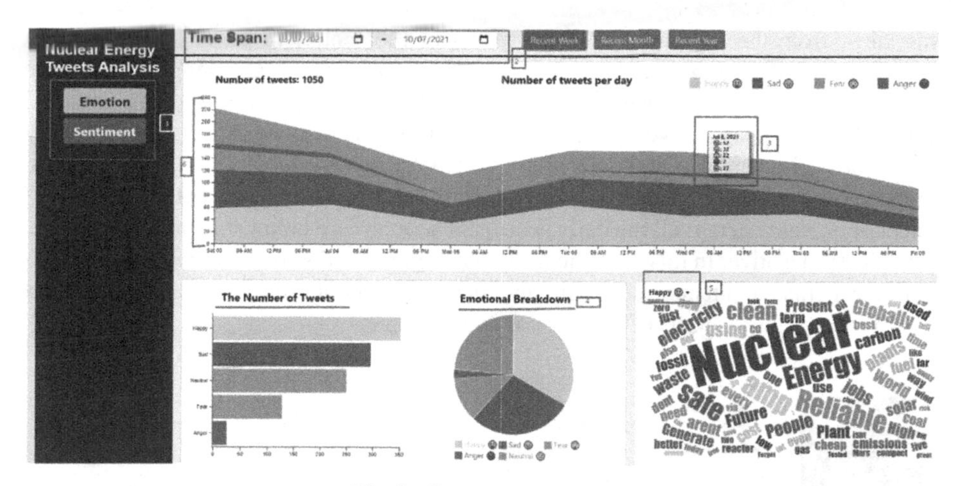

Fig. 2. Emotion dashboard

of all the aspects' tweets. Both the panels show the timeline chart of the selected time span (#6 in Fig. 2), the distributions of the tweets under various categories (#4), a word cloud showing frequent words appeared in the tweets (#5), and a panel of the tweets' content with their prediction categories (not shown in the figures). To display daily sentiment scores in the timeline chart of the sentiment panel, a daily sentiment score, ranged between 0 and 1, is calculated with the tweets of the day using the formula (1) for each aspect or all the aspects (i.e., "Overall").

$$\frac{((PositiveTweetCount * 1) + (NeutralTweetCount * 0.5))}{PositiveTweetCount + NegativeTweetCount + NeutralTweetCount} \quad (1)$$

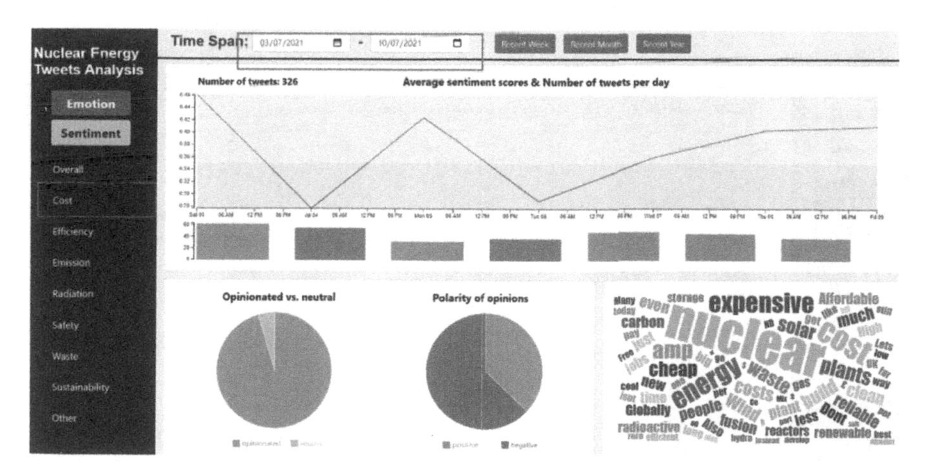

Fig. 3. Sentiment dashboard: the "cost" aspect is selected.

Using the dashboards, the trend of emotions and aspect-specific sentiments towards the nuclear energy topic can be easily explored. For example, the timeline chart in Fig. 4

was taken from the "sustainability" sub-panel of the sentiment panel during the one week between 03/07/2021 and 10/07/2021. As shown, in general, people were quite positive towards the "sustainability" aspect of nuclear energy, but in the two days, 03 and 07/07/2021, people gave negative and neutral feedback on the aspect. The "Polarity of opinions" pie chart also shows that there were more positive tweets than negative ones. Similarly in Fig. 5, we can see that people were quite negative towards the "radiation" aspect of nuclear energy; all the daily sentiment scores were below 0.35 and most of the tweets were negative. In this way, the dashboards generated with the automatic analysis of user-generated contents would help understand the public's emotions and sentiments towards the various aspects of the nuclear energy topic.

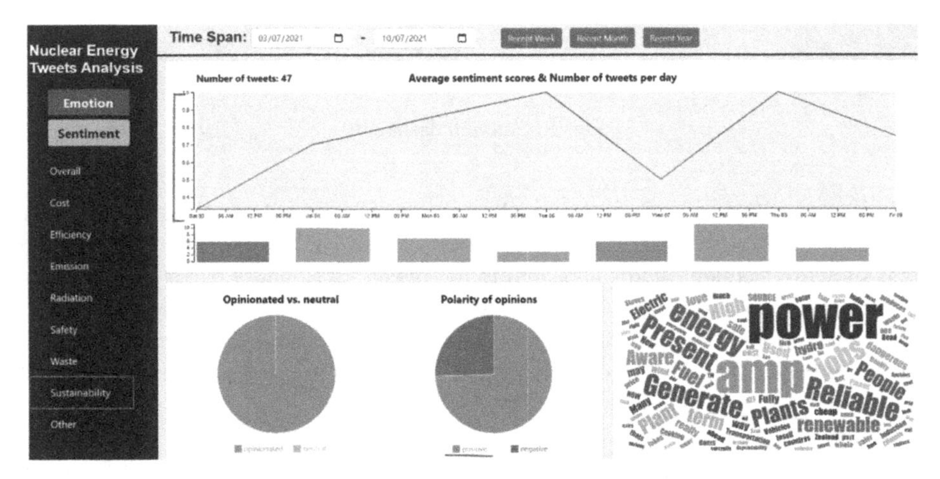

Fig. 4. Sentiment polarity trend with respect to the "sustainability" aspect.

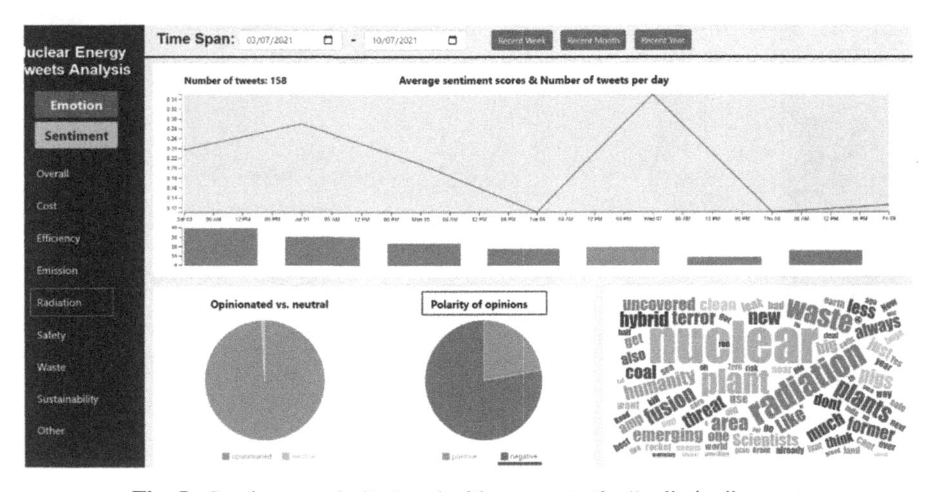

Fig. 5. Sentiment polarity trend with respect to the "radiation" aspect.

6 Conclusion

This brings to conclusion, satisfying our initial objective, that emotion and sentiment analysis models have been developed and hereby the models can predict the emotions of nuclear energy tweets and sentiment polarities are also predicted along with a daily sentiment score for each aspect. In addition, UI dashboards have been developed to visualize the analyzed results where data crawling, prediction and visualization processes are fully automated. Hence, this study demonstrates the potential of deep learning techniques for monitoring emotion and sentiment trends towards nuclear energy on Twitter. In future work, the visualization part will be improved further with usability evaluation.

References

1. Benedetto, F., Tedeschi, A.: Big data sentiment analysis for brand monitoring in social media streams by cloud computing. In: Pedrycz, W., Chen, S.-M. (eds.) Sentiment Analysis and Ontology Engineering. SCI, vol. 639, pp. 341–377. Springer, Cham (2016). https://doi.org/10.1007/978-3-319-30319-2_14
2. Devlin, J., Chang, M.-W., Lee, K., Toutanova, K.: BERT: pre-training of deep bidirectional transformers for language understanding. In: NAACL-HLT 2019, pp. 4171–4186 (2018)
3. Kant, N., Puri, R., Yakovenko, N., Catanzaro, B.: Practical text classification with large pre-trained language models (2018). https://arxiv.org/pdf/1812.01207.pdf
4. Liu, Z., Na, J.-C.: Aspect-based sentiment analysis of nuclear energy tweets with attentive deep neural network. In: Dobreva, M., Hinze, A., Žumer, M. (eds.) ICADL 2018. LNCS, vol. 11279, pp. 99–111. Springer, Cham (2018). https://doi.org/10.1007/978-3-030-04257-8_9
5. Olaleye, S.A., Sanusi, I.T., Salo, J.: Sentiment analysis of social commerce: a harbinger of online reputation management. Int. J. Electron. Bus. **14**(2) (2018). https://doi.org/10.1504/IJEB.2018.094864
6. Sayce, D.: The Number of tweets per day in 2020 (2020). https://www.dsayce.com/social-media/tweets-day/
7. Sun, C., Huang, L., Qiu, X.: Utilizing BERT for aspect-based sentiment analysis via constructing auxiliary sentence. In: NAACL-HLT 2019, pp. 380–385 (2019)
8. NLP-text-emotion-dataset. https://github.com/lukasgarbas/nlp-text-emotion/tree/master/data

Author Index

Printed in the United States
by Baker & Taylor Publisher Services